Springer Series on
SIGNALS AND COMMUNICATION TE

T0142261

Signals and Communication Technology

continued after index

Peter Seibt

Algorithmic Information Theory

Mathematics of Digital Information Processing

With 14 Figures

 Springer

Peter Seibt
Université de la Méditerranée
and
Centre de Physique Théorique
Campus de Luminy, Case 907
13288 Marseille cedex 9, France

ISSN 1860-4862

ISBN 978-3-642-06978-9   e-ISBN 978-3-540-33219-0

Springer is a part of Springer Science+Business Media.

springer.com

© Springer-Verlag Berlin Heidelberg 2006
Softcover reprint of the hardcover 1st edition 2006

Cover design: Design & Production, Heidelberg

Contents

Introduction

Shall we be destined to the days of eternity, on holy-days, as well as working days, to be shewing the RELICKS OF LEARNING, *as monks do the relicks of their saints – without working one – one single miracle with them?*

Laurence Sterne, *Tristram Shandy*

This book deals with information processing; so it is far from being a book on information theory (which would be built on description and estimation). The reader will be shown the horse, but not the saddle. At any rate, at the very beginning, there was a series of lectures on "Information theory, through the looking-glass of an algebraist", and, as years went on, a steady process of teaching and learning made the material evolve into the present form. There still remains an algebraic main theme: algorithms intertwining polynomial algebra and matrix algebra, in the shelter of signal theory.

A solid knowledge of elementary arithmetic and Linear Algebra will be the key to a thorough understanding of all the algorithms working in the various bit-stream landscapes we shall encounter. This priority of algebra will be the thesis that we shall defend. More concretely: We shall treat, in five chapters of increasing difficulty, five sensibly different subjects in Discrete Mathematics. The first two chapters on data compaction (lossless data compression) and cryptography are on an undergraduate level – the most difficult mathematical prerequisite will be a sound understanding of quotient rings, especially of finite fields (mostly in characteristic 2). The next two chapters are already on a graduate level; the reader should be slightly acquainted with arguments in signal theory – although Lebesque integration could remain the "grey box" that it usually is. We encounter sampling – an innocent operation of tremendous epistemological impact: the Platonic mathematician leaving his heaven of continuity (rule = truth) for the earth of discreteness (diversity = uncertainty) will be plainly comforted by the great interpolation theorems that lift him back to the heights. The chapter on error control codes which are designed according to signal theoretical ideas, complements – on a purely algebraic level – the invasion of signal theory. The fifth and final chapter is the most important, in length as well as in complexity. It deals with lossy (image) compression, and yields the mathematical background for the understanding of JPEG and JPEG 2000. Now, our Platonic mathematician will be expelled from paradise: The discrete world becomes absolute, and all continuous constructions are plainly auxiliary and relative.

But let us pass to a detailed description of the content.

The first chapter on data compaction is more or less an elementary introduction to algorithmic information theory. The central theme will be the nonredundant representation of information. Everything turns around the notion of entropy: What is the information content of a string of symbols (with given statistical behaviour), i.e. what is its minimal bit equivalent ? Entropy coding has its algorithmic stars: for memoryless sources, Huffman entropy coding is unbeatable, but from a dynamic viewpoint, arithmetic coding will be slightly better. Both methods are plainly integrated in advanced image compression standards – we shall give a "default" Huffman table for JPEG. The chapter will end with an (merely descriptive) exposition of the algorithm LZW which is universal in the sense that it compacts any character stream – without preliminary statistical evaluation – by establishing a dictionary that enumerates typical substrings (thereby creating its proper statistical evaluation). LZW is the perfect data compaction algorithm – but it needs large files in order to be efficient. That is why we do not meet it in image compression where the data units are too small.

The second chapter presents a set of rather austere lectures on cryptography. We aim to give the maximum of information in a minimum of space – there already exists a lot of highly coloured frescoes on the subject in print. The venerable algorithm DES – the cryptosystem the best understood on planet earth – will serve as an introduction to the subject. Things become more serious with the new standard AES-Rijndael, the mathematical basement of which is a nice challenge to the student's understanding of higher level (still) elementary arithmetic. He will learn to think in cyclic arithmetic – thus getting familiar with discrete logarithms in a very explicit way. This opens the door to digital signatures, i.e. to the practical realization of the public key paradigm: I tell you my position on an arithmetic circle, but I do not reveal the number of steps to get there. We shall treat the principal standard for digital signatures, the system DSA (*Digital Signature Algorithm*), as well as the variants rDSA (signatures via RSA) and ECDSA (signatures via elliptic curve arithmetic). As to RSA: This thirty-year-old algorithm has always been the cornerstone of academic zest to promote the public key idea. So we shall follow tradition – not without noting that RSA is a little bit old fashioned. Finally, the secure hash algorithm (SHA-1) will produce the message digests used in the various signature protocols. We shall need a lot of large prime numbers; hence we include a brief discussion on their efficient generation.

This completes the description of the easy part of this book. Teaching experience shows that students like data compaction for its simple elegance and are distant towards the iterative flatness of most cryptographic standards – are they to blame?

With the third chapter, we enter the mathematical world of signal theory. We have to answer the question: What is the discrete skeleton of a (continuous) signal? This means sampling, and reconstruction via interpolation. Putting aside all practical considerations, we shall treat the problem in vitro. Tough mathematical expositions are available; we have chosen a step-by-step approach. So, we begin with the discrete Fourier transform and its importance for trigonometric interpolation. Then we show ad hoc the classical interpolation theorem (of Whittaker–Shannon, Nyquist–Shannon, or simply Shannon, as you like it...) for precisely trigonometric polynomials. Finally, we attack the interpolation theorem in its usual form. There are some formal problems which need a short commentary. The natural mathematical framework for signal theory is the $\mathbf{L}^2$ Hilbert space formalism. Now, elements of an $\mathbf{L}^2$ space are not functions (which disappear in their clouds of equivalence) but function behaviour sketches. Precise numerical rules enter via duality. Thus, sampling – which is basically a Hilbert space nonsense – must be considered as a rule of behaviour (and should be duly formalized by a distribution). The equality in the Shannon interpolation formula (which means equality of distributions) is, in any down-to-earth exposition, considerably fragilized by the proof that establishes it. We shall try to be as simple as possible, and avoid easy "distribution tricks".

Logically, it is the fifth and last chapter on data compression that should now follow. Why this strange detour in the land of error control codes? There are at least two reasons. First, we get an equilibrium of complementary lectures, when alternating between non-algebraic and algebraic themes. Then, the fourth chapter logically reinforces our definite submission to signal theory. The codes of Reed–Solomon – our first subject – have a nice error- correcting algorithm that makes use of the Discrete Fourier Transform over finite fields of characteristic 2. And the convolutional codes – our second subject – are best understood via digital filtering in binary arithmetic. Our exposition there is non-standard, with a neat accent on algorithmic questions (no trellis nor finite automata formalisms).

Finally, we come to the fifth chapter, which is rather voluminous and treats data compression, i.e. the practice of intentionally reducing the information content of a data record – and this in such a way that the reproduction has as little distortion as possible. We shall concentrate on image compression, in particular on JPEG and JPEG 2000. The quality of compression depends on sifting out efficiently what is considered to be significant numerical information. Quantization towards bit representation will then annihilate everything that can be neglected.

Our main concern will be to find an intelligent information theoretic sieve method.

It is the *Discrete Cosine Transform* (DCT) in JPEG, and the *Discrete Wavelet Transform* (DWT) in JPEG 2000 that will resolve our problems. In both cases, a linear transformation will associate with regions of digital image samples (considered as matrices of pictural meaning) matrix transforms whose coefficients have no longer a pictorial but only a descriptive meaning. We must insist: Our transformations will not compress anything; they merely will arrange the numerical data in a transparent way, thus making it possible to define sound quantization criteria for efficient suppression of secondary numerical information.

We shall begin the fifth chapter with a slight non-thematic digression: the design of digital passband filters in a purely periodic context. This will be a sort of exercise for formally correct thinking in the sequel. Then we come up with the discrete cosine transform and its raison d'être in JPEG.

We shall first treat the 1D DCT (acting on vectors), then its 2D extension (acting on matrices), and finally its position in the Karhunen–Loève family. For the pragmatic reader who is only interested in image compression there is an easy argument in favour of the DCT that short circuits everything: The 2D DCT acts via conjugation on 8×8 matrices, preserves the energy (the Euclidian norm of matrices), and diagonalizes all constant matrices (which is a reasonable motive in image compression).

In the second part of the last chapter we shall encounter the discrete wavelet transform and its implantation into JPEG 2000. Exactly like the Karhunen–Loève transform, the discrete wavelet transform is rather an entire family of transforms. Our presentation will adopt the (two channel) filter bank approach, which is easy to explain – and plainly sufficient for the understanding of the way discrete wavelet transforms act in image compression. More concretely, we shall concentrate on three specific wavelet transforms: the DWT 5/3 spline, the DWT 7/5 Burt and the DWT 9/7 CDF. We shall also treat the reversible mode of JPEG 2000: how to get invertible transformations in integer arithmetic. The "lifting structure" that will guarantee non-linear integer-valued approximations of our initial matrix transformations bears a clear resemblance to the sequence of round transforms of the cryptosystem DES.

At the very end of our book we have to answer the question: Where are the wavelets (and why are there wavelets) behind all of that filter bank theory? This is a pretty mathematical subject – maybe a little bit too mathematical. But a thorough understanding of the criteria that govern the design of filter banks requires adopting the wavelet viewpoint.

Let us end this introduction with some remarks on teaching questions. This is a book on Mathematics. What about proofs? We have adopted a strategy that conciliates aesthetics with common sense: A proof should be interesting, not too long, and it should give sufficient information on the mathematical weight of the proposition. For example, the proof of the Kraft inequality (characterizing prefix codes) is of this kind. On a quite different level of reasoning, almost all proofs around the Viterbi decoding algorithm (for convolutional codes) are welcome, since they do not abound in common literature. In a certain sense, it is precisely our presentation of wavelet theory that shows the "sieve of rigour" that we have adopted.

A short remark on the nature of our exercises: Some readers will – perhaps – be shocked: Almost all exercises are dull, rather mechanical and lengthy. But we learn by repetition... Towards the end of the book, a good pocket calculator will be necessary. At any rate, a book in Concrete Mathematics should inflict the burden of complexity in a quantitative, rather than in a qualitative way. We have given a lot of hints, and many solutions. But we never aimed at completeness...

Let me end with a rather personal remark. This book is a result of a happy conjunction of teaching and learning. Learning has always been exciting for me. I hope the reader will feel it the same.

1

Data Compaction

This first, rather elementary chapter deals with *non-redundant* representation of information; in other words, we shall treat data compaction codes (i.e. algorithms for lossless data compression). More common, eventually lossy, data compression needs arguments and methods from signal theory, and will be considered in the last chapter of this book.

1.1 Entropy Coding

All coding methods that we shall encounter in this section are based on a preliminary statistical evaluation of our set of data. In a certain sense, the coding algorithms will treat the statistical profile of the data set rather than the data itself. Since we are only interested in coding methods, we shall always feel free to assume that the statistics we need are plainly at our disposal – so that our algorithms will run correctly.

Note that our probabilistic language is part of the tradition of information theory – which has always been considered as a peripheral discipline of probability theory. But you are perfectly allowed to think and argue in a purely deterministic way: the statistical evaluation of the data for compaction can be thought of as a specification of parameters – in the same way as the choice of the right number of nodes (or of the correct sampling frequency) in interpolation theory.

A historical remark: do not forget that almost all good ideas and clever constructions in this section have come to light between 1948 and 1952.

1.1.1 Discrete Sources and Their Entropy

We shall consider *memoryless* discrete sources, producing words (strings of letters, of symbols, of characters) in an alphabet $\{a_0, a_1, \ldots, a_{N-1}\}$ of N symbols.

We shall call

$p_j = p(a_j) \equiv$ the probability of (the production of) the letter $a_j, 0 \leq j \leq N - 1$.

Notation $\mathbf{p} = (p_0, p_1, \ldots, p_{N-1}) \equiv$ *the probability distribution which describes the production of our source.*

- *Regarding the alphabet*: think of $\{0, 1\}$ (a binary source: for example, a binary facsimile image) or of $\{00000000, 00000001, \ldots, 11111111\}$ (a source of 256 symbols, in 8-bit byte representation: for example, the ASCII character code).
- *Regarding the memoryless production*: this is a condition of probabilistic modelling which is *very strong*. Namely:

For a word $\mathbf{w} = a_{j_1} a_{j_2} \cdots a_{j_n}$ of length n, the *statistically independent* production of its letters at any moment is expressed by the identity

$$p(\mathbf{w}) = p(a_{j_1}) p(a_{j_2}) \cdots p(a_{j_n}).$$

This identity (the probability of a word is the product of the probabilities of its letters) models the production of our source by the iterated roll of a loaded dice, the faces of which are the letters of our alphabet – with the probability distribution $\mathbf{p}$ describing the outcome of our experience.

Note that this rather simple modelling has its virtues beyond simplicity: it describes the ugliest situation for data compression (which should improve according to the degree of correlation in the production of our data), thus meeting the demands of an austere and cautious design.

At any rate, we now dispose of an easy control for modelling – having a sort of "commutation rule" that permits us to decide what should be a *letter*, i.e. an atom of our alphabet. For a given binary source, for example, the words 01 and 10 may have sensibly different frequencies. It is evident that this source cannot be considered as a memoryless source for the alphabet $\{0, 1\}$; but a deeper statistical evaluation may show that we are permitted to consider it as a memoryless source over the alphabet of 8-bit bytes.

The Entropy of a Source

The entropy of a (discrete) source will be the *average information content* of a "generic" symbol produced by the source (measured in bits per symbol).

Let us insist on the practical philosophy behind this notion: you should think of entropy as a *scaling factor towards* (*minimal*) *bit-representation*: 1,000 symbols produced by the source (according to the statistics) "are worth" $1,000 \times$ entropy bits.

Prelude   *The information content of a message.*

Let $I(\mathbf{w})$ be the quantity of information contained in a word $\mathbf{w}$ that is produced by our source. We search a definition for $I(\mathbf{w})$ which satisfies the following two conditions:

(1) $I(\mathbf{w})$ is inversely proportional to the probability $p(\mathbf{w})$ of the production of $\mathbf{w}$ ("the less it is frequent, the more it is interesting").
Moreover, we want the information content of a *sure* event to be *zero*.

(2) $I(a_{j_1}a_{j_2}\cdots a_{j_n}) = I(a_{j_1}) + I(a_{j_2}) + \cdots + I(a_{j_n})$ (the information content of a word is the sum of the information contents of its letters – this stems from our hypothesis on the statistical independence in the production of the letters).

Passing to the synthesis of (1) and (2), we arrive at the following condition:
$I(\mathbf{w}) = F\left(\frac{1}{p(\mathbf{w})}\right)$ *where the real function F has to be strictly monotone (increasing) and must satisfy the identity $F(x \cdot y) = F(x) + F(y)$ as well as $F(1) = 0$.*

Now, there is essentially only one (continuous) function F which satisfies our conditions: the *logarithm*.

Thus the following definition comes up naturally.

Definition $I(\mathbf{w}) = \mathrm{Log}_2\left(\frac{1}{p(\mathbf{w})}\right) = -\mathrm{Log}_2 p(\mathbf{w}).$

$\left[\textit{Recall: } y = \mathrm{Log}_2 x \Longleftrightarrow x = 2^y \Longleftrightarrow x = e^{y\mathrm{Ln}\,2} \Longleftrightarrow y = \frac{\mathrm{Ln}\,x}{\mathrm{Ln}\,2}.\right]$

But why the logarithm to the base 2?

Answer   *We want the unity of the information content to be the bit.*
   Let us make things clearer with two examples.

(a) *Consider a source which produces a_0 = heads and a_1 = tails with the same probability $(\mathbf{p} = (p_0, p_1) = (\frac{1}{2}, \frac{1}{2}))$. We get $I(a_0) = I(a_1) = -\mathrm{Log}_2 2^{-1} = 1$.*
That is logical: when tossing coins with equal chance, heads is naturally coded by 0 and tails is naturally coded by 1.

(b) *Let us pursue this line of thought: now, our outcome will be the 256 integers between 0 and 255 (rolling a very, very big dice with 256 faces), all of equal chance: $\mathbf{p} = (p_0, p_1, \ldots, p_{255}) = \left(\frac{1}{256}, \frac{1}{256}, \ldots, \frac{1}{256}\right)$. $I(a_0) = I(a_1) = \cdots = I(a_{255}) = -\mathrm{Log}_2 2^{-8} = 8$. Once more: no surprise; assuming equal chance, the information content of any of the integers $0, 1, \ldots, 255$ has to be 8 bits: they are 8-bit bytes!*
But back to our source:

Let $\mathbf{p} = (p_0, p_1, \ldots, p_{N-1})$ be the probability distribution which describes the (memoryless) production of the letters of our alphabet.

Definition of the *entropy* of the source:

$H(\mathbf{p}) \equiv$ the *average quantity of information* per symbol (in bits per symbol)
$= p_0 I_0 + p_1 I_1 + \cdots + p_{N-1} I_{N-1} = -p_0 \mathrm{Log}_2 p_0 - p_1 \mathrm{Log}_2 p_1$
$- \cdots - p_{N-1} \mathrm{Log}_2 p_{N-1}.$

Exercises

(1) Compute the entropy of the source which produces the eight letters
$a_0, a_1, \ldots, a_7$, according to the probability distribution $\mathbf{p} = (p_0, p_1, \ldots, p_7)$
with $p_0 = \frac{1}{2}, p_1 = \frac{1}{4}, p_2 = p_3 = \frac{1}{16}, p_4 = p_5 = p_6 = p_7 = \frac{1}{32}$.

(2) Let us consider a memoryless source which produces four letters
a_0, a_1, a_2, a_3, according to the probability distribution $\mathbf{p} = (p_0, p_1, p_2, p_3)$.
Let us change our viewpoint. Consider the source as a producer of the 16
symbols $a_0a_0, a_0a_1, \ldots, a_3a_2, a_3a_3$, according to the product distribution
$\mathbf{p}^{(2)} = (p_{00}, p_{01}, \ldots, p_{23}, p_{33})$ with $p_{ij} = p_i p_j$, $0 \leq i, j \leq 3$.
Show that $H(\mathbf{p}^{(2)}) = 2H(\mathbf{p})$. Generalize.

Remarks

Our situation: the alphabet $\{a_0, a_1, \ldots, a_{N-1}\}$ will remain fixed; we shall vary
the probability distributions...

(1)
$$H(\mathbf{p}) = 0 \quad \Longleftrightarrow \quad \text{The source produces effectively only one letter}$$
(for example the letter a_0), with $p(a_0) = 1$.

Recall: a sure event has information content zero.

Hence: the entropy will be *minimal* (will be zero) as a characteristic of a
constant source production.

Thus, we extrapolate (and we are right):

(2) $H(\mathbf{p})$ is maximal $\quad \Longleftrightarrow \quad p_0 = p_1 = \cdots = p_{N-1} = \frac{1}{N}$.
In this case, we have $H(\mathbf{p}) = \text{Log}_2 N$.

Exercises

(1) A binary source produces $a_0 = $ white and $a_1 = $ black according to the
probability distribution $\mathbf{p} = (p_0, p_1)$.
Find the condition on the ratio white/black which characterizes $H(\mathbf{p}) < \frac{1}{2}$.

(2) Gibbs' inequality.
Consider $\mathbf{p} = (p_0, p_1, \ldots, p_{N-1})$ and $\mathbf{q} = (q_0, q_1, \ldots, q_{N-1})$, two strictly
positive probability distributions (no probability value is zero).
(a) Show that $-\sum_{j=0}^{N-1} p_j \text{Log}_2 p_j \leq -\sum_{j=0}^{N-1} p_j \text{Log}_2 q_j$.
(b) Show that the inequality above is an equality $\Longleftrightarrow \mathbf{p} = \mathbf{q}$.
(c) Deduce from (b): every probability distribution $\mathbf{p} = (p_0, p_1, \ldots, p_{N-1})$
satisfies $H(\mathbf{p}) \leq \text{Log}_2 N$ with equality $\Longleftrightarrow p_0 = p_1 = \cdots = p_{N-1} = \frac{1}{N}$.

$\Big[$ *Hint*:

(a) *Recall*: $\text{Ln} \, x \leq x - 1$ for all $x > 0$ with equality $\Longleftrightarrow x = 1$.

(b) You should get from (a) the following inequality

$$\sum_{j=0}^{N-1} p_j \left(\operatorname{Ln} \frac{q_j}{p_j} - \left(\frac{q_j}{p_j} - 1 \right) \right) \leq 0,$$

where all the terms of the sum are non-positive. This is the clue.]

Entropy Coding, A First Approach

Consider a memoryless source which produces the N symbols $a_0, a_1, \ldots, a_{N-1}$, according to the probability distribution $\mathbf{p} = (p_0, p_1, \ldots, p_{N-1})$.

We have seen: every letter a_j "is worth" $I(a_j)$ bits, $0 \leq j \leq N-1$.

This leads to the *natural idea* (Shannon (1948)): associate with the symbols of our alphabet binary code words of variable length in such a way that the length of a code word associated with a letter is precisely the information content of this letter (assume first that all probabilities are powers of 2, so that the information contents will be correctly integers).

More precisely:

Let l_j be the length (the number of bits) of the code word associated to the letter a_j, $0 \leq j \leq N-1$.

Our choice: $l_j = I(a_j)$, $0 \leq j \leq N-1$.

Let us look at the *average length* $\bar{l}$ of the code words:

$$\bar{l} = p_0 l_0 + p_1 l_1 + \cdots + p_{N-1} l_{N-1}.$$

Note that $\bar{l}$ is a *scaling factor*: our encoder will transform 1,000 symbols produced by the source (in conformity with the statistics used for the construction of the code) into $1,000 \times \bar{l}$ bits.

But, since we were able to choose $l_j = I(a_j)$, $0 \leq j \leq N-1$, we shall get

$$\bar{l} = H(\mathbf{p}).$$

Example Recall the source which produces the eight letters $a_0, a_1, \ldots, a_7$, according to the probability distribution $\mathbf{p} = (p_0, p_1, \ldots, p_7)$ with $p_0 = \frac{1}{2}, p_1 = \frac{1}{4}, p_2 = p_3 = \frac{1}{16}, p_4 = p_5 = p_6 = p_7 = \frac{1}{32}$.

This means: $I(a_0) = 1, I(a_1) = 2, I(a_2) = I(a_3) = 4, I(a_4) = I(a_5) = I(a_6) = I(a_7) = 5$.

We choose the following encoding:

$$
\begin{array}{ll}
a_0 \longmapsto 0 & a_4 \longmapsto 11100 \\
a_1 \longmapsto 10 & a_5 \longmapsto 11101 \\
a_2 \longmapsto 1100 & a_6 \longmapsto 11110 \\
a_3 \longmapsto 1101 & a_7 \longmapsto 11111
\end{array}
$$

Encoding *without statistics*, i.e. assuming equal chance, will oblige us to reserve three bits for any of the eight letters. On the other hand, with our code, we obtain $\bar{l} = H(\mathbf{p}) = 2.125$.

Let us insist: without statistical evaluation, 10,000 source symbols have to be transformed into 30,000 bits. With our encoder, based on the statistics of the source, we will transform 10,000 letters (produced in conformity with the statistics) into 21,250 bits. Manifestly, we have *compressed*.

Important remark concerning the choice of the code words in the example above.

Inspecting the list of our eight code words, we note that *no code word is the prefix of another code word*. We have constructed what is called a *binary prefix code*. In order to understand the practical importance of this notion, let us look at the following example:

$$A \longmapsto 0 \quad B \longmapsto 01 \quad C \longmapsto 10.$$

Let us try to decode 001010. We realize that there are three possibilities: *AACC*, *ABAC*, *ABBA*. The ambiguity of the decoding comes from the fact that the code word for A is the prefix of the code word for B. But look at our example: there is no problem to decode

01101001110111111110001000 back to $a_0 a_3 a_0 a_0 a_5 a_7 a_2 a_0 a_1 a_0 a_0$.

McMillan (1956) has shown that every variable length binary code that admits a *unique decoding algorithm* is isomorphic to a *prefix code*. This will be the reason for our loyalty to prefix codes in the sequel.

1.1.2 Towards Huffman Coding

In this section we shall recount the first explosion of ideas in information theory, between 1948 and 1952. Everything will begin with Claude Shannon, the founder of the theory, and will finally attain its "price of elegance" with the algorithm of Huffman, in 1952.

Do not forget that the theory we shall expose is built upon the rather restrictive hypothesis of a *memoryless source*.[1]

The Kraft Inequality and its Consequences

Let us consider a memoryless source producing N letters, $a_0, a_1, \ldots, a_{N-1}$, according to the probability distribution $\mathbf{p} = (p_0, p_1, \ldots, p_{N-1})$.

Shannon's coding paradigm. Associate to $a_0, a_1, \ldots, a_{N-1}$ words of a binary code, such that the lengths $l_0, l_1, \ldots, l_{N-1}$, of the code words *will correspond* to the information contents of the encoded symbols.

We need to make precise the term "*will correspond* to the information contents of the encoded symbols".

We aim at

$$l_j \approx I(a_j) = -\mathrm{Log}_2 p_j, \quad 0 \le j \le N - 1.$$

---

[1] One can do better – but there are convincing practical arguments for simple modelling.

More precisely, we put

$$l_j = \lceil I(a_j) \rceil = \lceil -\text{Log}_2 p_j \rceil, \quad 0 \le j \le N-1,$$

where $\lceil \; \rceil$ means rounding up to the next integer.

Our first problem will now be the following:

Is Shannon's programme soundly formulated: Suppose that we *impose N* lengths $l_0, l_1, \ldots, l_{N-1}$ for the words of a binary code to be constructed. What are the conditions that guarantee the existence of a binary *prefix code* which realizes these lengths? In particular, what about the soundness of the list of lengths derived from a probability distribution, following Shannon's idea? Is this list always realizable by the words of a binary prefix code?

Let us write down most explicitly the Shannon-conditions:

$$l_j - 1 < -\text{Log}_2 p_j \le l_j, \quad 0 \le j \le N-1, \quad \text{i.e.}$$

$$2^{-l_j} \le p_j < 2 \cdot 2^{-l_j}, \quad 0 \le j \le N-1.$$

Summing over all terms, we get:

$$\sum_{j=0}^{N-1} 2^{-l_j} = \frac{1}{2^{l_0}} + \frac{1}{2^{l_1}} + \cdots + \frac{1}{2^{l_{N-1}}} \le 1.$$

This innocent inequality will finally resolve all our problems.

We begin with a (purely combinatorial) result that has gloriously survived of a dissertation published in 1949:

Proposition (Kraft's Inequality) *Let $l_0, l_1, \ldots, l_{N-1}$ be imposed lengths (for N binary code words to construct). Then the following holds:*

There exists a binary prefix code which realizes these lengths $\Longleftrightarrow$ $\sum_{j=0}^{N-1} 2^{-l_j} \le 1.$

Proof Consider the binary tree of all binary words:

$$0 \qquad\qquad\qquad 1$$

$$00 \qquad 01 \qquad 10 \qquad 11$$

$$000 \quad 001 \quad 010 \quad 011 \; 100 \quad 101 \quad 110 \quad 111$$

On level l, there are 2^l binary words of length l, arranged according to their numerical values (every word, considered as the binary notation of an integer, indicates its position). The successors of a word (for the binary tree structure) are precisely its syntactical successors (i.e. the words which admit our word as a prefix).

This will be the convenient framework for the proof of our claim.

$\Longrightarrow$: Choose $l > l_j, 0 \le j \le N-1$. Every word of length l_j has 2^{l-l_j} successors on the level l of the binary tree of all binary words. The prefix property implies that these level-l-successor sets are all mutually *disjoint*.

Comparing the cardinality of their union with the number of all words on level l, we get: $\sum_{j=0}^{N-1} 2^{l-l_j} \leq 2^l$, i.e. $\sum_{j=0}^{N-1} 2^{-l_j} \leq 1$.

$\Longleftarrow$: Put $l = \max\{l_j : 0 \leq j \leq N-1\}$, and let $n_1, n_2, \ldots, n_l$ be the numbers of code words of length $1, 2, \ldots, l$ that we would like to construct. By our hypothesis we have:

$$n_1 \cdot 2^{-1} + n_2 \cdot 2^{-2} + \cdots + n_l \cdot 2^{-l} \leq 1,$$

i.e.

$$
\begin{array}{ll}
n_1 \cdot 2^{-1} \leq 1 & n_1 \leq 2 \\
n_1 \cdot 2^{-1} + n_2 \cdot 2^{-2} \leq 1 & n_2 \leq 2^2 - n_1 \cdot 2 \\
\qquad\qquad \vdots & \qquad\qquad \vdots \\
n_1 \cdot 2^{-1} + n_2 \cdot 2^{-2} + \cdots + n_l \cdot 2^{-l} \leq 1 & n_l \leq 2^l - n_1 \cdot 2^{l-1} - \cdots - n_{l-1} \cdot 2
\end{array}
$$

The first inequality shows that we can make our choice on level 1 of the binary tree of all binary words. The second inequality shows that the choice on level 2 is possible, after blockade of the $n_1 \cdot 2$ successors of the choice on level 1. And so on...

The last inequality shows that the choice on level l is possible, after blockade of the $n_1 \cdot 2^{l-1}$ successors of the choice on level 1, of the $n_2 \cdot 2^{l-2}$ successors of the choice on level 2, ..., of the $n_l \cdot 2$ successors of the choice on level $l-1$.

This finishes the proof of our proposition. $\square$

Exercises

(1) You would like to construct a binary prefix code with four words of length 3, and six words of length 4. How many words of length 5 can you add?

(2) Consider an alphabet of four letters: N, E, S, W.
Does there exist a prefix code on this alphabet which consists of two words of length 1, four words of length 2, 10 words of length 3 and 16 words of length 4?

(3) A memoryless source produces eight letters A, B, C, D, E, F, G, H according to the probability distribution $\mathbf{p} = (p(A), p(B), \ldots, p(H))$, with

$$
\begin{array}{lll}
p(A) = \frac{27}{64}, & p(B) = p(C) = \frac{3}{16}, & p(D) = \frac{1}{16}, \\
p(E) = p(F) = \frac{3}{64}, & p(G) = \frac{1}{32}, & p(H) = \frac{1}{64}.
\end{array}
$$

(a) Determine the information content of every letter and compute the entropy $H(\mathbf{p})$ of the source.

(b) Following Shannon's coding paradigm, find a binary prefix code associated with $\mathbf{p}$.

(c) Compute the average length $\bar{l}$ of the code words, and compare it with $H(\mathbf{p})$.

The most important consequence of the characterization of prefix codes via Kraft's inequality is the following theorem. Its small talk version could be: There is no lossless compression below entropy.

Theorem *Consider a memoryless source which produces N letters $a_0, a_1, \ldots,$ a_{N-1} according to the probability distribution $\mathbf{p} = (p_0, p_1, \ldots, p_{N-1})$.*
Let C be some associated binary prefix code, and

$$\bar{l} = \sum_{j=0}^{N-1} p_j l_j,$$

the average length of the code words (in bits per symbol).
Then: $H(\mathbf{p}) \leq \bar{l}$.
Moreover, the binary prefix codes constructed according to Shannon's idea satisfy the following inequality: $\bar{l} < H(\mathbf{p}) + 1$.

Proof (1) $H(\mathbf{p}) - \bar{l} \leq 0$:
$H(\mathbf{p}) - \bar{l} = -\sum_{j=0}^{N-1} p_j \mathrm{Log}_2 p_j - \sum_{j=0}^{N-1} p_j l_j = \frac{1}{\mathrm{Ln}\, 2} \cdot \sum_{j=0}^{N-1} p_j \mathrm{Ln} \left(\frac{2^{-l_j}}{p_j} \right)$
Now: $\mathrm{Ln}\, x \leq x - 1$ for $x > 0$, hence
$H(\mathbf{p}) - \bar{l} \leq \frac{1}{\mathrm{Ln}\, 2} \cdot \sum_{j=0}^{N-1} p_j \left(\frac{2^{-l_j}}{p_j} - 1 \right) = \frac{1}{\mathrm{Ln}\, 2} \cdot \sum_{j=0}^{N-1} \left(2^{-l_j} - p_j \right)$.
But, due to Kraft's inequality, we have: $\sum_{j=0}^{N-1} \left(2^{-l_j} - p_j \right) \leq 0$, and we are done.

(2) *Recall*: following Shannon's idea, one gets for the lengths of the code words associated with our symbols:

$$l_j - 1 < -\mathrm{Log}_2 p_j \leq l_j, \quad 0 \leq j \leq N - 1.$$

Summing up yields: $\sum_{j=0}^{N-1} (p_j l_j - p_j) < -\sum_{j=0}^{N-1} p_j \mathrm{Log}_2 p_j$, i.e.
$\bar{l} < H(\mathbf{p}) + 1$. □

Shannon Codes

Shannon coding is precisely the algorithmic realization of Shannon's coding paradigm:

Encode every source symbol into a binary word – the length of which equals the information content of the source symbol (rounded up to the next integer). We will obtain binary prefix codes. Unfortunately, Shannon codes are not always optimal (in a natural sense, which shall be made precise later) and were soon dethroned by Huffman coding. Why shall we altogether dwell on Shannon coding?

The principal reason is that *arithmetic coding* which is a very interesting "continuous" method of compaction (integrated in certain modes of JPEG and, more expressly, of JPEG 2000) is nothing but a dynamic version of Shannon coding. With the Shannon codes we are in the antechamber of arithmetic coding.

The idea of Shannon's algorithm is the following:

Consider a memoryless source producing N letters $a_0, a_1, \ldots, a_{N-1}$, according to the probability distribution $\mathbf{p} = (p_0, p_1, \ldots, p_{N-1})$.

Assume $p_0 \geq p_1 \geq \cdots \geq p_{N-1}$ (in order to guarantee that the following constructions will yield a *prefix* code). Associate with $\mathbf{p}$ a partition of the interval $[0, 1[$ in the following way:

$$
\begin{aligned}
A_0 &= 0, \\
A_1 &= p_0, \\
A_2 &= p_0 + p_1, \\
A_3 &= p_0 + p_1 + p_2, \\
&\;\;\vdots \\
A_N &= p_0 + p_1 + \cdots + p_{N-1} = 1.
\end{aligned}
$$

We note that, the length of every interval $[A_j, A_{j+1}[$ equals the probability of (the production of) the letter a_j:

$$
p_j = A_{j+1} - A_j, \quad 0 \leq j \leq N - 1.
$$

We shall associate with the letter a_j a binary word c_j, which will be *code word of the interval* $[A_j, A_{j+1}[$:

$$
c_j = c(A_j, A_{j+1}), \quad 0 \leq j \leq N - 1.
$$

Let us point out that the realization of Shannon's program demands that the length of the jth code word should be:

$$
l_j = \lceil I(a_j) \rceil = \lceil -\mathrm{Log}_2 p_j \rceil = \lceil -\mathrm{Log}_2 (A_{j+1} - A_j) \rceil, \quad 0 \leq j \leq N - 1.
$$

These considerations will oblige us to *define* the code word $c(A, B)$ of an interval $[A, B[\subset [0, 1[$ as follows:

$$
c(A, B) = \alpha_1 \alpha_2 \cdots \alpha_l \Longleftrightarrow A = 0 \cdot \alpha_1 \alpha_2 \cdots \alpha_l * \text{ (the beginning of the binary}
$$
$$
\text{notation of the real number } A\text{), with}
$$
$$
l = \lceil -\mathrm{Log}_2 (B - A) \rceil.
$$

We insist: the code word $c(A, B)$ of an interval $[A, B[$ is the initial segment of the binary notation of the left boundary A of this interval. One considers as much leading digits as the "information content of the interval" $-\mathrm{Log}_2(B - A)$ demands.

By the way: since the length $B - A$ of an interval $[A, B[\subset [0, 1[$ can actually be considered as the probability "of falling inside" – for certain evident geometric experiences – the value $\mathrm{Log}_2 \frac{1}{B-A}$ has indeed the flavour of an information content.

Exercises

Recall: binary notation of a real number A, $0 \leq A < 1$.

Assume that the development has already been established: $A = 0 \cdot \alpha_1 \alpha_2 \alpha_3 \alpha_4 \cdots$. Let us rediscover one by one the digits $\alpha_1, \alpha_2, \alpha_3, \alpha_4, \ldots$ Multiply by 2: $2A = \alpha_1 \cdot \alpha_2 \alpha_3 \alpha_4 \cdots$ (a comma-shift)

$$\text{If} \quad \begin{cases} 2A \geq 1, \text{ then } \alpha_1 = 1, \\ 2A < 1, \text{ then } \alpha_1 = 0. \end{cases}$$

First case: pass to $A^{(1)} = 2A - 1 = 0 \cdot \alpha_2 \alpha_3 \alpha_4 \cdots$,
Second case: pass to $A^{(1)} = 2A = 0 \cdot \alpha_2 \alpha_3 \alpha_4 \cdots$,
And so on...

Example Binary notation of $A = \frac{1}{11}$.

$$
\begin{aligned}
A &= 0 \cdot \alpha_1 \alpha_2 \alpha_3 \alpha_4 \cdots, \\
2A &= \frac{2}{11} < 1 \quad \Longrightarrow \quad \alpha_1 = 0, \\
4A &= \frac{4}{11} < 1 \quad \Longrightarrow \quad \alpha_2 = 0, \\
8A &= \frac{8}{11} < 1 \quad \Longrightarrow \quad \alpha_3 = 0, \\
16A &= \frac{16}{11} = 1 + \frac{5}{11} \quad \Longrightarrow \quad \alpha_4 = 1, \\
A^{(4)} &= \frac{5}{11} = 0 \cdot \alpha_5 \alpha_6 \alpha_7 \cdots, \\
2A^{(4)} &= \frac{10}{11} < 1 \quad \Longrightarrow \quad \alpha_5 = 0, \\
4A &= \frac{20}{11} = 1 + \frac{9}{11} \quad \Longrightarrow \quad \alpha_6 = 1, \\
A^{(6)} &= \frac{9}{11} = 0 \cdot \alpha_7 \alpha_8 \alpha_9 \cdots, \\
2A^{(6)} &= \frac{18}{11} = 1 + \frac{7}{11} \quad \Longrightarrow \quad \alpha_7 = 1, \\
A^{(7)} &= \frac{7}{11} = 0 \cdot \alpha_8 \alpha_9 \alpha_{10} \cdots, \\
2A^{(7)} &= \frac{14}{11} = 1 + \frac{3}{11} \quad \Longrightarrow \quad \alpha_8 = 1,
\end{aligned}
$$

$$A^{(8)} = \frac{3}{11} = 0 \cdot \alpha_9 \alpha_{10} \alpha_{11} \cdots,$$

$$2A^{(8)} = \frac{6}{11} < 1 \quad \Longrightarrow \quad \alpha_9 = 0,$$

$$4A^{(8)} = \frac{12}{11} = 1 + \frac{1}{11} \quad \Longrightarrow \quad \alpha_{10} = 1,$$

$$A^{(10)} = \frac{1}{11} = 0 \cdot \alpha_{11} \alpha_{12} \alpha_{13} \cdots = A.$$

The binary development of $A = \frac{1}{11} = 0 \cdot \overline{0001011101}$ (period length 10).

(1) Consider the subdivision of the interval $[0, 1[$ by iterated dichotomy:

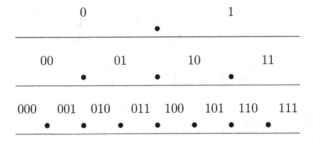

Let us encode these *standard-intervals* by the *paths* which point at them:

$$010 =\leftarrow\rightarrow\leftarrow \quad \text{points at} \quad \left[\frac{1}{4}, \frac{3}{8}\right[,$$

$$110 =\rightarrow\rightarrow\leftarrow \quad \text{points at} \quad \left[\frac{3}{4}, \frac{7}{8}\right[.$$

Show that the arithmetic code word $c(A, B)$ of a standard-interval $[A, B[$ equals the binary word that points at this interval (in the tree of dichotomy above).

Solution Let us show that the path $\alpha_1 \alpha_2 \cdots \alpha_l$ which points at the interval $[A, B[$ located on level l of our tree of dichotomy *is equal* to the l first bits (after the comma) of the binary notation of A.

Recursion on l:

$l = 1$: 0 and 1 are, respectively, the first bit of the binary notation of $0 = 0.0000 \ldots$ and of $\frac{1}{2} = 0.1000 \ldots$

$l \longmapsto l + 1$: consider $[A, B[$ on level $l + 1$ of our tree of dichotomy. $[A, B[$ is either the first half or the second half of an interval $[A^*, B^*[$ on level l.

By recursion hypothesis: $A^* = 0 \cdot \alpha_1 \alpha_2 \cdots \alpha_l$, and $\alpha_1 \alpha_2 \cdots \alpha_l$ points at $[A^*, B^*[$. If $[A, B[$ is to the left, then $\alpha_1 \alpha_2 \cdots \alpha_l 0$ points at $[A, B[$ and we have $A = A^* = 0 \cdot \alpha_1 \alpha_2 \cdots \alpha_l 0$.

If $[A, B[$ is to the right, then $\alpha_1 \alpha_2 \cdots \alpha_l 1$ points at $[A, B[$ and we have $A = A^* + \frac{1}{2}(B^* - A^*) = A^* + \frac{1}{2^{l+1}}$, i.e. the binary notation of A is $A = 0 \cdot \alpha_1 \alpha_2 \cdots \alpha_l 1$, as claimed.

(2) Find the following code words $c(A, B)$
   (a) $c\left(\frac{3}{8}, \frac{1}{2}\right)$
   (b) $c\left(\frac{1}{12}, \frac{7}{8}\right)$
   (c) $c\left(\frac{3}{5}, \frac{3}{4}\right)$
   (d) $c\left(\frac{1}{7}, \frac{1}{6}\right)$

(3) Determine all intervals $[A, B[$ such that $c(A, B) = 10101$.
   (Find first the standard-interval described by 10101, then think about the extent you can "deform" it without changing the code word.)

(4) Consider a memoryless source which produces N letters $a_0, a_1, \ldots, a_{N-1}$, according to the probability distribution $\mathbf{p} = (p_0, p_1, \ldots, p_{N-1})$.
   Assume $p_0 \geq p_1 \geq \cdots \geq p_{N-1}$.
   Show that the associated Shannon code is a *prefix code*.

Solution First, we have necessarily: $l_0 \leq l_1 \leq \cdots \leq l_{N-1}$ (where $l_j = \lceil I(a_j) \rceil = \lceil -\text{Log}_2 p_j \rceil$ is the length of the jth code word $0 \leq j \leq N - 1$). Let us show that the code word $c_j = c(A_j, A_{j+1})$ *cannot* be a prefix of the word $c_{j+1} = c(A_{j+1}, A_{j+2})$ for $0 \leq j \leq N - 2$. Otherwise we would have:

$$A_j = 0 \cdot \alpha_1 \alpha_2 \cdots \alpha_{l_j} *,$$
$$A_{j+1} = 0 \cdot \alpha_1 \alpha_2 \cdots \alpha_{l_j} *,$$

hence $p_j = A_{j+1} - A_j = 0 \cdot 0 \cdots 0*$ (with a block of at least l_j zeros after the comma) $\Longrightarrow p_j < 2^{-l_j} \Longrightarrow l_j < I(a_j)$, a contradiction.

Finally, if the code word c_j is a prefix of the code word c_k, $j < k$, then c_j must be necessarily a prefix of the word c_{j+1} (why?), and we can conclude.

(5) A memoryless source produces four letters A, B, C, D, with

$$p(A) = \frac{1}{2}, \quad p(B) = \frac{1}{4}, \quad p(C) = p(D) = \frac{1}{8}.$$

Write down the Shannon code word of $BADACABA$.

(6) Consider the source which produces the eight letters $a_0, a_1, \ldots, a_7$, according to the probability distribution $\mathbf{p} = (p_0, p_1, \ldots, p_7)$ where $p_0 = \frac{1}{2}, p_1 = \frac{1}{4}, p_2 = p_3 = \frac{1}{16}, p_4 = p_5 = p_6 = p_7 = \frac{1}{32}$.
   Find the associated Shannon code.

(7) Our memoryless source produces the eight letters A, B, C, D, E, F, G, H according to the probability distribution $\mathbf{p} = (p(A), p(B), \ldots, p(H))$ with

$$p(A) = \frac{27}{64} \qquad p(B) = p(C) = \frac{3}{16} \quad p(D) = \frac{1}{16},$$
$$p(E) = p(F) = \frac{3}{64} \qquad p(G) = \frac{1}{32} \qquad p(H) = \frac{1}{64}.$$

   (a) Find the associated Shannon code.
   (b) Compute the average word length $\bar{l}$ of the code words.

The Huffman Algorithm

Four years after Shannon's seminal papers, the Huffman algorithm appears, with universal acclaim. Being of utmost mathematical simplicity, it yields nevertheless the best – and thus definitive – algorithmic solution of the prefix coding problem for memoryless discrete sources.

Example Recall our source which produces the eight letters $a_0, a_1, \ldots, a_7$, according to the probability distribution $\mathbf{p} = (p_0, p_1, \ldots, p_7)$ with $p_0 = \frac{1}{2}, p_1 = \frac{1}{4}, p_2 = p_3 = \frac{1}{16}, p_4 = p_5 = p_6 = p_7 = \frac{1}{32}$.
We did encode as follows:

$$
\begin{array}{ll}
a_0 \longmapsto 0 & a_4 \longmapsto 11100 \\
a_1 \longmapsto 10 & a_5 \longmapsto 11101 \\
a_2 \longmapsto 1100 & a_6 \longmapsto 11110 \\
a_3 \longmapsto 1101 & a_7 \longmapsto 11111
\end{array}
$$

With the standard-interval coding of the preceding section in mind, where the code words are *paths* in a binary tree, one could come up with the following idea.

Let us interpret the code words above as paths in a binary tree which admits the symbols $a_0, a_1, \ldots, a_7$ as leaves (i.e. as terminal nodes).

We will obtain the following structure:

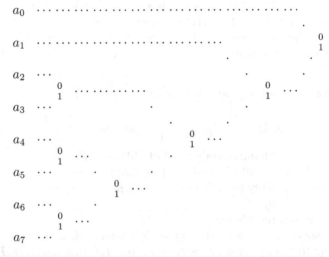

How can we generate, in general, this binary tree in function of the source symbols, more precisely: in function of the probability distribution $\mathbf{p}$ which describes the production of the source?

Let us adopt the following viewpoint.

We shall consider the given symbols as the leaves (as the terminal nodes) of a binary tree *which has to be constructed*. The *code words* associated with the symbols will be the *paths* towards the leaves (the symbols).

We note: the lesser the probability of a letter is, the longer should be the path towards this letter. The algorithm will have to create nodes (antecedents), conducted primarily by the *rare* letters; thus we shall need a numerical control by a *weighting* of the nodes.

The most primitive algorithm that we can invent – based on these design-patterns – will actually be the best one:

Algorithm of Huffman for the construction of a *weighted* binary tree:

Every step will create a new node, antecedent for two nodes taken from a list of candidates.

Start: Every source symbol is a weighted node (a candidate), the weight of which is its probability.

Step: The two nodes of minimal weight (in the actual list of candidates) create an antecedent whose weight will be the sum of the weights of its successors; it replaces them in the list of the candidates.

End: There remains a single node (of weight equal to 1) in the list of the candidates.

This is a recursive algorithm. Note that with every step the number of (couples of) nodes searching an antecedent becomes smaller and smaller. On the other hand, the sum over all weights remains always constant, i.e. equal to 1.

Attention: "the two nodes of minimal weight" are, in general, *not unique*. You frequently have to make a choice. So, the result of Huffman's algorithm is, in general, far from unique.

Back to our example:

a_0 $\frac{1}{2}$ ·

a_1 $\frac{1}{4}$ ·   1

a_2 $\frac{1}{16}$ · · ·

$\frac{1}{8}$ · · · · · · · · · · · · $\frac{1}{2}$ · · ·

a_3 $\frac{1}{16}$ · · ·

a_4 $\frac{1}{32}$ · · · $\frac{1}{4}$ · · ·

$\frac{1}{16}$ · · ·

a_5 $\frac{1}{32}$ · · ·

$\frac{1}{8}$ · · ·

a_6 $\frac{1}{32}$ · · ·

$\frac{1}{16}$ · · ·

a_7 $\frac{1}{32}$ · · ·

First, it is a_6 and a_7 which find an antecedent with weight $\frac{1}{16}$ (we make a choice at the end of the list). At the next step, we have no choice: it is a_4 and a_5 which have minimal weight and will thus find their antecedent of weight $\frac{1}{16}$. Now, we have six nodes as candidates, four of which have weight $\frac{1}{16}$. Once more, we shall choose at the end of the list, and we find this way a common antecedent for a_4, a_5, a_6 and a_7, the weight of which is $\frac{1}{8}$. And so on...

Exercises

(1) Our memoryless source producing the eight letters A, B, C, D, E, F, G, H according to the probability distribution $\mathbf{p} = (p(A), p(B), \ldots, p(H))$ with

$$p(A) = \tfrac{27}{64}, \qquad p(B) = p(C) = \tfrac{3}{16}, \qquad p(D) = \tfrac{1}{16},$$
$$p(E) = p(F) = \tfrac{3}{64}, \qquad p(G) = \tfrac{1}{32}, \qquad p(H) = \tfrac{1}{64}.$$

   (a) Find the associated Huffman code.
   (b) Compute the average word length $\bar{l}$ of the code words.

(2) Consider a source which produces the 12 letters $a_0, a_1, \ldots, a_{11}$ according to the probability distribution $\mathbf{p} = (p_0, p_1, \ldots, p_{11})$ with

$$p_0 = \tfrac{3}{16}, \qquad p_1 = \tfrac{5}{32}, \qquad p_2 = \tfrac{1}{8},$$
$$p_3 = p_4 = \tfrac{3}{32}, \quad p_5 = p_6 = p_7 = p_8 = \tfrac{1}{16}, \quad p_9 = p_{10} = p_{11} = \tfrac{1}{32}.$$

   (a) Compute $H(\mathbf{p})$.
   (b) Find the associated Huffman code, and compare $\bar{l}$, the average word length of the code words with $H(\mathbf{p})$.

(3) A memoryless source producing the three letters A, B, C with the probabilities $p(A) = \tfrac{3}{4}$, $p(B) = \tfrac{3}{16}$ and $p(C) = \tfrac{1}{16}$.
   (a) Compute the entropy of the source, find the associated Huffman code and the average word length for the (three) code words.
   (b) Consider now the same source, but as producer of the nine symbols AA, AB, AC, BA, BB, BC, CA, CB, CC, according to the product distribution (i.e. $p(AB) = p(BA) = \tfrac{9}{64}$).
   Generate the associated Huffman code, and compute the average word length of the code words *per initial symbol*. Compare with (a).

Remark The *compressed bit-rate* ρ of an encoder is defined as follows:

$$\rho = \frac{\text{average length of the code words}}{\text{average length of the source symbols}}.$$

It is clear that this definition makes sense only when complemented by an evaluation of the stream of source symbols: what is the average length of the source symbols – in *bits* per symbol?

(4) A binary source, which we consider as memoryless on words of length 4. We shall adopt the hexadecimal notation (example: $d = 1101$).
We observe the following probability distribution:
$p(0) = 0.40, p(4) = 0.01, p(8) = 0.01, p(c) = 0.05,$
$p(1) = 0.01, p(5) = 0.03, p(9) = 0.04, p(d) = 0.01,$
$p(2) = 0.01, p(6) = 0.04, p(a) = 0.03, p(e) = 0.01,$
$p(3) = 0.05, p(7) = 0.01, p(b) = 0.01, p(f) = 0.28.$
Generate the associated Huffman code, and compute the compressed bit-rate.

(5) A facsimile system for transmitting line-scanned documents uses black runlengths and white runlengths as the source symbols. We observe the following probability distribution:
$p(B1) = 0.05, p(B5) = 0.01, p(W1) = 0.02, p(W5) = 0.01,$
$p(B2) = 0.02, p(B6) = 0.02, p(W2) = 0.02, p(W6) = 0.01,$
$p(B3) = 0.01, p(B7) = 0.01, p(W3) = 0.01, p(W7) = 0.01,$
$p(B4) = 0.10, p(B8) = 0.25, p(W4) = 0.05, p(W8) = 0.40.$
As to the notation: $B3 \equiv 000, W5 \equiv 11111$.
(a) Find the Huffman code for this system.
(b) Compute the compressed bit-rate.

(6) A Huffman code associated with an alphabet of eight letters; we have the following eight code words (where two of them are masked):
00, 10, 010, 1100, 1101, 1111, $\mathbf{w}_1$, $\mathbf{w}_2$. Find $\mathbf{w}_1$ and $\mathbf{w}_2$.

(7) An alphabet of eight letters $a_0, a_1, a_2, a_3, a_4, a_5, a_6, a_7$.
A Huffman encoder has associated the following eight code words:
$c_0 = 00, c_1 = 01, c_2 = 100, c_3 = 101, c_4 = 1100, c_5 = 1101, c_6 = 1110,$
$c_7 = 1111.$
Find a probability distribution $\mathbf{p} = (p_0, p_1, p_2, p_3, p_4, p_5, p_6, p_7)$ such that the Shannon encoder yields the same list of code words.

(8) A binary source which is memoryless on the eight binary triples. A Huffman encoder associates the eight code words 00, 01, 100, 101, 1100, 1101, 1110, 1111.
(a) Find a probability distribution which fits with this code.
(b) Is it possible to choose a probability distribution which gives rise to a compressed bit-rate of 70%?

(9) Let a_0 be the symbol of highest probability p_0 of an alphabet which has N symbols ($N \geq 3$). A Huffman encoder associates a binary code word of length l_0. Show the following assertions:
(a) If $p_0 > \frac{4}{5}$, then $l_0 = 1$.
(b) If $p_0 < \frac{1}{3}$, then $l_0 \geq 2$.

(10) The optimal questionnaire.
You would like to participate in a TV game: you will have to find the profession of a person ("chosen at random") by three yes or no questions. You look at the statistics: there are 16 main professions $P1, P2, \ldots, P16$, occurring with the following frequencies:

$$p(P1) = 0.40, \ p(P5) = 0.05, \ p(P9) = 0.02, \ \ p(P13) = 0.01,$$
$$p(P2) = 0.18, \ p(P6) = 0.04, \ p(P10) = 0.02, \ p(P14) = 0.01,$$
$$p(P3) = 0.10, \ p(P7) = 0.03, \ p(P11) = 0.02, \ p(P15) = 0.01,$$
$$p(P4) = 0.06, \ p(P8) = 0.03, \ p(P12) = 0.01, \ p(P16) = 0.01.$$

(a) Find the strategy for the optimal questionnaire.

(b) Will you have a good chance (with only three questions)?

Huffman Coding in JPEG

Situation *JPEG treats a digital image as a sequence of blocks of 8×8 pixels. More precisely, a data unit will be a triple of 8×8 matrices. The first one for the pixel-values of luminance (Y), the two others for the pixel-values of chrominance (Cb, Cr).*

A linear invertible transformation (the 2D Discrete Cosine Transform) will transform each of these three matrices in a 8×8 matrix of the following type:

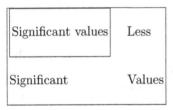

In lossy compression mode, an appropriate quantization procedure will finally set to zero most of the less significant values.

We ultimately come up with quantized schemes (of 64 *integers*) of the following type:

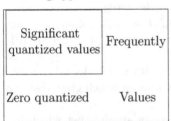

The value of the DC coefficient (*direct current*) in the left upper corner of the matrix will not be interesting – at least in the present context.

The Huffman coding deals with the 63 AC coefficients (*alternating current*), the absolute values of which are – in general – sensibly smaller than (the absolute value of) the dominant DC coefficient.

We shall make use of a sequential zigzag reading according to the scheme on the top of the next page.

The encoding concerns *the sequence of the non-zero coefficients* in the zigzag reading of the quantized scheme. It is clear that we also have to take into account the zero runlengths between the non-zero coefficients.

DC	1	5	6	14	15	27	28
2	4	7	13	16	26	29	42
3	8	12	17	25	30	41	43
9	11	18	24	31	40	44	53
10	19	23	32	39	45	52	54
20	22	33	38	46	51	55	60
21	34	37	47	50	56	59	61
35	36	48	49	57	58	62	63

Zigzag ordering of the quantized coefficients.

In order to prepare Huffman coding conveniently, we begin with a hierarchy of 10 categories for the non-zero coefficients:

1	-1	1
2	$-3, -2$	$2, 3$
3	$-7, -6, -5, -4$	$4, 5, 6, 7$
4	$-15, -14, \ldots, -9, -8$	$8, 9, \ldots, 14, 15$
5	$-31, -30, \ldots, -17, -16$	$16, 17, \ldots, 30, 31$
6	$-63, -62, \ldots, -33, -32$	$32, 33, \ldots, 62, 63$
7	$-127, -126, \ldots, -65, -64$	$64, 65, \ldots, 126, 127$
8	$-255, -254, \ldots, -129, -128$	$128, 129, \ldots, 254, 255$
9	$-511, 510, \ldots, -257, -256$	$256, 257, \ldots, 510, 511$
10	$-1,023, -1,022, \ldots, -513, -512$	$512, 513, \ldots, 1,022, 1,023$

Attention *There is a tacit convention concerning the encoding of all these integers; in category 4 for example, the code words will have four bits:*
$-15 \mapsto 0000$, $-14 \mapsto 0001, \ldots, -9 \mapsto 0110$, $-8 \mapsto 0111$, $8 \mapsto 1000$, $9 \mapsto 1001, \ldots, 14 \mapsto 1110$, $15 \mapsto 1111$.

We observe that a non-zero coefficient occurring in the sequential reading of a quantized scheme can be characterized by three parameters:

(1) The number of zeros which separate it from its non-zero predecessor.
(2) Its category.
(3) Its number within the category.

Example Consider the sequence $0\,8\,0\,0\,-2\,0\,4\,0\,0\,0\,1\ldots$
This means for

	Runlength/category	Value within the cat.
8	1/4	1000
−2	2/2	01
4	1/3	100
1	3/1	1

In order to be able to encode the sequential reading of the quantized coefficients, we need only a coding table for the symbols of the type runlength/category.

We shall give the table for the *luminance* AC coefficients.

The table has been developed by JPEG (*Joint Photographic Experts Group*) from the average statistics of a large set of images with 8 bit precision. It was not meant to be a default table, but actually it is.

Remark On two particular symbols.

(1) (EOB) ≡ *end of block* indicates the end of the non-zero coefficients in the sequence of the 63 AC coefficients to be encoded. The code word for this happy event will be 1010.
(2) (ZRL) ≡ *zero run list* indicates the outcome of the integer 0 preceded by a block of 15 zeros.[2]

---

[2] Attention, our zeros are zeros as integers – and not as bits!

Runlength/cat.	Code word	Runlength/cat.	Code word
0/0 (EOB)	1010	3/9	1111111110010100
0/1	00	3/a	1111111110010101
0/2	01	4/1	111011
0/3	100	4/2	1111111000
0/4	1011	4/3	1111111110010110
0/5	11010	4/4	1111111110010111
0/6	1111000	4/5	1111111110011000
0/7	11111000	4/6	1111111110011001
0/8	1111110110	4/7	1111111110011010
0/9	1111111110000010	4/8	1111111110011011
0/a	1111111110000011	4/9	1111111110011100
1/1	1100	4/a	1111111110011101
1/2	11011	5/1	1111010
1/3	1111001	5/2	11111110111
1/4	111110110	5/3	1111111110011110
1/5	11111110110	5/4	1111111110011111
1/6	1111111110000100	5/5	1111111110100000
1/7	1111111110000101	5/6	1111111110100001
1/8	1111111110000110	5/7	1111111110100010
1/9	1111111110000111	5/8	1111111110100011
1/a	1111111110001000	5/9	1111111110100100
2/1	11100	5/a	1111111110100101
2/2	11111001	6/1	1111011
2/3	1111110111	6/2	111111110110
2/4	111111110100	6/3	1111111110100110
2/5	1111111110001001	6/4	1111111110100111
2/6	1111111110001010	6/5	1111111110101000
2/7	1111111110001011	6/6	1111111110101001
2/8	1111111110001100	6/7	1111111110101010
2/9	1111111110001101	6/8	1111111110101011
2/a	1111111110001110	6/9	1111111110101100
3/1	111010	6/a	1111111110101101
3/2	111110111	7/1	11111010
3/3	111111110101	7/2	111111110111
3/4	1111111110001111	7/3	1111111110101110
3/5	1111111110010000	7/4	1111111110101111
3/6	1111111110010001	7/5	1111111110110000
3/7	1111111110010010	7/6	1111111110110001
3/8	1111111110010011	7/7	1111111110110010

Runlength/cat.	Code word	Runlength/cat.	Code word
7/8	1111111110110011	$b/7$	1111111111010101
7/9	1111111110110100	$b/8$	1111111111010110
7/a	1111111110110101	$b/9$	1111111111010111
8/1	111111000	b/a	1111111111011000
8/2	111111111000000	$c/1$	1111111010
8/3	1111111110110110	$c/2$	1111111111011001
8/4	1111111110110111	$c/3$	1111111111011010
8/5	1111111110111000	$c/4$	1111111111011011
8/6	1111111110111001	$c/5$	1111111111011100
8/7	1111111110111010	$c/6$	1111111111011101
8/8	1111111110111011	$c/7$	1111111111011110
8/9	1111111110111100	$c/8$	1111111111011111
8/a	1111111110111101	$c/9$	1111111111100000
9/1	111111001	c/a	1111111111100001
9/2	1111111110111110	$d/1$	11111111000
9/3	1111111110111111	$d/2$	1111111111100010
9/4	1111111111000000	$d/3$	1111111111100011
9/5	1111111111000001	$d/4$	1111111111100100
9/6	1111111111000010	$d/5$	1111111111100101
9/7	1111111111000011	$d/6$	1111111111100110
9/8	1111111111000100	$d/7$	1111111111100111
9/9	1111111111000101	$d/8$	1111111111101000
9/a	1111111111000110	$d/9$	1111111111101001
$a/1$	111111010	d/a	1111111111101010
$a/2$	1111111111000111	$e/1$	1111111111101011
$a/3$	1111111111001000	$e/2$	1111111111101100
$a/4$	1111111111001001	$e/3$	1111111111101101
$a/5$	1111111111001010	$e/4$	1111111111101110
$a/6$	1111111111001011	$e/5$	1111111111101111
$a/7$	1111111111001100	$e/6$	1111111111110000
$a/8$	1111111111001101	$e/7$	1111111111110001
$a/9$	1111111111001110	$e/8$	1111111111110010
a/a	1111111111001111	$e/9$	1111111111110011
$b/1$	1111111001	e/a	1111111111110100
$b/2$	1111111111010000	$f/0$ (ZRL)	11111111001
$b/3$	1111111111010001	$f/1$	1111111111110101
$b/4$	1111111111010010	$f/2$	1111111111110110
$b/5$	1111111111010011	$f/3$	1111111111110111
$b/6$	1111111111010100	$f/4$	1111111111111000
$f/5$	1111111111111001	$f/8$	1111111111111100
$f/6$	1111111111111010	$f/9$	1111111111111101
$f/7$	1111111111111011	f/a	1111111111111110

Example Consider a luminance data unit in gradation, its 2D DCT image, then the quantized scheme[3]

30	30	30	30	30	30	30	30
60	60	60	60	60	60	60	60
90	90	90	90	90	90	90	90
120	120	120	120	120	120	120	120
150	150	150	150	150	150	150	150
180	180	180	180	180	180	180	180
210	210	210	210	210	210	210	210
240	240	240	240	240	240	240	240

$\longmapsto$

1,080	0	0	0	0	0	0	0
−546.6	0	0	0	0	0	0	0
0	0	0	0	0	0	0	0
−57.1	0	0	0	0	0	0	0
0	0	0	0	0	0	0	0
−17	0	0	0	0	0	0	0
0	0	0	0	0	0	0	0
−4.3	0	0	0	0	0	0	0

$\longmapsto$

68	0	0	0	0	0	0	0
−46	0	0	0	0	0	0	0
0	0	0	0	0	0	0	0
−4	0	0	0	0	0	0	0
0	0	0	0	0	0	0	0
−1	0	0	0	0	0	0	0
0	0	0	0	0	0	0	0
0	0	0	0	0	0	0	0

$\Longleftarrow$

68 is the round of $\frac{1,080}{16}$

−46 is the round of $\frac{-546.6}{12}$

−4 is the round of $\frac{-57.1}{14}$

−1 is the round of $\frac{-17}{24}$

0 is the round of $\frac{-4.3}{72}$

The sequence of the 63 AC coefficients:

0 −46 0 0 0 0 0 0 −4 0 0 0 0 0 0 0 0 0 0 0 −1 0 0 0 0 0 0 0 0 0 0 0 0 0 0 0 0 0 0 0 0 0 0 0 0 0 0 0 0 0 0 0 0 0 0 0 0 0 0 0 0 0 0.

We have to encode the following symbols: $(1/6)[n^\circ 17](6/3)[n^\circ 3](a/1)[n^\circ 0]$ (EOB).

The code word: 1111111110000100 010001 1111111110100110 011 111111010 0 1010.[4]

As to the compressed bit-rate, we have associated with 64 8-bit bytes (the 8×8 luminance values) a code word of length 55. Hence: $\rho = \frac{55}{512} = 0.11$.

It is evident that the code word permits the reconstruction of the matrix of the quantized values (the quantized DC coefficient has its particular treatment...). Dequantization simply means multiplication with the divisorscheme that has been used for quantization. In our particular case, we get the following dequantized matrix:

---

[3] Quantization is done by means of a fixed scheme of 8×8 divisors, followed by a round to the next integer.

[4] *Attention*: we have got a prefix code – the separating blanks in our notation are redundant and exist only for the convenience of the reader...

$$
\begin{array}{ccccccc}
68\cdot16 & 0 & 0 & 0 & 0 & 0 & 0 & 0\\
-46\cdot12 & 0 & 0 & 0 & 0 & 0 & 0 & 0\\
0 & 0 & 0 & 0 & 0 & 0 & 0 & 0\\
-4\cdot14 & 0 & 0 & 0 & 0 & 0 & 0 & 0\\
0 & 0 & 0 & 0 & 0 & 0 & 0 & 0\\
-1\cdot24 & 0 & 0 & 0 & 0 & 0 & 0 & 0\\
0 & 0 & 0 & 0 & 0 & 0 & 0 & 0\\
0.72 & 0 & 0 & 0 & 0 & 0 & 0
\end{array}
=
\begin{array}{ccccccc}
1{,}088 & 0 & 0 & 0 & 0 & 0 & 0 & 0\\
-552 & 0 & 0 & 0 & 0 & 0 & 0 & 0\\
0 & 0 & 0 & 0 & 0 & 0 & 0 & 0\\
-56 & 0 & 0 & 0 & 0 & 0 & 0 & 0\\
0 & 0 & 0 & 0 & 0 & 0 & 0 & 0\\
-24 & 0 & 0 & 0 & 0 & 0 & 0 & 0\\
0 & 0 & 0 & 0 & 0 & 0 & 0 & 0\\
0 & 0 & 0 & 0 & 0 & 0 & 0 & 0
\end{array}
$$

This is the moment where the loss of information, due to quantization, becomes apparent. The decompression will be finished after transforming the matrix of dequantized values back, and making the necessary rounds (do not forget: we did start with 8-bit bytes). We obtain:

30	30	30	30	30	30	30	30
61	61	61	61	61	61	61	61
91	91	91	91	91	91	91	91
119	119	119	119	119	119	119	119
153	153	153	153	153	153	153	153
181	181	181	181	181	181	181	181
211	211	211	211	211	211	211	211
242	242	242	242	242	242	242	242

Exercises

(1) The Huffman coding described above has a two-level structure: there are code words for the type of the integer to be encoded, and there are the code words for its numerical value. Writing down the total Huffman tree for this coding, what will be the number of leaves (of code words)?

(2) We shall treat five 8×8 matrices of luminance values, together with their quantized 2D DCT matrices. Establish in every case the associated Huffman coding, and compute the compressed bit-rate.

(a)

$$
\begin{array}{cccccccc}
159 & 152 & 142 & 134 & 133 & 140 & 149 & 155\\
176 & 170 & 162 & 156 & 157 & 163 & 171 & 177\\
132 & 129 & 123 & 120 & 121 & 126 & 132 & 136\\
72 & 71 & 69 & 68 & 69 & 70 & 72 & 74\\
69 & 70 & 72 & 73 & 73 & 71 & 69 & 67\\
123 & 126 & 131 & 134 & 133 & 129 & 123 & 119\\
157 & 163 & 171 & 177 & 176 & 170 & 162 & 156\\
132 & 139 & 149 & 157 & 158 & 151 & 142 & 135
\end{array}
\longmapsto
\begin{array}{cccccccc}
64 & 0 & 0 & 0 & 0 & 0 & 0 & 0\\
0 & 0 & 4 & 0 & 0 & 0 & 0 & 0\\
17 & 0 & 0 & 0 & 0 & 0 & 0 & 0\\
0 & 0 & 0 & 0 & 0 & 0 & 0 & 0\\
-9 & 0 & 0 & 0 & 0 & 0 & 0 & 0\\
0 & 0 & 0 & 0 & 0 & 0 & 0 & 0\\
0 & 0 & 0 & 0 & 0 & 0 & 0 & 0\\
0 & 0 & 0 & 0 & 0 & 0 & 0 & 0
\end{array}
$$

(b)

$$
\begin{array}{cccccccc}
83 & 89 & 91 & 84 & 73 & 68 & 75 & 83\\
96 & 98 & 96 & 86 & 78 & 82 & 99 & 114\\
82 & 85 & 84 & 75 & 67 & 69 & 83 & 97\\
88 & 99 & 108 & 107 & 99 & 94 & 97 & 104\\
88 & 101 & 115 & 118 & 112 & 108 & 111 & 116\\
95 & 103 & 110 & 107 & 99 & 97 & 105 & 114\\
122 & 130 & 136 & 131 & 120 & 114 & 117 & 124\\
96 & 111 & 127 & 131 & 121 & 109 & 105 & 105
\end{array}
\longmapsto
\begin{array}{cccccccc}
50 & 0 & 0 & -3 & 0 & 0 & 0 & 0\\
-9 & 0 & 2 & 0 & 0 & 0 & 0 & 0\\
0 & 1 & 0 & 0 & 0 & 0 & 0 & 0\\
1 & 0 & 0 & 0 & 0 & 0 & 0 & 0\\
0 & 0 & -1 & 0 & 0 & 0 & 0 & 0\\
0 & 0 & 0 & 0 & 0 & 0 & 0 & 0\\
-1 & 0 & 0 & 0 & 0 & 0 & 0 & 0\\
0 & 0 & 0 & 0 & 0 & 0 & 0 & 0
\end{array}
$$

(c)

25	32	41	46	47	48	51	53
45	57	75	90	94	89	80	73
26	34	43	48	46	41	37	34
82	89	92	84	72	69	78	90
87	101	109	98	80	79	102	126
61	78	91	84	67	65	87	110
112	122	131	129	123	126	141	156
94	88	80	77	85	106	131	149

$\longmapsto$

41	−5	1	−3	0	0	0	0
−15	2	−4	1	0	0	0	0
0	−3	0	1	0	0	0	0
1	0	0	−1	0	0	0	0
0	0	1	0	0	0	0	0
0	0	0	0	0	0	0	0
−3	0	0	0	0	0	0	0
0	0	0	0	0	0	0	0

(d)

133	139	89	198	114	91	151	114
126	127	95	207	136	110	148	124
134	128	186	73	32	158	93	127
104	128	75	40	246	129	143	141
141	143	129	246	40	75	128	104
127	93	158	32	73	186	128	134
124	148	110	136	207	95	127	126
114	151	91	114	198	89	139	133

$\longmapsto$

63	0	0	0	1	0	−1	0
0	1	0	−1	0	1	0	−1
2	0	−2	0	1	0	−1	0
0	1	0	−1	0	1	0	−1
0	0	−1	0	1	0	−1	0
0	−1	0	1	0	−1	0	1
−1	0	1	0	−1	0	1	0
0	1	0	−1	0	1	0	−1

(e)

65	61	68	59	69	60	67	63
61	72	52	78	50	76	56	67
68	52	81	44	84	47	76	60
59	78	44	88	40	84	50	69
69	50	84	40	88	44	78	59
60	76	47	84	44	81	52	68
67	56	76	50	78	52	72	61
63	67	60	69	59	68	61	65

$\longmapsto$

32	0	0	0	0	0	0	0
0	0	0	0	0	0	0	0
0	0	0	0	0	0	0	0
0	0	0	0	0	0	0	0
0	0	0	0	0	0	0	0
0	0	0	0	0	0	0	0
0	0	0	0	0	0	0	0
0	0	0	0	0	0	0	1

We observe a stubborn propagation of non-zero values in the quantized scheme (d). This comes from a (pictorially) perverse distribution of the luminance values in the initial scheme. Note that in general the zigzag ordering has the property that the probability of coefficients being zero is an approximately monotonic increasing function of the index.

Huffman Coding is Optimal

Situation *A memoryless source, producing N letters $a_0, a_1, \ldots, a_{N-1}$ according to the fixed probability distribution $\mathbf{p} = (p_0, p_1, \ldots, p_{N-1})$.*

Now consider *all* associated binary *prefix* codes.
Such a binary prefix code C is *optimal* : $\Longleftrightarrow$
The average word length $\bar{l} = p_0 l_0 + p_1 l_1 + \cdots + p_{N-1} l_{N-1}$ of its words is *minimal*. (Note that $\bar{l} = \bar{l}(l_0, l_1, \ldots, l_{N-1})$ is a function of the lengths of the N code words associated with $a_0, a_1, \ldots, a_{N-1}$; the probabilities $p_0, p_1, \ldots, p_{N-1}$ are the *constants* of the problem.)

  Our goal: we shall show that *the Huffman algorithm necessarily produces an optimal binary prefix code.*

But first several characteristic properties of optimal codes:

Proposition *Let C be an optimal binary prefix code, associated with* $\mathbf{p} = (p_0, p_1, \ldots, p_{N-1})$.

Then we necessarily have:

1. $p_j > p_k \Longrightarrow l_j \le l_k$.
2. *The code will have an even number of words of maximal length.*
3. *Whenever several code words have the same length, two of them will be equal except for the last bit.*

Proof 1. Assume $p_j > p_k$ *and* $l_j > l_k$; then

$$(p_j - p_k)(l_j - l_k) > 0,$$

i.e.

$$p_j l_j + p_k l_k > p_j l_k + p_k l_j.$$

This *shows*: if we exchange the code words of a_j and of a_k, then we get a better code.

2. Otherwise there would exist a code word of maximal length without a partner which differs only in the last bit. We can suppress its last bit, and will obtain a word which is not a code word (prefix property!). This permits us to change to a better code.

3. Consider all code words of length l (arbitrary, but fixed). Assume that all these words remain distinct when skipping everywhere the last bit. Due to the prefix property, we would thus get a better code.   □

The Huffman codes are optimal: this is an immediate consequence of the following proposition.

Proposition *Consider a source $\mathbb{S}$ of N states, controlled by the probability distribution* $\mathbf{p} = (p_0, p_1, \ldots, p_{N-1})$.

Replace the two symbols a_{j_1} and a_{j_2} of smallest probabilities by a single symbol $a_{(j_1, j_2)}$ with probability $p_{(j_1, j_2)} = p_{j_1} + p_{j_2}$. Let $\mathbb{S}'$ be the source of $N-1$ states we get this way.

Let C' be an optimal binary prefix code for $\mathbb{S}'$, and let $\mathbf{x}$ be the code word of $a_{(j_1, j_2)}$.

Let C be the ensuing binary prefix code for $\mathbb{S}$:

$$a_{j_1} \longmapsto \mathbf{x}0,$$
$$a_{j_2} \longmapsto \mathbf{x}1.$$

Then C is optimal for $\mathbb{S}$.

Proof The lengths of the code words (L for C', l for C):

$$l_j = \begin{cases} L_{(j_1, j_2)} + 1, & \text{if } j = j_1, j_2, \\ L_j, & \text{else.} \end{cases}$$

One gets for the *average lengths* ($\overline{L}$ for C', $\overline{l}$ for C):

$$\overline{l} = \sum_{j \ne j_1, j_2} p_j l_j + p_{j_1} l_{j_1} + p_{j_2} l_{j_2} = \sum_{j \ne j_1, j_2} p_j L_j + p_{(j_1, j_2)} L_{(j_1, j_2)} + p_{j_1} + p_{j_2} = \overline{L} + p_{j_1} + p_{j_2}.$$

But $p_{j_1} + p_{j_2}$ is a *constant* for our optimality arguments (we make the word lengths vary), and it is the constant of the smallest possible difference between $\overline{l}$ and $\overline{L}$ (due to the choice of p_{j_1} and p_{j_2}!). Thus, $\overline{L}$ minimal $\Longrightarrow \overline{l}$ minimal.   □

Corollary *The Huffman algorithm produces optimal binary prefix codes.*

In particular, the average word length $\bar{l}$ of the code words is constant for all Huffman codes associated with a fixed probability distribution $\mathbf{p}$ (note that you will be frequently obliged to make choices when constructing Huffman trees).

Observation   *For every Huffman code associated with a fixed probability distribution $\mathbf{p}$ we have the estimation $H(\mathbf{p}) \leq \bar{l} < H(\mathbf{p}) + 1$ (this is true for Shannon coding, hence a forteriori for Huffman coding).*

Exercises

(1) Does there exist a binary prefix code which consists of three words of length 3, of five words of length 4 and of nine words of length 5?
   Does there exist a Huffman code which consists of three words of length 3, of five words of length 4 and of nine words of length 5?

(2) Show: every Huffman code is a Shannon code.
   More precisely: let C be a Huffman code (given by its binary tree); then there exists a probability distribution $\mathbf{p}$ such that the set of code words of C is the associated Shannon code.

(3) Let $\{A, B, C, D\}$ be an alphabet of four letters, with $p(A) \geq p(B) \geq p(C) \geq p(D)$.
   (a) Find all associated Huffman codes.
   (b) Give an example of a Shannon code (in choosing appropriate probabilities) which is not a Huffman code.

(4) Is an *optimal* binary prefix code necessarily a Huffman code?

Approximation of the Entropy via Block Encoding

Consider a memoryless source, producing the N letters $a_0, a_1, \ldots, a_{N-1}$, according to the probability distribution $\mathbf{p} = (p_0, p_1, \ldots, p_{N-1})$.
Let us change our outlook:
We shall take the *words of length n* ($n \geq 1$) as our new production units:
$\mathbf{x} = a_{j_1} a_{j_2} \cdots a_{j_n}$ will be the notation for these "big letters".
The product probability distribution:

$$\mathbf{p}^{(n)} = \left(\prod p_{j_1} p_{j_2} \cdots p_{j_n} \right)_{0 \leq j_1, j_2, \ldots, j_n \leq N-1}$$

- (there are N^n words of length n on an alphabet of N elements).

   Recall

$$H(\mathbf{p}^{(n)}) = n \cdot H(\mathbf{p})$$

(this was a previous exercise).

Proposition *A memoryless source, producing the N letters $a_0, a_1, \ldots, a_{N-1}$, according to the probability distribution $\mathbf{p} = (p_0, p_1, \ldots, p_{N-1})$.*

Let us pass to an encoding of blocks in n letters.

Let C be an associated Huffman code, and let $\bar{l}_i$ be the average word length of the code words per initial symbol.

Then we have:

$$H(\mathbf{p}) \le \bar{l}_i < H(\mathbf{p}) + \frac{1}{n}.$$

Proof Let $\bar{l}_n = \sum p(\mathbf{x}) l(\mathbf{x})$ be the average length of the code words of C. We have the following estimation:

$$H(\mathbf{p}^{(n)}) \le \bar{l}_n < H(\mathbf{p}^{(n)}) + 1.$$

But $H(\mathbf{p}^{(n)}) = n \cdot H(\mathbf{p})$, $\bar{l}_n = n \cdot \bar{l}_i$; whence the final result. □

Exercises

(1) Consider a binary source which produces one bit per unit of time (for example, every μs). We know: $p_0 = \frac{3}{4}$, $p_1 = \frac{1}{4}$.
Suppose that our bitstream has to pass through a channel which accepts only 0.82 bits per unit of time. Construct an adapter by means of Huffman block encoding.
(N.B.: $H(\mathbf{p}) = 0.81$).

(2) A memoryless source, producing the N letters $a_0, a_1, \ldots, a_{N-1}$, according to the probability distribution $\mathbf{p} = (p_0, p_1, \ldots, p_{N-1})$.
Let $C = C_1$ be an associated binary prefix code, and let $\bar{l}$ be the average word length of its code words.
For $n \ge 1$ let us encode as follows:

$$\mathbf{x} = a_{j_1} a_{j_2} \cdots a_{j_n} \longmapsto c(\mathbf{x}) = c(a_{j_1}) c(a_{j_2}) \cdots c(a_{j_n}).$$

Let C_n be the ensuing code; $n \ge 1$.
(a) Show that C_n is a binary prefix code associated with $\mathbf{p}^{(n)}$, $n \ge 1$.
(b) Show that we have, for each C_n: $\bar{l}_i = \bar{l}$.
(c) Give a necessary and sufficient condition for all the C_n to be optimal (with respect to $\mathbf{p}^{(n)}$).

1.1.3 Arithmetic Coding

The arithmetic coding is a dynamic version (presented by a recursive algorithm) of Shannon block encoding (with continually increasing block lengths). Actually, all of these block codes will be visited very shortly: for every block length $n \ge 1$, we shall encode only a single word of length n: the initial segment of length n of a given stream of source symbols.

Arithmetic coding is much easier to explain than to understand. That is why we shall adopt – at least at the beginning – a very pedantic viewpoint. Towards the end, we shall be concerned with more practical aspects of arithmetic coding.

The Elias Encoder

The Elias encoder was initially meant to be a purely academic construction. Its first (and rather discreet) presentation dates from 1968. It was between 1976 and 1979 (Pasco, Jones, Rubin) that arithmetic coding began to be considered as a practically interesting method for lossless compression.

The situation A memoryless source, producing the N letters $a_0, a_1, \ldots,$ a_{N-1}, according to the probability distribution $\mathbf{p} = (p_0, p_1, \ldots, p_{N-1})$.

We shall always suppose $p_0 \geq p_1 \geq \cdots \geq p_{N-1}$.

The arithmetic encoder will associate with a stream of source symbols $a_{j_1} a_{j_2} \cdots a_{j_n} \cdots$ (which could be theoretically unlimited), a bitstream $\alpha_1 \alpha_2 \alpha_3 \cdots \alpha_l \cdots$ (which would then also be unlimited).

But let us stop after n encoding steps:

The code word $\alpha_1 \alpha_2 \alpha_3 \cdots \alpha_l$ of l bits associated with the n first source symbols $a_{j_1} a_{j_2} \cdots a_{j_n}$ will be the code word $c(a_{j_1} a_{j_2} \cdots a_{j_n})$ of a Shannon block encoding formally adapted to recursiveness according to the device: "every step yields a tree-antecedent to the next step".

We shall, in particular, inherit from the preceding section: if the actual production of the source is statistically correct, then l will be the *average length* of the code words for a Shannon block encoding, and consequently

$$H(\mathbf{p}) \leq \frac{l}{n} < H(\mathbf{p}) + \frac{1}{n}.$$

In this sense, the arithmetic encoder will work optimally, i.e. very close to the entropy of the source. The number of bits produced by the arithmetic encoder, counted per source symbol, will be asymptotically equal to the entropy of the source. As a first initiation in arithmetic coding, let us consider a "superfluous" version.

Example A (memoryless) source, producing the four letters a, b, c, d according to the probability distribution given by $p(a) = \frac{1}{2}$, $p(b) = \frac{1}{4}$, $p(c) = p(d) = \frac{1}{8}$.

The Shannon code:

$a \longmapsto 0 \qquad 0 = 0.\underline{0}000\ldots$

$b \longmapsto 10 \qquad \frac{1}{2} = 0.\underline{10}000\ldots$

$c \longmapsto 110 \qquad \frac{3}{4} = 0.\underline{110}000\ldots$

$d \longmapsto 111 \qquad \frac{7}{8} = 0.\underline{111}000\ldots$

Now, look at the source word *badacaba*. Its code word is 10011101100100. Let us write down this concatenation in "temporal progression":

b	10
ba	100
bad	100111
$bada$	1001110
$badac$	1001110110
$badaca$	10011101100
$badacab$	1001110110010
$badacaba$	10011101100100

Recall our binary tree of all *standard-intervals* obtained by iterated dichotomy.

10	is the code word of the interval	$\left[\frac{1}{2}, \frac{3}{4}\right[$,
100	is the code word of the interval	$\left[\frac{1}{2}, \frac{5}{8}\right[$,
100111	is the code word of the interval	$\left[\frac{39}{64}, \frac{5}{8}\right[$,
1001110	is the code word of the interval	$\left[\frac{39}{64}, \frac{79}{128}\right[$,
1001110110	is the code word of the interval	$\left[\frac{315}{512}, \frac{631}{1024}\right[$,
10011101100	is the code word of the interval	$\left[\frac{315}{512}, \frac{1,261}{2,048}\right[$,
1001110110010	is the code word of the interval	$\left[\frac{2,521}{4,096}, \frac{5,043}{8,192}\right[$,
10011101100100	is the code word of the interval	$\left[\frac{2,521}{4,096}, \frac{10,085}{16,384}\right[$.

We have obtained a *chain* of eight intervals: $\left[\frac{1}{2}, \frac{3}{4}\right[$ is the interval of b for the Shannon partition of the interval $[0, 1[$.

The interval $\left[\frac{1}{2}, \frac{5}{8}\right[$ is the interval of a for the Shannon partition of the interval $\left[\frac{1}{2}, \frac{3}{4}\right[$:

The interval $\left[\frac{39}{64}, \frac{5}{8}\right[$ is the interval of d for the Shannon partition of the interval $\left[\frac{1}{2}, \frac{5}{8}\right[$:

and so on. . .

Let us insist:
b points at its interval $\left[\frac{1}{2}, \frac{3}{4}\right[$.
ba points at the interval of a within the interval of b.
bad points at the interval of d within the interval of a within the interval of b, and so on. . .

The recursive algorithm for the construction of the chain of intervals corresponding to the successive arrivals of the source symbols:

Start: $A_0 = 0 \; B_0 = 1$.

Step: The initial segment $s_1 s_2 \cdots s_m$ of the source stream points at $[A_m, B_m[$.

Compute three division points D_1, D_2 and D_3 for the interval $[A_m, B_m[$:

$D_1 = A_m + p(a)(B_m - A_m)$,
$D_2 = A_m + (p(a) + p(b))(B_m - A_m)$,
$D_3 = A_m + (p(a) + p(b) + p(c))(B_m - A_m)$.

Let s_{m+1} be the $(m+1)$st source symbol.

$$
\text{If} \quad s_{m+1} = \begin{cases} a \\ b \\ c \\ d \end{cases} \quad \text{then} \quad \begin{cases} A_{m+1} = A_m, \; B_{m+1} = D_1, \\ A_{m+1} = D_1, \; B_{m+1} = D_2, \\ A_{m+1} = D_2, \; B_{m+1} = D_3, \\ A_{m+1} = D_3, \; B_{m+1} = B_m. \end{cases}
$$

End: $m = n$. The code word of $s_1 s_2 \cdots s_n$ is the word $c(A_n, B_n) \equiv$ the code word of the interval $[A_n, B_n[$.

Note that the least upper bounds B_n of our intervals are not very important (and are often completely neglected): the code word is a prefix of the binary notation of A_n, and its length l is determined by the information content $I(s_1 s_2 \cdots s_n)$.

Let us sum up our observations:

The simple Shannon coding $c(badacaba) = c(b)c(a)c(d)c(a)c(c)c(a)c(b)c(a)$ becomes considerably more complicated from our viewpoint of "dynamic block coding" and by the arithmetic generating the correct strings of intervals. Our first impression is that there is no advantage in adopting this new vision of computing code words.

But now let us change the data.

Our source will still produce the four letters a, b, c, d, but now according to a new probability distribution given by $p(a) = \frac{3}{4}$, $p(b) = \frac{1}{8}$, $p(c) = p(d) = \frac{1}{16}$.

We compute: $I(a) = 0.42$, $I(b) = 3$, $I(c) = I(d) = 4$.

The Shannon code:

$a \longmapsto 0$

$b \longmapsto 110$

$c \longmapsto 1110$

$d \longmapsto 1111$

Let us choose a source word in conformity with the statistics: *daaabaaa-caaabaaa*.

The associated Shannon code word is 11110001100001110000110000 and has 26 bits.

But if we adopt an arithmetic encoding, making use of the recursive algorithm explained earlier, we will end with a code word of l bits, where $l = \lceil I(daaabaaacaaabaaa) \rceil = \lceil 12I(a) + 2I(b) + I(c) + I(d) \rceil = \lceil 5.04 + 6 + 8 \rceil = 20$.

So, the arithmetic code word of $daaabaaacaaabaaa$ is *shorter* than the (concatenated) Shannon code word. This is a general fact: whenever the probabilities are not powers of $\frac{1}{2}$, arithmetic coding is better than any block coding (of fixed block length).

Exercises

(1) The memoryless source which produces the four letters a, b, c, d, according to the probability distribution given by $p(a) = \frac{3}{4}$, $p(b) = \frac{1}{8}$, $p(c) = p(d) = \frac{1}{16}$.
Compute the arithmetic code word of $daaabaaacaaabaaa$ (thus completing the example above).

(2) A memoryless binary source such that $p_0 = \frac{3}{4}, p_1 = \frac{1}{4}$.
Compute the arithmetic code word of 00101000.

(3) A memoryless source producing the three letters a, b, c according to the probability distribution given by $p(a) = \frac{3}{4}$, $p(b) = p(c) = \frac{1}{8}$.
Find the arithmetic code word of $aabaacaa$.

(4) Write down the general version of the recursive algorithm for arithmetic coding. *Recall*: we have to do with a memoryless source producing N letters $a_0, a_1, \ldots, a_{N-1}$, according to the probability distribution $\mathbf{p} = (p_0, p_1, \ldots, p_{N-1})$, and $p_0 \geq p_1 \geq \cdots \geq p_{N-1} > 0$.

(5) The situation as in exercise (4). Suppose that all probabilities are powers of $\frac{1}{2}$: $p_j = 2^{-l_j}, 0 \leq j \leq N - 1$.
Show that in this case the arithmetic code word of a source word $s_1 s_2 \cdots s_n$ is *equal* to the Shannon code word (obtained by simple concatenation of the code words for $s_1, s_2, \ldots, s_n$).

(6) A memoryless source, producing the N letters $a_0, a_1, \ldots, a_{N-1}$ according to the ("decreasing") probability distribution $\mathbf{p} = (p_0, p_1, \ldots, p_{N-1})$.
Let $\mathbf{s}_1$ and $\mathbf{s}_2$ be two source words such that $c(\mathbf{s}_1) = c(\mathbf{s}_2)$ (they have the *same* arithmetic code word).
Show that either $\mathbf{s}_1$ is a prefix of $\mathbf{s}_2$ or $\mathbf{s}_2$ is a prefix of $\mathbf{s}_1$.

[**Solution** $\mathbf{s}_1$ points at $[A_m, B_m[$.
$\mathbf{s}_2$ points at $[C_n, D_n[$.
Our hypothesis: $A_m = 0 \cdot \alpha_1 \alpha_2 \cdots \alpha_l *$, $C_n = 0 \cdot \alpha_1 \alpha_2 \cdots \alpha_l *$, where $l = \lceil -\text{Log}_2(B_m - A_m) \rceil = \lceil -\text{Log}_2(D_n - C_n) \rceil$.
This means:
$2^{-l} \leq B_m - A_m < 2 \cdot 2^{-l}$,
$2^{-l} \leq D_n - C_n < 2 \cdot 2^{-l}$.
Suppose $m \leq n$.

We have only to show that $[C_n, D_n[\subseteq [A_m, B_m[$ (this inclusion implies that s_1 is a prefix of s_2).

Otherwise, we would have $[C_n, D_n[\cap [A_m, B_m[= \emptyset$

Hence, $|C_n - A_m| \geq 2^{-l}$, i.e. the binary notations of A_m and of C_n would already differ (at least) at the lth position after the comma. But this is a contradiction to our hypothesis above.]

The Arithmetic Decoder

The arithmetic decoder will reconstruct, step by step, the source stream from the code stream. It has to discern the encoding decisions by the means of the information which arrive with the code stream. Every decision of the encoder is dedicated to the identification of a source symbol – a process which has to be imitated by the decoder.

This is the occasion where our modelling in terms of (chains of) intervals will be helpful.

First, consider the situation from a *static* viewpoint. The decoder can take into account the *complete* code stream $\alpha_1 \alpha_2 \cdots \alpha_L$.

This means that we face the code word $c(A_n, B_n)$ of an interval
$A_n = 0 \cdot \alpha_1 \alpha_2 \cdots \alpha_L * (* \equiv$ masked part),
$L = \lceil -\text{Log}_2(B_n - A_n) \rceil$.

We search: the source stream $s_1 s_2 \cdots s_n$, which points at this interval (in the Shannon partition tree whose structure comes from the lexicographic ordering of the pointers).

Let us look at the following.

Example A (memoryless) source producing the three letters a, b, c according to the probability distribution $\mathbf{p}$ given by $p(a) = \frac{3}{4}$, $p(b) = p(c) = \frac{1}{8}$.

The code word: 100110111.

Our terminal interval $[A_n, B_n[$ is described by
$A_n = 0.100110111*$,
$\frac{1}{2^9} \leq B_n - A_n < \frac{1}{2^8}$.

The decoding will consist of keeping track of the *encoder*'s decisions.

The logical *main argument* for decoding is the following: the hierarchy of the Shannon partition tree accepts only *chains of intervals* or *empty intersections*.

This means: whenever a *single* point of such an interval is to the left or to the right of a division point, then the *entire* interval will be to the left or to the right of this point (recall the solution of exercise (6) at the end of the preceding section).

Let us begin with the decoding.

First step:
Compute $D_1 = \frac{3}{4} = 0.11$,
$\qquad\qquad D_2 = \frac{7}{8} = 0.111$,
$A_n < D_1 \Longrightarrow [A_n, B_n[\subset [0, D_1[\equiv$ the interval of a.

$\Longrightarrow s_1 = a.$

Thus, $A_1 = 0$, $B_1 = \frac{3}{4}$.

Second step:

Compute $D_1 = \frac{9}{16} = 0.1001$,

$\qquad\qquad D_2 = \frac{21}{32} = 0.10101$,

$D_1 < A_n < D_2 \Longrightarrow [A_n, B_n[\subset [D_1, D_2[\equiv$ the interval of ab.

$\Longrightarrow s_1 s_2 = ab.$

Thus, $A_2 = \frac{9}{16}$, $B_2 = \frac{21}{32}$.

Third step:

Compute $D_1 = \frac{81}{128} = 0.1010001$,

$\qquad\qquad D_2 = \frac{93}{128} = 0.1011101$,

$A_n < D_1 \Longrightarrow [A_n, B_n[\subset [A_2, D_1[\equiv$ the interval of aba.

$\Longrightarrow s_1 s_2 s_3 = aba.$

Thus, $A_3 = \frac{9}{16}$, $B_3 = \frac{81}{128}$.

Fourth step:

Compute $D_1 = \frac{315}{512} = 0.100111011$,

$\qquad\qquad D_2 = \frac{639}{1024} = 0.1001111111$,

$A_n < D_1 \Longrightarrow [A_n, B_n[\subset [A_3, D_1[\equiv$ the interval of $abaa$.

$\Longrightarrow s_1 s_2 s_3 s_4 = abaa.$

Thus, $A_4 = \frac{9}{16}$, $B_4 = \frac{315}{512}$.

Fifth step:

Compute $D_1 = \frac{1233}{2048} = 0.10011010001$,

$\qquad\qquad D_2 = \frac{2493}{4096} = 0.100110111101$,

A_n and D_2 have the *same* binary notation – until the masked part of A_n.

Question How shall we continue?

Answer $A_n = D_2$ and $s_5 = c$ (i.e. $[A_n, B_n[\subset [D_2, B_4[)$.

Justification *If we suppose that $A_n = D_2$ and $s_5 = c$, then we have found a source word* $\mathbf{s} = s_1 s_2 s_3 s_4 s_5 = abaac$ *with the code word* 10011011 *(note that $I(abaac) = 3 \times 0.42 + 3 + 3 = 7.26$).*

Suppose, furthermore, that only the letter a was produced in the sequel ($\frac{2493}{4096}$ remains then the left end point $A_5 = A_6 = A_7$, etc. of the code interval), so we will have $s_1 s_2 s_3 s_4 s_5 s_6 s_7 = abaacaa$ with $I(abaacaa) = 5 \times 0.42 + 3 + 3 = 8.1$. This source word will produce our code word, but also

$s_1 s_2 s_3 s_4 s_5 s_6 s_7 s_8 = abaacaaa \quad (I(s_1 \cdots s_8) = 8.52)$,

$s_1 s_2 s_3 s_4 s_5 s_6 s_7 s_8 s_9 = abaacaaaa \quad (I(s_1 \cdots s_9) = 8.94)$.

We obtain this way three source words

$s_1 s_2 s_3 s_4 s_5 s_6 s_7$,

$s_1 s_2 s_3 s_4 s_5 s_6 s_7 s_8$,

$s_1 s_2 s_3 s_4 s_5 s_6 s_7 s_8 s_9$,

which produce the same code word 100110111.

Back to the question: why $A_n = \frac{2,493}{4,096}$?

The answer: Under this hypothesis, we obtain a coherent and logical decoding. But, according to exercise (6) of the preceding section, *every* string of source symbols with the code word 100110111 must be a *prefix* of *abaacaaaa*.

Attention *The argument concerning the end of our decoding algorithm depends on a complete message: the string of the received code bits has to be the code word of an interval.*

Now consider an *incomplete message*.

In the situation of our example, assume that the first three bits of the code stream are 101. This is not a code word of an interval in our partition tree associated with the given probability distribution. Let us try to determine altogether the first two source symbols. 101 is the prefix of the binary notation of the greatest lower bound A of an interval $[A, B[$ coming up at some step of the encoding:

$$A = 0.101 * .$$

We note: $A < \frac{3}{4} = D_1 \equiv$ the inferior division point of the first step of the encoder,

$$\Longrightarrow s_1 = a.$$

The division points of the second step (inside the interval $[A_1, B_1[= [0, \frac{3}{4}[)$:
$D_1 = \frac{9}{16} = 0.1001$, $D_2 = \frac{21}{32} = 0.10101$,
$A > D_1 \Longrightarrow s_2 = b$ or $s_2 = c$.
The case $s_1 s_2 = ac$ gives $A_2 = 0.\underline{101}01$, and this is ok.
The case $s_1 s_2 = ab$ needs a closer inspection of the possible continuations.
And indeed: the code word of $s_1 s_2 s_3 = abc$ is $\underline{101}0010$, since the second division point D_2 inside $[A_2, B_2[= [\frac{9}{16}, \frac{21}{32}[$ is $D_2 = \frac{165}{256} = 0.\underline{101}00101$.

We see: our rules about the decoding at the end of the code stream can only be applied in a *terminal* situation. We have to face the code word of an interval.

Exercises

(1) Write down the decoding algorithm in general, i.e. for a memoryless source, producing the N letters $a_0, a_1, \ldots, a_{N-1}$ according to the probability distribution $\mathbf{p} = (p_0, p_1, \ldots, p_{N-1})$ with $p_0 \geq p_1 \geq \cdots \geq p_{N-1} > 0$.

(2) A (memoryless) binary source $\mathbf{p} = (p_0, p_1)$ with $p_0 = \frac{3}{4}$, $p_1 = \frac{1}{4}$. Decode 1101010.

(3) The binary source of exercise (2). The decoder receives a code stream beginning with 1001. Find the first three bits of the source stream.

(4) A memoryless source which produces the four letters a, b, c, d according to the probability distribution $\mathbf{p}$ given by $p(a) = \frac{3}{4}$, $p(b) = \frac{1}{8}$, $p(c) = p(d) = \frac{1}{16}$. Decode 11101111000101010000.

(5) A (binary) arithmetic encoder produces zeros and ones; hence, it is nothing but a binary source. What then is its entropy?

Practical Considerations

It is clear that the spirit of arithmetic coding demands a continual bitstream output of the encoder. On the other hand, the constraints of a bounded arithmetic call for perpetual renormalization in order to avoid an asphyxiation in more and more complex computations. So, we need a practicable version of the Elias encoder and decoder, i.e. a sort of prototype for effective implementation.

Some (almost) trivial observations

Let $[A, B[$ be a (small) interval taking place in our computations. We are interested in the division point

$D = A + p \cdot (B - A)$ ($p \equiv$ a cumulative probability).

If the difference $B - A$ is *small*, then we will have (generically – there are the well-known exceptions)

$$A = a + r,$$
$$B = a + s,$$

where a describes the *identical* prefix of the binary notation[5] of A and of B.

Compute $d = r + p \cdot (s - r)$ (subdivision of the remainder parts ...).

Then $D = a + d$.

On the other hand (stability of our computations under "arithmetic zoom"): consider, for $m \geq 1$,

$R = 2^m r,$

$S = 2^m s.$

Then $R + p \cdot (S - R) = 2^m (r + p \cdot (s - r))$.

Consequence *Suppose that the binary notations of A and of B are identical up to a certain position t:*

$$A = \alpha_1 \cdots \alpha_t \alpha_{t+1} \cdots ,$$
$$B = \alpha_1 \cdots \alpha_t \beta_{t+1} \cdots .$$

Then the computation of $D = A + p \cdot (B - A)$ is equivalent to the computation of $T = R + p \cdot (S - R)$

with $R = 0 \cdot \alpha_{t+1} \cdots$

$S = 0 \cdot \beta_{t+1} \cdots$

More precisely, if $T = 0 \cdot \tau_1 \tau_2 \cdots$, then $D = 0 \cdot \alpha_1 \cdots \alpha_t \tau_1 \tau_2 \cdots$

This allows a renormalizable arithmetic: our computations remain in the higher levels of appropriate partition trees of simple intervals inside $[0, 1[$.

The criterion for sending off a string of code bits will come from a simple syntactical comparison.

Let $[A_n, B_n[$ be the interval of the source word $s_1 s_2 \cdots s_n$.

Write $A_n = 0 \cdot \alpha_1 \alpha_2 \cdots \alpha_t \alpha_{t+1} \cdots ,$

$B_n = 0 \cdot \beta_1 \beta_2 \cdots \beta_t \beta_{t+1} \cdots .$

---

[5] We shall never consider "false" infinite notations with period "1".

If $\alpha_1 = \beta_1$ $\alpha_2 = \beta_2 \cdots \alpha_t = \beta_t$, then our encoder will produce (send off) the string of code bits $\alpha_1 \alpha_2 \cdots \alpha_t$ and will continue the encoding with

$$A'_n = 0 \cdot \alpha_{t+1} \cdots ,$$
$$B'_n = 0 \cdot \beta_{t+1} \cdots .$$

Example The (memoryless) source which produces the three letters a, b, c according to the probability distribution $\mathbf{p}$ given by $p(a) = \frac{3}{4}$, $p(b) = p(c) = \frac{1}{8}$.

Consider the source word *abaacaaaa* (an old friend – look at our decoding example in the last section!).

First step of the encoder:

$D_1 = \frac{3}{4} = 0.11$,

$D_2 = \frac{7}{8} = 0.111$,

$s_1 = a \Longrightarrow A_1 = 0 = 0.000\ldots,$

$\qquad\qquad B_1 = \frac{3}{4} = 0.11.$

No common prefix can be discarded.

Second step of the encoder:

$D_1 = \frac{9}{16} = 0.1001$,

$D_2 = \frac{21}{32} = 0.10101$,

$s_2 = b \Longrightarrow A_2 = \frac{9}{16} = 0.\underline{1}001,$

$\qquad\qquad B_2 = \frac{21}{32} = 0.\underline{1}0101.$

We send off: $\alpha_1 \alpha_2 = 10$.

We renormalise: $A'_2 = 0.01 = \frac{1}{4}$,

$\qquad\qquad\quad B'_2 = 0.101 = \frac{5}{8}.$

Third step of the encoder:

$D_1 = \frac{1}{4} + \frac{3}{4} \cdot \frac{3}{8} = \frac{17}{32} = 0.10001$,

$D_2 = \frac{1}{4} + \frac{7}{8} \cdot \frac{3}{8} = \frac{37}{64} = 0.100101$,

$s_3 = a \Longrightarrow A_3 = \frac{1}{4} = 0.01,$

$\qquad\qquad B_3 = \frac{17}{32} = 0.10001.$

No common prefix can be discarded.

Fourth step of the encoder:

$D_1 = \frac{1}{4} + \frac{3}{4} \cdot \frac{9}{32} = \frac{59}{128} = 0.0111011$,

$D_2 = \frac{1}{4} + \frac{7}{8} \cdot \frac{9}{32} = \frac{127}{256} = 0.01111111$,

$s_4 = a \Longrightarrow A_4 = \frac{1}{4} = 0.01,$

$\qquad\qquad B_4 = \frac{59}{128} = 0.\underline{01}11011.$

We send off: $\alpha_3 \alpha_4 = 01$.

We renormalise: $A'_4 = 0$,

$\qquad\qquad\quad B'_4 = 0.11011 = \frac{27}{32}.$

Fifth step of the encoder:

$D_1 = \frac{3}{4} \cdot \frac{27}{32} = \frac{81}{128} = 0.1010001$,

$D_2 = \frac{7}{8} \cdot \frac{27}{32} = \frac{189}{256} = 0.10111101$,

$s_5 = c \Longrightarrow A_5 = \frac{189}{256} = 0.10111101$

$\qquad\qquad B_5 = \frac{27}{32} = 0.\underline{1}1011.$

We send off: $\alpha_5 = 1$.

We renormalise: $A'_5 = 0.0111101 = \frac{61}{128}$,

$\qquad\qquad\quad B'_5 = 0.1011 = \frac{11}{16}.$

Sixth step of the encoder:

$D_1 = \frac{61}{128} + \frac{3}{4} \cdot \frac{27}{128} = \frac{325}{512} = 0.101000101,$

$s_6 = a \Longrightarrow A_6 = A_5 = 0.0111101,$

$\qquad B_6 = D_1 = 0.101000101.$

No common prefix can be discarded.

Seventh step of the encoder:

$D_1 = \frac{61}{128} + \frac{3}{4} \cdot \frac{81}{512} = \frac{1,219}{2,048} = 0.10011000011,$

$s_7 = a \Longrightarrow A_7 = A_6 = 0.0111101,$

$\qquad B_7 = D_1 = 0.10011000011.$

No common prefix can be discarded.

Eighth step of the encoder:

$D_1 = \frac{61}{128} + \frac{3}{4} \cdot \frac{243}{2,048} = \frac{4,633}{8,192} = 0.1001000011001,$

$s_8 = a \Longrightarrow A_8 = A_7 = 0.0111101,$

$\qquad B_8 = D_1 = 0.1001000011001.$

No common prefix can be discarded.

Ninth step of the encoder:

$D_1 = \frac{61}{128} + \frac{3}{4} \cdot \frac{729}{8,192} = \frac{17,803}{32,768} = 0.100010110001011,$

$s_9 = a \Longrightarrow A_9 = A_8 = 0.0111101,$

$\qquad B_9 = D_1 = 0.100010110001011.$

No common prefix can be discarded.

Exercises

(1) Continue the previous example: find the shortest source word $s_1 s_2 \cdots$ $s_9 s_{10} \cdots$ such that the encoder will effectively send off (after the convenient syntactical tests) $\alpha_6 \alpha_7 \alpha_8 \alpha_9 = 0111$.

(2) True or false: if $s_1 s_2 \cdots s_n$ is the beginning of the source stream, then its code word $c(s_1 s_2 \cdots s_n)$ is the beginning of the code stream?

(3) Our example cited earlier seems to indicate the danger of a *blocking* due to inefficiency of the syntactical test: it is the constellation $[A_n, B_n[=$ $[0.011 \cdots 1*, 0.100 \cdots 0 * [$ which looks dangerous for renormalization. Try to control the situation numerically. Will we need an algorithmic solution (exceptional case)?

Remark In arithmetic coding, the sequence of source symbols $s_1 s_2 \cdots s_n$ will be encoded and decoded progressively; this allows us to implement an *adaptive version* of arithmetic coding which will learn progressively the actual probability distribution $\mathbf{p}^{(n)}$ (after the production of the first n source symbols).

Let us make this more precise: at the beginning, there is no statistical information concerning the production of the N symbols $a_0, a_1, \ldots, a_{N-1}$. We are obliged to put $\mathbf{p}^{(0)} = (\frac{1}{N}, \frac{1}{N}, \ldots, \frac{1}{N})$ (uniform distribution).

Now let $\mathbf{p}^{(n)} = (p_0^{(n)}, \ldots, p_{N-1}^{(n)})$ be the actual distribution at the nth pulse.

$\mathbf{p}^{(n)}$ was computed at the last node n_k of a control pulsation $(n_k \leq n)$ according to the histogram of $s_0 s_1 \cdots s_{n_k}$.

Let $s_{n+1} = a_j$ $(0 \leq j \leq N-1)$.

Then we shall put: $A_{n+1} = A_n + \left(\sum_{k=0}^{j-1} p_k^{(n)} \right) (B_n - A_n),$

$\qquad B_{n+1} - A_{n+1} = p_j^{(n)} \cdot (B_n - A_n).$

The decoder will recover s_{n+1}, since it knows $\mathbf{p}^{(n)}$ – the actual probability distribution after the production of the nth character – due to the histogram established with the information of anterior decoding.

1.2 Universal Codes: The Example LZW

The algorithms for data compaction which we shall treat now are "universal" in the following sense: the idea of a memoryless source – which is a nice but very rare object – will be sacrificed. We shall joyfully accept source productions of a high statistical complexity; this means that explicit statistical evaluations of the source production will disappear. It is a new way of coding (by means of a *dictionary*) that creates an *implicit* histogram.

The birth of the central algorithm Lempel–Ziv–Welch (LZW) dates to around 1977. There exist a lot of clever variants. We shall restrict ourselves to the presentation of a single version which is seductively clear and simple.

1.2.1 LZW Coding

Situation *A source produces a stream of letters, taken from a finite alphabet.*

The encoder will establish a *dictionary* of strings of characters (of motives) which are *characteristic* for the source stream (and it will create this way *implicit statistics* of the source production). The compressed message is given by the stream of *pointers* ($\equiv$ the numbers attributed to the strings in the dictionary). Note that we meet here a method of coding where the code words have *fixed* length.

The Encoding Algorithm

The *principal aim* of the encoding algorithm is the writing of a *dictionary*:

Strings of characters $\longleftrightarrow$ Numbers for these strings

The production of the code stream (i.e. of the sequence of pointers) is (logically) secondary.

The encoding algorithm works in *three steps*:

(1) Read the next character x (arriving from the source).
(2) Complete a *current string* s (which is waiting in a buffer for admission to the dictionary) by concatenation: $s \longmapsto sx$.
(3) Write the current string into the dictionary as soon as it will be admissible (i.e. unknown to the dictionary) – otherwise go back to (1). Produce (i.e. send off) the number of s at the moment where you write sx into the dictionary; initialize $s = x$.

Remark What is the reason for this enigmatic delay at the output of the encoder? Let us insist: the encoder produces the code word of s at the moment when it writes sx into the dictionary.

The reason is the following. The decoder will have to *reconstruct* the dictionary by means of the code stream which it receives (the dictionary of the encoder will not survive the end of the compaction). Without this delay between writing and producing, the reconstruction of the dictionary by the decoder would be impossible.

Let us recapitulate the encoding algorithm in a more formalized version.

STEP: read the next character x of the source stream.

> If   no character x (end of message),
>> then produce the code word (the pointer) of the current string s;
>> end.
>
> If   the string sx exists already in the table,
>> then replace s by the new current string sx;
>>> repeat STEP.
>
> If   the string sx is not yet in the table,
>> then write sx in the table, produce the code of s and put $s = x$;
>>> repeat STEP.

Example Initialization of the dictionary: (1) a
(2) b
(3) c

Encode *baca caba baba caba.*

Read	Produce	Write	Current string
		(1) a	
		(2) b	
		(3) c	
b			b
a	(2)	(4) ba	a
c	(1)	(5) ac	c
a	(3)	(6) ca	a
c			ac
a	(5)	(7) aca	a
b	(1)	(8) ab	b
a			ba
b	(4)	(9) bab	b
a			ba
b			bab
a	(9)	(10) $baba$	a
c			ac
a			aca
b	(7)	(11) $acab$	b
a			ba
	(4)		

Look, once more, how the source stream is transformed into the code stream:

$$b \quad a \quad c \quad ac \quad a \quad ba \quad bab \quad aca \quad ba$$

$$(2)\ (1)\ (3)\ (5)\ (1)\ (4)\ (9)\ (7)\ (4)$$

Exercises

(1) The situation of the preceding example.
Encode now *baba caba baca caba baba caba*.
(2) A binary source. We initialize: (0) 0
    (1) 1
    Encode 010101010101010101010101.
(3) (a) Give an example of an LZW encoding which produces a sequence of pointers of the type ... (♯)(♯) ... (i.e. with a pointer doubled).
    (b) Does there exist sequences of LZW code words of the type ... (♯)(♯)(♯) ... (i.e. with a pointer tripled)?

1.2.2 The LZW Decoder

A First Approach

The principal goal of the decoder is the *reconstruction of the dictionary* of the encoder. It has to correctly interpret the stream of code words (pointers) that it receives. The down-to-earth decoding (the identification of the code words) is a part of this task.

The *current string s*, candidate for admission to the dictionary, will still remain in the centre of the algorithm. Let us begin immediately with an example.

Example Initialized dictionary: (1) *a*
                                   (2) *b*
                                   (3) *c*
Decode (3)(1)(2)(5)(1)(4)(6)(6).

Read	Produce	Write	Current string
		(1) *a*	
		(2) *b*	
		(3) *c*	
(3)	*c*		*c*
(1)	*a*	(4) *ca*	*a*
(2)	*b*	(5) *ab*	*b*
(5)	*ab*	(6) *ba*	*ab*
(1)	*a*	(7) *aba*	*a*
(4)	*ca*	(8) *ac*	*ca*
(6)	*ba*	(9) *cab*	*ba*
(6)	*ba*	(10) *bab*	*ba*

Attention   We note that the two first columns are the result of a mechanical identification according to the input information of the decoder, whereas the two last columns come from a backward projection. What did the encoder do? Recall that the encoder can only write into the dictionary when appending a single character to the current string. Every word in the dictionary is preceded by the "dispersed pyramid" of all its prefixes. The steps of non-writing of the encoder disappear during the decoding. We write at every step.

We observe:

(1) *At the beginning of every decoding step, the current string will be the prefix of the next writing into the dictionary. We append the first character of the decoded string.*

(2) *At the end of every decoding step, the current string will be equal to the string that we just decoded (the column "produce" and the column "current string" of our decoding model are identical: at the moment when we identify a code word, the demasked string appears in the "journal" of the encoder – a consequence of the small delay for the output during the encoding).*

The Exceptional Case

Example   The situation is as in the first example. Decode (2)(1)(4)(6).

Read	Produce	Write	Current string
		(1) a	
		(2) b	
		(3) c	
(2)	b		b
(1)	a	(4) ba	a
(4)	ba	(5) ab	ba
(6)	bax	(6) bax	bax

Look a little bit closer at the last line of our decoding scheme. We have to identify a pointer which points at nothing in the dictionary, but which shall get its meaning precisely at this step. But recall our previous observations. We need to write a string of the form bax (current string plus a new character); but this must be, at the same time, the decoded string. On the other hand, the character x must be the *first* character of the decoded string. Thus, $x = b$ (= the first character of the current string).

All these considerations generalize:

We have to identify a code word (a pointer) the source word of which is not yet in the dictionary. We proceed precisely as in the foregoing example.

Let s be the current string at the end of the last step. Arrives the fatal pointer ($\natural$). If we take our decoding observations (1) and (2) as rules, then we are obliged to produce: sx write: ($\natural$) : sx put: $s = sx$, x is an (a priori) unknown character, which will be identified according to the rule: $x =$ the first character of the current string s.

The question remains: what is the encoding constellation which provokes the exceptional case of the decoder?

Exercise Show that the exceptional decoding case comes from the following encoding situation: the encoder sends off the last pointer available in the dictionary.

The Decoding Algorithm

Let us sum up, in pseudo-program version, the functioning of our decoder:

    STEP: read the next pointer (n) of the code stream.
      if   no such (n), <u>end</u>.
      if   the string of (n) is not in the dictionary,
          <u>then</u> produce sx, where x is the first character of the
               current string s, write sx into the dictionary,
               replace s by the new current string sx
               repeat STEP
      <u>else</u> the string u of (n) exists in the dictionary,
          <u>then</u> produce u, write sx into the dictionary, where s is the
               current string, and x is the first character of u, replace
               the current string by u,
               repeat STEP.

Exercises

(1) Initialized dictionary: (1) a (2) b (3) c (4) d
    Decode (1)(3)(4)(3)(5)(7)(2)(11)(2)(1)(2)(6)(1)(17)(1).
(2) Initialized dictionary: (1) a (2) b (3) c (4) d
    Decode (3)(1)(2)(6)(8)(4)(8)(7)(10)(5)(10)(7)(12)(11)(9)(1)(5)(14)(18)
        (5)(16)(19)(25)(13)(18)(28)(10)(15)(18)(7)(21)(20)(1)(32)(25)
        (39)(33)(18).
(3) A LZW decoder, initialized: (1) a (2) b ... (26) z.
    At its 76th decoding step the decoder encounters an exceptional case.
    What is the number of this pointer?
(4) Initialized dictionary: (0) 0 (1) 1
    (a) What is the bitstream which is encoded in (0)(1)(2)(3)(4)(5)(6)(7)(8)
        (9)(10)(11)?
    (b) Let us continue: we shall produce in this way all the integers, in natural
        order, from 0 to 65,535.
    Let us write the pointers in 16 bits. What will be the compressed bit-rate

$$\rho = \frac{\text{number of code bits}}{\text{number of source bits}}?$$

(5) The situation of exercise (4). The binary source producing a bitstream such that the stream of LZW-codes will be $(0)(1)(2)\dots(65{,}534)(65{,}535)$. Model this source as a memoryless (!) source for the alphabet of the 256 8-bit bytes $00000000, 00000001, \dots, 11111111$. What will be (roughly) the source statistics? What will be its entropy?

(6) Initialized dictionary: (1) a (2) b (3) c (4) d
    Decode $(4)(1)(3)(6)(1)(5)(7)(7)(5)(5)(2)(1)(15)(14)(9)(19)(17)(1)(19)(12)$ $(16)(19)(13)(22)(10)(20)(26)(25)(15)(33)(32)(35)(15)(24)(38)(22)$ $(7)(1)$.

2

Cryptography

Cryptography is the theory (and the practice) of the computational aspects in the domain of *data security*. It deals with the design and validation of algorithmic methods which are meant to guarantee the integrity, the authenticity and the confidentiality in data transmission (and in data storage).

The palette of cryptosystems is rich. On the other hand, the fundamental algorithmic ideas in most of the currently canonized ciphers are almost exclusively the result of a lucky raid in elementary number theory. This has determined our selection. Thus, after having looked at Data Encryption Standard (DES – which is of a comforting mathematical simplicity), and having rather extensively treated the new data encryption standard: the cipher *AES-Rijndael*, where a deeper mathematical understanding will be needed, we shall finally limit our attention to the investigation of a hard core of cryptosystems: The "arithmetical" systems, based principally on the idea of encryption by means of exponentiation in certain multiplicative groups. Beginning with a traditional and rigorously academic approach, we feel happy to present the (aging) system *RSA*, with which elementary arithmetic entered definitely and with unanimous acclaim into cryptography. Then, public key oblige, the chapter will hoist to a relatively broad exposition of digital signature algorithms; the algorithm *DSA*, the mathematical framework of which is relatively close to that of *RSA*, should prove exciting for all fans of technical details. We also supply the variant *rDSA* and the algorithm *ECDSA*, which features elliptic curves – thus proving the practical interest of the purest mathematics. Finally, at the end of this ride across the "land of standards", we will also have looked closely to the hash algorithm *SHA-1*, which is a sort of twisted parody of our old friend, the DES.

2.1 The Data Encryption Standard

In 1973, the *National Bureau of Standards* of the United States invited tenders for a cryptosystem, abutting finally onto the DES – an amplified version of a former "family cipher" of IBM, called *LUCIFER*. After a long series of public discussions and controversies, *DES* was finally adopted as standard for data encryption in January 1977. Since, *DES* has been reevaluated almost every five years. The most recent version dates from 1994.[1]

2.1.1 The DES Scheme

The algorithm DES transforms plaintext blocks of 64 bits into ciphertext blocks of 64 bits. The cipher keys (attributed to the users) will consist of 56 bits, in 64-bit presentation, with eight parity-check bits at the positions 8, 16, 24, 32, 40, 48, 56, 64. The *same* algorithm is used as well for encryption as for decryption.[2]

Global Structure of the Algorithm

The algorithm works in three principal steps:

1. The plaintext block $T = t_1 t_2 \cdots t_{64}$ is first treated by a fixed *initial permutation IP*, which shuffles, in a highly regular manner, the binary positions of the eight 8-bit bytes.
2. Then follow 16 iterations (16 rounds) of a transformation which depends, at each round, on another *round key*, extracted from the cipher key of the user by an auxiliary algorithm (the *key schedule*).
3. The result of the 16 rounds has still to undergo the permutation IP^{-1}, the inverse of the initial permutation, in order to become the ciphertext block.

---

[1] Omitting all preliminary definitions and generalities, we shall consider DES as an introductory example for cryptographic language.
[2] A strange statement – to be explained in a moment.

Scheme of the Algorithm DES

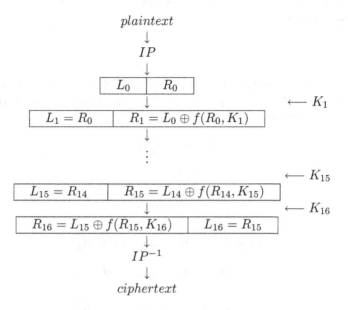

The Functioning of the 16 Iterations

Let T_i be the result of the ith iteration, $1 \leq i \leq 16$.
$$T_i = L_i R_i = t_1 t_2 \cdots t_{32} t_{33} \cdots t_{64}$$

Then
$$L_i = R_{i-1}$$
$$R_i = L_{i-1} \oplus f(R_{i-1}, K_i) \qquad 1 \leq i \leq 16$$

where $\oplus$ means addition of vectors with 32 entries over the field $\mathbb{F}_2 = \{0, 1\}$ (in another language: we use 32 times the Boolean operation XOR). f is a "mixing function", the first argument of which (the right half of the current state) has 32 bits, and the second argument of which (the current round key) has 48 bits. Its values will have 32 bits.

Attention The result of the last iteration is $R_{16} L_{16}$, and IP^{-1} will operate on $R_{16} L_{16}$.

Remark We immediately see that the DES scheme is "generically" *invertible*. More precisely: Every round is invertible – for any cipher key K and *every possible choice* of the "mixing" function f. You have only to write the round transformation scheme "upside down":

$$L_{i-1} = R_i \oplus f(L_i, K_i), \quad R_{i-1} = L_i, \qquad 1 \leq i \leq 16.$$

This basic observation will give rise to an important construction in filter bank theory: The *lifting structures*. We shall treat this subject in the last chapter of our book.

Exercise

We consider a mini-DES of four rounds, which transforms 8-bit bytes $x_1 x_2 \cdots x_8$ into 8-bit bytes $y_1 y_2 \cdots y_8$. We shall make use of keys $K = u_1 u_2 \cdots u_8$ of eight bits.

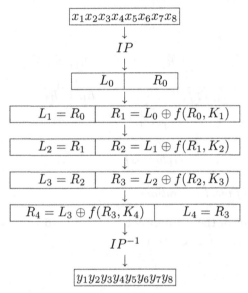

(a) $IP = \begin{pmatrix} 1\,2\,3\,4\,5\,6\,7\,8 \\ 3\,8\,7\,5\,1\,4\,6\,2 \end{pmatrix}$ (assignment of positions) i.e.
$IP(x_1 x_2 x_3 x_4 x_5 x_6 x_7 x_8) = x_5 x_8 x_1 x_6 x_4 x_7 x_3 x_2$.

(b) The algorithm which computes the four round keys: $K = u_1 u_2 \cdots u_8$ will give rise to

$$K_1 = u_7 u_1 u_3 u_5, \quad K_3 = u_1 u_4 u_7 u_2,$$
$$K_2 = u_8 u_2 u_4 u_6, \quad K_4 = u_2 u_5 u_8 u_3.$$

(c) The mixing function: $f(r_1 r_2 r_3 r_4, t_1 t_2 t_3 t_4) = z_1 z_2 z_3 z_4$, where
$z_1 = r_1 + t_2 + r_3 + t_4 \bmod 2,$
$z_2 = r_2 + t_2 + r_3 + t_3 \bmod 2,$
$z_3 = r_1 + t_1 + r_4 + t_4 \bmod 2,$
$z_4 = r_2 + t_1 + r_4 + t_3 \bmod 2.$

- Encrypt $x = 10011010$ with $K = 11001100$.

The Symmetry of the Scheme DES

For the decryption, we shall use the *same* algorithm as for the encryption, but with *the order of the round keys inverted*: K_{16} will parameterise the first iteration, K_{15} the second one, ..., K_1 the last one.

Let us show that, indeed, $D_K(E_K(M)) = M$ for every plaintext M (where E_K and D_K are the encryption and decryption transformations for the cipher key K).

Observation: $L_{i-1} = R_i \oplus f(L_i, K_i)$ $R_{i-1} = L_i$ $1 \le i \le 16$.

We have

$$C = E_K(M) = IP^{-1} \circ T_{16}^{K_{16}} \circ T_{15}^{K_{15}} \circ \cdots \circ T_1^{K_1} \circ IP(M)$$
$$M' = D_K(C) = IP^{-1} \circ T_{16}^{K_1} \circ T_{15}^{K_2} \circ \cdots \circ T_1^{K_{16}} \circ IP(C)$$

We have to show: $M' = M$.

First:

$$M' = IP^{-1} \circ T_{16}^{K_1} \circ T_{15}^{K_2} \circ \cdots \circ T_1^{K_{16}} \circ IP \circ IP^{-1} \circ T_{16}^{K_{16}} \circ T_{15}^{K_{15}} \circ \cdots \circ T_1^{K_1}$$
$$\circ IP(M) = IP^{-1} \circ T_{16}^{K_1} \circ T_{15}^{K_2} \circ \cdots \circ T_1^{K_{16}}(R_{16}, L_{16})$$

Then

$$T_1^{K_{16}}(R_{16}, L_{16}) = (L_{16}, R_{16} \oplus f(L_{16}, K_{16})) = (R_{15}, L_{15}),$$
$$T_2^{K_{15}}(R_{15}, L_{15}) = (L_{15}, R_{15} \oplus f(L_{15}, K_{15})) = (R_{14}, L_{14})$$

$$\vdots$$

$$T_{15}^{K_2}(R_2, L_2) = (L_2, R_2 \oplus f(L_2, K_2)) = (R_1, L_1),$$
$$T_{16}^{K_1}(R_1, L_1) = (R_1 \oplus f(L_1, K_1), L_1) = (L_0, R_0)$$

and we have finished...

Remark The symmetry of the DES scheme, as we have shown it above, does neither depend on the size of the plaintext block, nor on the number of rounds, nor on the "mixing" function f, nor on the key schedule. We dispose of a broad range for variation.

2.1.2 The Cipher DES in Detail

1. *The initial permutation IP*:

It is described by the following table:

IP
58 50 42 34 26 18 10 2
60 52 44 36 28 20 12 4
62 54 46 38 30 22 14 6
64 56 48 40 32 24 16 8
57 49 41 33 25 17 9 1
59 51 43 35 27 19 11 3
61 53 45 37 29 21 13 5
63 55 47 39 31 23 15 7

You have to read line by line: $IP(t_1 t_2 \cdots t_{63} t_{64}) = t_{58} t_{50} \cdots t_{15} t_7$.

IP will shuffle, in a very regular manner, the bit-positions of the eight 8-bit bytes which constitute the block T of 64 bits: every new byte contains precisely one single bit of every old byte.

The inverse permutation IP^{-1}:

IP^{-1}			
40 8 48	16	56 24 64 32
39 7 47	15	55 23 63 31
38 6 46	14	54 22 62 30
37 5 45	13	53 21 61 29
36 4 44	12	52 20 60 28
35 3 43	11	51 19 59 27
34 2 42	10	50 18 58 26
33 1 41	9	49 17 57 25

2. *The function f and the S-boxes:*

The first argument R of the function f is a 32-bit string, the second J has length 48, and the result $f(R, J)$ will have 32 bits. The operations are organized according to the following scheme:

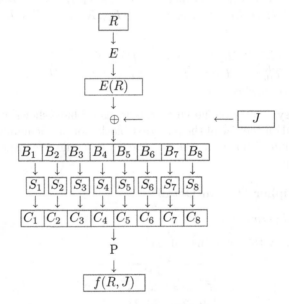

- At the beginning, the first argument of the function f is "stretched" into a string of 48 bits, according to an *expansion table* E. $E(R)$ is made of all bits of R in a natural order, with 16 repetitions.
- $E(R) \oplus J$ is computed, and the result B is treated as a block of eight substrings made of six bits: $B = B_1 B_2 B_3 B_4 B_5 B_6 B_7 B_8$.
- The following step makes use of eight *S-boxes* S_1, S_2, ... S_8. Every S_j transforms its block B_j of six bits into a block C_j of four bits, $1 \le j \le 8$.

– The string $C = C_1C_2C_3C_4C_5C_6C_7C_8$ of length 32 is reorganized according to a fixed permutation P. The result $P(C)$ will be $f(R, J)$.

The operation of expansion E is defined by the following table:

E					
32	1	2	3	4	5
4	5	6	7	8	9
8	9	10	11	12	13
12	13	14	15	16	17
16	17	18	19	20	21
20	21	22	23	24	25
24	25	26	27	28	29
28	29	30	31	32	1

Read it line by line: $E(r_1r_2r_3 \cdots r_{32}) = r_{32}r_1r_2 \cdots r_{32}r_1$.

Before explaining the functioning of the S-boxes, we shall present them all:

S_1															
14	4	13	1	2	15	11	8	3	10	6	12	5	9	0	7
0	15	7	4	14	2	13	1	10	6	12	11	9	5	3	8
4	1	14	8	13	6	2	11	15	12	9	7	3	10	5	0
15	12	8	2	4	9	1	7	5	11	3	14	10	0	6	13

S_2															
15	1	8	14	6	11	3	4	9	7	2	13	12	0	5	10
3	13	4	7	15	2	8	14	12	0	1	10	6	9	11	5
0	14	7	11	10	4	13	1	5	8	12	6	9	3	2	15
13	8	10	1	3	15	4	2	11	6	7	12	0	5	14	9

S_3															
10	0	9	14	6	3	15	5	1	13	12	7	11	4	2	8
13	7	0	9	3	4	6	10	2	8	5	14	12	11	15	1
13	6	4	9	8	15	3	0	11	1	2	12	5	10	14	7
1	10	13	0	6	9	8	7	4	15	14	3	11	5	2	12

S_4															
7	13	14	3	0	6	9	10	1	2	8	5	11	12	4	15
13	8	11	5	6	15	0	3	4	7	2	12	1	10	14	9
10	6	9	0	12	11	7	13	15	1	3	14	5	2	8	4
3	15	0	6	10	1	13	8	9	4	5	11	12	7	2	14

S_5															
2	12	4	1	7	10	11	6	8	5	3	15	13	0	14	9
14	11	2	12	4	7	13	1	5	0	15	10	3	9	8	6
4	2	1	11	10	13	7	8	15	9	12	5	6	3	0	14
11	8	12	7	1	14	2	13	6	15	0	9	10	4	5	3

						S_6									
12	1	10	15	9	2	6	8	0	13	3	4	14	7	5	11
10	15	4	2	7	12	9	5	6	1	13	14	0	11	3	8
9	14	15	5	2	8	12	3	7	0	4	10	1	13	11	6
4	3	2	12	9	5	15	10	11	14	1	7	6	0	8	13

						S_7									
4	11	2	14	15	0	8	13	3	12	9	7	5	10	6	1
13	0	11	7	4	9	1	10	14	3	5	12	2	15	8	6
1	4	11	13	12	3	7	14	10	15	6	8	0	5	9	2
6	11	13	8	1	4	10	7	9	5	0	15	14	2	3	12

						S_8									
13	2	8	4	6	15	11	1	10	9	3	14	5	0	12	7
1	15	13	8	10	3	7	4	12	5	6	11	0	14	9	2
7	11	4	1	9	12	14	2	0	6	10	13	15	3	5	8
2	1	14	7	4	10	8	13	15	12	9	0	3	5	6	11

The functioning of the S-boxes:

Consider a 6-bit string $B_j = b_1 b_2 b_3 b_4 b_5 b_6$. We shall compute a 4-bit string $S_j(B_j)$ in the following way:

The two bits $b_1 b_6$ are considered to be the binary notation of the index r of one of the four lines of S_j, $0 \leq r \leq 3$.

The four bits $b_2 b_3 b_4 b_5$ constitute the binary notation of the index s of one of the 16 columns of the table, $0 \leq s \leq 15$.

$S_j(B_j)$ will be the 4-bit notation of the entry $S_j(r, s)$.

Example $S_5(101010) = 1101$.
$\quad (b_1 b_6 = 10 \leftrightarrow 2 \qquad b_2 b_3 b_4 b_5 = 0101 \leftrightarrow 5 \qquad S_5(2,5) = 13 \leftrightarrow 1101)$

Remark The known criteria (made public in 1976) for the design of the S-boxes are the following:

0. Every line of an S-box is a permutation of the integers $0, 1, \ldots, 15$.
1. No S-box is an affine transformation.
2. The modification of a single input bit of an S-box causes the modification of at least two output bits.
3. For every S-box S and all $B = b_1 b_2 b_3 b_4 b_5 b_6$, $S(B)$ and $S(B \oplus 001100)$ differ in at least two bits.
4. For every S-box S and all $B = b_1 b_2 b_3 b_4 b_5 b_6$ we have $S(B) \neq S(B \oplus 11\alpha\beta00)$ for $\alpha, \beta \in \{0, 1\}$.

5. For every S-box, if an input bit is kept constant and if we are only interested in a fixed output position, the number of quintuples which produce 0 and the number of quintuples which produce 1 are near to the mean, i.e. they vary between 13 and 19 (note that whenever the fixed bit is b_1 or b_6, there are precisely 16 values which produce 0 and 16 values which produce 1, according to the criterion **0**).

Finally, the permutation P (which seems to be meant as a remedy to the local treatment of the data):

P			
16	7	20	21
29	12	28	17
1	15	23	26
5	18	31	10
2	8	24	14
32	27	3	9
19	13	30	6
22	11	4	25

Read line by line: $P(c_1c_2c_3\cdots c_{32}) = c_{16}c_7c_{20}\cdots c_4c_{25}.$

3. *The key schedule*:

The cipher key K (of some user) is a 64-bit string, 56 positions of which constitute the real key, and eight positions support parity-check bits. The bits on the positions $8,16,\ldots,64$ are such that every 8-bit byte contains an odd number of ones. This way one can detect an error of a 1-bit fault per 8-bit byte. In the computation of the round keys, the parity-check bits are clearly neglected.

1. Starting with the 64 bits of K, one skips first the parity-check bits, and permutes the key bits; the operation $PC-1$ (*permuted choice*) finally gives $C_0D_0 = PC-1(K)$ where C_0 is the block of the 28 firsts bits of $PC-1(K)$ and D_0 the block of the remaining 28 bits.
2. Then, one computes 16 times

$$C_i = LS_i(C_{i-1}),$$
$$D_i = LS_i(D_{i-1})$$

and $K_i = PC-2(C_i, D_i)$. LS_i is a cyclic left shift of one or of two positions, depending on the value of i: it is of one position if $i = 1,2,9$ or 16, otherwise it is of two positions. $PC-2$ is another "permuted choice", which has to skip eight positions in order to obtain finally 48-bit strings.

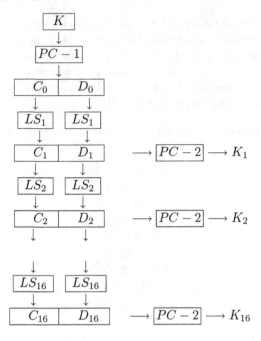

The two selective permutations $PC-1$ and $PC-2$, operating for the computation of the round keys, are:

PC.1						
57	49	41	33	25	17	9
1	58	50	42	34	26	18
10	2	59	51	43	35	27
19	11	3	60	52	44	36
63	55	47	39	31	23	15
7	62	54	46	38	30	22
14	6	61	53	45	37	29
21	13	5	28	20	12	4

PC.2						
14	17	11	24	1	5
3	28	15	6	21	10
23	19	12	4	26	8
16	7	27	20	13	2
41	52	31	37	47	55
30	40	51	45	33	48
44	49	39	56	34	53
46	42	50	36	29	32

Exercises

(1) A mini-DES with four iterations, transforming 8-bit bytes into 8-bit bytes. The keys will have also eight bits.

(a) The initial permutation: $IP(x_1x_2x_3x_4x_5x_6x_7x_8) = x_8x_5x_2x_7x_4x_1x_6x_3$.

(b) The key schedule: Let $K = u_1u_2u_3u_4u_5u_6u_7u_8$, then $K_1 = u_4u_1u_3u_8u_5u_7$, $K_2 = u_3u_4u_2u_7u_8u_6$, $K_3 = u_2u_3u_1u_6u_7u_5$, $K_4 = u_5u_6u_8u_1u_2u_4$.

(c) The function $f(R_{i-1}, K_i)$:
Compute first the binary word $b_1b_2b_3b_4b_5b_6$ from $R_{i-1} = r_1r_2r_3r_4$ and $K_i = t_1t_2t_3t_4t_5t_6$ in the following way:
$b_1b_2b_3b_4b_5b_6 = r_1r_2r_3r_4r_1r_3 \oplus t_1t_2t_3t_4t_5t_6$
($\oplus$: Boolean XOR, componentwise). Then, we shall use the following S-box:

	0	1	2	3	4	5	6	7	8	9	10	11	12	13	14	15
0	4	11	2	14	15	0	8	13	3	12	9	7	5	10	6	1
1	13	0	11	7	4	9	1	10	14	3	5	12	2	15	8	6
2	1	4	11	13	12	3	7	14	10	15	6	8	0	5	9	2
3	6	11	13	8	1	4	10	7	9	5	0	15	14	2	3	12

$b_1b_2b_3b_4b_5b_6$ points at a line b_1b_6 and at a column $b_2b_3b_4b_5$ of the S-box. $f(R_{i-1}, K_i)$ will be the 4-bit notation of the integer which occupies the entry.

(d) Encrypt $M = 00100010$ with $K = 01110111$.

(2) Now we deal with usual DES.
The plaintext: $M \equiv 8af1ee80a6b7c385$. Write down the fifth 8-bit byte of $IP(M)$.

(3) Let $B = 0fe31e4fb0f0$ (48 bits; you have to think in 8 times six bits). Find $S(B)$ (32 bits – in hexadecimal notation).

(4) The plaintext: 00 00 00 00 00 00 00 00 (64 zeros).
The cipher key: 01 01 01 01 01 01 01 01 (56 zeros – the ones are the parity-check bits).
$(L_1, R_1) =$?

(5) Let $c(.)$ denote the operation of complementation for binary words (p.ex.: $c(0110) = 1001$).
Let $y = \text{DES}(x, K)$ be the ciphertext obtained from the plaintext block x and the cipher key K. Show that $c(y) = \text{DES}(c(x), c(K))$ (complementation of the plaintext and of the key will yield the complement of the ciphertext).

(6) The round keys may be obtained from the cipher key of the user by means of "selection tables". Find the selection table for K_6.

Solution

K_6
3 44 27 17 42 10 26 50 60 2 41 35
25 57 19 18 1 51 52 59 58 49 11 34
13 23 30 45 63 62 38 21 31 12 14 55
20 47 29 54 6 15 4 5 39 53 46 22

(7) Let $K = abababababababab$ (64 bits) be your cipher key. Which is your sixth round key K_6 (48 bits, in hexadecimal notation)?

(8) Differential cryptanalysis – level 0.

S_1	0	1	2	3	4	5	6	7	8	9	10	11	12	13	14	15
0	14	4	13	1	2	15	11	8	3	10	6	12	5	9	0	7
1	0	15	7	4	14	2	13	1	10	6	12	11	9	5	3	8
2	4	1	14	8	13	6	2	11	15	12	9	7	3	10	5	0
3	15	12	8	2	4	9	1	7	5	11	3	14	10	0	6	13

Notation   *The inputs: B, B^*, etc. (six bits)*
            The outputs: C, C^*, etc. (four bits).

(a) For an input modification B' and an output variation C' of (S_1) let support$(B' \longrightarrow C') = \{B : S_1(B \oplus B') \oplus S_1(B) = C'\}$. Find support$(110100 \longrightarrow 0100)$ – a set of two elements.

(b) *Observation*: The first six bits of the round key K_1 are the bits k_{10}, k_{51}, k_{34}, k_{60}, k_{49}, k_{17} of the cipher key. Let $R_0 = 00001 \cdots 0$ and $R_0^* = 10101 \cdots 1$ be two chosen "plaintexts".
We observe: The two outputs of the S_1-box C and C^* (coming from R_0 and from R_0^*) differ only at the second position ($C \oplus C^* = 0100$).
Find k_{17}, k_{34} and k_{49} (we shall make use of the result of (a)).

2.2 The Advanced Encryption Standard: The Cipher Rijndael

It was the cipher *Rijndael*, designed by Vincent Rijmen and Joan Daemen, which was the surprise winner of the contest for the new Advanced Encryption Standard of the United States. This contest was organized and run by the National Institute for Standards and Technology (NIST), at the end of the 1990s. In December 2001, the cipher Rijndael became the successor of DES, i.e. the *AES* ≡ *Advanced Encryption Standard*. Much more soaked in Mathematics than its predecessor, the system AES remains altogether in the tradition: you will find *S*-boxes, rounds and round transforms, and round keys.

2.2.1 Some Elementary Arithmetic

The *letters* of the system AES will be *8-bit bytes*, with an intrinsic arithmetic that stems from their identification with the 256 elements of a *field of residues* (modulo a certain irreducible binary polynomial of degree 8).

In order to be perfectly at ease with this kind of arithmetic, we shall recall the essential facts about the most frequent specimen of quotient rings, first the integer case (i.e. we shall treat the quotient rings of $\mathbb{Z}$), where everything is rather natural (and familiar ...), then the case of binary polynomial arithmetic, where the algebraic skeleton is simply obtained by formal analogy with the integer case.

Note that the quotient rings of these two *Euclidian* rings – the ring of integers and the ring of binary polynomials – are omnipresent as algorithmic tools for information security (in cryptography as well as in theory and practice of error control codes).

Recall: The Arithmetic of the Rings of Residues $\mathbb{Z}/n\mathbb{Z}$

Let us keep $n \geq 1$ fixed. Definition of an equivalence relation on $\mathbb{Z}$:
$x \equiv y \bmod n \quad \Longleftrightarrow \quad x$ and y leave the same remainder when divided by $n \quad \Longleftrightarrow \quad x - y$ is divisible by n.

Notation $[x] = x \bmod n$ *for the equivalence class of x. $x \bmod n$ will also denote the canonical representative of this class: the (non-negative) remainder when dividing x by n.*

Fundamental Observation

The so-defined equivalence relation is compatible with addition and with multiplication on $\mathbb{Z}$:

$$\begin{array}{ll} x \equiv x' \bmod n & \\ y \equiv y' \bmod n \end{array} \Rightarrow \begin{array}{ll} x + y \equiv x' + y' \bmod n \\ x \cdot y \equiv x' \cdot y' \bmod n \end{array}$$

Consequence:

The set $\mathbb{Z}/n\mathbb{Z}$ of the n equivalence classes $\{[0], [1], \ldots [n-1]\}$ bears a natural ring structure (of a finite commutative ring):

$$[x] + [y] = [x + y] \qquad [x] \cdot [y] = [x \cdot y],$$

where $[0]$ is the zero element for addition, $[1]$ is the unit for multiplication. We have, of course: $-[x] = [-x]$: the opposite class is the class of the opposite representative.

Example $\mathbb{Z}/12\mathbb{Z} = \{[0], [1], \ldots [11]\}$

Let us look closer at the following products:

$$\begin{array}{lll} [1] \cdot [1] = [1], & [2] \cdot [6] = [0], & [3] \cdot [4] = [0], \\ [5] \cdot [5] = [1], & [7] \cdot [7] = [1], & [8] \cdot [3] = [0], \\ [9] \cdot [4] = [0], & [10] \cdot [6] = [0], & [11] \cdot [11] = [1]. \end{array}$$

We note: $[a]$ is invertible for the multiplication in $\mathbb{Z}/12\mathbb{Z}$ $\Longleftrightarrow$ a is
prime to 12   (i.e. $a \equiv 1, 5, 7, 11 \bmod 12$).

Let us turn back to the general case:

We denote by $(\mathbb{Z}/n\mathbb{Z})^*$ the group of multiplicatively invertible elements
in $\mathbb{Z}/n\mathbb{Z}$.

Proposition
$$[a] \in (\mathbb{Z}/n\mathbb{Z})^* \quad \Longleftrightarrow \quad gcd(a, n) = 1.$$

Proof $[a] \in (\mathbb{Z}/n\mathbb{Z})^*$ $\Leftrightarrow$ there exists $x \in \mathbb{Z}$: $[a] \cdot [x] = [1]$ $\Leftrightarrow$ there
exist $x, y \in \mathbb{Z}$: $ax - 1 = ny$ $\Leftrightarrow$ the equation $aX + nY = 1$ admits a
solution $(x, y) \in \mathbb{Z} \times \mathbb{Z}$.

We still have to show:
The equation $aX + nY = 1$ admits a solution $(x, y) \in \mathbb{Z} \times \mathbb{Z}$ $\Longleftrightarrow$
$gcd(a, n) = 1$. The reasoning goes as follows:

On the one hand, a common factor p of a and n must divide $ax + ny$,
which we suppose to be equal to 1; on the other hand, if $gcd(a, n) = 1$,   then
we compute a solution $(x, y) \in \mathbb{Z} \times \mathbb{Z}$ for the equation $aX + nY = 1$ by
means of the Euclidean algorithm which determines the $gcd(a, n)$ (=1).   □

Example Compute an integer solution of the equation $37X + 100Y = 1$.
Let us recover $gcd(37, 100) = 1$ by iterated Euclidian divisions:

$$100 = 2 \cdot 37 + 26, \quad \text{i.e.} \quad 100 = 2 \cdot 37 + r_1,$$
$$37 = 1 \cdot 26 + 11, \quad\quad\quad 37 = r_1 + r_2,$$
$$26 = 2 \cdot 11 + 4, \quad\quad\quad r_1 = 2 \cdot r_2 + r_3,$$
$$11 = 2 \cdot 4 + 3, \quad\quad\quad r_2 = 2 \cdot r_3 + r_4,$$
$$4 = 1 \cdot 3 + 1, \quad\quad\quad r_3 = r_4 + 1.$$

Now, we have to eliminate the remainders:
$$1 = r_3 - r_4 = r_3 - (r_2 - 2r_3) = 3r_3 - r_2 = 3(r_1 - 2r_2) - r_2 = 3r_1 - 7r_2 =$$
$$3r_1 - 7(37 - r_1) = 10r_1 - 7 \cdot 37 = 10(100 - 2 \cdot 37) - 7 \cdot 37 = 37 \cdot (-27) + 100 \cdot 10.$$
Our solution: $(x, y) = (-27, 10)$.
Once more: $[37]^{-1} = [-27] = [73]$ in $(\mathbb{Z}/100\mathbb{Z})^*$.

Corollary $\mathbb{Z}/n\mathbb{Z}$ *is a field (every non-zero residue is invertible for the*
multiplication in $\mathbb{Z}/n\mathbb{Z}$) $\quad \Longleftrightarrow \quad$ *n is a prime number.*

In this case   • $n = p$ is a prime number • $(\mathbb{Z}/p\mathbb{Z})^*$ is a *cyclic* group of
order $p - 1$:

There exists an element $\omega \in \mathbb{Z}/p\mathbb{Z}$ such that the list of the $p - 1$ powers
$\omega, \omega^2, \dots \omega^{p-1} = 1$ contains precisely *all* the elements of $(\mathbb{Z}/p\mathbb{Z})^*$.
An element ω with this property is called *primitive* modulo p.

Example $(\mathbb{Z}/17\mathbb{Z})^*$:

$$
\begin{array}{llll}
\omega = 3, & \omega^2 = 9, & \omega^3 = 10, & \omega^4 = 13, \\
\omega^5 = 5, & \omega^6 = 15, & \omega^7 = 11, & \omega^8 = 16, \\
\omega^9 = 14, & \omega^{10} = 8, & \omega^{11} = 7, & \omega^{12} = 4, \\
\omega^{13} = 12, & \omega^{14} = 2, & \omega^{15} = 6, & \omega^{16} = 1.
\end{array}
$$

Let us insist: If p is a prime number, then the multiplicative group $(\mathbb{Z}/p\mathbb{Z})^*$ is a cyclic group of $(p-1)$th roots of unity, i.e. it can be written as

$$(\mathbb{Z}/p\mathbb{Z})^* = \{\omega, \omega^2, \ldots, \omega^{p-1} = 1\}.$$

In particular, for every divisor n of $p-1$, the group $(\mathbb{Z}/p\mathbb{Z})^*$ contains a subgroup Γ_n of nth roots of unity:

$$\Gamma_n = \left\{ \omega^{\frac{p-1}{n}}, \omega^{2\frac{p-1}{n}}, \ldots \omega^{n\frac{p-1}{n}} = 1 \right\}.$$

Remark Starting with a fixed prime number p, the smallest (positive) integer which is a primitive root modulo p is *not* obtained in a natural way: Take, for example, the following five consecutive prime numbers, and the associated (minimal) primitive roots:

$$
\begin{array}{ll}
p = 760301 : & \omega = 2, \\
p = 760321 : & \omega = 73, \\
p = 760343 : & \omega = 5, \\
p = 760367 : & \omega = 5, \\
p = 760373 : & \omega = 2,
\end{array}
$$

By the way, already the candidate 2 gives rise to a remarkable problem: One does not know, whether 2 is a primitive root modulo p for an infinity of prime numbers p.

Exercises

(1) Find the smallest positive integer ω which is a primitive root modulo $p = 3, 5, 7, 11, 13, 17$.
Decide, in all cases, if ω remains primitive modulo p^2 (ω will then remain primitive modulo p^k for every $k \geq 2$).

The following exercises deal with the important *Chinese Remainder Theorem*.

(2) Show the Chinese Remainder Theorem for the following particular case (which is perfectly generic...):

Let $N = n_1 n_2 n_3$ be a product of three (mutually) coprime integers. Then the natural mapping

$$\pi : \mathbb{Z}/N\mathbb{Z} \longrightarrow \mathbb{Z}/n_1\mathbb{Z} \times \mathbb{Z}/n_2\mathbb{Z} \times \mathbb{Z}/n_3\mathbb{Z}$$
$$x \bmod N \longmapsto (x \bmod n_1, x \bmod n_2, x \bmod n_3)$$

is an *isomorphism* of rings.

Solution Since the source and the target set have the same cardinality $N = n_1 n_2 n_3$, we have only to show the surjectivity of π:
For given $x_1, x_2, x_3 \in \mathbb{Z}$, there exists $x \in \mathbb{Z}$ with $x \equiv x_1 \bmod n_1$, $x \equiv x_2 \bmod n_2$, $x \equiv x_3 \bmod n_3$.

We search x as a linear combination $x = a_1 x_1 + a_2 x_2 + a_3 x_3$ (where a_1, a_2, a_3 will *not* depend of x_1, x_2, x_3, i.e. they will *define* the inverse mapping π^{-1}).
We need:

$$
\begin{array}{lll}
a_1 \equiv 1 \bmod n_1, & a_1 \equiv 0 \bmod n_2, & a_1 \equiv 0 \bmod n_3, \\
a_2 \equiv 0 \bmod n_1, & a_2 \equiv 1 \bmod n_2, & a_2 \equiv 0 \bmod n_3, \\
a_3 \equiv 0 \bmod n_1, & a_3 \equiv 0 \bmod n_2, & a_3 \equiv 1 \bmod n_3,
\end{array}
$$

We find:
$$
\begin{array}{lll}
a_1 = \alpha_1 n_2 n_3 & & \alpha_1 \equiv (n_2 n_3)^{-1} \bmod n_1, \\
a_2 = n_1 \alpha_2 n_3 & \text{with} & \alpha_2 \equiv (n_1 n_3)^{-1} \bmod n_2, \\
a_3 = n_1 n_2 \alpha_3 & & \alpha_3 \equiv (n_1 n_2)^{-1} \bmod n_3.
\end{array}
$$

(the multiplicative inverses exist precisely because of our coprimality hypothesis on n_1, n_2, n_3). It is clear, that everything generalizes to coprime decompositions of arbitrary length.

(3) In the situation of exercise (2), let $N = 1001$, with $n_1 = 7$, $n_2 = 11$, $n_3 = 13$.

Make explicit $\pi^{-1} : \mathbb{Z}/7\mathbb{Z} \times \mathbb{Z}/11\mathbb{Z} \times \mathbb{Z}/13\mathbb{Z} \longrightarrow \mathbb{Z}/1001\mathbb{Z}$
(i.e. find a_1, a_2, a_3).

(4) Resolve the following congruences (separately):

$$
\begin{array}{ll}
3x \equiv 4 \bmod 12 & 27x \equiv 25 \bmod 256 \\
9x \equiv 12 \bmod 21 & 103x \equiv 612 \bmod 676
\end{array}
$$

(5) Show that $n^5 - n$ is always divisible by 30.

(6) Find the smallest non-negative solution for each of the following three systems of congruences:

(a) $x \equiv 2 \bmod 3$

(b) $x \equiv 12 \bmod 31$

$x \equiv 3 \bmod 5$

(c) $19x \equiv 103 \bmod 900$

$x \equiv 87 \bmod 127$

$x \equiv 4 \bmod 11$

$10x \equiv 511 \bmod 841$

$x \equiv 91 \bmod 255$

$x \equiv 5 \bmod 16$

The Arithmetic of the Field of Residues $\mathbb{F}_2[x]/(p(x))$

Let $\mathbb{F}_2[x]$ be the ring of polynomials with coefficients in the field $\mathbb{F}_2 = \mathbb{Z}/2\mathbb{Z}$. This is an Euclidian ring (we dispose of the Euclidian algorithm), exactly like $\mathbb{Z}$,the ring of integers.

As a first step in analogies, we shall be interested in "prime numbers" for $\mathbb{F}_2[x]$.

Definition $p(x)$ *is an irreducible polynomial* $\iff$ $p(x)$ *has no non-trivial divisor (i.e.* 1 *and* $p(x)$ *are the only divisors of* $p(x)$).

Example The irreducible polynomials of degree ≤ 4:

(1) degree 1: $x, x+1$
(2) degree 2: $x^2 + x + 1$ (note that $x^2 + 1 = (x+1)^2$, since $1 + 1 = 0$)
(3) degree 3: $x^3 + x + 1$, $x^3 + x^2 + 1$ (a cubic polynomial is irreducible $\iff$ it has no linear factor $\iff$ it admits neither 0 nor 1 as a root)
(4) degree 4: $x^4 + x + 1$, $x^4 + x^3 + 1$, $x^4 + x^3 + x^2 + x + 1$

(the absence of a linear factor forces the polynomial to have a constant term $= 1$ and an odd number of additive terms; on the other hand: $(x^2 + x + 1)^2 = x^4 + x^2 + 1$).

Exercise

Show that the binary polynomial $p(x) = x^8 + x^4 + x^3 + x^2 + 1$ is irreducible (this is an exercise in Euclidian division: you have to test whether $p(x)$ is divisible by any of the irreducible polynomials of degree ≤ 4).

Our next step: the **rings of residues** $\mathbb{F}_2[x]/(p(x))$.

Let $p(x) \in \mathbb{F}_2[x]$ be a *fixed* polynomial .

The ring of residues (for the division by $p(x)$) $\mathbb{F}_2[x]/(p(x))$ is defined in precisely the same manner as in the case of the integers.

We note: If the degree of $p(x)$ is equal to n, then $\mathbb{F}_2[x]/(p(x))$ admits 2^n elements: there are 2^n remainders for the division by $p(x)$ (i.e. there are 2^n binary polynomials of degree $< n$).

Let us look at a simple but instructive

Example $p(x) = x^4 + x + 1$.

$\mathbb{F}_2[x]/(x^4 + x + 1) = \{ [0], \quad [1], \quad [x], \quad [x+1], \quad [x^2], \quad [x^2 + 1], \quad [x^2 + x],$ $[x^2 + x + 1], [x^3], \quad [x^3 + 1], \quad [x^3 + x], \quad [x^3 + x + 1], \quad [x^3 + x^2], \quad [x^3 + x^2 + 1],$ $[x^3 + x^2 + x], \quad [x^3 + x^2 + x + 1] \quad \}$.

The 16 binary words of length 4 will be coded in remainders of division (by $p(x) = x^4 + x + 1$) in the following way:

$$0000 \quad \leftrightarrow \quad [0],$$
$$0001 \quad \leftrightarrow \quad [1],$$
$$0010 \quad \leftrightarrow \quad [x]$$
$$\vdots$$
$$1111 \quad \leftrightarrow \quad [x^3 + x^2 + x + 1].$$

The addition of residues is simple:

$[x^3 + 1] + [x^3 + x^2 + 1] = [x^2]$,
$\quad$ 1001 $\quad \oplus \quad$ 1101 = $\quad$ 0100

(addition will always come from the Boolean XOR, position by position).

As to multiplication, one should – in theory – proceed this way:

$$[x^3 + x + 1] \cdot [x^3 + x^2] = [x^6 + x^5 + x^4 + x^2].$$

Then, one searches the remainder of division (by $x^4 + x + 1$):

$$
\begin{array}{l}
x^6 + x^5 + x^4 \qquad\ + x^2 : (x^4 + x + 1) = x^2 + x + 1 \\
\underline{x^6 \qquad\qquad\quad + x^3 + x^2} \\
\quad\ x^5 + x^4 + x^3 \\
\quad\ \underline{x^5 \qquad\qquad\quad\ \ + x^2 + x} \\
\qquad\quad x^4 + x^3 + x^2 + x \\
\qquad\quad \underline{x^4 \qquad\qquad\qquad + x + 1} \\
\qquad\qquad\quad x^3 + x^2 \qquad + 1
\end{array}
$$

This yields: $\quad [x^3 + x + 1] \cdot [x^3 + x^2] = [x^3 + x^2 + 1]$

$$1011 \quad \cdot 1100 \quad = \quad 1101.$$

How to avoid these (boring) Euclidian divisions when computing products?
We observe: $[x^4] = [x + 1]$ (since $[x^4 + x + 1] = [0]$). Let us give up the bracket notation (i.e. we replace equivalence by equality):

$x^4 = x + 1$

We can continue:

$x^5 = x^2 + x$,
$x^6 = x^3 + x^2$,

Finally:

$x^6 + x^5 + x^4 + x^2 = (x^3 + x^2) + (x^2 + x) + (x + 1) + x^2 = x^3 + x^2 + 1$, as expected.

Remember, for the general case:

Let $\quad p(x) = x^n + \beta_{n-1}x^{n-1} + \ldots \beta_1 x + \beta_0 \in \mathbb{F}_2[x]$. Then the arithmetic of residues modulo $p(x)$ (the arithmetic in $\mathbb{F}_2[x]/(p(x))$) will obey the following rules:

(1) The addition of remainders is done "coefficient by coefficient", respecting $1 + 1 = 0$.
(2) The multiplication of remainders is done in two steps:

(i) Multiplication of the remainders as simple binary polynomials.
(ii) Reduction of the result to a remainder (i.e. a polynomial of degree $< n$), according to the reduction rule $x^n = \beta_{n-1}x^{n-1} + \cdots + \beta_1 x + \beta_0$.

But back to $\mathbb{F}_2[x]/(x^4 + x + 1)$:
  Let us consider the successive powers of $\omega = x$:

$$\begin{array}{lll}
\omega = x, & \omega^6 = x^3 + x^2, & \omega^{11} = x^3 + x^2 + x, \\
\omega^2 = x^2, & \omega^7 = x^3 + x + 1, & \omega^{12} = x^3 + x^2 + x + 1, \\
\omega^3 = x^3, & \omega^8 = x^2 + 1, & \omega^{13} = x^3 + x^2 + 1, \\
\omega^4 = x + 1, & \omega^9 = x^3 + x, & \omega^{14} = x^3 + 1, \\
\omega^5 = x^2 + x, & \omega^{10} = x^2 + x + 1, & \omega^{15} = 1,
\end{array}$$

We remark: The powers of $\omega = x$ yield a complete list of all 15 non-zero remainders in $\mathbb{F}_2[x]/(x^4 + x + 1)$.

$$\mathbb{F}_2[x]/(x^4 + x + 1) \text{ is a } field \text{ (of 16 elements)}$$

Every non-zero residue admits clearly a multiplicative inverse:

$$(x^3 + x^2)^{-1} = (\omega^6)^{-1} = \omega^{-6} = \omega^9 = x^3 + x.$$

  For the incredulous reader:

$$(x^3 + x^2)(x^3 + x) = x^6 + x^5 + x^4 + x^3 = (x^3 + x^2) + (x^2 + x) + (x + 1) + x^3 = 1$$

But back to the general situation:

Proposition *The ring of residues $\mathbb{F}_2[x]/(p(x))$ is a field $\iff$ $p(x)$ is an irreducible polynomial.*

The argument is the same as in the case of the integers: You have to use the equivalence: $[a(x)]$ is invertible in the ring $\mathbb{F}_2[x]/(p(x))$ $\iff$ $\gcd(a(x), p(x)) = 1$.

Complement *Let $p(x) = x^n + \cdots + 1$ be an irreducible polynomial of degree n.*
  Then the multiplicative group $G = (\mathbb{F}_2[x]/(p(x)))^*$ is a cyclic group of order $2^n - 1$ (this is a non-trivial fact ...). If $\omega = x$ is a generator of this group (i.e. if the powers x^k, $1 \le k \le 2^n - 1$, produce all the non-zero residues modulo $p(x)$), then we say that the (irreducible) polynomial $p(x)$ is primitive.

Exercise

Show that the polynomial $p(x) = x^4 + x^3 + x^2 + x + 1$ is irreducible, but that $p(x)$ is *not* primitive.

Some irreducible and *primitive* (binary) polynomials:

$x^2 + x + 1,$	$x^7 + x^3 + 1,$	$x^{12} + x^6 + x^4 + x + 1,$	$x^{17} + x^3 + 1,$
$x^3 + x + 1,$	$x^8 + x^4 + x^3 + x^2 + 1,$	$x^{13} + x^4 + x^3 + x + 1,$	$x^{18} + x^7 + 1,$
$x^4 + x + 1,$	$x^9 + x^4 + 1,$	$x^{14} + x^{10} + x^6 + x + 1,$	$x^{19} + x^5 + x^2 + x + 1,$
$x^5 + x^2 + 1,$	$x^{10} + x^3 + 1,$	$x^{15} + x + 1,$	$x^{20} + x^3 + 1,$
$x^6 + x + 1,$	$x^{11} + x^2 + 1,$	$x^{16} + x^{12} + x^3 + x + 1.$	

Remark Let $p(x) = x^n + \cdots + 1$ be a primitive (irreducible) polynomial. Then $\omega = x$ is a primitive $(2^n - 1)$th root of unity in $\mathbb{F}_2[x]/(p(x))$

$$\omega^{2^n - 1} = 1 \quad \text{and} \quad \omega^k \neq 1 \quad \text{for} \quad 1 \leq k \leq 2^n - 1.$$

For every divisor N of $2^n - 1$, put $N' = \frac{2^n - 1}{N}$. Then $\omega_N = \omega^{N'}$ is a (primitive) Nth root of unity in $\mathbb{F}_2[x]/(p(x))$.

Notation $\mathbb{F}_{2^n} = \mathbb{F}_2[x]/(p(x))$ ($p(x)$ is a distinguished irreducible binary polynomial of degree n – the notational laxity is partly justified by the fact that all finite fields of the *same cardinality* are *isomorphic*.[3])

Let us fix the notation (and the corresponding arithmetic) for the sequel:

$\mathbb{F}_2 = \mathbb{Z}/2\mathbb{Z} = \{0, 1\},$

$\mathbb{F}_4 = \mathbb{F}_2[x]/(x^2 + x + 1),$

$\mathbb{F}_8 = \mathbb{F}_2[x]/(x^3 + x + 1),$

$\mathbb{F}_{16} = \mathbb{F}_2[x]/(x^4 + x + 1),$

$\mathbb{F}_{64} = \mathbb{F}_2[x]/(x^6 + x + 1),$

$\mathbb{F}_{256} = \mathbb{F}_2[x]/(x^8 + x^4 + x^3 + x^2 + 1).$

Exercises

We shall begin with the field $\mathbb{F}_{64} = \mathbb{F}_2[x]/(x^6 + x + 1)$, then turn to the field $\mathbb{F}_{256} = \mathbb{F}_2[x]/(x^8 + x^4 + x^3 + x^2 + 1)$. We shall always identify binary words (of length 6 or 8) with remainders of division, according to the scheme $100101 = x^5 + x^2 + 1$ in $\mathbb{F}_{64}$, $00100111 = x^5 + x^2 + x + 1$ in $\mathbb{F}_{256}$.

(1) Write down the table of the 63 non-zero residues in $\mathbb{F}_{64}$, in function of the successive powers of $\omega = \omega_{63} = x$.

---

[3] There are 30 irreducible binary polynomials of degree 8, hence 30 realizations for "the" Galois field of 256 elements in terms of residue-arithmetic. In the following exercises, we shall compare two of them. The reader will rapidly note the important differences in down-to-earth computations.

Solution The cyclic group of 63th roots of unity $(\mathbb{F}_{64})^*$:

$$
\begin{aligned}
x^6 &= x + 1, & x^{25} &= x^5 + x, & x^{44} &= x^5 + x^3 + x^2 + 1, \\
x^7 &= x^2 + x, & x^{26} &= x^2 + x + 1, & x^{45} &= x^4 + x^3 + 1, \\
x^8 &= x^3 + x^2, & x^{27} &= x^3 + x^2 + x, & x^{46} &= x^5 + x^4 + x, \\
x^9 &= x^4 + x^3, & x^{28} &= x^4 + x^3 + x^2, & x^{47} &= x^5 + x^2 + x + 1, \\
x^{10} &= x^5 + x^4, & x^{29} &= x^5 + x^4 + x^3, & x^{48} &= x^3 + x^2 + 1, \\
x^{11} &= x^5 + x + 1, & x^{30} &= x^5 + x^4 + x + 1, & x^{49} &= x^4 + x^3 + x, \\
x^{12} &= x^2 + 1, & x^{31} &= x^5 + x^2 + 1, & x^{50} &= x^5 + x^4 + x^2, \\
x^{13} &= x^3 + x, & x^{32} &= x^3 + 1, & x^{51} &= x^5 + x^3 + x + 1, \\
x^{14} &= x^4 + x^2, & x^{33} &= x^4 + x, & x^{52} &= x^4 + x^2 + 1, \\
x^{15} &= x^5 + x^3, & x^{34} &= x^5 + x^2, & x^{53} &= x^5 + x^3 + x, \\
x^{16} &= x^4 + x + 1, & x^{35} &= x^3 + x + 1, & x^{54} &= x^4 + x^2 + x + 1, \\
x^{17} &= x^5 + x^2 + x, & x^{36} &= x^4 + x^2 + x, & x^{55} &= x^5 + x^3 + x^2 + x, \\
x^{18} &= x^3 + x^2 + x + 1, & x^{37} &= x^5 + x^3 + x^2, & x^{56} &= x^4 + x^3 + x^2 + x + 1, \\
x^{19} &= x^4 + x^3 + x^2 + x, & x^{38} &= x^4 + x^3 + x + 1, & x^{57} &= x^5 + x^4 + x^3 + x^2 + x, \\
x^{20} &= x^5 + x^4 + x^3 + x^2, & x^{39} &= x^5 + x^4 + x^2 + x, & x^{58} &= x^5 + x^4 + x^3 + x^2 + x + 1, \\
x^{21} &= x^5 + x^4 + x^3 + x + 1, & x^{40} &= x^5 + x^3 + x^2 + x + 1, & x^{59} &= x^5 + x^4 + x^3 + x^2 + 1, \\
x^{22} &= x^5 + x^4 + x^2 + 1, & x^{41} &= x^4 + x^3 + x^2 + 1, & x^{60} &= x^5 + x^4 + x^3 + 1, \\
x^{23} &= x^5 + x^3 + 1, & x^{42} &= x^5 + x^4 + x^3 + x, & x^{61} &= x^5 + x^4 + 1, \\
x^{24} &= x^4 + 1, & x^{43} &= x^5 + x^4 + x^2 + x + 1, & x^{62} &= x^5 + 1.
\end{aligned}
$$

(2) (The discrete logarithm problem in $\mathbb{F}_{64}$):

Let $\qquad \alpha = 11100,$
$\qquad \qquad \beta = 10101.$
Find the unique exponent e , $0 \le e \le 62$, such that

$$\alpha^e \equiv \beta \bmod (x^6 + x + 1)$$

(Notation: $e = log_\alpha \beta$)

(3) Write down the table for the 255 non-zero residues in $\mathbb{F}_{256}$, in function of the successive powers of $\omega = \omega_{255} = x$.

Solution The cyclic group of 255th roots of unity $(\mathbb{F}_{256})^*$:

$$
\begin{aligned}
x^8 &= x^4 + x^3 + x^2 + 1, & x^{19} &= x^6 + x^4 + x^3 + x, \\
x^9 &= x^5 + x^4 + x^3 + x, & x^{20} &= x^7 + x^5 + x^4 + x^2, \\
x^{10} &= x^6 + x^5 + x^4 + x^2, & x^{21} &= x^6 + x^5 + x^4 + x^2 + 1, \\
x^{11} &= x^7 + x^6 + x^5 + x^3, & x^{22} &= x^7 + x^6 + x^5 + x^3 + x, \\
x^{12} &= x^7 + x^6 + x^3 + x^2 + 1, & x^{23} &= x^7 + x^6 + x^3 + 1, \\
x^{13} &= x^7 + x^2 + x + 1, & x^{24} &= x^7 + x^3 + x^2 + x + 1, \\
x^{14} &= x^4 + x + 1, & x^{25} &= x + 1, \\
x^{15} &= x^5 + x^2 + x, & x^{31} &= x^7 + x^6, \\
x^{16} &= x^6 + x^3 + x^2, & x^{32} &= x^7 + x^4 + x^3 + x^2 + 1, \\
x^{17} &= x^7 + x^4 + x^3, & x^{33} &= x^5 + x^2 + x + 1, \\
x^{18} &= x^5 + x^3 + x^2 + 1, & x^{34} &= x^6 + x^3 + x^2 + x,
\end{aligned}
$$

$x^{35} = x^7 + x^4 + x^3 + x^2,$

$x^{36} = x^5 + x^2 + 1,$

$x^{38} = x^7 + x^4 + x^2,$

$x^{39} = x^5 + x^4 + x^2 + 1,$

$x^{40} = x^6 + x^5 + x^3 + x,$

$x^{41} = x^7 + x^6 + x^4 + x^2,$

$x^{42} = x^7 + x^5 + x^4 + x^2 + 1,$

$x^{43} = x^6 + x^5 + x^4 + x^2 + x + 1,$

$x^{44} = x^7 + x^6 + x^5 + x^3 + x^2 + x,$

$x^{45} = x^7 + x^6 + 1,$

$x^{46} = x^7 + x^4 + x^3 + x^2 + x + 1,$

$x^{47} = x^5 + x + 1,$

$x^{49} = x^7 + x^3 + x^2,$

$x^{50} = x^2 + 1,$

$x^{55} = x^7 + x^5,$

$x^{56} = x^6 + x^4 + x^3 + x^2 + 1,$

$x^{57} = x^7 + x^5 + x^4 + x^3 + x,$

$x^{58} = x^6 + x^5 + x^3 + 1,$

$x^{59} = x^7 + x^6 + x^4 + x,$

$x^{60} = x^7 + x^5 + x^4 + x^3 + 1,$

$x^{61} = x^6 + x^5 + x^3 + x^2 + x + 1,$

$x^{62} = x^7 + x^6 + x^4 + x^3 + x^2 + x,$

$x^{63} = x^7 + x^5 + 1,$

$x^{64} = x^6 + x^4 + x^3 + x^2 + x + 1,$

$x^{65} = x^7 + x^5 + x^4 + x^3 + x^2 + x,$

$x^{66} = x^6 + x^5 + 1,$

$x^{67} = x^7 + x^6 + x,$

$x^{68} = x^7 + x^4 + x^3 + 1,$

$x^{69} = x^5 + x^3 + x^2 + x + 1,$

$x^{70} = x^6 + x^4 + x^3 + x^2 + x,$

$x^{71} = x^7 + x^5 + x^4 + x^3 + x^2,$

$x^{72} = x^6 + x^5 + x^2 + 1,$

$x^{73} = x^7 + x^6 + x^3 + x,$

$x^{74} = x^7 + x^3 + 1,$

$x^{75} = x^3 + x^2 + x + 1,$

$x^{79} = x^7 + x^6 + x^5 + x^4,$

$x^{80} = x^7 + x^6 + x^5 + x^4 + x^3 + x^2 + 1,$

$x^{81} = x^7 + x^6 + x^5 + x^2 + x + 1,$

$x^{82} = x^7 + x^6 + x^4 + x + 1,$

$x^{83} = x^7 + x^5 + x^4 + x^3 + x + 1,$

$x^{84} = x^6 + x^5 + x^3 + x + 1,$

$x^{85} = x^7 + x^6 + x^4 + x^2 + x,$

$x^{86} = x^7 + x^5 + x^4 + 1,$

$x^{87} = x^6 + x^5 + x^4 + x^3 + x^2 + x + 1,$

$x^{88} = x^7 + x^6 + x^5 + x^4 + x^3 + x^2 + x,$

$x^{89} = x^7 + x^6 + x^5 + 1,$

$x^{90} = x^7 + x^6 + x^4 + x^3 + x^2 + x + 1,$

$x^{91} = x^7 + x^5 + x + 1,$

$x^{92} = x^6 + x^4 + x^3 + x + 1,$

$x^{93} = x^7 + x^5 + x^4 + x^2 + x,$

$x^{94} = x^6 + x^5 + x^4 + 1,$

$x^{95} = x^7 + x^6 + x^5 + x,$

$x^{96} = x^7 + x^6 + x^4 + x^3 + 1,$

$x^{97} = x^7 + x^5 + x^3 + x^2 + x + 1,$

$x^{98} = x^6 + x + 1,$

$x^{99} = x^7 + x^2 + x,$

$x^{100} = x^4 + 1,$

$x^{103} = x^7 + x^3,$

$x^{104} = x^3 + x^2 + 1,$

$x^{108} = x^7 + x^6 + x^4,$

$x^{109} = x^7 + x^5 + x^4 + x^3 + x^2 + 1,$

$x^{110} = x^6 + x^5 + x^2 + x + 1,$

$x^{111} = x^7 + x^6 + x^3 + x^2 + x,$

$x^{112} = x^7 + 1,$

$x^{113} = x^4 + x^3 + x^2 + x + 1,$

$x^{116} = x^7 + x^6 + x^5 + x^4 + x^3,$

$x^{117} = x^7 + x^6 + x^5 + x^3 + x^2 + 1,$

$x^{118} = x^7 + x^6 + x^2 + x + 1,$

$x^{119} = x^7 + x^4 + x + 1,$

$x^{120} = x^5 + x^4 + x^3 + x + 1,$

$x^{122} = x^7 + x^6 + x^5 + x^3 + x^2,$

$x^{123} = x^7 + x^6 + x^2 + 1,$

$x^{124} = x^7 + x^4 + x^2 + x + 1,$

$x^{125} = x^5 + x^4 + x + 1,$

$x^{127} = x^7 + x^6 + x^3 + x^2,$

$x^{128} = x^7 + x^2 + 1,$

$x^{129} = x^4 + x^2 + x + 1,$

$x^{132} = x^7 + x^5 + x^4 + x^3,$

$x^{133} = x^6 + x^5 + x^3 + x^2 + 1,$

$x^{134} = x^7 + x^6 + x^4 + x^3 + x,$

$x^{135} = x^7 + x^5 + x^3 + 1,$

$x^{136} = x^6 + x^3 + x^2 + x + 1,$

$x^{137} = x^7 + x^4 + x^3 + x^2 + x,$

$x^{138} = x^5 + 1,$

$x^{140} = x^7 + x^2,$

$x^{141} = x^4 + x^2 + 1,$

$x^{144} = x^7 + x^5 + x^3,$

$x^{145} = x^6 + x^3 + x^2 + 1,$

$x^{146} = x^7 + x^4 + x^3 + x,$

$x^{147} = x^5 + x^3 + 1,$

$x^{149} = x^7 + x^5 + x^2,$

$x^{150} = x^6 + x^4 + x^2 + 1,$

$x^{151} = x^7 + x^5 + x^3 + x,$

$x^{152} = x^6 + x^3 + 1,$

$x^{153} = x^7 + x^4 + x,$

$x^{154} = x^5 + x^4 + x^3 + 1,$

$x^{155} = x^6 + x^5 + x^4 + x,$

$x^{156} = x^7 + x^6 + x^5 + x^2,$

$x^{157} = x^7 + x^6 + x^4 + x^2 + 1,$

$x^{158} = x^7 + x^5 + x^4 + x^2 + x + 1,$

$x^{159} = x^6 + x^5 + x^4 + x + 1,$

$x^{160} = x^7 + x^6 + x^5 + x^2 + x,$

$x^{161} = x^7 + x^6 + x^4 + 1,$

$x^{162} = x^7 + x^5 + x^4 + x^3 + x^2 + x + 1,$

$x^{163} = x^6 + x^5 + x + 1,$

$x^{164} = x^7 + x^6 + x^2 + x,$

$x^{165} = x^7 + x^4 + 1,$

$x^{166} = x^5 + x^4 + x^3 + x^2 + x + 1,$

$x^{168} = x^7 + x^6 + x^5 + x^4 + x^3 + x^2,$

$x^{169} = x^7 + x^6 + x^5 + x^2 + 1,$

$x^{170} = x^7 + x^6 + x^4 + x^2 + x + 1,$

$x^{171} = x^7 + x^5 + x^4 + x + 1,$

$x^{172} = x^6 + x^5 + x^4 + x^3 + x + 1,$

$x^{173} = x^7 + x^6 + x^5 + x^4 + x^2 + x,$

$x^{174} = x^7 + x^6 + x^5 + x^4 + 1,$

$x^{175} = x^7 + x^6 + x^5 + x^4 + x^3 + x^2 + x + 1,$

$x^{176} = x^7 + x^6 + x^5 + x + 1,$

$x^{177} = x^7 + x^6 + x^4 + x^3 + x + 1,$

$x^{178} = x^7 + x^5 + x^3 + x + 1,$

$x^{179} = x^6 + x^3 + x + 1,$

$x^{180} = x^7 + x^4 + x^2 + x,$

$x^{181} = x^5 + x^4 + 1,$

$x^{183} = x^7 + x^6 + x^2,$

$x^{184} = x^7 + x^4 + x^2 + 1,$

$x^{185} = x^5 + x^4 + x^2 + x + 1,$

$x^{186} = x^6 + x^5 + x^3 + x^2 + x,$

$x^{187} = x^7 + x^6 + x^4 + x^3 + x^2,$

$x^{188} = x^7 + x^5 + x^2 + 1,$

$x^{189} = x^6 + x^4 + x^2 + x + 1,$

$x^{190} = x^7 + x^5 + x^3 + x^2 + x,$

$x^{191} = x^6 + 1,$

$x^{192} = x^7 + x,$

$x^{193} = x^4 + x^3 + 1,$

$x^{196} = x^7 + x^6 + x^3,$

$x^{197} = x^7 + x^3 + x^2 + 1,$

$x^{198} = x^2 + x + 1,$

$x^{203} = x^7 + x^6 + x^5,$

$x^{204} = x^7 + x^6 + x^4 + x^3 + x^2 + 1,$

$x^{205} = x^7 + x^5 + x^2 + x + 1,$

$x^{206} = x^6 + x^4 + x + 1,$

$x^{207} = x^7 + x^5 + x^2 + x,$

$x^{208} = x^6 + x^4 + 1,$

$x^{209} = x^7 + x^5 + x,$

$x^{210} = x^6 + x^4 + x^3 + 1,$

$x^{211} = x^7 + x^5 + x^4 + x,$

$x^{212} = x^6 + x^5 + x^4 + x^3 + 1,$

$x^{213} = x^7 + x^6 + x^5 + x^4 + x,$

$x^{214} = x^7 + x^6 + x^5 + x^4 + x^3 + 1,$

$x^{215} = x^7 + x^6 + x^5 + x^3 + x^2 + x + 1,$

$x^{216} = x^7 + x^6 + x + 1,$

$x^{217} = x^7 + x^4 + x^3 + x + 1,$

$x^{218} = x^5 + x^3 + x + 1,$

$x^{220} = x^7 + x^5 + x^3 + x^2,$

$x^{221} = x^6 + x^2 + 1,$

$x^{222} = x^7 + x^3 + x,$

$x^{223} = x^3 + 1,$

$x^{227} = x^7 + x^4,$

$x^{228} = x^5 + x^4 + x^3 + x^2 + 1,$

$x^{230} = x^7 + x^6 + x^5 + x^4 + x^2,$

$x^{231} = x^7 + x^6 + x^5 + x^4 + x^2 + 1,$

$x^{232} = x^7 + x^6 + x^5 + x^4 + x^2 + x + 1,$

$x^{233} = x^7 + x^6 + x^5 + x^4 + x + 1,$

$$x^{234} = x^7 + x^6 + x^5 + x^4 + x^3 + x + 1,$$
$$x^{235} = x^7 + x^6 + x^5 + x^3 + x + 1,$$
$$x^{236} = x^7 + x^6 + x^3 + x + 1,$$
$$x^{237} = x^7 + x^3 + x + 1,$$
$$x^{238} = x^3 + x + 1,$$
$$x^{242} = x^7 + x^5 + x^4,$$
$$x^{243} = x^6 + x^5 + x^4 + x^3 + x^2 + 1,$$
$$x^{245} = x^7 + x^6 + x^5 + x^3 + 1,$$
$$x^{246} = x^7 + x^6 + x^3 + x^2 + x + 1,$$
$$x^{247} = x^7 + x + 1,$$
$$x^{248} = x^4 + x^3 + x + 1,$$
$$x^{251} = x^7 + x^6 + x^4 + x^3,$$

$$x^{252} = x^7 + x^5 + x^3 + x^2 + 1,$$
$$x^{253} = x^6 + x^2 + x + 1,$$
$$x^{254} = x^7 + x^3 + x^2 + x,$$
$$x^{255} = 1.$$

(4) Solve the following two linear systems (over $\mathbb{F}_{256} = \mathbb{F}_2[x]/(x^8 + x^4 + x^3 + x^2 + 1)$):

(a)
$$xT_0 + (x+1)T_1 + T_2 + T_3 = 1,$$
$$T_0 + xT_1 + (x+1)T_2 + T_3 = 0,$$
$$T_0 + T_1 + xT_2 + (x+1)T_3 = 0,$$
$$(x+1)T_0 + T_1 + T_2 + xT_3 = 0.$$

(b)
$$x^{89}T_0 + x^{248}T_1 + x^{182}T_2 + x^{166}T_3 = x^{52},$$
$$x^{52}T_0 + x^{89}T_1 + x^{248}T_2 + x^{182}T_3 = x^{73},$$
$$x^{73}T_0 + x^{52}T_1 + x^{89}T_2 + x^{248}T_3 = x^{146},$$
$$x^{146}T_0 + x^{73}T_1 + x^{52}T_2 + x^{89}T_3 = x^{158}.$$

(5) Show that $(x+1)^{51} \equiv 1 \bmod (x^8 + x^4 + x^3 + x^2 + 1)$.

(6) Consider the involutive automorphism
$$\Phi : \mathbb{F}_2[x] \longrightarrow \mathbb{F}_2[x]$$
$$x \longmapsto x + 1$$

(a) Put $p(x) = x^8 + x^4 + x^3 + x^2 + 1$
$m(x) = x^8 + x^4 + x^3 + x + 1$
Show that $\Phi(p(x)) = m(x)$.
(b) Deduce a natural isomorphism $\phi : \mathbb{F}_{256} = \mathbb{F}_2[x]/(p(x)) \longrightarrow \mathbf{R}_{256} = \mathbb{F}_2[x]/(m(x))$, $x \longmapsto x + 1$.
(c) Show that the polynomial $m(x) = x^8 + x^4 + x^3 + x + 1$ is irreducible, but that $m(x)$ is *not* primitive. Find the order of x in $\mathbf{R}_{256}^* = \mathbb{F}_2[x]/(m(x))^*$, i.e. find the smallest positive exponent e such that $x^e \equiv 1 \bmod m(x)$.

(7) Compute the following multiplicative inverses in $\mathbf{R}_{256} = \mathbb{F}_2[x]/(m(x))$:

(a) $11101010^{-1} = ?$

(b) $01010101^{-1} = ?$

(c) $11111111^{-1} = ?$

(8) The reference code in C of Rijndael contains two tables to support multiplication *) in $\mathbf{R}_{256} = \mathbb{F}_2[x]/(x^8 + x^4 + x^3 + x + 1)$:

word 8 Logtable [256] = {
0, 0, 25, 1, 50, 2, 26, 198, 75, 199, 27, 104, 51, 238, 223, 3, $\cdots$ } ;
word 8 Alogtable [256] = {
1, 3, 5, 15, 17, 51, 85, 255, 26, 46, 114, 150, 161, 248, 19, 53, $\cdots$ } ;

(a) Find the last four entries of every table. Justify.

(b) Replace now $\mathbf{R}_{256}$ by $\mathbb{F}_{256} = \mathbb{F}_2[x]/(x^8 + x^4 + x^3 + x^2 + 1)$.

Write the first 16 and the last 16 entries of each of the two tables, now for the arithmetic of $\mathbb{F}_{256}$.

---

*) word8 mul(word8 a, word8 b) {
　　/* multiply two elements of $\mathbf{R}_{256}$
　　*/
　　if (a && b) return Alogtable[(Logtable[a]+Logtable[b])%255];
　　else return 0;
}

(9) Write down the table of the 255 non-zero residues in $\mathbf{R}_{256}$, in function of the successive powers of $\xi = x + 1$.

Solution the cyclic group of 255th roots of unity $(\mathbf{R}_{256})^*$:

$\xi^2 = x^2 + 1$,
$\xi^3 = x^3 + x^2 + x + 1$,
$\xi^4 = x^4 + 1$,
$\xi^5 = x^5 + x^4 + x + 1$,
$\xi^6 = x^6 + x^4 + x^2 + 1$,
$\xi^7 = x^7 + x^6 + x^5 + x^4 + x^3 + x^2 + x + 1$,
$\xi^8 = x^4 + x^3 + x$,
$\xi^9 = x^5 + x^3 + x^2 + x$,

$\xi^{10} = x^6 + x^5 + x^4 + x$,
$\xi^{11} = x^7 + x^4 + x^2 + x$,
$\xi^{12} = x^7 + x^5 + 1$,
$\xi^{13} = x^7 + x^6 + x^5 + x^4 + x^3$,
$\xi^{14} = x^4 + x + 1$,
$\xi^{15} = x^5 + x^4 + x^2 + 1$,
$\xi^{16} = x^6 + x^4 + x^3 + x^2 + x + 1$,
$\xi^{17} = x^7 + x^6 + x^5 + 1$,

$\xi^{18} = x^5 + x^4 + x^3,$

$\xi^{19} = x^6 + x^3,$

$\xi^{20} = x^7 + x^6 + x^4 + x^3,$

$\xi^{21} = x^6 + x^5 + x^4 + x + 1,$

$\xi^{22} = x^7 + x^4 + x^2 + 1,$

$\xi^{23} = x^7 + x^5 + x^2,$

$\xi^{24} = x^7 + x^6 + x^5 + x^4 + x^2 + x + 1,$

$\xi^{25} = x,$

$\xi^{26} = x^2 + x,$

$\xi^{27} = x^3 + x,$

$\xi^{28} = x^4 + x^3 + x^2 + x,$

$\xi^{29} = x^5 + x,$

$\xi^{30} = x^6 + x^5 + x^2 + x,$

$\xi^{31} = x^7 + x^5 + x^3 + x,$

$\xi^{32} = x^7 + x^6 + x^5 + x^2 + 1,$

$\xi^{33} = x^5 + x^4 + x^2,$

$\xi^{34} = x^6 + x^4 + x^3 + x^2,$

$\xi^{35} = x^7 + x^6 + x^5 + x^2,$

$\xi^{36} = x^5 + x^4 + x^2 + x + 1,$

$\xi^{37} = x^6 + x^4 + x^3 + 1,$

$\xi^{38} = x^7 + x^6 + x^5 + x^3 + x + 1,$

$\xi^{39} = x^5 + x^2 + x,$

$\xi^{40} = x^6 + x^5 + x^3 + x,$

$\xi^{41} = x^7 + x^5 + x^4 + x^3 + x^2 + x,$

$\xi^{42} = x^7 + x^6 + x^4 + x^3 + 1,$

$\xi^{43} = x^6 + x^5 + x^4,$

$\xi^{44} = x^7 + x^4,$

$\xi^{45} = x^7 + x^5 + x^3 + x + 1,$

$\xi^{46} = x^7 + x^6 + x^5 + x^2 + x,$

$\xi^{47} = x^5 + x^4 + 1,$

$\xi^{48} = x^6 + x^4 + x + 1,$

$\xi^{49} = x^7 + x^6 + x^5 + x^4 + x^2 + 1,$

$\xi^{50} = x^2,$

$\xi^{51} = x^3 + x^2,$

$\xi^{52} = x^4 + x^2,$

$\xi^{53} = x^5 + x^4 + x^3 + x^2,$

$\xi^{54} = x^6 + x^2,$

$\xi^{55} = x^7 + x^6 + x^3 + x^2,$

$\xi^{56} = x^6 + x^3 + x^2 + x + 1,$

$\xi^{57} = x^7 + x^6 + x^4 + 1,$

$\xi^{58} = x^6 + x^5 + x^2,$

$\xi^{59} = x^7 + x^5 + x^4 + x^3,$

$\xi^{60} = x^7 + x^6 + x^4 + x + 1,$

$\xi^{61} = x^6 + x^5 + x^3 + x^2 + x,$

$\xi^{62} = x^7 + x^5 + x^4 + x,$

$\xi^{63} = x^7 + x^6 + x^3 + x^2 + 1,$

$\xi^{64} = x^6 + x^3 + x^2,$

$\xi^{65} = x^7 + x^6 + x^4 + x^2,$

$\xi^{66} = x^6 + x^5 + x^2 + x + 1,$

$\xi^{67} = x^7 + x^5 + x^3 + 1,$

$\xi^{68} = x^7 + x^6 + x^5,$

$\xi^{69} = x^5 + x^4 + x^3 + x + 1,$

$\xi^{70} = x^6 + x^3 + x^2 + 1,$

$\xi^{71} = x^7 + x^6 + x^4 + x^2 + x + 1,$

$\xi^{72} = x^6 + x^5 + x,$

$\xi^{73} = x^7 + x^5 + x^2 + x,$

$\xi^{74} = x^7 + x^6 + x^5 + x^4 + 1,$

$\xi^{75} = x^3,$

$\xi^{76} = x^4 + x^3,$

$\xi^{77} = x^5 + x^3,$

$\xi^{78} = x^6 + x^5 + x^4 + x^3,$

$\xi^{79} = x^7 + x^3,$

$\xi^{80} = x^7 + x + 1,$

$\xi^{81} = x^7 + x^4 + x^3 + x^2 + x,$

$\xi^{82} = x^7 + x^5 + x^4 + x^3 + 1,$

$\xi^{83} = x^7 + x^6 + x^4,$

$\xi^{84} = x^6 + x^5 + x^3 + x + 1,$

$\xi^{85} = x^7 + x^5 + x^4 + x^3 + x^2 + 1,$

$\xi^{86} = x^7 + x^6 + x^4 + x^3 + x^2,$

$\xi^{87} = x^6 + x^5 + x^4 + x^3 + x^2 + x + 1,$

$\xi^{88} = x^7 + 1,$

$\xi^{89} = x^7 + x^4 + x^3,$

$\xi^{90} = x^7 + x^5 + x^4 + x + 1,$

$\xi^{91} = x^7 + x^6 + x^3 + x^2 + x,$

$\xi^{92} = x^6 + x^3 + 1,$

$\xi^{93} = x^7 + x^6 + x^4 + x^3 + x + 1,$

$\xi^{94} = x^6 + x^5 + x^4 + x^2 + x,$

$\xi^{95} = x^7 + x^4 + x^3 + x,$

$\xi^{96} = x^7 + x^5 + x^4 + x^2 + 1,$

$\xi^{97} = x^7 + x^6 + x^2,$

$\xi^{98} = x^6 + x^4 + x^2 + x + 1,$

$\xi^{99} = x^7 + x^6 + x^5 + x^4 + x^3 + 1,$

$\xi^{100} = x^4,$

$\xi^{101} = x^5 + x^4,$

$\xi^{102} = x^6 + x^4,$

$\xi^{103} = x^7 + x^6 + x^5 + x^4,$

$\xi^{104} = x^3 + x + 1,$

$\xi^{105} = x^4 + x^3 + x^2 + 1,$

$\xi^{106} = x^5 + x^2 + x + 1,$

$\xi^{107} = x^6 + x^5 + x^3 + 1,$

$\xi^{108} = x^7 + x^5 + x^4 + x^3 + x + 1,$

$\xi^{109} = x^7 + x^6 + x^4 + x^2 + x,$

$\xi^{110} = x^6 + x^5 + 1,$

$\xi^{111} = x^7 + x^5 + x + 1,$

$\xi^{112} = x^7 + x^6 + x^5 + x^4 + x^3 + x^2 + x,$

$\xi^{113} = x^4 + x^3 + 1,$

$\xi^{114} = x^5 + x^3 + x + 1,$

$\xi^{115} = x^6 + x^5 + x^4 + x^3 + x^2 + 1,$

$\xi^{116} = x^7 + x^2 + x + 1,$

$\xi^{117} = x^7 + x^4 + x,$

$\xi^{118} = x^7 + x^5 + x^3 + x^2 + 1,$

$\xi^{119} = x^7 + x^6 + x^5 + x^3 + x^2,$

$\xi^{120} = x^5 + x^3 + x^2 + x + 1,$

$\xi^{121} = x^6 + x^5 + x^4 + 1,$

$\xi^{122} = x^7 + x^4 + x + 1,$

$\xi^{123} = x^7 + x^5 + x^3 + x^2 + x,$

$\xi^{124} = x^7 + x^6 + x^5 + x^3 + 1,$

$\xi^{125} = x^5,$

$\xi^{126} = x^6 + x^5,$

$\xi^{127} = x^7 + x^5,$

$\xi^{128} = x^7 + x^6 + x^5 + x^4 + x^3 + x + 1,$

$\xi^{129} = x^4 + x^2 + x,$

$\xi^{130} = x^5 + x^4 + x^3 + x,$

$\xi^{131} = x^6 + x^3 + x^2 + x,$

$\xi^{132} = x^7 + x^6 + x^4 + x,$

$\xi^{133} = x^6 + x^5 + x^3 + x^2 + 1,$

$\xi^{134} = x^7 + x^5 + x^4 + x^2 + x + 1,$

$\xi^{135} = x^7 + x^6 + x,$

$\xi^{136} = x^6 + x^4 + x^3 + x^2 + 1,$

$\xi^{137} = x^7 + x^6 + x^5 + x^2 + x + 1,$

$\xi^{138} = x^5 + x^4 + x,$

$\xi^{139} = x^6 + x^4 + x^2 + x,$

$\xi^{140} = x^7 + x^6 + x^5 + x^4 + x^3 + x,$

$\xi^{141} = x^4 + x^2 + 1,$

$\xi^{142} = x^5 + x^4 + x^3 + x^2 + x + 1,$

$\xi^{143} = x^6 + 1,$

$\xi^{144} = x^7 + x^6 + x + 1,$

$\xi^{145} = x^6 + x^4 + x^3 + x^2 + x,$

$\xi^{146} = x^7 + x^6 + x^5 + x,$

$\xi^{147} = x^5 + x^4 + x^3 + x^2 + 1,$

$\xi^{148} = x^6 + x^2 + x + 1,$

$\xi^{149} = x^7 + x^6 + x^3 + 1,$

$\xi^{150} = x^6,$

$\xi^{151} = x^7 + x^6,$

$\xi^{152} = x^6 + x^4 + x^3 + x + 1,$

$\xi^{153} = x^7 + x^6 + x^5 + x^3 + x^2 + 1,$

$\xi^{154} = x^5 + x^3 + x^2,$

$\xi^{155} = x^6 + x^5 + x^4 + x^2,$

$\xi^{156} = x^7 + x^4 + x^3 + x^2,$

$\xi^{157} = x^7 + x^5 + x^4 + x^3 + x^2 + x + 1,$

$\xi^{158} = x^7 + x^6 + x^4 + x^3 + x,$

$\xi^{159} = x^6 + x^5 + x^4 + x^2 + 1,$

$\xi^{160} = x^7 + x^4 + x^3 + x^2 + x + 1,$

$\xi^{161} = x^7 + x^5 + x^4 + x^3 + x,$

$\xi^{162} = x^7 + x^6 + x^4 + x^2 + 1,$

$\xi^{163} = x^6 + x^5 + x^2,$

$\xi^{164} = x^7 + x^5 + x^3 + x^2,$

$\xi^{165} = x^7 + x^6 + x^5 + x^3 + x^2 + x + 1,$

$\xi^{166} = x^5 + x^3 + x,$

$\xi^{167} = x^6 + x^5 + x^4 + x^3 + x^2 + x,$

$\xi^{168} = x^7 + x,$

$\xi^{169} = x^7 + x^4 + x^3 + x^2 + 1,$

$\xi^{170} = x^7 + x^5 + x^4 + x^3 + x^2,$

$\xi^{171} = x^7 + x^6 + x^4 + x^3 + x^2 + x + 1,$

$\xi^{172} = x^6 + x^5 + x^4 + x^3 + x,$

$\xi^{173} = x^7 + x^3 + x^2 + x,$

$\xi^{174} = x^7 + x^3 + 1,$

$\xi^{175} = x^7,$

$\xi^{176} = x^7 + x^4 + x^3 + x + 1,$

$\xi^{177} = x^7 + x^5 + x^4 + x^2 + x,$

$\xi^{178} = x^7 + x^6 + 1,$

$\xi^{179} = x^6 + x^4 + x^3,$

$\xi^{180} = x^7 + x^6 + x^5 + x^3,$

$\xi^{181} = x^5 + x + 1,$

$\xi^{182} = x^6 + x^5 + x^2 + 1,$

$\xi^{183} = x^7 + x^5 + x^3 + x^2 + x + 1,$

$\xi^{184} = x^7 + x^6 + x^5 + x^3 + x,$

$\xi^{185} = x^5 + x^2 + 1,$

$\xi^{186} = x^6 + x^5 + x^3 + x^2 + x + 1,$

$\xi^{187} = x^7 + x^5 + x^4 + 1,$

$\xi^{188} = x^7 + x^6 + x^3,$

$\xi^{189} = x^6 + x + 1,$

$\xi^{190} = x^7 + x^6 + x^2 + 1,$

$\xi^{191} = x^6 + x^4 + x^2,$

$\xi^{192} = x^7 + x^6 + x^5 + x^4 + x^3 + x^2,$

$\xi^{193} = x^4 + x^3 + x^2 + x + 1,$

$\xi^{194} = x^5 + 1,$

$\xi^{195} = x^6 + x^5 + x + 1,$

$\xi^{196} = x^7 + x^5 + x^2 + 1,$

$\xi^{197} = x^7 + x^6 + x^5 + x^4 + x^2,$

$\xi^{198} = x^2 + x + 1,$

$\xi^{199} = x^3 + 1,$

$\xi^{200} = x^4 + x^3 + x + 1,$

$\xi^{201} = x^5 + x^3 + x^2 + 1,$

$\xi^{202} = x^6 + x^5 + x^4 + x^2 + x + 1,$

$\xi^{203} = x^7 + x^4 + x^3 + 1,$

$\xi^{204} = x^7 + x^5 + x^4,$

$\xi^{205} = x^7 + x^6 + x^3 + x + 1,$

$\xi^{206} = x^6 + x^2 + x,$

$\xi^{207} = x^7 + x^6 + x^3 + x,$

$\xi^{208} = x^6 + x^2 + 1,$

$\xi^{209} = x^7 + x^6 + x^3 + x^2 + x + 1,$

$\xi^{210} = x^6 + x^3 + x,$

$\xi^{211} = x^7 + x^6 + x^4 + x^3 + x^2 + x,$

$\xi^{212} = x^6 + x^5 + x^4 + x^3 + 1,$

$\xi^{213} = x^7 + x^3 + x + 1,$

$\xi^{214} = x^7 + x^2 + x,$

$\xi^{215} = x^7 + x^4 + 1,$

$\xi^{216} = x^7 + x^5 + x^3,$

$\xi^{217} = x^7 + x^6 + x^5 + x + 1,$

$\xi^{218} = x^5 + x^4 + x^3 + x^2 + x,$

$\xi^{219} = x^6 + x,$

$\xi^{220} = x^7 + x^6 + x^2 + x,$

$\xi^{221} = x^6 + x^4 + 1,$

$\xi^{222} = x^7 + x^6 + x^5 + x^4 + x + 1,$

$\xi^{223} = x^3 + x^2 + x,$

$\xi^{224} = x^4 + x,$

$\xi^{225} = x^5 + x^4 + x^2 + x,$

$\xi^{226} = x^6 + x^4 + x^3 + x,$

$\xi^{227} = x^7 + x^6 + x^5 + x^3 + x^2 + x,$

$\xi^{228} = x^5 + x^3 + 1,$

$\xi^{229} = x^6 + x^5 + x^4 + x^3 + x + 1,$

$\xi^{230} = x^7 + x^3 + x^2 + 1,$

$\xi^{231} = x^7 + x^3 + x^2,$

$\xi^{232} = x^7 + x^3 + x^2 + x + 1,$

$\xi^{233} = x^7 + x^3 + x,$

$\xi^{234} = x^7 + x^2 + 1,$

$\xi^{235} = x^7 + x^4 + x^2,$

$\xi^{236} = x^7 + x^5 + x^2 + x + 1,$

$\xi^{237} = x^7 + x^6 + x^5 + x^4 + x,$

$\xi^{238} = x^3 + x^2 + 1,$

$\xi^{239} = x^4 + x^2 + x + 1,$

$\xi^{240} = x^5 + x^4 + x^3 + 1,$

$\xi^{241} = x^6 + x^3 + x + 1,$

$\xi^{242} = x^7 + x^6 + x^4 + x^3 + x^2 + 1,$

$\xi^{243} = x^6 + x^5 + x^4 + x^3 + x^2,$

$\xi^{244} = x^7 + x^2,$

$\xi^{245} = x^7 + x^4 + x^2 + x + 1,$

$\xi^{246} = x^7 + x^5 + x,$

$\xi^{247} = x^7 + x^6 + x^5 + x^4 + x^3 + x^2 + 1,$

$\xi^{248} = x^4 + x^3 + x^2,$

$\xi^{249} = x^5 + x^2,$

$\xi^{250} = x^6 + x^5 + x^3 + x^2,$

$\xi^{251} = x^7 + x^5 + x^4 + x^2,$

$\xi^{252} = x^7 + x^6 + x^2 + x + 1,$

$\xi^{253} = x^6 + x^4 + x,$

$\xi^{254} = x^7 + x^6 + x^5 + x^4 + x^2 + x,$

$\xi^{255} = 1.$

2.2.2 Specification of Rijndael

Generalities

AES encrypts a plaintext block of 128 bits (of 16 bytes) into a ciphertext block of 128 bits (of 16 bytes).

AES exists in three versions:

AES-128:   the cipher keys have 128 bits,
AES-192:   the cipher keys have 192 bits,
AES-256:   the cipher keys have 256 bits.

The plaintext block undergoes $\begin{cases} 10\ (\text{AES} - 128) \\ 12\ (\text{AES} - 192) \\ 14\ (\text{AES} - 256) \end{cases}$ iterations of a transformation which depends each round on another round key extracted from the user's cipher key.

Note that the round key K_i, $\quad 0 \le i \le Nr (\equiv$ the number of rounds) will have 128 bits, i.e. it will have the same size as the message block during the encryption process.

The round keys are produced by an auxiliary algorithm – the key schedule – the input of which is the cipher key of the user, exactly as in the DES.

Remarks on Formal Questions

The letters of the cipher AES–Rijndael are the (8-bit) bytes

- Syntactical notation: 10100111
- Hexadecimal notation: $a7$ $(1010 \equiv a \quad 0111 \equiv 7)$
- Polynomial notation: $x^7 + x^5 + x^2 + x + 1$ (this is an element of the field $\mathbf{R}_{256}$, i.e. a remainder of division by $m(x) = x^8 + x^4 + x^3 + x + 1$).

Summing up: $\beta_7 \beta_6 \cdots \beta_1 \beta_0 = \{(\beta_7 \cdots \beta_4)_{16} (\beta_3 \cdots \beta_0)_{16}\} = \beta_7 x^7 + \beta_6 x^6 + \cdots + \beta_1 x + \beta_0$

External Presentation and Internal Presentation of a Block of 128 Bits

The external presentation of the 128-bit plaintext block (and of the ciphertext block) is sequential:

$$\beta_0 \beta_1 \beta_2 \cdots \beta_{127} \equiv a_0 a_1 a_2 \cdots a_{15} \qquad \text{with} \qquad \begin{aligned} a_0 &= \beta_0 \beta_1 \cdots \beta_7 \\ a_1 &= \beta_8 \beta_9 \cdots \beta_{15} \\ &\vdots \\ a_{15} &= \beta_{120} \beta_{121} \cdots \beta_{127} \end{aligned}$$

Caution In individual treatment, every 8-bit byte will have a decreasing indexing of the positions: 7.6.5.4.3.2.1.0.

The internal presentation of the 128-bit message block during the encryption (or of the ciphertext block during the decryption) will be in *matrix form* (and will be called a *state* or *state array*).

$$
\begin{matrix}
a_0 & a_4 & a_8 & a_{12} \\
a_1 & a_5 & a_9 & a_{13} \\
a_2 & a_6 & a_{10} & a_{14} \\
a_3 & a_7 & a_{11} & a_{15}
\end{matrix}
\quad = \quad
\begin{matrix}
s_{0,0} & s_{0,1} & s_{0,2} & s_{0,3} \\
s_{1,0} & s_{1,1} & s_{1,2} & s_{1,3} \\
s_{2,0} & s_{2,1} & s_{2,2} & s_{2,3} \\
s_{3,0} & s_{3,1} & s_{3,2} & s_{3,3}
\end{matrix}
$$

Every state is a quadruple of (vertical) *words*, each consisting of 4 letters (4 bytes):

$$
w_0 = \begin{matrix} a_0 \\ a_1 \\ a_2 \\ a_3 \end{matrix} = \begin{matrix} s_{0,0} \\ s_{1,0} \\ s_{2,0} \\ s_{3,0} \end{matrix}
\qquad
w_1 = \begin{matrix} a_4 \\ a_5 \\ a_6 \\ a_7 \end{matrix} = \begin{matrix} s_{0,1} \\ s_{1,1} \\ s_{2,1} \\ s_{3,1} \end{matrix}
$$

$$
w_2 = \begin{matrix} a_8 \\ a_9 \\ a_{10} \\ a_{11} \end{matrix} = \begin{matrix} s_{0,2} \\ s_{1,2} \\ s_{2,2} \\ s_{3,2} \end{matrix}
\qquad
w_3 = \begin{matrix} a_{12} \\ a_{13} \\ a_{14} \\ a_{15} \end{matrix} = \begin{matrix} s_{0,3} \\ s_{1,3} \\ s_{2,3} \\ s_{3,3} \end{matrix}
$$

> In the sequel, we only consider the version *AES-128*.

The Round Transformation

The round transformation is composed of four constituent transformations, called the *steps*:

- The *S*-box: It consists of 16 *identical* copies of an *S*-box acting on letters; the 16 bytes of the state are treated independently.
- The ShiftRows step: This is a byte transposition that cyclically shifts the rows of the state over different offsets.
- The MixColumns step: The four columns of the state undergo simultaneously a circular convolution (as elements of $\mathbf{R}_{256}^4$) with a simple impulse response.
- Addition of the round key: The round key adds to the state (addition of two 4×4 matrixes over $\mathbf{R}_{256}$).

The S-box

The *S*-box is the *non-linear* component of the round transformation (exactly like the *S*-boxes of the DES). It acts in parallel on the letters, i.e. as 16 identical *S*-boxes:

$$
\begin{matrix}
s_{0,0} \; s_{0,1} \; s_{0,2} \; s_{0,3} \\
s_{1,0} \; s_{1,1} \; s_{1,2} \; s_{1,3} \\
s_{2,0} \; s_{2,1} \; s_{2,2} \; s_{2,3} \\
s_{3,0} \; s_{3,1} \; s_{3,2} \; s_{3,3}
\end{matrix}
\longmapsto
\begin{matrix}
s'_{0,0} \; s'_{0,1} \; s'_{0,2} \; s'_{0,3} \\
s'_{1,0} \; s'_{1,1} \; s'_{1,2} \; s'_{1,3} \\
s'_{2,0} \; s'_{2,1} \; s'_{2,2} \; s'_{2,3} \\
s'_{3,0} \; s'_{3,1} \; s'_{3,2} \; s'_{3,3}
\end{matrix}
=
\begin{matrix}
S(s_{0,0}) \; S(s_{0,1}) \; S(s_{0,2}) \; S(s_{0,3}), \\
S(s_{1,0}) \; S(s_{1,1}) \; S(s_{1,2}) \; S(s_{1,3}), \\
S(s_{2,0}) \; S(s_{2,1}) \; S(s_{2,2}) \; S(s_{2,3}), \\
S(s_{3,0}) \; S(s_{3,1}) \; S(s_{3,2}) \; S(s_{3,3}).
\end{matrix}
$$

Let us describe the S-box as a byte transformation: $S(\beta_7 \beta_6 \cdots \beta_0) = \hat{\beta}_7 \hat{\beta}_6 \cdots \hat{\beta}_0$, where the final result comes from the following affine transformation of the space $(\mathbb{F}_2)^8$

$$
\begin{pmatrix}
\hat{\beta}_0 \\
\hat{\beta}_1 \\
\hat{\beta}_2 \\
\hat{\beta}_3 \\
\hat{\beta}_4 \\
\hat{\beta}_5 \\
\hat{\beta}_6 \\
\hat{\beta}_7
\end{pmatrix}
=
\begin{pmatrix}
1 0 0 0 1 1 1 1 \\
1 1 0 0 0 1 1 1 \\
1 1 1 0 0 0 1 1 \\
1 1 1 1 0 0 0 1 \\
1 1 1 1 1 0 0 0 \\
0 1 1 1 1 1 0 0 \\
0 0 1 1 1 1 1 0 \\
0 0 0 1 1 1 1 1
\end{pmatrix}
\begin{pmatrix}
\beta'_0 \\
\beta'_1 \\
\beta'_2 \\
\beta'_3 \\
\beta'_4 \\
\beta'_5 \\
\beta'_6 \\
\beta'_7
\end{pmatrix}
+
\begin{pmatrix}
1 \\
1 \\
0 \\
0 \\
0 \\
1 \\
1 \\
0
\end{pmatrix}.
$$

The intermediate result $\beta'_7 \beta'_6 \cdots \beta'_0$ is computed as follows:

$$
\beta'_7 x^7 + \beta'_6 x^6 + \cdots + \beta'_1 x + \beta'_0 = (\beta_7 x^7 + \beta_6 x^6 + \cdots + \beta_1 x + \beta_0)^{-1}
$$

in

$$
\mathbf{R}_{256} = \mathbb{F}_2[x]/(x^8 + x^4 + x^3 + x + 1)
$$

(the byte 00000000 remains fixed).

The complexity of the S-box is equivalent to the complexity of multiplicative inversion in $\mathbf{R}_{256}$; the philosophy for the ensuing affine transformation is merely to create a Boolean transformation *without fixed points*.

Exercises

(1) Compute the *inverse* transformation of the affine transformation above.

Solution

$$
\begin{pmatrix}
1 0 0 0 1 1 1 1 \\
1 1 0 0 0 1 1 1 \\
1 1 1 0 0 0 1 1 \\
1 1 1 1 0 0 0 1 \\
1 1 1 1 1 0 0 0 \\
0 1 1 1 1 1 0 0 \\
0 0 1 1 1 1 1 0 \\
0 0 0 1 1 1 1 1
\end{pmatrix}^{-1}
=
\begin{pmatrix}
0 0 1 0 0 1 0 1 \\
1 0 0 1 0 0 1 0 \\
0 1 0 0 1 0 0 1 \\
1 0 1 0 0 1 0 0 \\
0 1 0 1 0 0 1 0 \\
0 0 1 0 1 0 0 1 \\
1 0 0 1 0 1 0 0 \\
0 1 0 0 1 0 1 0
\end{pmatrix}
$$

i.e.

$$
\begin{pmatrix} \beta_0' \\ \beta_1' \\ \beta_2' \\ \beta_3' \\ \beta_4' \\ \beta_5' \\ \beta_6' \\ \beta_7' \end{pmatrix} = \begin{pmatrix} 0\,0\,1\,0\,0\,1\,0\,1 \\ 1\,0\,0\,1\,0\,0\,1\,0 \\ 0\,1\,0\,0\,1\,0\,0\,1 \\ 1\,0\,1\,0\,0\,1\,0\,0 \\ 0\,1\,0\,1\,0\,0\,1\,0 \\ 0\,0\,1\,0\,1\,0\,0\,1 \\ 1\,0\,0\,1\,0\,1\,0\,0 \\ 0\,1\,0\,0\,1\,0\,1\,0 \end{pmatrix} \begin{pmatrix} \hat{\beta}_0 + 1 \\ \hat{\beta}_1 + 1 \\ \hat{\beta}_2 \\ \hat{\beta}_3 \\ \hat{\beta}_4 \\ \hat{\beta}_5 + 1 \\ \hat{\beta}_6 + 1 \\ \hat{\beta}_7 \end{pmatrix}.
$$

(2) Show that the polynomial description of the affine transformation which is the final constituent of the S-box is given by

$$
\hat{\beta}_7 x^7 + \hat{\beta}_6 x^6 + \cdots + \hat{\beta}_0 = [(\beta_7' x^7 + \beta_6' x^6 + \cdots + \beta_0')(x^4 + x^3 + x^2 + x + 1)
$$
$$
+ x^6 + x^5 + x + 1] \bmod (x^8 + 1).
$$

(3) The S-box finally becomes, after some computational effort, the following hexadecimal table:

	0	1	2	3	4	5	6	7	8	9	a	b	c	d	e	f
0	63	7c	77	7b	f2	6b	6f	c5	30	01	67	2b	fe	d7	ab	76
1	ca	82	c9	7d	fa	59	47	f0	ad	d4	a2	af	9c	a4	72	c0
2	b7	fd	93	26	36	3f	f7	cc	34	a5	e5	f1	71	d8	31	15
3	04	c7	23	c3	18	96	05	9a	07	12	80	e2	eb	27	b2	75
4	09	83	2c	1a	1b	6e	5a	a0	52	3b	d6	b3	29	e3	2f	84
5	53	d1	00	ed	20	fc	b1	5b	6a	cb	be	39	4a	4c	58	cf
6	d0	ef	aa	fb	43	4d	33	85	45	f9	02	7f	50	3c	9f	a8
7	51	a3	40	8f	92	9d	38	f5	bc	b6	da	21	10	ff	f3	d2
8	cd	0c	13	ec	5f	97	44	17	c4	a7	7e	3d	64	5d	19	73
9	60	81	4f	dc	22	2a	90	88	46	ee	b8	14	of	5e	0b	db
a	e0	32	3a	0a	49	06	24	5c	c2	d3	ac	62	91	95	e4	79
b	e7	c8	37	6d	8d	d5	4e	a9	6c	56	f4	ea	65	7a	ae	08
c	ba	78	25	2e	1c	a6	b4	c6	e8	dd	74	1f	4b	bd	8b	8a
d	70	3e	b5	66	48	03	f6	0e	61	35	57	b9	86	c1	1d	9e
e	e1	f8	98	11	69	d9	8e	94	9b	1e	87	e9	ce	55	28	df
f	8c	a1	89	0d	bf	e6	42	68	41	99	2d	0f	b0	54	bb	16

Read *If $s_{i,j} = xy$, then $s_{i,j}' = S(s_{i,j}) \equiv$ the letter (8-bit byte) of the table at the position (x, y), $0 \le i, j \le 3$.*

Example $S(53) = ed$, i.e. $S(01010011) = 11101101$.

(a) Verify, following the algorithm defining the S-box, that we indeed get: $S(88) = c4$, i.e. $S(10001000) = 11000100$.
(You should avoid computing the multiplicative inverse of $x^7 + x^3$ modulo $m(x)$ in painfully solving the equation $(x^7 + x^3)U + (x^8 + x^4 + x^3 + x + 1)V = 1$. It is easier to use the table of the 255 powers of $\xi = x + 1$ in $\mathbf{R}_{256}^*$.)

(b) Write down the table of the *inverse S*-box.

Solution

	0	1	2	3	4	5	6	7	8	9	a	b	c	d	e	f
0	52	09	6a	d5	30	36	a5	38	bf	40	a3	9e	81	f3	d7	fb
1	7c	e3	39	82	9b	2f	ff	87	34	8e	43	44	c4	of	e9	cb
2	54	7b	94	32	a6	c2	23	3d	ee	4c	95	0b	42	fa	c3	4e
3	08	2e	a1	66	28	d9	24	b2	76	5b	a2	49	6d	8b	d1	25
4	72	f8	f6	64	86	68	98	16	d4	a4	5c	cc	5d	65	b6	92
5	6c	70	48	50	fd	ed	b9	da	5e	15	46	57	a7	8d	9d	84
6	90	d8	ab	00	8c	bc	d3	0a	f7	e4	58	05	b8	b3	45	06
7	d0	2c	1e	8f	ca	3f	0f	02	c1	af	bd	03	01	13	8a	6b
8	3a	91	11	41	4f	67	dc	ea	97	f2	cf	ce	f0	b4	e6	73
9	96	ac	74	22	e7	ad	35	85	e2	f9	37	e8	1c	75	df	6e
a	47	f1	1a	71	1d	29	c5	89	6f	b7	62	0e	aa	18	be	1b
b	fc	56	3e	4b	c6	d2	79	20	9a	db	c0	fe	78	cd	5a	f4
c	1f	dd	a8	33	88	07	c7	31	b1	12	10	59	27	80	ec	5f
d	60	51	7f	a9	19	b5	4a	0d	2d	e5	7a	9f	93	c9	9c	ef
e	a0	e0	3b	4d	ae	2a	f5	b0	c8	eb	bb	3c	83	53	99	61
f	17	2b	04	7e	ba	77	d6	26	e1	69	14	63	55	21	0c	7d

(4) Give an example of a state of four distinct words which is *reproduced* after two iterations of the transformation S:

$$
\begin{array}{cccc}
s_{0,0} & s_{0,1} & s_{0,2} & s_{0,3} \\
s_{1,0} & s_{1,1} & s_{1,2} & s_{1,3} \\
s_{2,0} & s_{2,1} & s_{2,2} & s_{2,3} \\
s_{3,0} & s_{3,1} & s_{3,2} & s_{3,3}
\end{array}
\quad
\begin{array}{c}
S \circ S \\
\longmapsto
\end{array}
\quad
\begin{array}{cccc}
s_{0,0} & s_{0,1} & s_{0,2} & s_{0,3} \\
s_{1,0} & s_{1,1} & s_{1,2} & s_{1,3} \\
s_{2,0} & s_{2,1} & s_{2,2} & s_{2,3} \\
s_{3,0} & s_{3,1} & s_{3,2} & s_{3,3}.
\end{array}
$$

(5) Modify the *S*-box of the cipher AES in replacing the multiplicative inversion in $\mathbf{R}_{256} = \mathbb{F}_2[x]/(x^8 + x^4 + x^3 + x + 1)$ by that in $\mathbb{F}_{256} = \mathbb{F}_2[x]/(x^8 + x^4 + x^3 + x^2 + 1)$ – while keeping unchanged the affine transformation which constitutes the second step of the *S*-box.
What shall now be the value $S(\mathrm{aa}) = S(10101010)$?

The ShiftRows Step

The first line is kept fixed, the second line is cyclically shifted one byte to the left, the following two bytes to the left, and the last line three byte-positions to the left. This gives:

$$
\begin{array}{cccc}
s_{0,0} & s_{0,1} & s_{0,2} & s_{0,3} \\
s_{1,0} & s_{1,1} & s_{1,2} & s_{1,3} \\
s_{2,0} & s_{2,1} & s_{2,2} & s_{2,3} \\
s_{3,0} & s_{3,1} & s_{3,2} & s_{3,3}
\end{array}
\quad
\longmapsto
\quad
\begin{array}{cccc}
s_{0,0} & s_{0,1} & s_{0,2} & s_{0,3} \\
s_{1,1} & s_{1,2} & s_{1,3} & s_{1,0} \\
s_{2,2} & s_{2,3} & s_{2,0} & s_{2,1} \\
s_{3,3} & s_{3,0} & s_{3,1} & s_{3,2}.
\end{array}
$$

The MixColumns Step

Each of the four columns of the state is considered as a vector of length 4 over the field $\mathbf{R}_{256}$.

Define the circular convolution product of two vectors:

Let $x = \begin{pmatrix} x_0 \\ x_1 \\ x_2 \\ x_3 \end{pmatrix} \in \mathbf{R}_{256}^4$. The associated circular matrix $C(x)$ is defined by

$$C(x) = \begin{pmatrix} x_0 & x_3 & x_2 & x_1 \\ x_1 & x_0 & x_3 & x_2 \\ x_2 & x_1 & x_0 & x_3 \\ x_3 & x_2 & x_1 & x_0 \end{pmatrix}.$$

Example

$$x = \begin{pmatrix} 02 \\ 01 \\ 01 \\ 03 \end{pmatrix}, \quad \text{then} \quad C(x) = \begin{pmatrix} 02 & 03 & 01 & 01 \\ 01 & 02 & 03 & 01 \\ 01 & 01 & 02 & 03 \\ 03 & 01 & 01 & 02 \end{pmatrix}.$$

The multiplication of a circular matrix with a vector presents some important particularities: Put

$$x = \begin{pmatrix} x_0 \\ x_1 \\ x_2 \\ x_3 \end{pmatrix}, \quad y = \begin{pmatrix} y_0 \\ y_1 \\ y_2 \\ y_3 \end{pmatrix}, \quad \text{and} \quad z = \begin{pmatrix} z_0 \\ z_1 \\ z_2 \\ z_3 \end{pmatrix} = C(x) \cdot y.$$

Then:

$$z_0 = x_0 y_0 + x_3 y_1 + x_2 y_2 + x_1 y_3,$$
$$z_1 = x_1 y_0 + x_0 y_1 + x_3 y_2 + x_2 y_3,$$
$$z_2 = x_2 y_0 + x_1 y_1 + x_0 y_2 + x_3 y_3,$$
$$z_3 = x_3 y_0 + x_2 y_1 + x_1 y_2 + x_0 y_3.$$

We observe: $z_j = \sum_{\mu+\nu \equiv j \bmod 4} x_\mu y_\nu$, $0 \le j \le 3$.

Consequences   *(1) Commutativity: $C(x) \cdot y = C(y) \cdot x$.*

(2) Invariance under cyclic permutation: $C(x) \cdot \sigma(y) = \sigma(C(x) \cdot y)$, where

$$\sigma \begin{pmatrix} z_0 \\ z_1 \\ z_2 \\ z_3 \end{pmatrix} = \begin{pmatrix} z_3 \\ z_0 \\ z_1 \\ z_2 \end{pmatrix}.$$

Combining (1) and (2), we obtain

(3) If $z = C(x) \cdot y$, then $C(z) = C(x)C(y) = C(y)C(x)$ (multiplication of circular matrices).

It is evident that the properties (1),(2) and (3) generalize for arbitrary n.

Definition *The (circular) convolution product* $* : \mathbf{R}_{256}^4 \times \mathbf{R}_{256}^4 \longrightarrow \mathbf{R}_{256}^4$:
Let us define, for

$$x = \begin{pmatrix} x_0 \\ x_1 \\ x_2 \\ x_3 \end{pmatrix}, \quad y = \begin{pmatrix} y_0 \\ y_1 \\ y_2 \\ y_4 \end{pmatrix} \in \mathbf{R}_{256}^4$$

their (circular) convolution product $z = x * y$ *by* $z = C(x) \cdot y = C(y) \cdot x$.

We immediately get:

(1) $x * y = y * x$
(2) $\sigma(x * y) = (\sigma(x)) * y = x * (\sigma(y))$
(3) $C(x * y) = C(x)C(y)$

Observation (Polynomial notation) *Replace the vectors*

$$x = \begin{pmatrix} x_0 \\ x_1 \\ x_2 \\ x_3 \end{pmatrix}, \quad y = \begin{pmatrix} y_0 \\ y_1 \\ y_2 \\ y_3 \end{pmatrix}, \quad z = x * y = \begin{pmatrix} z_0 \\ z_1 \\ z_2 \\ z_3 \end{pmatrix}$$

by the polynomials

$p_x(T) = x_0 + x_1 T + x_2 T^2 + x_3 T^3,$
$p_y(T) = y_0 + y_1 T + y_2 T^2 + y_3 T^3,$
$p_z(T) = z_0 + z_1 T + z_2 T^2 + z_3 T^3,$

then $p_z(T) = p_x(T)p_y(T) \bmod (T^4 + 1)$, *i.e.* $p_{x*y}(T)$ *is the remainder of division of the polynomial product* $p_x(T)p_y(T)$ *by the polynomial* $T^4 + 1$.

Now consider the arithmetical filtering (of the words) of the state array given by circular convolution with the "impulse response"

$$a = \begin{pmatrix} 02 \\ 01 \\ 01 \\ 03 \end{pmatrix} = \begin{pmatrix} x \\ 1 \\ 1 \\ x+1 \end{pmatrix}$$

More explicitely, we shall have

$$\begin{matrix} s_{0,0} & s_{0,1} & s_{0,2} & s_{0,3} \\ s_{1,0} & s_{1,1} & s_{1,2} & s_{1,3} \\ s_{2,0} & s_{2,1} & s_{2,2} & s_{2,3} \\ s_{3,0} & s_{3,1} & s_{3,2} & s_{3,3} \end{matrix} \longmapsto \begin{matrix} s'_{0,0} & s'_{0,1} & s'_{0,2} & s'_{0,3} \\ s'_{1,0} & s'_{1,1} & s'_{1,2} & s'_{1,3} \\ s'_{2,0} & s'_{2,1} & s'_{2,2} & s'_{2,3} \\ s'_{3,0} & s'_{3,1} & s'_{3,2} & s'_{3,3} \end{matrix}$$

with

$$
w_0' = \begin{pmatrix} s_{0,0}' \\ s_{1,0}' \\ s_{2,0}' \\ s_{3,0}' \end{pmatrix} = \begin{pmatrix} 02 \\ 01 \\ 01 \\ 03 \end{pmatrix} * \begin{pmatrix} s_{0,0} \\ s_{1,0} \\ s_{2,0} \\ s_{3,0} \end{pmatrix}, \quad
w_1' = \begin{pmatrix} s_{0,1}' \\ s_{1,1}' \\ s_{2,1}' \\ s_{3,1}' \end{pmatrix} = \begin{pmatrix} 02 \\ 01 \\ 01 \\ 03 \end{pmatrix} * \begin{pmatrix} s_{0,1} \\ s_{1,1} \\ s_{2,1} \\ s_{3,1} \end{pmatrix},
$$

$$
w_2' = \begin{pmatrix} s_{0,2}' \\ s_{1,2}' \\ s_{2,2}' \\ s_{3,2}' \end{pmatrix} = \begin{pmatrix} '02 \\ 01 \\ 01 \\ 03 \end{pmatrix} * \begin{pmatrix} s_{0,2} \\ s_{1,2} \\ s_{2,2} \\ s_{3,2} \end{pmatrix}, \quad
w_3' = \begin{pmatrix} s_{0,3}' \\ s_{1,3}' \\ s_{2,3}' \\ s_{3,3}' \end{pmatrix} = \begin{pmatrix} 02 \\ 01 \\ 01 \\ 03 \end{pmatrix} * \begin{pmatrix} s_{0,3} \\ s_{1,3} \\ s_{2,3} \\ s_{3,3} \end{pmatrix}.
$$

Exercises

(1) Show that a is *invertible* for the circular convolution product. You have to find b with $a * b = e_0 = \begin{pmatrix} 01 \\ 00 \\ 00 \\ 00 \end{pmatrix}$ [4].

Solution

$$
b = \begin{pmatrix} 0e \\ 09 \\ 0d \\ 0b \end{pmatrix} = \begin{pmatrix} x^3 + x^2 + x \\ x^3 + 1 \\ x^3 + x^2 + 1 \\ x^3 + x + 1 \end{pmatrix}.
$$

(2) True or false:

$$
\begin{pmatrix} x \\ 1 \\ 1 \\ x+1 \end{pmatrix} * \begin{pmatrix} x^3 + x^2 + x \\ x^3 + 1 \\ x^3 + x^2 + 1 \\ x^3 + x + 1 \end{pmatrix} = \begin{pmatrix} 1 \\ 0 \\ 0 \\ 0 \end{pmatrix}
$$

in $(\mathbb{F}_2[x]/(p(x))^4$ for *any* irreducible polynomial $p(x)$ over $\mathbb{F}_2$.

(3) The vector $\mathbf{b}$ as in exercise (1). Then $\mathbf{b} * \begin{pmatrix} ae \\ ae \\ ae \\ ae \end{pmatrix} = ?$

(4) $\mathbf{a} = \begin{pmatrix} 02 \\ 01 \\ 01 \\ 03 \end{pmatrix} = \begin{pmatrix} x \\ 1 \\ 1 \\ x+1 \end{pmatrix} \qquad \mathbf{b} = \begin{pmatrix} 0e \\ 09 \\ 0d \\ 0b \end{pmatrix} = \begin{pmatrix} x^3 + x^2 + x \\ x^3 + 1 \\ x^3 + x^2 + 1 \\ x^3 + x + 1 \end{pmatrix}.$

It has been observed that actually $\mathbf{b} = \mathbf{a} * \mathbf{p}$ with a rather simple vector $\mathbf{p}$. Consequently, it is possible to implement the decryption operation $\mathbf{b} * ()$ as a composition of a simple preprocessing step $\mathbf{p} * ()$, followed by the *encryption* operation $\mathbf{a} * ()$. Find $\mathbf{p}$.

---

[4] $C(e_0)$ is the identity matrix, hence e_0 is the unit for circular convolution.

Addition of the Round Key to the State Array

Here you are in a DES-like situation: You apply 128 times the Boolean operation XOR, position by position (bit by bit). Algebraically, one computes the sum of two matrices over $\mathbf{R}_{256}$:

$$\begin{matrix} s_{0,0} & s_{0,1} & s_{0,2} & s_{0,3} \\ s_{1,0} & s_{1,1} & s_{1,2} & s_{1,3} \\ s_{2,0} & s_{2,1} & s_{2,2} & s_{2,3} \\ s_{3,0} & s_{3,1} & s_{3,2} & s_{3,3} \end{matrix} \quad \oplus \quad \begin{matrix} k_{0,0} & k_{0,1} & k_{0,2} & k_{0,3} \\ k_{1,0} & k_{1,1} & k_{1,2} & k_{1,3} \\ k_{2,0} & k_{2,1} & k_{2,2} & k_{2,3} \\ k_{3,0} & k_{3,1} & k_{3,2} & k_{3,3} \end{matrix}$$

Exercises

(1) Find the initial state

$$\begin{matrix} s_{0,0} & s_{0,1} & s_{0,2} & s_{0,3} \\ s_{1,0} & s_{1,1} & s_{1,2} & s_{1,3} \\ s_{2,0} & s_{2,1} & s_{2,2} & s_{2,3} \\ s_{3,0} & s_{3,1} & s_{3,2} & s_{3,3} \end{matrix}$$

of a round which guarantees that at the end of the round *state = round key*.

(2) Give an example of a *non-constant* initial state (of a round) such that its round transform (before the final addition of the round key) is nothing but a simple permutation of its 16 letters.
(Help: Search first for the 8-bit bytes x with $S(S(x)) = x$.)

(3) Consider the following polynomial, with coefficients in $\mathbf{R}_{256}$:

$$S(X) = 63 + 8fX^{127} + b5X^{191} + 01X^{223} + f4X^{239} + 25X^{247} + f9X^{251}$$
$$+09X^{253} + 05X^{254}$$

(a) Compute $S(00), S(01), S(02), S(04)$.
(b) Let us take for granted:The polynomial $S(X)$ *is* (as a function on $\mathbf{R}_{256}$) the S-box of AES-Rijndael.
Deduce the polynomial $T(X) \in \mathbf{R}_{256}[X]$ which is (as a function on $\mathbf{R}_{256}$) the affine transformation

$$\begin{pmatrix} \hat{\beta}_0 \\ \hat{\beta}_1 \\ \hat{\beta}_2 \\ \hat{\beta}_3 \\ \hat{\beta}_4 \\ \hat{\beta}_5 \\ \hat{\beta}_6 \\ \hat{\beta}_7 \end{pmatrix} = \begin{pmatrix} 1&0&0&0&1&1&1&1 \\ 1&1&0&0&0&1&1&1 \\ 1&1&1&0&0&0&1&1 \\ 1&1&1&1&0&0&0&1 \\ 1&1&1&1&1&0&0&0 \\ 0&1&1&1&1&1&0&0 \\ 0&0&1&1&1&1&1&0 \\ 0&0&0&1&1&1&1&1 \end{pmatrix} \begin{pmatrix} \beta_0 \\ \beta_1 \\ \beta_2 \\ \beta_3 \\ \beta_4 \\ \beta_5 \\ \beta_6 \\ \beta_7 \end{pmatrix} + \begin{pmatrix} 1 \\ 1 \\ 0 \\ 0 \\ 0 \\ 1 \\ 1 \\ 0 \end{pmatrix}.$$

(c) How can you finally find the polynomial $S^{-1}(X) \in \mathbf{R}_{256}[X]$ which is (as a function on $\mathbf{R}_{256}$) the inverse S-box?

2.2.3 The Key Schedule

The round keys will be extracted from the user's cipher key by means of the following algorithm:

- The cipher key K will generate an *expanded key array* K^* which is a vector of N words (i.e. a $4 \times N$ matrix of bytes).
- The length N of the expanded key array K^*: $N = 4 \times (n+1)$, where

$$
n = Nr \equiv \text{the number of rounds} = \begin{cases} 10 & AES - 128 \\ 12 & AES - 192 \\ 14 & AES - 256 \end{cases}
$$

- We shall need $n+1$ round keys: $K_0, K_1, \ldots K_n$ $(n = 10, 12, 14)$, which are cut out sequentially from the expanded key array:

$K_0 \equiv$ the first four words of the expanded key,
$K_1 \equiv$ the following four words of the expanded key,
$$\vdots$$
$K_n \equiv$ the last four words of the expanded key,
$K^* = (K_0, K_1, \ldots, K_n)$.

But let us begin with the presentation of the algorithm which computes the expanded key array from the user's cipher key.

First we need two simple operations on words:

$$
S \begin{pmatrix} a_0 \\ a_1 \\ a_2 \\ a_3 \end{pmatrix} = \begin{pmatrix} S(a_0) \\ S(a_1) \\ S(a_2) \\ S(a_3) \end{pmatrix} \qquad \text{and} \qquad \tau \begin{pmatrix} a_0 \\ a_1 \\ a_2 \\ a_3 \end{pmatrix} = \begin{pmatrix} a_1 \\ a_2 \\ a_3 \\ a_0 \end{pmatrix}
$$

$S \equiv$ the operation of the S-box on the bytes

$\tau = \sigma^{-1}$ for the cyclic permutation σ introduced together with the circular convolution

$$
RC[i] = \begin{pmatrix} x^{i-1} \\ 0 \\ 0 \\ 0 \end{pmatrix} \in \mathbf{R}_{256}^4 \quad i \geq 1.
$$

(a translation vector, with variable first component).
Recursive computation of the words $w[i]$, $0 \leq i < 4(n+1)$

$$
n = \begin{cases} 10 & AES - 128, \\ 12 & AES - 192, \\ 14 & AES - 256, \end{cases}
$$

which are the components of the *expanded key array*.

$$K^* = \begin{cases} w[0]w[1]\cdots w[43] & \text{if AES} - 128 \\ w[0]w[1]\cdots w[51] & \text{if AES} - 192 \\ w[0]w[1]\cdots w[59] & \text{if AES} - 256 \end{cases}$$

START:

$w[0]w[1]\cdots w[Nk-1]$ is simply the cipher key $Nk = \begin{cases} 4 & \text{en AES} - 128 \\ 6 & \text{en AES} - 192 \\ 8 & \text{en AES} - 256 \end{cases}$

STEP:

For $i \geq Nk$ we have $\qquad w[i] = w[i - Nk] \oplus w[i-1]^{\sharp}$, where

$$w[i-1]^{\sharp} = \begin{cases} w[i-1] & \text{for } \begin{cases} i \not\equiv 0 \bmod Nk & \text{if Nk} = 4,6 \\ i \not\equiv 0,4 \bmod Nk & \text{if Nk} = 8 \end{cases} \\ S(\tau(w[i-1])) \oplus RC[\frac{i}{Nk}] & \text{for } i \equiv 0 \bmod Nk \\ S(w[i-1]) & \text{for } i \equiv 4 \bmod Nk \text{ and } Nk = 8 \end{cases}$$

END: $i = 4n + 3$.

Scheme *Beginning of the computation of K^* in AES-128.*

$$w[0]w[1]w[2]w[3] = K = \begin{matrix} k_{0,0}\ k_{0,1}\ k_{0,2}\ k_{0,3} \\ k_{1,0}\ k_{1,1}\ k_{1,2}\ k_{1,3} \\ k_{2,0}\ k_{2,1}\ k_{2,2}\ k_{2,3} \\ k_{3,0}\ k_{3,1}\ k_{3,2}\ k_{3,3} \end{matrix}$$

$$w[4] = w[0] \oplus \begin{pmatrix} S(k_{1,3}) \\ S(k_{2,3}) \\ S(k_{3,3}) \\ S(k_{0,3}) \end{pmatrix} \oplus \begin{pmatrix} 01 \\ 00 \\ 00 \\ 00 \end{pmatrix},$$

$w[5] = w[1] \oplus w[4],$
$w[6] = w[2] \oplus w[5],$
$w[7] = w[3] \oplus w[6],$

$$w[8] = w[4] \oplus S(\tau(w[7])) \oplus \begin{pmatrix} 02 \\ 00 \\ 00 \\ 00 \end{pmatrix},$$

$w[9] = w[5] \oplus w[8],$
$w[10] = w[6] \oplus w[9],$
$w[11] = w[7] \oplus w[10],$

$$w[12] = w[8] \oplus S(\tau(w[11])) \oplus \begin{pmatrix} 04 \\ 00 \\ 00 \\ 00 \end{pmatrix}$$

$\vdots$

Example of the computation of an expanded key array: The cipher key $K = 2b\ 7e\ 15\ 16\ 28\ ae\ d2\ a6\ ab\ f7\ 15\ 88\ 09\ cf\ 4f\ 3c$. This yields:

$$w_0 = 2b7e1516, \quad w_1 = 28aed2a6, \quad w_2 = abf71588, \quad w_3 = 09cf4f3c.$$

index	$w[i-1]$	$\tau()$	$S()$	$RC\left[\frac{i}{Nk}\right]$	$\oplus RC$	$w[i-Nk]$	$w[i]$
4	09cf4f3c	cf4f3c09	8a84eb01	01000000	8b84eb01	2b7e1516	a0fafe17
5	a0fafe17					28aed2a6	88542cb1
6	88542cb1					abf71588	23a33939
7	23a33939					09cf4f3c	2a6c7605
8	2a6c7605	6c76052a	50386be5	02000000	52386be5	a0fafe17	f2c295f2
9	f2c295f2					88542cb1	7a96b943
10	7a96b943					23a33939	5935807a
11	5935807a					2a6c7605	7359f67f
12	7359f67f	59f67f73	cb42d28f	04000000	cf42d28f	f2c295f2	3d80477d
13	3d80477d					7a96b943	4716fe3e
14	4716fe3e					5935807a	1e237e44
15	1e237e44					7359f67f	6d7a883b
16	6d7a883b	7a883b6d	dac4e23c	08000000	d2c4e23c	3d80477d	ef44a541
17	ef44a541					4716fe3e	a8525b7f
18	a8525b7f					1e237e44	b671253b
19	b671253b					6d7a883b	db0bad00
20	db0bad00	0bad00db	2b9563b9	10000000	3b9563b9	ef44a541	d4d1c6f8
21	d4d1c6f8					a8525b7f	7c839d87
22	7c839d87					b671253b	caf2b8bc
23	caf2b8bc					db0bad00	11f915bc
24	11f915bc	f915bc11	99596582	20000000	b9596582	d4d1c6f8	6d88a37a
25	6d88a37a					7c839d87	110b3efd
26	110b3efd					caf2b8bc	dbf98641
27	dbf98641					11f915bc	ca0093fd
28	ca0093fd	0093fdca	63dc5474	40000000	23dc5474	6d88a37a	4e54f70e
29	4e54f70e					110b3efd	5f5fc9f3

index	$w[i-1]$	$\tau()$	$S()$	$RC\left[\frac{i}{Nk}\right]$	$\oplus RC$	$w[i-Nk]$	$w[i]$
30	5f5fc9f3					dbf98641	84a64fb2
31	84a64fb2					ca0093fd	4ea6dc4f
32	4ea6dc4f	a6dc4f4e	2486842f	80000000	a486842f	4e54f70e	ead27321
33	ead27321					5f5fc9f3	b58dbad2
34	b58dbad2					84a64fb2	312bf560
35	312bf560					4ea6dc4f	7f8d292f
36	7f8d292f	8d292f7f	5da515d2	1b000000	46a515d2	ead27321	ac7766f3
37	ac7766f3					b58dbad2	19fadc21
38	19fadc21					312bf560	28d12941
39	28d12941					7f8d292f	575c006e
40	575c006e	5c006e57	4a639f5b	36000000	7c639f5b	ac7766f3	d014f9a8
41	d014f9a8					19fadc21	c9ee2589
42	c9ee2589					28d12941	e13f0cc8
43	e13f0cc8					575c006e	b6630ca6

The Encryption Protocol

The encryption with AES–Rijndael is done in 11, 13 or 15 rounds, depending on the size of the keys (128, 192 or 256 bits). There are

- *The initial round*: Addition of K_0 to the initial state ($\equiv$ the plaintext block)
- *The rounds* $n°1, \ldots n°$ $\begin{cases} 9 \\ 11 \\ 13 \end{cases}$: as described in (c)
- *The final round*: Without the MixColumns step

Example of an encryption with AES-128:

- The plaintext block: 32 43 f6 a8 88 5a 30 8d 31 31 98 a2 e0 37 07 34
- The cipher key: 2b 7e 15 16 28 ae d2 a6 ab f7 15 88 09 cf 4f 3c

round	initial state	after S()	after ←	after a*()		round key
0	32 88 31 e0 43 5a 31 37 f6 30 98 07 a8 8d a2 34				⊕	2b 28 ab 09 7e ae f7 cf 15 d2 15 4f 16 a6 88 3c =
1	19 a0 9a e9 3d f4 c6 f8 e3 e2 8d 48 be 2b 2a 08	d4 e0 b8 1e 27 bf b4 41 11 98 5d 52 ae f1 e5 30	d4 e0 b8 1e bf b4 41 27 5d 52 11 98 30 ae f1 e5	04 e0 48 28 66 cb f8 06 81 19 d3 26 e5 9a 7a 4c	⊕	a0 88 23 2a fa 54 a3 6c fe 2c 39 76 17 b1 39 05 =
2	a4 68 6b 02 9c 9f 5b 6a 7f 35 ea 50 f2 2b 43 49	49 45 7f 77 0f db 39 02 d2 96 87 53 89 f1 1a 3b	49 45 7f 77 db 39 02 0f 87 53 d2 96 3b 89 f1 1a	58 1b db 1b 4d 4b e7 6b ca 5a ca b0 f1 ac a8 e5	⊕	f2 7a 59 73 c2 96 35 59 95 b9 80 f6 f2 43 7a 7f =
3	aa 61 82 68 8f dd d2 32 5f e3 4a 46 03 ef d2 9a	ac ef 13 45 73 c1 b5 23 cf 11 d6 5a 7b df b5 b8	ac ef 13 45 c1 b5 23 73 d6 5a cf 11 b8 7b df b5	75 20 53 bb ec 0b c0 25 09 63 cf d0 93 33 7c dc	⊕	3d 47 1e 6d 80 16 23 7a 47 fe 7e 88 7d 3e 44 3b =
4	48 67 4d d6 6c 1d e3 5f 4e 9d b1 58 ee 0d 38 e7	52 85 e3 f6 50 a4 11 cf 2f 5e c8 6a 28 d7 07 94	52 85 e3 f6 a4 11 cf 50 c8 6a 2f 5e 94 28 d7 07	0f 60 6f 5e d6 31 c0 b3 da 38 10 13 a9 bf 6b 01	⊕	ef a8 b6 db 44 52 71 0b a5 5b 25 ad 41 7f 3b 00 =
5	e0 c8 d9 85 92 63 b1 b8 7f 63 35 be e8 c0 50 01	e1 e8 35 97 4f fb c8 6c d2 fb 96 ae 9b ba 53 7c	e1 e8 35 97 fb c8 6c 4f 96 ae d2 fb 7c 9b ba 53	25 bd b6 4c d1 11 3a 4c a9 d1 33 c0 ad 68 8e b0	⊕	d4 7c ca 11 d1 83 f2 f9 c6 9d b8 15 f8 87 bc bc =
6	f1 c1 7c 5d 00 92 c8 b5 6f 4c 8b d5 55 ef 32 0c	a1 78 10 4c 63 4f e8 d5 d8 29 3d 03 fc df 23 fe	a1 78 10 4c 4f e8 d5 63 3d 03 a8 29 fe fc df 23	4b 2c 33 37 86 4a 9d d2 8d 89 f4 18 6d 80 e8 d8	⊕	6d 11 db ca 88 0b f9 00 a3 3e 86 93 7a fd 41 fd =
7	26 3d e8 fd 0e 41 64 d2 2e b7 72 8b 17 7d a9 25	f7 27 9b 54 ab 83 43 b5 31 a9 40 3d f0 ff d3 3f	f7 27 9b 54 83 43 b5 ab 40 3d 31 a9 3f f0 ff d3	14 46 27 34 15 16 46 2a b5 15 56 d8 bf ec d7 43	⊕	4e 5f 84 4e 54 5f a6 a6 f7 c9 4f dc 0e f3 b2 4f =
8	5a 19 a3 7a 41 49 e0 8c 42 dc 19 04 b1 1f 65 0c	be d4 0a da 83 3b e1 64 2c 86 d4 f2 c8 c0 4d fe	be d4 0a da 3b e1 64 83 d4 f2 2c 86 fe c8 c0 4d	00 b1 54 fa 51 c8 76 1b 2f 89 6d 99 d1 ff cd ea	⊕	ea b5 31 7f d2 8d 2b 8d 73 ba f5 29 21 d2 60 2f =
9	ea 04 65 85 83 45 5d 96 5c 33 98 b0 f0 2d ad c5	87 f2 4d 97 ec 6e 4c 90 4a c3 46 e7 8c d8 95 a6	87 f2 4d 97 6e 4c 90 ec 46 e7 4a c3 a6 8c d8 95	47 40 a3 4c 37 d4 70 9f 94 e4 3a 42 ed a5 a6 bc	⊕	ac 19 28 57 77 fa d1 5c 66 dc 29 00 f3 21 41 6e =
10	eb 59 8b 1b 40 2e a1 c3 f2 38 13 42 1e 84 e7 d2	e9 cb 3d af 09 31 32 2e 89 07 7d 2c 72 5f 94 b5	e9 cb 3d af 31 32 2e 09 7d 2c 89 07 b5 72 5f 94		⊕	d0 c9 e1 b6 14 ee 3f 63 f9 25 0c 0c a8 89 c8 a6 =

```
39 02 dc 19
25 dc 11 6a
84 09 85 0b
1d fb 97 32
```

- The ciphertext: 39 25 84 1d 02 dc 09 fb dc 11 85 97 19 6a 0b 32

Exercises

(1) The plaintext unit: 00 00 00 00 00 00 00 00 00 00 00 00 00 00 00 00.
The cipher key: 00 00 00 00 00 00 00 00 00 00 00 00 00 00 00 00.
The state array at the beginning of round n^0 2?

(2) Your cipher key: 43 c6 84 53 33 25 0c 80 1d 2b c3 97 e2 cc 40 b3.
Your plaintext block: 4a a3 3c 69 4f 4f 3b ad 59 7f f3 d9 ec e8 32 0c.
The state array at the *beginning* of the second round?

(3) In the example of encryption with AES-128 (on the preceding pages) the
third round transformation transforms the state array

$$I = \begin{matrix} \text{aa } 61 \ 82 \ 68 \\ \text{8f dd d2 } 32 \\ \text{5f e3 } 4a \ 46 \\ 03 \ \text{ef d2 } 9a \end{matrix} \quad \text{into} \quad F = \begin{matrix} 48 \ 67 \ 4d \ d6 \\ 6c \ 1d \ e3 \ 5f \\ 4e \ 9d \ b1 \ 58 \\ \text{ee } 0d \ 38 \ e7 \end{matrix}$$

A modification of four bits in the initial state I gives

$$F' = \begin{matrix} 11 \ 67 \ 4d \ d6 \\ 4b \ 1d \ e3 \ 5f \\ \text{ed } 9d \ b1 \ 58 \\ \text{fd } 0d \ 38 \ e7 \end{matrix}$$

Find the error.

(4) The eighth round key: $K_8 =$

01	04	85	88
91	04	46	44
81	04	27	22
c0	0c	18	11

The ninth round key: $K_9 = ?$

(5) What was the *initial state* of a (standard) round that produces as the
result of the third step – i.e. before the addition of the round key – the
following state array:

7c	77	7b	f2
7c	77	7b	f2
7c	77	7b	f2
7c	77	7b	f2

(6) The plaintext block: 00 11 22 33 44 55 66 77 88 99 aa bb cc dd ee ff.

The cipher key: 00 01 02 03 04 05 06 07 08 09 0a 0b 0c 0d 0e 0f.
The initial state of the second round will be

89 d8 10 e8 ** ** ** ** 2d 18 43 d8 cb 12 8f e4.

Compute the four masked bytes.

2.2.4 Decryption with Rijndael

First consider the decryption algorithm in its logical structure: We have simply to apply the inverse transformations of the encryption algorithm, with *the order* of the transformations inverted.

This will give, for a standard round of *inverse* AES-Rijndael:

(1) The addition of the round key
(2) The inverse convolution on the words of the state array
(3) The inverse "rotation" of the lines of the state array
(4) The inverse S-box.

But the cipher Rijndael seems to have been designed with a tender look towards the old DES, where the decryption is nothing but an encryption of the ciphertext (with the order of the round keys inverted). Here, the situation is less symmetrical; but altogether, it is possible to create something similar: the decryption operations may be executed *in the order of encryption*, in replacing only each constituent transformation by its inverse. The order of the round keys will be inverted, but the round keys will be "inverted" themselves.

To better understand this, consider a simplified version of AES:

Example AES–Rijndael inverse in two rounds:

(1) Addition of the round key K_2
(2) (Inversion of the) rotation of the lines of the state array
(3) Operation of the inverse S-box
(4) Addition of the round key K_1
(5) Inverse convolution operation on the words of the state array
(6) (Inversion of the) rotation of the lines of the state array
(7) Operation of the inverse S-box
(8) Addition of the round key K_0

Observation (1) The operation of the S-box *commutes* with the operation of the line rotations (since the first simply acts on the letters, in parallel).
(2) The composition (addition of the round key K_i) *with* (inverse convolution operation on the columns of the state)
  equals the composition (inverse convolution operation on the columns of the state) *with* (addition of the inverted round key $K_i^\sharp$) where $K_i^\sharp = b * K_i$ (column by column).

(this is an immediate consequence of the linearity of the operation of circular convolution

$$Recall: \quad b = \begin{pmatrix} 0e \\ 09 \\ 0d \\ 0b \end{pmatrix} = \begin{pmatrix} x^3 + x^2 + x \\ x^3 + 1 \\ x^3 + x^2 + 1 \\ x^3 + x + 1 \end{pmatrix}.$$

Consequences The decryption by AES–Rijndael in two rounds, in encryption-adapted order:

(1) Addition of the round key K_2
(2) Operation of the inverse S-box
(3) (Inversion of the) rotation of the lines of the state array
(4) Inverse convolution operation on the words of the state array
(5) Addition of the round key $K_1^\sharp$
(6) Operation of the inverse S-box
(7) (Inversion of the) rotation of the lines of the state array
(8) Addition of the round key K_0

The structure *initial round – normal round – final round* has been respected. The order of the operations corresponds to their order in the encryption algorithm – up to the inversion of the particular transformations.

This will clearly also be true in the general case.

The round keys $K_i^\sharp$ for the decryption algorithm are obtained as follows:

(1) Compute the expanded key array $K^* = (K_0, K_1, \ldots, K_n)$, $\quad n = \begin{cases} 10 \\ 12 \\ 14 \end{cases}$

(2) Let Nr (= 10, 12, 14) be the number of rounds. then

$$K_i^\sharp = \begin{cases} K_i & \text{for } i = 0, Nr \\ b * K_i & \text{otherwise} \end{cases}$$

The round keys $K_i^\sharp$ will operate in inverse order: $K_{Nr}^\sharp, K_{Nr-1}^\sharp, \ldots, K_1^\sharp, K_0^\sharp$.

2.3 The Public Key Paradigm and the Cryptosystem RSA

The idea of a public key system dates from 1976, and is due to Diffie and Hellman. The first realization of such a system is the cipher *RSA*, published in 1978 by *R*ivest, *S*hamir and *A*dleman.

One observes that almost all technical ideas around public key ciphers refer to encryption methods by exponentiation (in certain multiplicative groups supporting the encoded information).

2.3.1 Encryption and Decryption via Exponentiation

Still more Arithmetic

Let us introduce first *Euler's totient function* $\varphi(n)$:

$\varphi(n) \equiv$ the number of integers a between 1 and n which are prime to n.

Example

$$\varphi(2) = 1, \quad \varphi(8) = 4,$$
$$\varphi(3) = 2, \quad \varphi(9) = 6,$$
$$\varphi(4) = 2, \quad \varphi(10) = 4,$$
$$\varphi(5) = 4, \quad \varphi(11) = 10,$$
$$\varphi(6) = 2, \quad \varphi(12) = 4,$$
$$\varphi(7) = 6, \quad \varphi(13) = 12.$$

Important observation. $\varphi(n)$ *counts* the number of residues which are invertible for the multiplication in $\mathbb{Z}/n\mathbb{Z}$ (*Recall:* $[a]$ is invertible in $\mathbb{Z}/n\mathbb{Z} \iff \gcd(a, n) = 1$). This means: $\varphi(n)$ is *the order of the multiplicative group*$(\mathbb{Z}/n\mathbb{Z})^*$.

Computation of $\varphi(n)$:

(1) $\varphi(p^k) = p^{k-1}(p - 1)$,   if p is prime, $k \geq 1$.
(2) $\varphi(m \cdot n) = \varphi(m) \cdot \varphi(n)$,   if m and n are coprime.

In particular: $\varphi(p \cdot q) = (p-1)(q-1)$ for two distinct prime numbers p and q.

Let us insist: $\varphi(n)$ is calculable *if* one *knows* the prime factorization of n.

Example $\varphi(792) = \varphi(2^3) \cdot \varphi(3^2) \cdot \varphi(11) = 4 \cdot 6 \cdot 10 = 240$.

It is surprising that the mathematical basis for most of the arithmetical ciphers is a very, very elementary result. It is the

Theorem (Lagrange) *Let G be a finite (multiplicative) group of order φ. Then every $a \in G$ satisfies $a^\varphi = 1$.*

Proof (for commutative groups) Choose your a (arbitrary, but fixed), and an indexing of the group elements: $x_1, x_2, \ldots, x_\varphi$. Multiply all of them by a : $\{ax_1, ax_2, \ldots, ax_\varphi\} = \{x_1, x_2, \ldots, x_\varphi\}$. Hence: $(ax_1)(ax_2)\cdots(ax_\varphi) = x_1 x_2 \cdots x_\varphi$. This yields $a^\varphi = 1$, as claimed. $\square$

Consequence (Euler) $a^{\varphi(n)} \equiv 1 \bmod n$, whenever a is prime to n.

In particular (Fermat): $a^p \equiv a \bmod p$ for every $a \in \mathbb{Z}$ and every prime number p.

The situation in which we are particularly interested is the following:

Let G be a multiplicative (abelian) group of φ elements (which encode the blocks of our information). We aim at encrypting (decrypting) via exponentiation in this group. How can the exponents e of encryption and d of decryption be harmonized?

The answer is simple and basic for all arithmetical cryptography:

Let φ be the number of elements of G.

Suppose that $ed \equiv 1 \bmod \varphi$

Then: $(a^e)^d = a$ for every $a \in G$.

The argument: Write $ed = 1 + k \cdot \varphi$. Then $(a^e)^d = a \cdot (a^\varphi)^k = a$
since $a^\varphi = 1$ (Lagrange - Euler).

Systems of Encryption by Exponentiation

At this point, our first idea should be the following:

Let us digitize our message blocks (in the broad sense) as *remainders of division* in "natural" multiplicative groups of the type

(1) $G = (\mathbb{Z}/p\mathbb{Z})^* \equiv$ the group of the $p-1$ non-zero residues of the field $\mathbb{Z}/p\mathbb{Z}$, p prime.
(2) $G = (\mathbb{F}_2[x]/(p(x)))^* \equiv$ the group of the $2^n - 1$ non-zero residues of the field $\mathbb{F}_2[x]/(p(x))$, $p(x) = x^n + \cdots + 1$ irreducible over $\mathbb{F}_2 = \{0, 1\}$.

Note that at first approach we have counted on *cyclicity*: The relevant groups are all generated by a single element (which is not unique): We consider *discrete circles*.

The arithmetical cryptography encounters a typical situation here: Whenever these groups are sufficiently big, then the arithmetic will be blind in face of geometry; the cyclicity will be algorithmically inaccessible (*the problem of the discrete logarithm will be difficult*).

On the other hand, if the "discrete circle" is algorithmically well controlled (cf. the log tables for $\mathbb{F}_{256}$ and for $\mathbf{R}_{256}$, serving as arithmetical support for AES), then the multiplicative structure of G will be trivialized down to an additive structure, by the discrete logarithms (the "angles" replacing the "points"), and the cryptographic interest of G will vanish. We shall make all this more precise in a moment.

But let us return to the groups G that we proposed above. The relation between the encryption exponents e and the decryption exponents d is then the following:

(1) If $ed \equiv 1 \bmod (p-1)$ then $(a^e)^d \equiv a \bmod p$.
(2) If $ed \equiv 1 \bmod (2^n - 1)$ then $(r(x)^e)^d \equiv r(x) \bmod p(x)$.

We shall call the cryptosystems based on exponentiation in the groups $(\mathbb{Z}/p\mathbb{Z})^*$ or $(\mathbb{F}_2[x]/(p(x)))^*$ uniformly *Pohlig–Hellman systems*.

Exercises

First a technical complement:
  Fast exponentiation.

- Fundamental idea: Compute a^e mod n in such a way that the arithmetic never exceeds the size of n^2 (compute $r(x)^e$ mod $p(x)$ in such a way that the arithmetic never exceeds 2 times the degree of $p(x)$).
- *Realization of the idea*: Write the exponent e in binary notation, i.e. as a sum of powers of 2; the exponentiation will then essentially reduce to the (iterated) computation of squares.

Example 3509^{17} mod $4717 = ?$

$$[3509]^{17} = [3509]^{16}[3509]$$
but
$$[3509]^2 = [1711]$$

$$[3509]^4 = [1711]^2 = [2981]$$

$$[3509]^8 = [2981]^2 = [4250]$$

$$[3509]^{16} = [4250]^2 = [1107]$$

Hence: 3509^{17} mod $4717 = 2372$.

(1) A Pohlig–Hellman system, in binary polynomial arithmetic modulo $p(x) = x^4 + x + 1$. Your encryption exponent: $e = 13$.
   Compute the ciphertext C corresponding to $M = 0101 \equiv x^2 + 1$, then d, your decryption exponent, and verify, in decrypting, the restoration of M.
(2) A Pohlig–Hellman system, in binary polynomial arithmetic modulo $p(x) = x^8 + x^4 + x^3 + x^2 + 1$. Your encryption exponent: $e = 32$. You receive the ciphertext $C = 10011101$. Decrypt!
(3) The same system as in exercise (2).
   (a) A encrypts the plaintext $M = 11010101$ into the ciphertext $C = 01000011$.
      Find her encryption key e and her decryption key d.
   (b) The same question for B, who encrypts the plaintext $M = 11010010$ into the ciphertext $C = 11001001$.
   (This exercise is a first illustration to our remark on the trivialization of an arithmetical cryptosystem in case *the problem of the discrete logarithm is easy*. Here, we control perfectly the discrete circle of the 255 non-zero residues modulo $p(x) = x^8 + x^4 + x^3 + x^2 + 1$).
(4) Still the same system. The user F tells you that he encrypted the plaintext 11111111 into the ciphertext 10101010. You find, he is a practical joker. Why?

2.3.2 The Cryptosystem RSA

The Algorithm RSA

The Pohlig–Hellman system is *symmetrical*, exactly like the ciphers DES and AES: The knowledge of the keys for encryption is equivalent to the knowledge of the keys for decryption. Let us make this more precise:

Let (e_A, p_A) be the encryption key of A (A encrypts her plaintext blocks, digitalized in remainders of division modulo p_A – a big prime number – via exponentiation with e_A in $\mathbb{Z}/p_A\mathbb{Z}$). Then her decryption exponent d_A is simply the solution (in X) of the equation $e_A X + (p_A - 1)Y = 1$.

RSA creates an *asymmetrical* situation with an arithmetical triteness: One replaces the prime number p_A which defines the arithmetic of the user A by a *composite* number $n_A = p_A q_A$, where p_A and q_A are two (big) distinct prime numbers.

The exponents of encryption e_A and of decryption d_A are connected by

$$e_A d_A \equiv 1 \bmod \varphi(n_A), \quad \varphi(n_A) = (p_A - 1)(q_A - 1)$$

(*Recall*: It is the theorem of Euler relative to $G_A = (\mathbb{Z}/n_A\mathbb{Z})^*$ which controls the relation between the exponents e_A and d_A).

Remark As soon as you have found and attributed e_A and d_A, you are allowed to encrypt and to decrypt in *all of* $\mathbb{Z}/n_A\mathbb{Z}$: We will have $(a^{e_A})^{d_A} \equiv a \bmod n_A$ for *every* a, invertible for the multiplication modulo n_A or not. This is an immediate consequence of the Chinese Remainder Theorem (Exercise).

We note: We succeeded in complicating the Pohlig–Hellman situation. But what are now our benefits?

The answer:

The advantage of the new situation is the possibility of making *public* the encryption keys whilst keeping *secret* the decryption keys.

The Functioning of the Cipher RSA

Every user A of the cryptosystem RSA enciphers and deciphers in its *individual arithmetic*, i.e. it treats its plaintext (ciphertext) blocks as remainders of division modulo n_A, where $n_A = p_A q_A$ is a product of two big distinct prime numbers.

The exponents of encryption e_A and of decryption d_A are multiplicative inverses modulo $\varphi(n_A) = (p_A - 1)(q_A - 1)$.

The encryption keys (e_A, n_A) are *public* (in some directory).

The decryption keys (d_A, n_A) are *private* (secret).

Problem How is it possible *not* to reveal the decryption exponents, while rendering public the encryption keys?

Let us try to compute d_A in function of (e_A, n_A).

We know: d_A is the solution (in X) of the linear equation $e_A X + \varphi(n_A) Y = 1$. d_A is calculable *under the assumption that* $\varphi(n_A)$ will be calculable in function of n_A. But this is equivalent to finding the two prime factors p_A and q_A of n_A.

But the factorization algorithms for big integers are slow.

If the size of n_A is sufficiently large, it is practically impossible to find (by algorithm) the two prime factors p_A and q_A.

The historical challenge came from the authors of the system RSA themselves, who proposed, in 1978, to all of the planet earth the factorization of an integer of 129 decimal positions (the integer RSA-129). Seventeen years later, in 1994, Atkins, Graff, Lenstra and Leyland succeeded, with the aid of 1,600 computers connected via Internet, in finding the two prime factors (of 64 and 65 decimal positions).

Today the size of 1,024 binary positions (for the n_A) is considered to be sufficient to guarantee the security of the system RSA. Note that the key attribution demands the generation of big prime numbers p_A and q_A, of 512 binary positions each. This has created an algorithmic brotherhood of pseudorandom generators (in order to produce the candidates) and of good primality tests (in order to choose the winners). We shall treat this subject in a moment.

Attention • *Conceptually, the keys of the system RSA are attributed for a relatively long period (cf the RSA-129 challenge above).*

• *Practically, there is an important exception: When using the algorithm RSA for digital signatures (in the standard rDSA), the encryption exponent e may even be shared by an entire group of users, whereas the private keys (d_A, n_A) will change with every new signature.*

Finally, let us point out that RSA is slow. For example: DES is at least 1,000 times faster than RSA.

As a consequence, RSA will mainly be used for "little confidences" – like, for example, the transmission of the cipher key of a huge document encrypted with AES.

RSA as a Public Key System

We shall denote by $E_X(\cdot)$ and $D_X(\cdot)$ the encryption and decryption transformations of the user X.

Personal Envelopes

The user A would like to send a (secret) message M to the user B.

She encrypts $M \longmapsto C = E_B(M)$ with the *public* encryption transformation of the receiver B (she shuts the envelope for B). Only B can open: He recovers the plaintext using his private transformation: $C \longmapsto M = D_B(C)$. The confidentiality is thus guaranteed.

Signatures Inside Personal Envelopes

In a first approach, we shall adopt a simplifying and somehow academic viewpoint, which reflects nevertheless exactly how the functioning of RSA has been perceived during the first 15 years of its existence.

The sender A reserves a particular patch P of her message M for the signature (P would be, nowadays, the "message digest" of the full plaintext – we shall treat the standardized digital signatures in Sect. 2.4). The receiver B should then be able to verify whether the integrity has been respected. How shall we proceed?

The logic of the affair will not change if we suppose that $P = M$.

(i) *A* *signs* in treating M as a ciphertext: she computes $D_A(M)$ with her private key,
(ii) Then she sends $C = E_B(D_A(M))$ to the receiver B.
(iii) B opens the envelope, in decrypting with his private key; he obtains $D_B(E_B(D_A(M))) = D_A(M)$.
(iv) Finally, B verifies the signature of A: he applies the public transformation of the sender and gets correctly $M = E_A(D_A(M))$.

Attention   *The procedure described above only works for $n_A < n_B$! (why?) But in the opposite case (and this will be a public relation!) we may pass to an inversion: A will send to B the ciphertext $C' = D_A(E_B(M))$ and B will first apply E_A and then D_B.*

Example of a simple encryption:

$$p_A = 53, \quad q_A = 61, \quad n_A = 3233$$

$$\varphi(n_A) = 52 \cdot 60 = 3120.$$

$$e_A = 71 \quad d_A = 791.$$

$$M \equiv RENAISSANCE$$

Let us encode in decimal 4-digit blocks (of two letters) according to $A \equiv 00$,

$B \equiv 01$,

$\vdots$

$Z \equiv 25$,

$space \equiv 26$,

$$M \equiv RE\ NA\ IS\ SA\ NC\ E \quad \equiv 170413000818180013020426,$$

$(e_A, n_A) = (71, 3233),$

$1704^{71} \equiv 3106 \bmod 3233$ (Fast exponentiation!), etc.

We obtain finally: $C = E_A(M) = 310601000931269119842927.$

Exercises

(1) We consider a mini-RSA, with plaintext (and ciphertext) units consisting of decimal 4-digit blocks (i.e. of two letter blocks: $A \equiv 00, B \equiv 01, \ldots, Z \equiv 25$, *space* $\equiv 26$).
The user B receives the ciphertext $C \equiv 2433$ which is a *signed* message of the user A.
The public keys:

$$(e_A, n_A) = (53, 2923)$$
$$(e_B, n_B) = (233, 4087)$$

Decrypt the message.

(2) The same situation as above.
The user B receives the ciphertext $C \equiv 1441$ which is a *signed* message of the user A.

$$(e_A, n_A) = (11, 2867)$$

The public keys: $(e_B, n_B) = (689, 2987)$ Decrypt the message.

(3) Let $E : (\mathbb{Z}/n\mathbb{Z})^* \longrightarrow (\mathbb{Z}/n\mathbb{Z})^*, x \mapsto x^e$ be the operation of encryption in RSA.
Show: There exists $m \geq 1$ such that $E^m(x) = x$ for every $x \in (\mathbb{Z}/n\mathbb{Z})^*$. $E^m \equiv m$-times iteration of the encryption operation. The plaintext can (surprisingly) be recovered from the ciphertext by iterated *encryption*! (Help: Use the theorem of Euler, "on the right level".)

(4) Still another mini-RSA, with encryption and decryption in decimal 4-digit blocks (i.e. in two letter blocks: $A = 10, B = 11, C = 12, \ldots, Z = 35$, *espace* $= 99$). The user B receives the ciphertext $C \equiv QNSE$, which is a *signed* message of the user A.
The public keys: $(e_A, n_A) = (37, 4331)$, $(e_B, n_B) = (1493, 5561)$. Decrypt!

(5) The situation of the exercise (4).
The user B receives the ciphertext $C \equiv 0141.2685$, which is a *signed* message of the user A.
The public keys: $(e_A, n_A) = (37, 5183)$, $(e_B, n_B) = (2491, 5293)$. Decrypt!

(6) Once more the situation of exercise (4). The user B receives the ciphertext $C \equiv 2176.3509$, which is a *signed* message of the user A.
The public keys:
$(e_A, n_A) = (17, 4559)$, $(e_B, n_B) = (2961, 4717)$. Decrypt!

(7) Still the situation of exercise (4). The user B receives the ciphertext $C \equiv$ 3997.3141, which is a signed message of the user A. The public keys: $(e_A, n_A) = (67, 4307)$, $(e_B, n_B) = (3973, 4331)$. Decrypt!

2.4 Digital Signatures

The algorithms of signature are standardized[5] in Digital Signature Standard (DSS), the last version of which dates from January 2000. The standard is given in three variants: the *DSA*, which is the principal signature algorithm, the *rDSA*, which precises the use of the cipher RSA for digital signatures, and the *ECDSA*, which is the most fashionable signature algorithm, using triumphantly the arithmetic of elliptic curves. We shall present all three algorithms. But, to begin appropriately, we have first to speak of the "support" for the signatures, i.e. we have to speak of message digests.

2.4.1 Message Digests via SHA-1

The algorithm Secure Hash Algorithm (SHA-1) creates a message digest of 160 bits of a message (of a data file) of any length less than 2^{64} bits.

The *security* of the algorithm is understood as follows: it is computationally infeasible to create a message which will be transformed into a *given* digest. Similarly, is it computationally infeasible to find two *different* messages with the *same* message digest.

A modification of the message text during its transmission will give – with a very high probability – a new message digest that is *different* from the digest which has been extracted by the sender. If the signatory of the message and the verifier of the signature use both the message digest of the document as the "algorithmic support" of the signature, then the signature will be "attached to the document".

The algorithm SHA-1 is modeled after the MD4 (Message Digest Algorithm) of Ronald L. Rivest (1990).

SHA-1 became a standard in 1995.

Message Padding

Language: A *word* will be a binary string of 32 bits; we shall use hexadecimal notation almost exclusively:

1010 0001 0000 0011 1111 1110 0010 0011 =a103fe23.

A *block* will be a binary string of 512 bits, i.e. a block consists of 16 words. Now let M be the message, considered as a binary string of length l ($l < 2^{64}$). The purpose of message padding is to make the *total length* of a padded message a

---

[5] At least for the United States...

multiple of 512. The "blowing-up" of M into the padded message M^* – which will be a sequence of n *blocks*:

$M^* = M[1]M[2] \cdots M[n], \quad n \geq 1.$

The length l^* of M^* thus becomes a *multiple of 512*. The construction of M^* is simple: $M^* = M10 \cdots 0w_1w_2$. The (long) prefix is M, followed by the bit 1, then by a certain number of zeros, finally by 64 bits w_1w_2 which are the binary notation (over 64 bits) of the length l of M. The number of intermediate zeros that are appended will be minimal for the length of M^* to be a multiple of 512.

In computing the message digest of M, the algorithm SHA-1 will treat, one by one, the blocks $M[1]$, $M[2]$, ..., $M[n]$ of M^*.

Example $M = $01100001 01100010 01100011 01100100 01100101
the length is $l = 40$; this yields for M^*, written as a block of 16 words (in hexadecimal notation):
M^* = 61626364 65800000 00000000 00000000 00000000 00000000 00000000 00000000 00000000 00000000 00000000 00000000 00000000 00000000 00000000 00000028.

The Algorithm SHA-1, Global Structure

SHA-1 computes the message digest $m = SHA - 1(M)$ in function of $M^* = M[1]M[2] \cdots M[n]$. We shall have n *exterior rounds*, one for (the treatment of) each block of M^*. Every exterior round consists of *80 interior rounds*. The algorithm is (structurally) best understood if you look at it this way: The message digest $m = SHA - 1(M)$ is the ciphertext of a *fixed* plaintext

$$m_0 = 67452301efcdab8998badcfe10325476c3d2e1f0$$

(which is *the same* for every run of the algorithm). M enters into the algorithm via $M^* = M[1]M[2] \cdots M[n]$ which is a *cipher key vector*. Every $M[i]$ is blown up into an expanded key array (for the i-th exterior round) of 80 words $W_0W_1 \cdots W_{79}$ (recall the key schedule of AES–Rijndael). Each of these 80 words (of 32 bits) will serve as a round key for an interior round.

Scheme of an exterior round (treatment of a block $M[i]$):

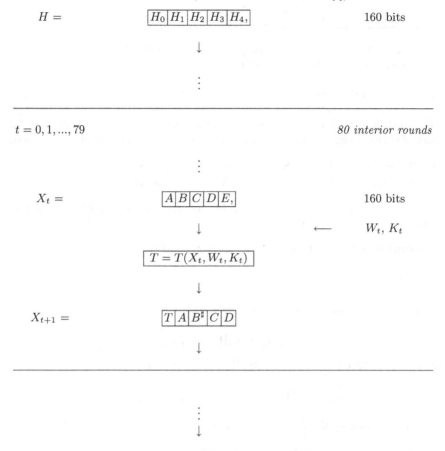

$H = $ $\boxed{H_0|H_1|H_2|H_3|H_4,}$ 160 bits

$\downarrow$

$\vdots$

$t = 0, 1, ..., 79$ *80 interior rounds*

$\vdots$

$X_t = $ $\boxed{A|B|C|D|E,}$ 160 bits

$\downarrow$ $\longleftarrow$ W_t, K_t

$\boxed{T = T(X_t, W_t, K_t)}$

$\downarrow$

$X_{t+1} = $ $\boxed{T|A|B^\sharp|C|D}$

$\downarrow$

$\vdots$

$\downarrow$

$H = $ $\boxed{H_0 + A|H_1 + B|H_2 + C|H_3 + D|H_4 + E}$ 160 bits

Let's sum up:

The algorithm SHA-1 computes the message digest $m = $ SHA-1(M) using the padded version $M^* = M[1]M[2] \cdots M[n]$ of M to transform a universal plaintext m_0 (160 bits) into the ciphertext m (160 bits).

The $M[i]$ serve as keys for "exterior rounds". A key schedule generates, for each $M[i]$, 80 round keys $W_0 W_1 \cdots W_{79}$ for 80 "interior rounds".

We remark the presence of two five-word buffers:

(i) $H_0 H_1 H_2 H_3 H_4$ which will be the *exterior round state array*.
(ii) $ABCDE$ which will be the *interior round state array*.

A buffer T of 32 bits is to be added; its content corresponds to $f(R_{i-1}, K_i)$ in DES. The content of T changes with every interior round.

The Algorithm SHA-1 in Detail

(1) Initialization of the Buffer H

$H_0 = 67452301,$ $H_2 = 98badcfe,$ $H_4 = c3d2e1f0,$
$H_1 = efcdab89,$ $H_3 = 10325476,$

(2) Two Operations on Words:

The algorithm SHA-1 uses several operations on (binary, length 32) words; most of them are rather familiar. Two of them should be, however, underlined:

(i) The cyclic left shift: $S(x_1 x_2 \cdots x_{32}) = x_2 x_3 \cdots x_{32} x_1$
(ii) The sum: $Z = X + Y \Longleftrightarrow | Z | \equiv | X | + | Y | \mod 2^{32}$

(where $| \ . \ |$ means "numerical value of the binary word". Example: $| c3d2e1f0 | = 3285377520$)

Sequential treatment of $M^* = M[1]M[2] \cdots M[n]$:

for $i = 0, 1, \ldots, n$:

(3) Expansion of $M[i]$ into $W_0 W_1 \cdots W_{79}$:

$$M[i] = W_0 W_1 \cdots W_{15}.$$

Recursive computation of the word W_t, $16 \leq t \leq 79$:

$$W_t = S(W_{t-3} \oplus W_{t-8} \oplus W_{t-14} \oplus W_{t-16}).$$

(4) Initialization of the Interior Round Buffer:

$A = H_0,$ $C = H_2,$ $E = H_4,$
$B = H_1,$ $D = H_3,$

(5) Processing of an Interior Round:

$t = 0, 1, \ldots, 79$:

$T = S^5(A) + f_t(B, C, D) + E + W_t + K_t,$
where

$$f_t(B,C,D) = \begin{cases} (B \wedge C) \vee ((\neg B) \wedge D) & 0 \leq t \leq 19, \\ B \oplus C \oplus D & 20 \leq t \leq 39, \\ (B \wedge C) \vee (B \wedge D) \vee (C \wedge D) & 40 \leq t \leq 59, \\ B \oplus C \oplus D & 60 \leq t \leq 79. \end{cases}$$

Attention *The Boolean operations on the words are executed bit by bit (i.e. 32 times in parallel).*

$$K_t = \begin{cases} \text{5a827999} & 0 \le t \le 19, \\ \\ \text{6ed9eba1} & 20 \le t \le 39, \\ \\ \text{8f1bbcdc} & 40 \le t \le 59, \\ \\ \text{ca62c1d6} & 60 \le t \le 79. \end{cases}$$

Then $A = T$, $B = A$, $C = S^{30}(B)$, $D = C$, $E = D$, (We note the formal similarity of an interior round with a round of a "5-word generalized DES").

(6) End of the Exterior Round:

$$\begin{aligned} H_0 &= H_0 + A, \\ H_1 &= H_1 + B, \\ H_2 &= H_2 + C, \\ H_3 &= H_3 + D, \\ H_4 &= H_4 + E. \end{aligned}$$

Example Consider the text $M =$ "abcdbcdecdefdefgefghfghighijhijkijkljklm-klmnlmnomnopnopq" in ASCII.

This text of 56 8-bit characters has length $l = 448$. The hexadecimal representation of 448 (over 64 bits) is 00000000 000001c0. We will have $M^* = M[1]M[2] = M1\underbrace{00\cdots0}_{511\,\text{bits}}00000000\ 000001c0$.

Let us begin with the first exterior round. The words of the first block are

$W[0] = $ 61626364,
$W[1] = $ 62636465,
$W[2] = $ 63646466,
$W[3] = $ 64656667,
$W[4] = $ 65666768,
$W[5] = $ 66676869,
$W[6] = $ 6768696a,
$W[7] = $ 68696a6b,
$W[8] = $ 696a6b6c,
$W[9] = $ 6a6b6c6d,
$W[10] = $ 6b6c6d6e,
$W[11] = $ 6c6d6e6f,
$W[12] = $ 6d6e6f70,
$W[13] = $ 6e6f7071,

$W[14] = 80000000,$
$W[15] = 00000000,$

The values for A, B, C, D, E after the tth interior round are the following:

	A	B	C	D	E
$t = 0$:	0116fc17	67452301	7bf36ae2	98badcfe	10325476
$t = 1$:	ebf3b452	0116fc17	59d148c0	7bf36ae2	98badcfe
$t = 2$:	5109913a	ebf3b452	c045bf05	59d148c0	7bf36ae2
$t = 3$:	2c4f6eac	5109913a	bafced14	c045bf05	59d148c0
$t = 4$:	33f4ae5b	2c4f6eac	9442644e	bafced14	c045bf05
$t = 5$:	96b85189	33f4ae5b	0b13dbab	9442644e	bafced14
$t = 6$:	db04cb58	96b85189	ccfd2b96	0b13dbab	9442644e
$t = 7$:	45833f0f	db04cb58	65ae1462	ccfd2b96	0b13dbab
$t = 8$:	c565c35e	45833f0f	36c132d6	65ae1462	ccfd2b96
$t = 9$:	6350afda	c565c35e	d160cfc3	36c132d6	65ae1462
$t = 10$:	8993ea77	6350afda	b15970d7	d160cfc3	36c132d6
$t = 11$:	e19ecaa2	8993ea77	98d42bf6	b15970d7	d160cfc3
$t = 12$:	8603481e	e19ecaa2	e264fa9d	98d42bf6	b15970d7
$t = 13$:	32f94a85	8603481e	b867b2a8	e264fa9d	98d42bf6
$t = 14$:	b2e7a8be	32f94a85	a180d207	b867b2a8	e264fa9d
$t = 15$:	42637e39	b2e7a8be	4cbe52a1	a180d207	b867b2a8
$t = 16$:	6b068048	42637e39	acb9ea2f	4cbe52a1	a180d207
$t = 17$:	426b9c35	6b068048	5098df8e	acb9ea2f	4cbe52a1
$t = 18$:	944b1bd1	426b9c35	1ac1a012	5098df8e	acb9ea2f
$t = 19$:	6c445652	944b1bd1	509ae70d	1ac1a012	5098df8e
$t = 20$:	95836da5	6c445652	6512c6f4	509ae70d	1ac1a012
$t = 21$:	09511177	95836da5	9b111594	6512c6f4	509ae70d
$t = 22$:	e2b92dc4	09511177	6560db69	9b111594	6512c6f4
$t = 23$:	fd224575	e2b92dc4	c254445d	6560db69	9b111594
$t = 24$:	eeb82d9a	fd224575	38ae4b71	c254445d	6560db69
$t = 25$:	5a142c1a	eeb82d9a	7f48915d	38ae4b71	c254445d
$t = 26$:	2972f7c7	5a142c1a	bbae0b66	7f48915d	38ae4b71
$t = 27$:	d526a644	2972f7c7	96850b06	bbae0b66	7f48915d
$t = 28$:	e1122421	d526a644	ca5cbdf1	96850b06	bbae0b66
$t = 29$:	05b457b2	e1122421	3549a991	ca5cbdf1	bbae0b66
$t = 30$:	a9c84bec	05b457b2	78448908	3549a991	ca5cbdf1
$t = 31$:	52e31f60	a9c84bec	816d15ec	78448908	3549a991
$t = 32$:	5af3242c	52e31f60	2a7212fb	816d15ec	78448908
$t = 33$:	31c756a9	5af3242c	14b8c7d8	2a7212fb	816d15ec
$t = 34$:	e9ac987c	31c756a9	16bcc90b	14b8c7d8	2a7212fb
$t = 35$:	ab7c32ee	e9ac987c	4c71d5aa	16bcc90b	14b8c7d8
$t = 36$:	5933fc99	ab7c32ee	3a6b261f	4c71d5aa	16bcc90b
$t = 37$:	43f87ae9	5933fc99	aadf0cbb	3a6b261f	4c71d5aa
$t = 38$:	24957f22	43f87ae9	564cff26	aadf0cbb	3a6b261f

$t = 39$: adeb7478   24957f22   50fe1eba   564cff26   aadf0cbb
$t = 40$: d70e5010   adeb7478   89255fc8   50fe1eba   564cff26
$t = 41$: 79bcfb08   d70e5010   2b7add1e   89255fc8   50fe1eba
$t = 42$: f9bcb8de   79bcfb08   35c39404   2b7add1e   89255fc8
$t = 43$: 633e9561   f9bcb8de   1e6f3ec2   35c39404   2b7add1e
$t = 44$: 98c1ea64   633e9561   be6f2e37   1e6f3ec2   35c39404
$t = 45$: c6ea241e   98c1ea64   58cfa558   be6f2e37   1e6f3ec2
$t = 46$: a2ad4f02   c6ea241e   26307a99   58cfa558   be6f2e37
$t = 47$: c8a69090   a2ad4f02   b1ba8907   26307a99   58cfa558
$t = 48$: 88341600   c8a69090   a8ab53c0   b1ba8907   26307a99
$t = 49$: 7e846f58   88341600   3229a424   a8ab53c0   b1ba8907
$t = 50$: 86e358ba   7e846f58   220d0580   3229a424   a8ab53c0
$t = 51$: 8d2e76c8   86e358ba   1fa11bd6   220d0580   3229a424
$t = 52$: ce892e10   8d2e76c8   a1b8d62e   1fa11bd6   220d0580
$t = 53$: edea95b1   ce892e10   234b9db2   a1b8d62e   1fa11bd6
$t = 54$: 36d1230a   edea95b1   33a24b84   234b9db2   a1b8d62e
$t = 55$: 776c3910   36d1230a   7b7aa56c   33a24b84   234b9db2
$t = 56$: a681b723   776c3910   8db448c2   7b7aa56c   33a24b84
$t = 57$: ac0a794f   a681b723   1ddb0e44   8db448c2   7b7aa56c
$t = 58$: f03d3782   ac0a794f   e9a06dc8   1ddb0e44   8db448c2
$t = 59$: 9ef775c3   f03d3782   eb029e53   e9a06dc8   1ddb0e44
$t = 60$: 36254b13   9ef775c3   bc0f4de0   eb029e53   e9a06dc8
$t = 61$: 4080d4dc   36254b13   e7bddd70   bc0f4de0   eb029e53
$t = 62$: 2bfaf7a8   4080d4dc   cd8952c4   e7bddd70   bc0f4de0
$t = 63$: 513f9ca0   2bfaf7a8   10203537   cd8952c4   e7bddd70
$t = 64$: e5895c81   513f9ca0   0afebdea   10203537   cd8952c4
$t = 65$: 1037d2d5   e5895c81   144fe728   0afebdea   10203537
$t = 66$: 14a82da9   1037d2d5   79625720   144fe728   0afebdea
$t = 67$: 6d17c9fd   14a82da9   440df4b5   79625720   144fe728
$t = 68$: 2c7b07bd   6d17c9fd   452a0b6a   440df4b5   79625720
$t = 69$:   fdf6efff   2c7b07bd   5b45f27f   452a0b6a   440df4b5
$t = 70$: 112b96e3   fdf6efff   4b1ec1ef   5b45f27f   452a0b6a
$t = 71$: 84065712   112b96e3   ff7dbbff   4b1ec1ef   5b45f27f
$t = 72$: ab89fb71   84065712   c44ae5b8   ff7dbbff   4b1ec1ef
$t = 73$: c5210e35   ab89fb71   a10195c4   c44ae5b8   ff7dbbff
$t = 74$: 352d9f4b   c5210e35   6ae27edc   a10195c4   c44ae5b8
$t = 75$: 1a0e0e0a   352d9f4b   7148438d   6ae27edc   a10195c4
$t = 76$: d0d47349   1a0e0e0a   cd4b67d2   7148438d   6ae27edc
$t = 77$: ad38620d   d0d47349   86838382   cd4b67d2   7148438d
$t = 78$: d3ad7c25   ad38620d   74351cd2   86838382   cd4b67d2
$t = 79$: 8ce34517   d3ad7c25   6b4e1883   74351cd2   86838382

We just finished the processing of $M[1]$. The exterior round state array becomes

$$H_0 = 67452301 + 8ce34517 = f4286818,$$
$$H_1 = efcdab89 + d3ad7c25 = c37b27ae,$$
$$H_2 = 98badcfe + 6b4e1883 = 0408f581,$$
$$H_3 = 10325476 + 74351cd2 = 84677148,$$
$$H_4 = c3d2e1f0 + 86838382 = 4a566572.$$

Let us attack now the second exterior round. The words of the block $M[2]$ are

$$W[0] = 00000000,$$
$$W[1] = 00000000,$$
$$W[2] = 00000000,$$
$$W[3] = 00000000,$$
$$W[4] = 00000000,$$
$$W[5] = 00000000,$$
$$W[6] = 00000000,$$
$$W[7] = 00000000,$$
$$W[8] = 00000000,$$
$$W[9] = 00000000,$$
$$W[10] = 00000000,$$
$$W[11] = 00000000,$$
$$W[12] = 00000000,$$
$$W[13] = 00000000,$$
$$W[14] = 00000000,$$
$$W[15] = 000001c0,$$

The values for A, B, C, D, E after the t-th interior round are now the following:

	A	B	C	D	E
$t = 0$:	2df257e9	f4286818	b0dec9eb	0408f581	84677148
$t = 1$:	4d3dc58f	2df257e9	3d0a1a06	b0dec9eb	0408f581
$t = 2$:	c352bb05	4d3dc58f	4b7c95fa	3d0a1a06	b0dec9eb
$t = 3$:	eef743c6	c352bb05	d34f7163	4b7c95fa	3d0a1a06
$t = 4$:	41e34277	eef743c6	70d4aec1	d34f7163	4b7c95fa
$t = 5$:	5443915c	41e34277	bbbdd0f1	70d4aec1	d34f7163
$t = 6$:	e7fa0377	5443915c	d078d09d	bbbdd0f1	70d4aec1
$t = 7$:	c6946813	e7fa0377	1510e457	d078d09d	bbbdd0f1
$t = 8$:	fdde1de1	c6946813	f9fe80dd	1510e457	d078d09d
$t = 9$:	b8538aca	fdde1de1	f1a51a04	f9fe80dd	1510e457

$t = 10:$	6ba94f63	b8538aca	7f778778	f1a51a04	f9fe80dd
$t = 11:$	43a2792f	6ba94f63	ae14e2b2	7f778778	f1a51a04
$t = 12:$	fecd7bbf	43a2792f	daea53d8	ae14e2b2	7f778778
$t = 13:$	a2604ca8	fecd7bbf	d0e89e4b	daea53d8	ae14e2b2
$t = 14:$	258b0baa	a2604ca8	ffb35eef	d0e89e4b	daea53d8
$t = 15:$	d9772360	258b0baa	2898132a	ffb35eef	d0e89e4b
$t = 16:$	5507db6e	d9772360	8962c2ea	2898132a	ffb35eef
$t = 17:$	a51b58bc	5507db6e	365dc8d8	8962c2ea	2898132a
$t = 18:$	c2eb709f	a51b58bc	9541f6db	365dc8d8	8962c2ea
$t = 19:$	d8992153	c2eb709f	2946d62f	9541f6db	365dc8d8
$t = 20:$	37482f5f	d8992153	f0badc27	2946d62f	9541f6db
$t = 21:$	ee8700bd	37482f5f	f6264854	f0badc27	2946d62f
$t = 22:$	9ad594b9	ee8700bd	cdd20bd7	f6264854	f0badc27
$t = 23:$	8fbaa5b9	9ad594b9	7ba1c02f	cdd20bd7	f6264854
$t = 24:$	88fb5867	8fbaa5b9	66b5652e	7ba1c02f	cdd20bd7
$t = 25:$	eec50521	88fb5867	63eea96e	66b5652e	7ba1c02f
$t = 26:$	50bce434	eec50521	e23ed619	63eea96e	66b5652e
$t = 27:$	5c416daf	50bce434	7bb14148	e23ed619	63eea96e
$t = 28:$	2429be5f	5c416daf	142f390d	7bb14148	e23ed619
$t = 29:$	0a2fb108	2429be5f	d7105b6b	142f390d	7bb14148
$t = 30:$	17986223	0a2fb108	c90a6f97	d7105b6b	142f390d
$t = 31:$	8a4af384	17986223	028bec42	c90a6f97	d7105b6b
$t = 32:$	6b629993	8a4af384	c5e61888	028bec42	c90a6f97
$t = 33:$	f15f04f3	6b629993	2292bce1	c5e61888	028bec42
$t = 34:$	295cc25b	f15f04f3	dad8a664	2292bce1	c5e61888
$t = 35:$	696da404	295cc25b	fc57c13c	dad8a664	2292bce1
$t = 36:$	cef5ae12	696da404	ca573096	fc57c13c	dad8a664
$t = 37:$	87d5b80c	cef5ae12	1a5b6901	ca573096	fc57c13c
$t = 38:$	84e2a5f2	87d5b80c	b3bd6b84	1a5b6901	ca573096
$t = 39:$	03bb6310	84e2a5f2	21f56e03	b3bd6b84	1a5b6901
$t = 40:$	c2d8f75f	03bb6310	a138a97c	21f56e03	b3bd6b84
$t = 41:$	bfb25768	c2d8f75f	00eed8c4	a138a97c	21f56e03
$t = 42:$	28589152	bfb25768	f0b63dd7	00eed8c4	a138a97c
$t = 43:$	ec1d3d61	28589152	2fec95da	f0b63dd7	00eed8c4
$t = 44:$	3caed7af	ec1d3d61	8a162454	2fec95da	f0b63dd7
$t = 45:$	c3d033ea	3caed7af	7b074f58	8a162454	2fec95da
$t = 46:$	7316056a	c3d033ea	cf2bb5eb	7b074f58	8a162454
$t = 47:$	46f93b68	7316056a	b0f40cfa	cf2bb5eb	7b074f58
$t = 48:$	dc8e7f26	46f93b68	9cc5815a	b0f40cfa	cf2bb5eb
$t = 49:$	850d411c	dc8e7f26	11be4eda	9cc5815a	b0f40cfa

$t = 50$:   7e4672c0   850d411c   b7239fc9   11be4eda   9cc5815a
$t = 51$:   89fbd41d   7e4672c0   21435047   b7239fc9   11be4eda
$t = 52$:   1797e228   89fbd41d   1f919cb0   21435047   b7239fc9
$t = 53$:   431d65bc   1797e228   627ef507   1f919cb0   21435047
$t = 54$:   2bdbb8cb   431d65bc   05e5f88a   627ef507   1f919cb0
$t = 55$:   6da72e7f   2bdbb8cb   10c7596f   05e5f88a   627ef507
$t = 56$:   a8495a9b   6da72e7f   caf6ee32   10c7596f   05e5f88a
$t = 57$:   e785655a   a8495a9b   db69cb9f   caf6ee32   10c7596f
$t = 58$:   5b086c42   e785655a   ea1256a6   db69cb9f   caf6ee32
$t = 59$:   a65818f7   5b086c42   b9e15956   ea1256a6   db69cb9f
$t = 60$:   7aab101b   a65818f7   96c21b10   b9e15956   ea1256a6
$t = 61$:   93614c9c   7aab101b   e996063d   96c21b10   b9e15956
$t = 62$:   f66d9bf4   93614c9c   deaac406   e996063d   96c21b10
$t = 63$:   d504902b   f66d9bf4   24d85327   deaac406   e996063d
$t = 64$:   60a9da62   d504902b   3d9b66fd   24d85327   deaac406
$t = 65$:   8b687819   60a9da62   f541240a   3d9b66fd   24d85327
$t = 66$:   083e90c3   8b687819   982a7698   f541240a   3d9b66fd
$t = 67$:   f6226bbf   083e90c3   62da1e06   982a7698   f541240a
$t = 68$:   76c0563b   f6226bbf   c20fa430   62da1e06   982a7698
$t = 69$:   989dd165   76c0563b   fd889aef   c20fa430   62da1e06
$t = 70$:   8b2c7573   989dd165   ddb0158e   fd889aef   c20fa430
$t = 71$:   ae1b8e7b   8b2c7573   66277459   ddb0158e   fd889aef
$t = 72$:   ca1840de   ae1b8e7b   e2cb1d5c   66277459   ddb0158e
$t = 73$:   16f3babb   ca1840de   eb86e39e   e2cb1d5c   66277459
$t = 74$:   d28d83ad   16f3babb   b2861037   eb86e39e   e2cb1d5c
$t = 75$:   6bc02dfe   d28d83ad   c5bceeae   b2861037   eb86e39e
$t = 76$:   d3a6e275   6bc02dfe   74a360eb   c5bceeae   b2861037
$t = 77$:   da955482   d3a6e275   9af00b7f   74a360eb   c5bceeae
$t = 78$:   58c0aac0   da955482   74e9b89d   9af00b7f   74a360eb
$t = 79$:   906fd62c   58c0aac0   b6a55520   74e9b89d   9af00b7f

This ends the treatment of $M[2]$. The (final) state of the buffer H after the second round will be

$H_0 = $ f4286818 $+$ 906fd62c $=$ 84983e44,
$H_1 = $ c37b27ae $+$ 58c0aac0 $=$ 1c3bd26e,
$H_2 = $ 0408f581 $+$ b6a55520 $=$ baae4aa1,
$H_3 = $ 84677148 $+$ 74e9b89d $=$ f95129e5,
$H_4 = $ 4a566572 $+$ 9af00b7f $=$ e54670f1.

We sum up:

If $M = $ "abcdbcdecdefdefgefghfghighijhijkijkljklmklmnlmnomnopnopq" in ASCII, then SHA-1(M)=84983e44 1c3bd26e baae4aa1 f95129e5 e54670f1 (in hex).

Exercises

(1) *Recall*: The Boolean operations on words are executed bitwise – i.e. 32 times in parallel.

    (a) Compute $f(B,C,D) = (B \wedge C) \vee ((\neg B) \wedge D)$ for B = 69, C = f2, D = 57.

    (b) Compute $g(B,C,D) = (B \wedge C) \vee (B \wedge D) \vee (C \wedge D)$ for $B = $ a3, $C = $ 02, $D = $ fe.

(2) Write S^5, the 5-position cyclic left shift on 32-bit words, by means of arithmetical operations (on the integer values of the words) and of Boolean operations (position by position).

(3) Consider the following hash algorithm μ-sha which is an academic pocket version of SHA-1:

The *words* will have four bits (instead of 32 bits); the *blocks* will have 16 bits (instead of 512 bits).

We shall treat messages M of length $l < 2^{16}$.

$M^* = M[1]M[2] \cdots M[n]$ is obtained as in SHA-1; here, the last block $M[n]$ *is* the length of M (in binary notation, over 16 bits).

The message digest $m = $ μ-sha(M) will have 20 bits (5 words).

We shall have exterior rounds for the successive treatment of the $M[i]$, and interior rounds which carry out this treatment. Every exterior round will be composed of four interior rounds ($t = 0, 1, 2, 3$).

The algorithm imitates SHA-1, with the following modifications:

(i) Initialization: $H_0 = 9$, $H_1 = 5$, $H_2 = a$, $H_3 = c$, $H_4 = 3$.

(ii) $M[i] = W_0 W_1 W_2 W_3$ (16 bits = 4 words)

    There are no other words to compute.

(iii) The logical functions:

$$f_t(B,C,D) = \begin{cases} (B \wedge C) \vee ((\neg B) \wedge D), & t = 0, \\ B \oplus C \oplus D, & t = 1, \\ (B \wedge C) \vee (B \wedge D) \vee (C \wedge D), & t = 2, \\ B \oplus C \oplus D, & t = 3, \end{cases}$$

(iv) The interior round constants:

$$K_0 = 2, \qquad K_1 = 6, \qquad K_2 = 10, \qquad K_3 = 14.$$

(v) The interior round transformation is defined as in SHA-1; note only that $S^5 = S$ and that $S^{30} = S^2$.

    (a) Let $M = $ bebad (in hex). Compute $m = $ μ-sha(M).

    (b) Let $M = $ faced (in hex). Compute $m = $ μ-sha(M).

(4) Back to the algorithm SHA-1. Consider $M = $ "abc" in ASCII:
$M = 011000010110001001100011 = 616263$.

(a) Write down $M^* = M[1] = W_0 W_1 \cdots W_{15}$.
(b) Compute the content of the buffer ABCDE after two interior rounds $(t = 0, 1)$.

(5) In the situation of exercise (4), introduce an error in the leading position:
$$M = 616263 \quad \longmapsto \quad M' = \text{e}16263.$$

What are the binary positions of the buffer ABCDE which are affected by this error after four interior rounds (i.e. after $t = 3$)?

2.4.2 DSA: Digital Signature Algorithm

Generalities

DSA is the principal standard for digital signatures. It is a kind of public key cryptosystem on message digests. Every user will dispose of a private (secret) multi-key and a public key. The private keys have a constant component (identification of the user) and a variable component (identification of the act of signature), which changes with every new signature. The public keys are constant (during a certain period) and may even be shared by some group of users. The private keys act (as parameters) on the signature algorithm; the public keys act on the verification algorithm (of the signatures). The signature generation (for a message M) consists in computing a "ciphertext" which is a couple of 160-bit words.

The first component of the signature is "the (secret) identifier of the act of signature, in an envelope". The second component of the signature is "the signature attached to the document"; more precisely, it is a ciphertext computed from the *message digest* of the document M, which will be masked by means of the action of the private key of the signatory (i.e. of the signatory's identifier as well as the identifier of the act of signature).

The verification process is an *implicit decryption*: One verifies if the first component of the signature is correctly an (implicit) function of the second component.

The scheme of functioning for DSA

Signature Generation	Signature Verification
Message ⇓ SHA-1 ⇓ Message Digest ⇓	Received Message ⇓ SHA-1 ⇓ Message Digest ⇓

Private ⇒ Key → DSA Sign Operation → Digital ⇒ Signature | Digital ⇒ Signature → DSA Signature Verification → Public ⇐ Key

⇓

Yes – Signature Verified
No – Verification Failed

The Functioning of DSA

The Individual Parameters.

The private keys and the public keys of a user A stem from the following parameters:

(i) p: a prime number such that $2^{1023} < p < 2^{1024}$
(ii) q: a prime number dividing $p - 1$, with $2^{159} < q < 2^{160}$
(iii) g: an integer $1 < g < p$ such that $g^q \equiv 1 \bmod p$

• p, q, g are public and are often common to a group of users   •

(iv) $x = x_A$: $1 < x < q$, the private key of the user A
     $y = y_A = g^x \bmod p$: the public key of the user A
(v) $k = k_A$: $1 < k < q$, the private key of the user A, designed to mark the act of signature

• $x = x_A$ and $y = y_A$ are in general kept fixed for a period of time. $k = k_A$ must be *regenerated* for each signature   •

The algorithms of (pseudorandom) generation of the values p, q, x and k will be presented in a moment.

Remark

(1) How is it possible to publish $y = g^x \bmod p$ whilst keeping secret the value x? We note that we are confronted with the problem of *computing the discrete logarithm* (to the base g) in $\mathbb{Z}/p\mathbb{Z}$. The situation is similar to the problem of factoring big integers in RSA: the algorithms which

compute discrete logarithms have the same slowness as those which factor. This explains the *size* of p (familiar to us by the discussions on RSA) which has to guarantee the *algorithmic infeasibility* of finding (in a reasonable time) x from $y = g^x \bmod p$.

(2) Let's insist: The security of DSA (as a public key system) depends on the choice of the size of p. The appearance of q – which is of sensibly smaller size – comes from the needs of economy (in space and in time).[6] We shall see in the sequel that a lot of computations (and of storage) will be thus restricted to 160-bit format (instead of 1024-bit format).

The Signature Generation

The user A would like to sign a message (a document) M (of length $l < 2^{64}$ bits).

- A first generates the private key of the act of signature $k = k_A$.
- A computes $m = \text{SHA} - 1(M)$, the message digest of M.
- Then it computes the signature[7] $\sigma = (r, s)$, where $r = (g^k \bmod p) \bmod q$, $s = (k^{-1}(m + xr)) \bmod q$ [8]
- If $r = 0$ or if $s = 0$, then A has to generate a new value for $k = k_A$.

The Signature Verification

For a thorough understanding of the authentication procedure, we have to look first a little bit closer at the mathematical structure of the signature $\sigma = (r, s)$. Actually, there is a *implicit function* relation

$$\boxed{r = F(r, s, m)}$$

which does not depend on the knowledge of the private parameters (x, k) of the signatory A.

Encouraging *observation*: Compute simply in $\mathbb{R}$:

$$r = g^k, \quad s = \frac{m + xr}{k}, \quad y = g^x.$$

---

[6] To the intuitive reader: The complexity of the situation depends altogether on p. Compare with the geometry of a plane defined by 26 linear equations in $\mathbb{R}^{28}$.

[7] There is no permanent place in the world for ugly mathematics. – G.H. Hardy.

[8] We identify the binary word $m = m_1 m_2 \cdots m_{160}$ with the integer $m_1 2^{159} + m_2 2^{158} + \cdots + m_{160}$; k^{-1} will mean the integer u: $1 < u < q$, such that $ku \equiv 1 \bmod q$.

Put

$$u_1 = \frac{m}{s} = \frac{km}{m+xr},$$
$$u_2 = \frac{r}{s} = \frac{kr}{m+xr}.$$

Then: $g^{u_1} y^{u_2} = g^{\frac{km}{m+xr}} \cdot g^{\frac{kxr}{m+xr}} = g^k = r$. This observation indicates the right way for the signature validation procedure. In fact:

Proposition *Let $\sigma = (r, s)$ be the signature of the message M by the user A, and let $m = SHA-1(M)$. Let us put*

$w = s^{-1} \bmod q$

$u_1 = mw \bmod q$

$u_2 = rw \bmod q$

$v = (g^{u_1} y^{u_2} \bmod p) \bmod q$

Then, $v = r$.

Proof We observe first that $g^{e_1} \equiv g^{e_2} \bmod p \iff e_1 \equiv e_2 \bmod q$.

This is an immediate consequence of the fact that the order of g modulo p is precisely equal to q. The remainder is easy: $g^{u_1} y^{u_2} \bmod p = g^{(m+xr)w} \bmod p$. But: $(m + xr)w \equiv ksw \equiv k \bmod q$ i.e. $g^{(m+xr)w} \equiv g^k \bmod p$. This yields the claim. $\square$

We get *the procedure of signature verification*:

Let M be the message (the document) signed by the user A, $m = $ SHA-1(M), and let $\sigma = (r, s)$ be the signature. Let M' be the received message (the received document), and let $\sigma' = (r', s')$ be the transmitted signature. Finally, let (y, p, q) be the *public key of A*.

One computes

- $m' = SHA-1(M')$,
- $w = (s')^{-1} \bmod q$,
- $u_1 = m'w \bmod q$,
- $u_2 = r'w \bmod q$,
- $v = (g^{u_1} y^{u_2} \bmod p) \bmod q$.

If $v = r'$, then the signature is verified. If $v \neq r'$, then the message is considered to be invalid.

Exercises

(1) How can g be found (an element of order q in $(\mathbb{Z}/p\mathbb{Z})^*$)?
   Suppose that p and q are already at hand, and let $e = \frac{p-1}{q}$.
   Show that the following algorithm will find an appropriate value for g.

   > a. Let h: $1 < h < p - 1$ be the next test value.
   > b. Compute $g = h^e \bmod p$
   > c. If $g = 1$, return to a. ; otherwise, g will do.

(2) A class room version of DSA.

We will choose $q = 101$, $p = 78q + 1 = 7879$.

(a) Show that 3 is a generator of $(\mathbb{Z}/7879\mathbb{Z})^*$.

(b) We may take $g = 3^{78}$ mod $7879 = 170$. Let us choose $x = 75$. Suppose that the message digest of M is $m = 1234$. Our signature key will be $k = 50$. Compute the signature $\sigma = (r, s)$.

(c) Verify the signature.

(3) The situation of the preceding exercise. Now, consider $m = 5,001$ and $k = 49$. Compute, once more, the signature $\sigma = (r, s)$, and carry out the validation.

2.4.3 Auxiliary Algorithms for DSA

The algorithm DSA is a great consumer of big numbers – a lot of them prime numbers (like the public p and q), but a great portion not necessarily so (like the private x and k) – which have all to be supplied by pseudorandom generators, i.e. in simulating methodically a kind of "chance under control". We shall outline briefly the indispensable details of these algorithmic providers of big numbers, while respecting the logic of attribution: the public keys are – in this sense – primary, the private keys secondary.

The Miller–Rabin Primality Test

With the birth of the arithmetical cryptosystems, where the manipulation of big prime numbers became commonplace, the efforts of the mathematicians to provide a formidable palette of primality tests were plainly successful: The inheritance of the last centuries was too rich. The test of Miller–Rabin, which we shall present in a moment, is the little grey duck in a family of white swans: On the one hand, it is very easy to understand (modulo a statistical information which can be established via several exercises in elementary number theory), on the other hand, since it refers to a probabilistic argument, it will not be joyously accepted by the mathematical pencil-and-paper community. But ultimately, for a pragmatic and unscrupulous practitioner, it is one of the handiest available primality tests.

(1) The test $SPP(n, b) = \begin{cases} \text{yes,} \\ \text{no.} \end{cases}$

$[SPP(n, b) \equiv$ n is a strong pseudoprime for the base b. $]$

Let us fix $n = 2^a m + 1$, with $a \geq 1$, m odd. n will be the *candidate* – a (big) odd number that we would like to be prime. The integer b is chosen *randomly* between 1 and n: $1 < b < n$.

Then: $SPP(n, b) = $ yes $\iff$ either $b^m \equiv 1 \bmod n$ or $b^m \equiv -1 \bmod n$
or $b^{2m} \equiv -1 \bmod n$ or $b^{4m} \equiv -1 \bmod n$ ... or $b^{2^{a-1}m} \equiv -1 \bmod n$

---

(2) The justification of the test SPP:

If p is a *prime* number, and if b is an *arbitrary* integer between 1 and p, then we necessarily get: $SPP(p, b) = yes$.

The argument:

In the given situation, the (little) theorem of Fermat yields: $b^{p-1} \equiv 1 \bmod p$. Write $p - 1 = 2^a m$ with $a \geq 1$ and m odd.

$$\left[\textbf{\textit{Lemma:}} \text{ The polynomial } X^{p-1} - 1 \text{ admits the following factorization:} \right.$$
$$\left. X^{p-1} - 1 = (X^m - 1)(X^m + 1)(X^{2m} + 1)(X^{4m} + 1) \cdots (X^{2^{a-1}m} + 1). \right]$$

$$\left. \textit{Proof} \text{ Recursion on } a \geq 1. \right]$$

Thus, $b^{p-1} - 1 = (b^m - 1)(b^m + 1)(b^{2m} + 1)(b^{4m} + 1) \cdots (b^{2^{a-1}m} + 1)$.

p divides $b^{p-1} - 1$, hence p divides at least one of the $a + 1$ factors above: $b^m \equiv 1 \bmod n$ or $b^m \equiv -1 \bmod n$ or $b^{2m} \equiv -1 \bmod n$ or $b^{4m} \equiv -1 \bmod n$ ... or $b^{2^{a-1}m} \equiv -1 \bmod n$; i.e. $SPP(p, b) = yes$ for *every* b, $1 < b < p$.

(3) The fundamental result for the Miller–Rabin test:

If n is *composite*, then $SPP(n, b) = yes$ for *not more than* 25% of the b between 0 and n. In other words:

> The answer of the test is against its intent with a probability $\leq \frac{1}{4}$.

Let then n be our candidate (which we want to be prime). We shall randomly choose b between 1 and n. If $SPP(n, b) = yes$, then the probability of n *not* being prime is $\leq \frac{1}{4}$.

(4) *The Miller–Rabin test* (1980).

Let n (a positive odd integer) be the candidate of our primality test. We choose b randomly between 1 and n. If $SPP(n, b) = yes$, then n has passed the first round of the test: n is not prime with a probability $\leq \frac{1}{4}$. Let us choose, a second time, b randomly between 1 and n. If $SPP(n, b) = yes$, then n has passed the second round of the test: n is not prime with a probability $\leq \frac{1}{16}$ and so on. If n passes N rounds of the test SPP, then the probability of n not being prime is $\leq (\frac{1}{4})^N$. One fixes $N \geq 50$. If n passes N times the test SPP, one considers n to be a prime number.

Exercises

(1) For $N \geq 2$, let $G_N = \{a \in (\mathbb{Z}/N\mathbb{Z})^* : a^{N-1} = 1\}$.

    (a) Show that G_N is a subgroup of $(\mathbb{Z}/N\mathbb{Z})^*$.

    (b) Determine the order (the cardinality) of G_N in case $N = 1763$.

    (c) Find explicit representatives for G_{1763}.

(2) $N = 20051533153 = 100129 \cdot 200257$.

    Find two integers b, $1 < b < N$, such that $SPP(N, b) = yes$ and also two integers b, $1 < b < N$, such that $SPP(N, b) = no$.

Generation of the Prime Numbers p and q for DSA

Let us begin with some remarks on the density of the prime numbers (in $\mathbb{N}$).

Remark (1) There are 78,497 odd prime numbers up to 10^6. This gives a density of $\frac{1}{12,7}$. (Note that $\frac{1}{Ln10^6} = \frac{1}{6Ln10} = \frac{1}{13,8}$.)

(2) In general, we dispose of the following famous result (Hadamard, de la Vallée-Poussin, 1896):

Let $\pi(x)$ be the number of primes $\leq x$, and let

$$Li(x) = \int_2^x \frac{dt}{Ln\,t}$$

Then:

$$\lim_{x \to \infty} \frac{Li(x)}{\pi(x)} = 1.$$

(3) Consider a (big) integer $z \in \mathbb{N}$; then the density of the prime numbers of the size of z is approximately equal to $\frac{1}{Ln\,z}$. Example: For $z \approx 10^{100}$, the corresponding density will be $\frac{1}{230}$. This means that we must – in average – look at 115 consecutive odd candidates in order to find a prime number of 100 decimal positions.

Back to DSA.

We have to generate – as public parameters of a user or of a group of users – two prime numbers p and q such that

(1) $2^{159} < q < 2^{160}$,
(2) $2^{1023} < p < 2^{1024}$,
(3) q divides $p - 1$.

– (1)The generation of q is done by SHA-1 (perverted into an auxiliary algorithm), using the numerical content of a buffer Σ (for SEED), defined by the user.
– (2) and (3) The same value of Σ (as that used in (1)) will serve to compute some "prototype" X of p: $2^{1023} < X < 2^{1024}$.

Then, we have to "trim" X in order to obtain the prime number $p \equiv 1 \bmod 2q$.

Recall   As always until now, we shall make no notational distinction between a binary word $x_1 x_2 \cdots x_\sigma$ of length σ and the integer $x = x_1 2^{\sigma-1} + x_2 2^{\sigma-2} + \cdots + x_\sigma$ $(0 \leq x < 2^\sigma)$ which admits the binary word as its notation.

We insist: The lecture is like the usual decimal lecture; the most significant bit is to the left, the least significant bit is to the right.

(1) The small prime number q:
   (1a) The user initializes Σ ($\sigma = $ the length (of the content) of Σ; $\sigma \geq 160$).
   (1b) $U = $ SHA-1$(\Sigma) \oplus $ SHA-1$((\Sigma + 1) \bmod 2^\sigma)$

(1c) $q = 2^{159} \vee U \vee 1$ (the first and the last bit are now equal to 1)
$$2^{159} < q < 2^{160}$$

(1d) Finally, q has to pass a *robust* primality test (the probability of a non-prime number passing the test must be $\leq \frac{1}{2^{80}}$). If q is not prime, return to (1a).

(2) The big prime number p:

We begin with a crude approach which neglects the divisibility constraint (q has to divide $p - 1$):

(2a) Let C be the counter for the iterated runs to find p using the specific seed Σ which helped to generate q. We initialize $C = 0$. At the same time, we initialize a shift constant (for the computation to come) $D = 2$.

(2b) Now, we compute seven digits of a 2^{160}-adic development:

$$V_k = SHA - 1[(\Sigma + D + k) \bmod 2^\sigma] \qquad 0 \leq k \leq 6.$$

This gives $W = V_6 \cdot 2^{960} + V_5 \cdot 2^{800} + V_4 \cdot 2^{640} + V_3 \cdot 2^{480} + V_2 \cdot 2^{320} + V_1 \cdot 2^{160} + V_0$. Then $X = 2^{1023} + W$ (so we get $2^{1023} < X < 2^{1024}$). Finally, we must try to get p "with the help of X":

(2c) Let then $c \equiv X \bmod 2q$ and let $p = X - (c - 1)$ (hence $p \equiv 1 \bmod 2q$). If $p \leq 2^{1023}$, we go to (2d).

Otherwise, we perform a robust primality test on p. If p passes the test, we save the values of Σ and of C (for use in certifying the proper generation of p and of q). Else, we go to (2d):

(2d) Let us increment: $C \mapsto C + 1$, $D \mapsto D + 7$.

If $C \geq 2^{12} = 4096$ (we have tested in vain too many candidates on the base of the seed Σ, etc.), then we start again – beginning with (1a).

Else, we go back to (2b) – q remains valid, still a little effort for p ...

Random Number Generators for the DSA

The Auxiliary Function $G(t, c)$

The function $G(t, c)$ will accept arguments t of 160 bits and arguments c of 160 bits (version DES) or of 160–512 bits (version SHA-1) in order to produce values that are 160-bit words.

The SHA-1 version. There will be only *one* exterior round of the algorithm SHA-1: $t = t_0 t_1 t_2 t_3 t_4$ (5 times 32 bits) initializes the buffer $H = H_0 H_1 H_2 H_3 H_4$ of SHA-1. The second argument c (which will have b bits: $160 \leq b \leq 512$) defines the 16 first words (the "round keys" for the 16 first interior rounds) in the following way: $W_0 W_1 \cdots W_{15} = c \; \underbrace{0 \cdots 0}_{512 - b \, \text{times}}$.

Note, by the way, that the block of 512 bits above is not of the form $M^* = M[1]$. $G(t, c)$ will then be the final state of $H = H_0 H_1 H_2 H_3 H_4$.

The DES version. Let's begin with a (notational)

Definition *Let a_1, a_2, b_1, b_2 be four 32-bit words.*
$DES_{b_1,b_2}(a_1,a_2) = DES_K(A)$, *where*

$$A = a_1 a_2 \quad \text{(plaintext of 64 bits)}$$

$$K = b_1' b_2 \quad \text{(key of 56 bits)}$$
$$b_1' \equiv \text{the 24 last bits of } b_1$$

Let now t and c be given (of 160 bits each); compute $G(t,c)$ (of 160 bits):

(i) $t = t_1 t_2 t_3 t_4 t_5$, $c = c_1 c_2 c_3 c_4 c_5$ (segmentation into five words, of 32 bits each)

(ii) $x_i = t_i \oplus c_i \quad 1 \le i \le 5$.

(iii) For $1 \le i \le 5$ we put:
$a_1 = x_i$,
$a_2 = x_{(i \bmod 5)+1} \oplus x_{((i+3) \bmod 5)+1}$,
$b_1 = c_{((i+3) \bmod 5)+1}$,
$b_2 = c_{((i+2) \bmod 5)+1}$,
$y_{i,1} y_{i,2} = DES_{b_1,b_2}(a_1, a_2)$.

(iv) $G(t,c) = z_1 z_2 z_3 z_4 z_5$ with $z_i = y_{i,1} \oplus y_{((i+1) \bmod 5)+1,2} \oplus y_{((i+2) \bmod 5)+1,1}$
$1 \le i \le 5$ (for example $z_3 = y_{3,1} \oplus y_{5,2} \oplus y_{1,1}$).

The algorithm for computing m values of the private key x. b will denote the size of c – the second argument of the auxiliary function $G(t,c)$ ($160 \le b \le 512$). We shall proceed this way:

1. Choose a new, secret value for $x_{(b)}$ ($\equiv$ the *seed-key* – of length b).
2. $t = 67452301$ efcdab89 98badcfe 10325476 c3d2e1f0   (the "universal" initial value for SHA-1).
3. Computation of m copies for x: $x_0, x_1, \ldots, x_{m-1}$:   For $j = 0, 1, 2, \ldots, m-1$ do:
   (i) Define σx_j ($\equiv x$-seed) by an optional user input (of length b).
   (ii) For $i = 0, 1$ do (computing twin-values, for every j):
       a. $c_i = (x_{(b)} + \sigma x_j) \bmod 2^b$,
       b. $w_i = G(t, c_i)$,
       c. $x_{(b)} = (1 + x_{(b)} + w_i) \bmod 2^b$ (the seed-key will change $2m$ times).
   (iii) $x_j = (w_0 \| w_1) \bmod q$ (concatenation of w_0 and w_1).

The algorithm for precomputing m values of the key of signature k for several documents. Let us sum up the situation: p, q and g, the public parameters of the user, have already been generated. b will have the same meaning as before. One wants to precompute k, k^{-1} and r for m messages at a time. The steps of the algorithm are the following:

1. Choose a secret initial value $k_{(b)}$ (once more the *seed-key* – of length b).
2. $t =$ efcdab89 98badcfe 10325476 c3d2e1f0 67452301 (a cyclic word-shift of the standard initial value for SHA-1)

3. For $j = 0, 1, 2, \ldots, m - 1$ do:
   a. $w_0 = G(t, k_{(b)})$, $k_{(b)} = (1 + k_{(b)} + w_0) \bmod 2^b$,
      $w_1 = G(t, k_{(b)})$, $k_{(b)} = (1 + k_{(b)} + w_1) \bmod 2^b$.
   b. $k = (w_0 \| w_1) \bmod q$.
   c. Compute $k_j^{-1} = k^{-1} \bmod q$.
   d. Compute $r_j = (g^k \bmod p) \bmod q$.
4. Let now M_0, M_1, M_2, ..., M_{m-1} be the next m messages to sign. We continue this way:
   e. $h = SHA - 1(M_j)$
   f. $s_j = (k_j^{-1}(h + x r_j)) \bmod q$
   g. $(r_j, s_j) \equiv$ the signature for M_j.
5. $t = h$.
6. Go to 3.

We note: Step 3 permits pre-computation of the quantities needed to sign the next m messages. Step 4 can begin whenever the first of these m messages is ready (and the execution of step 4 can be suspended whenever the next of the m messages is not ready). As soon as steps 4 and 5 have been completed, step 3 can be executed, and the result saved until the first member of the next group of m messages is ready.

An Example of DSA

We shall consider a "soft version", with $2^{511} < p < 2^{512}$.

First some remarks on the initializations chosen for the auxiliary algorithms:

The seed-value Σ of the algorithm which computes p and q was chosen to be $\Sigma = $ d5014e4b 60ef2ba8 b6211b40 62ba3224 e0427dd3.

With this value for Σ, the algorithm did produce p and q when the counter C was at 150 (recall that we save the values of Σ and of C in order to certify the correct generation of p and q).

The auxiliary function $G(t, c)$ was constructed in version SHA-1, with $b = 160$.

All the *seed-keys* had the same length of 160 bits.

$x_{(b)} = $ bd029bbe 7f51960b cf9edb2b 61f06f0f eb5a38b6,
$t = $ 67452301 efcdab89 98badcfe 10325476 c3d2e1f0,
$x = G(t, x_{(b)}) \bmod q$,
$k_{(b)} = $ 687a66d9 0648f993 867e121f 4ddf9ddb 01205584,
$t = $ efcdab89 98badcfe 10325476 c3d2e1f0 67452301,
$k = G(t, k_{(b)}) \bmod q$[9]. Now we are ready to present our example:
$h = 2$ (h is the value introduced in exercise (1) on DSA),

[9] The watchful reader will realize that the auxiliary algorithms appear here in an older version – without splitting into $w_0 w_1$-twins and some eager reductions modulo q.

p = 8df2a494 492276aa 3d25759b b06869cb eac0d83a fb8d0cf7 cbb8324f 0d7882e5 d0762fc5 b7210eaf c2e9adac 32ab7aac 49693dfb f83724c2 ec0736ee 31c80291,

q = c773218c 737ec8ee 993b4f2d ed30f48e dace915f.

Σ = 626d0278 39ea0a13 413163a5 5b4cb500 299d5522 956cefcb 3bff10f3 99ce2c2e 71cb9de5 fa24babf 58e5b795 21925c9c c42e9f6f 464b088c c572af53 e6d78802.

x = 2070b322 3dba372f de1c0ffc 7b2e3b49 8b260614.

k = 358dad57 1462710f 50e254cf 1a376b2b deaadfbf.

k^{-1} = 0d516729 8202e49b 4116ac10 4fc3f415 ae52f917.

The message M was the ASCII equivalent of "abc" (cf exercise (4) on message digests).

$m = SHA - 1(M)$ = a9993e36 4706816a ba3e2571 7850c26c 9cd0d89d,

y = 19131871 d75b1612 a819f29d 78d1b0d7 346f7aa7 7bb62a85 9bfd6c56 75da9d21 2d3a36ef 1672ef66 0b8c7c25 5cc0ec74 858fba33 f44c0669 9630a76b 030ee333,

r = 8bac1ab6 6410435c b7181f95 b16ab97c 92b341c0,

s = 41e2345f 1f56df24 58f426d1 55b4ba2d b6dcd8c8,

w = 9df4ece5 826be95f ed406d41 b43edc0b 1c18841b,

u_1 = bf655bd0 46f0b35e c791b004 804afcbb 8ef7d69d,

u_2 = 821a9263 12e97ade abcc8d08 2b527897 8a2df4b0,

$g^{u_1} \bmod p$ = 51b1bf86 7888e5f3 af6fb476 9dd016bc fe667a65 aafc2753 9063bd3d 2b138b4c e02cc0c0 2ec62bb6 7306c63e 4db95bbf 6f96662a 1987a21b e4ec1071 010b6069,

$y^{u_2} \bmod p$ = 8b510071 2957e950 50d6b8fd 376a668e 4b0d633c 1e46e665 5c611a72 e2b28483 be52c74d 4b30de61 a668966e dc307a67 c19441f4 22bf3c34 08aeba1f 0a4dbec7,

v = 8bac1ab6 6410435c b7181f95 b16ab97c 92b341c0.

2.4.4 The Signature Algorithm rDSA

The use of the cryptosystem RSA for digital signatures was standardized in September 1998 (by the ANSI≡American National Standards Institute).

We shall present a "minimal" version, skipping details on the auxiliary algorithms and on options of refinement.

The Key Generation

We have to

- Choose the public exponent e (for the verification of the signature)
- Generate two big prime numbers p and q (which must be kept secret), then compute
  $n = pq$ (which will be public)

– Compute the private exponent d, which will serve for the signature generation.

First, we shall fix the length of the binary notation of n equal to 1,024 (this is the smallest possible value accepted by the standard). Then, we put $e = 3$ (or $e = 2^{16} + 1 = 65537$), uniformly for *all* users (we are also permitted to produce e individually and randomly ...).

The pseudorandom generation of p and of q (of 512 bits each) is controlled by a list of constraints similar to those which control the auxiliary algorithms of DSA. Note that the exponent e must be prime to $p - 1$ and to $q - 1$.

For protection of n against certain classical factorization algorithms, we have to respect two rules:

(i) $p \pm 1$ and $q \pm 1$ must each have a big prime factor of size between 100 and 120 bits.

(ii) p and q must differ at least in one of the first 100 binary positions (i.e. $|p - q| > 2^{412}$).

For the private exponent d we demand that $d > 2^{512}$ (if one knows that the exponent d is small, an attack against RSA is possible, etc.)

The Signature Generation

Let M be the message to be signed.

First, we compute $m = \text{SHA-1}(M)$, the message digest of the message. Then, we proceed to carry out an *encapsulation* of the message digest m:

$$m^* \equiv \boxed{\text{header} \mid \text{padding} \mid \text{digest} \mid \text{trailer}}$$

m^* will have 1024 bits ($\equiv$ the size of n).
Header: (4 bits) $6 \equiv 0110$

Padding: (844 bits) $\begin{array}{ll} 210 \text{ times} & \text{b} \equiv 1011 \\ 1 \text{ time} & \text{a} \equiv 1010 \end{array}$

Digest: (160 bits) $m = SHA - 1(M)$

Trailer: (16 bits) $33\text{cc} \equiv 0011001111001100$
("cc" will always be the end of m^*; "33" identifies SHA-1, the hash algorithm producing the message digest)
Note that numerically we have $2^{1022} < m^* < 2^{1023}$ and $m^* \equiv 12 \bmod 16$.
We can now produce the *signature*:

$$\sigma = \min\{m^{*d} \bmod n, n - (m^{*d} \bmod n)\}$$

Remark Numerically, we always get $2^{1022} < \sigma < 2^{1023}$. (Why? – an exercise !)

The Signature Validation

Let M' be the transmitted message, and let σ' be the transmitted signature.

- Decryption:
  Compute first $\tau' = (\sigma')^e \bmod n$.
  Then $m'^* = \begin{cases} \tau' & \text{if } \tau' \equiv 12 \bmod 16 \\ n - \tau' & \text{if } n - \tau' \equiv 12 \bmod 16 \end{cases}$
  If neither case is true, one rejects σ'.
- σ' is rejected, if m'^* is not of the form   6 bbb...ba m' 33cc.
  In case we are formally satisfied, we recover m'.
- Validation of the signature:

Compute SHA-1(M') and verify if $m' = $ SHA-1(M').
If the answer is affirmative, the signature is verified. Otherwise, we reject it.

An Example

We consider the case where $e = 3$. Our two prime numbers p and q, of 512 binary positions each, are:

p	d8cd81f0 35ec57ef e8229551 49d3bff7 0c53520d 769d6d76 646c7a79 2e16ebd8 9fe6fc5b 606b56f6 3eb11317 a8dccdf2 03650ef2 8d0cb9a6 d2b2619c 52480f51
q	cc109249 5d867e64 065dee3e 7955f2eb c7d47a2d 7c995338 8f97dddc 3e1ca19c 35ca659e dc3d6c08 f64068ea fedbd911 27f9cb7e dc174871 1b624e30 b857caad
d	is computed in solving the congruence $3X \equiv 1 \bmod ppcm(p-1, q-1)$[10] 1ccda20b cffb8d51 7ee96668 66621b11 822c7950 d55f4bb5 bee37989 a7d17312 e326718b e0d79546 eaae87a5 6623b919 b1715ffb d7f16028 fc400774 1961c88c 5d7b4daa ac8d36a9 8c9efbb2 6c8a4a0e 6bc15b35 8e528a1a c9d0f042 beb93bca 16b541b3 3f80c933 a3b76928 5c462ed5 677bfe89 df07bed5 c127fd13 241d3c4b
n	acd1cc46 dfe54fe8 f9786672 664ca269 0d0ad7e5 003bc642 7954d939 eee8b271 52e6a947 450d7fa9 80172de0 64d6569a 28a83fe7 0fa840f5 e9802cb8 984ab34b d5c1e639 9ec21e4d 3a3a69be 4e676f39 5aafef7c 4925fd4f aee9f9e5 e48af431 5df0ec2d b9ad7a35 0b3df2f4 d15dc003 9846d1ac a3527b1a 75049e3f e34f43bd

Let us pass now to the signature of the document $M = $ "abc" $= 616263$ (in ASCII).

---

[10] By the Chinese Remainder Theorem, one immediately gets, for every $a \in \mathbb{Z}$ which is prime to $n = pq$: $a^{ppcm(p-1,q-1)} \equiv 1 \bmod pq$ – and that's what we need, etc.

$m = \text{SHA-1}(M)$ a9993e36 4706816a ba3e2571 7850c26c 9cd0d89d

m^* 6bbbbbbb bbbbbbbb bbbbbbbb bbbbbbbb bbbbbbbb
bbbbbbbb bbbbbbbb bbbbbbbb bbbbbbbb bbbbbbbb bbbbbbbb
bbbbbbbb bbbbbbbb bbbbbbbb bbbbbbbb bbbbbbbb bbbbbbbb
bbbbbbbb bbbbbbbb bbbbbbbb bbbbbbbb bbbbbbbb bbbbbbbb
bbbbbbbb bbbbbbbb bbbbbbbb bbbaa999
3e364706 816aba3e 25717850 c26c9cd0 d89d33cc

$m^{*d} \bmod n$ a6b496f4 a802af90 92f1f561 931d84db d0b943ef 34c102b9
4dd51ab0 1e1054bc 0e0572a1 fb2db034 569883f3 82b74e44
9f6c80c4 060fbc0f fbd3a9ca 9d66685b 90873007 d207c1d6
4c692d01 11157bb9 76a4551e 72ddc83c 767a9d75 a4746c51
9b73ce52 c2bfbd1c 3c431d25 4fe8bb43 08fea486 787f239f
d2944390 da49de45

This value exceeds $\frac{n}{2}$; the signature will be the opposite value mod n.

σ 61d35523 7e2a0586 6867110d 32f1d8d3 c5193f5c b7ac3892 b7fbe89d
0d85db54 4e136a54 9dfcf752 97ea9ece 21f08558 93bbf230 99884e5e
dac82edf ae44af04 53ab631c cba5c76e dd13cbd3 d51f37fe 40b9a5dd
64835133 86f5c704 01687dfc 27d1ddaf 6edbd18c efad5cf8 17504c08
f482d262 ad3577aa 2705aaf0 9056578

Finally, we can verify the validity of the signature:

τ' We compute now $\tau' = \sigma^3 \bmod n$.
4116108b 2429942d 3dbcaab6 aa90e6ad 514f1c29 44800a86
bd991d7e 332cf6b5 972aed8b 8951c3ed c45b7224 a91a9ade
6cec842b 53ec853a 2dc470fc dc8ef790 1a062a7d e3066291 7e7eae02
92abb37d 9ef433c0 8d6a4193 f32e3e2a 28cf3875 a2353071 fdf1be79
4f83495b 932778fd 16dc176e 7de102c9 b298016f 0ab20ff1

It is $n - \tau'$, which is congruent to 12 mod 16; we thus obtain:

m'^* 6bbbbbbb bbbbbbbb bbbbbbbb bbbbbbbb bbbbbbbb bbbbbbbb
bbbbbbbb bbbbbbbb bbbbbbbb bbbbbbbb bbbbbbbb bbbbbbbb
bbbbbbbb bbbbbbbb bbbbbbbb bbbbbbbb bbbbbbbb bbbbbbbb
bbbbbbbb bbbbbbbb bbbbbbbb bbbbbbbb bbbbbbbb bbbbbbbb
bbbbbbbb bbbbbbbb bbbaa999 3e364706 816aba3e 25717850
c26c9cd0 d89d33cc

which contains $m = \text{SHA-1}(M)$ at the right place. We can now accomplish
the validation.

2.4.5 ECDSA – Elliptic Curve Digital Signatures

The theory of elliptic curves is perhaps the most distinguished domain of
modern Pure Mathematics. Its nobility comes from a fascinating interaction
between Analysis, Geometry, Algebra and Number Theory, which has created
a formidable tapestry of deep and beautiful results.

There also remains some precious crumbs for cryptography.

More precisely: Classically, an elliptic curve (over $\mathbb{C}$) is defined by an equation of the form

$$y^2 = x^3 + ax + b$$

where the cubic polynomial in x has only simple roots.

The important property of elliptic curves for cryptography is the following: There exists a natural method – in *geometric language* – to put on such a curve the structure of an *abelian group* $\mathcal{E}$.

As a consequence, if you now consider the same situation over a finite field k, the corresponding (finite) group E will be a kind of complicated *additive substitute* for the multiplicative group k^*, but with a less decipherable structure for the lorgnettes of cryptanalysis.

If we then replace exponentiation by iterated addition, it will be easy to formulate a signature protocol in complete analogy with that of DSA.

Our exposition will focus on the case of characteristic 2 (hence: $1+1 = 0$). Why?

The arithmetic of the fields $\mathbb{F}_{2^m} = \mathbb{F}_2[x]/(p(x))$ – where $p(x)$ is an irreducible binary polynomial of degree m – harmonizes perfectly with the binary computer technology and simplifies considerably in case of irreducible trinomials (pentanomials) $p(x)$.

But, until now, we have never seen groups of type $(\mathbb{F}_{2^m})^*$ in public key cryptography. The reason is simple: In order to get the discrete logarithm problem to be difficult in the multiplicative group $(\mathbb{F}_{2^m})^*$, one is obliged to raise considerably the degree m. On the other hand, the groups $\mathcal{E} = \mathcal{E}(\mathbb{F}_{2^m})$ of elliptic curves over the fields $\mathbb{F}_{2^m}$ become pretty complicated – in the sense we need it – for *already reasonable* degrees m. Thus, the marriage will be logical: *If* we want to work, for digital signatures, in binary polynomial arithmetic, *then* we *need* the elliptic curves over the various fields $\mathbb{F}_{2^m}$.

The prize to pay is a certain alienation of the "elliptic formulary": the case of characteristic 2 is not natural with respect to classical theory.

The algorithm ECDSA will work precisely like the two other variants of DSS:

First, there is the *signatory* who generates a digital signature of the message M; then, there is the *verifier* who certifies the validity of the signature. Every signatory disposes of a private key (for the generation of the signature) and of a public key (for the verification of the signature). For the *two* procedures (the generation and the verification of the signature) a digest of the message M will be extracted (by SHA-1, in general) in order to supply the document-support for the signature.

Elliptic Curves Over $\mathbb{F}_{2^m}$

Our field of constants will be the field of residues $\mathbb{F}_{2^m} = \mathbb{F}_2[x]/(p(x))$, where $p(x) = x^m + \cdots + 1$ is an irreducible binary polynomial of degree m. For fixed m, all extensions of $\mathbb{F}_2$ of degree m are isomorphic. A (multiplicative) arithmetic which is easy to implement will be furnished by those $p(x)$ which are trinomials or (if irreducible trinomials do not exist) by pentanomials.

Example Irreducible trinomials over $\mathbb{F}_2$ – *some examples*:

$x^{161} + x^{18} + 1,$
$x^{191} + x^{9} + 1,$
$x^{255} + x^{52} + 1,$
$x^{300} + x^{5} + 1,$
$x^{319} + x^{36} + 1,$
$x^{399} + x^{26} + 1,$
$x^{511} + x^{10} + 1,$
$x^{620} + x^{9} + 1,$
$x^{1023} + x^{7} + 1,$
$x^{1366} + x + 1,$
$x^{1596} + x^{3} + 1,$
$x^{1980} + x^{33} + 1,$
$x^{1994} + x^{15} + 1.$

There are no irreducible trinomials of degree m when $m \equiv 0 \bmod 8$ – a surprising hostility of mathematics towards the classical formats in computer science. But irreducible pentanomials will exist, for all degrees $m \geq 4$.

Irreducible pentanomials over $\mathbb{F}_2$ – *some examples*:

$x^{160} + x^{117} + x^{2} + x + 1,$
$x^{192} + x^{7} + x^{2} + x + 1,$
$x^{256} + x^{155} + x^{2} + x + 1,$
$x^{320} + x^{7} + x^{2} + x + 1,$
$x^{400} + x^{245} + x^{2} + x + 1,$
$x^{512} + x^{51} + x^{2} + x + 1,$
$x^{640} + x^{253} + x^{3} + x + 1,$
$x^{800} + x^{463} + x^{2} + x + 1,$
$x^{1024} + x^{515} + x^{2} + x + 1,$
$x^{1536} + x^{881} + x^{2} + x + 1,$
$x^{1600} + x^{57} + x^{2} + x + 1,$
$x^{1920} + x^{67} + x^{2} + x + 1,$
$x^{2000} + x^{981} + x^{2} + x + 1.$

We shall now pass to the principal subject of this paragraph: the elliptic curves.

Definition *An elliptic curve*[11] $\mathcal{E}(\mathbb{F}_{2^m})$ *over* $\mathbb{F}_{2^m}$, *defined by the couple* $(a, b) \in (\mathbb{F}_{2^m})^2$, $b \neq 0$, *is the set of solutions* $(X, Y) \in (\mathbb{F}_{2^m})^2$ *of the equation*

$$Y^2 + XY = X^3 + aX^2 + b$$

plus one distinguished point $\mathbb{O}$, *which is meant to be the point at infinity*.

Notation $\#\mathcal{E}(\mathbb{F}_{2^m}) \equiv$ the number of points of $\mathcal{E}(\mathbb{F}_{2^m})$.

---

[11] Our definition is a little bit restrictive: The condition $b \neq 0$ happily eliminates the supersingular case which haunts the odd-prime-characteristic theory.

Remark (1) Pass to homogeneous coordinates:

$$Y^2 Z + XYZ = X^3 + aX^2 Z + bZ^3$$

The projective curve which is defined by this equation is the correct object of the foregoing definition:

In homogeneous coordinates (a projective point is nothing but a 1D linear subspace minus the origin) we have:

$\mathbb{O} = [0, 1, 0] \equiv$ the (0-dotted) Y-axis as a projective point .

The finite points are then of the form $[X, Y, 1]$, where (X, Y) is a solution of the initial non-homogeneous equation.[12]

(2) $\#E(\mathbb{F}_{2^m})$ is even.

By a celebrated theorem of Hasse's we have the estimation

$$2^m + 1 - 2\sqrt{2^m} \leq \#\mathcal{E}(\mathbb{F}_{2^m}) \leq 2^m + 1 + 2\sqrt{2^m}.$$

For an elliptic curve over $\mathbb{F}_{16}$ this means: $10 \leq \#\mathcal{E}(\mathbb{F}_{16}) \leq 24$ and for an elliptic curve over $\mathbb{F}_{256}$ this yields: $226 \leq \#\mathcal{E}(\mathbb{F}_{256}) \leq 288$.

Fundamental observation. $\mathcal{E}(\mathbb{F}_{2^m})$ bears the structure of an additive abelian group according to the following laws:

1. $\mathbb{O} \equiv$ the point at infinity, is the zero-element for the addition.

2. The opposite point for the addition:
$$\boxed{-(X, Y) = (X, X + Y)}$$

3. The sum of two distinct (finite) points (which are not opposite):

$$(X_1, Y_1), (X_2, Y_2) \in \mathcal{E}(\mathbb{F}_{2^m}), \quad X_1 \neq X_2.$$

$$(X_1, Y_1) + (X_2, Y_2) = (X_3, Y_3)$$

with

$$\boxed{\begin{array}{l} X_3 = \lambda^2 + \lambda + X_1 + X_2 + a, \quad Y_3 = \lambda(X_1 + X_3) + X_3 + Y_1 \\[2mm] \text{where} \quad \lambda = \frac{Y_1 + Y_2}{X_1 + X_2} \in \mathbb{F}_{2^m}. \end{array}}$$

4. Doubling a point (of order >2):

$$(X_1, Y_1) \in \mathcal{E}(\mathbb{F}_{2^m}), \quad X_1 \neq 0.$$

---

[12] For those who are not accustomed to homogeneous coordinates: In affine thinking, the last coordinate is nothing but a *common denominator* for the first two coordinates. So, $\mathbb{O}$ is indeed at infinity – in direction of the y-axis.

$$2(X_1, Y_1) = (X_3, Y_3)$$

with

$$X_3 = \lambda^2 + \lambda + a \qquad\qquad Y_3 = X_1^2 + (\lambda + 1)X_3$$

where $\quad \lambda = X_1 + \frac{Y_1}{X_1} \in \mathbb{F}_{2^m}$.

Remark The (arithmetical) laws for the addition and the doubling of (finite) points on an elliptic curve have a precise *geometrical* meaning:

First consider the case of two points $P_1 = (X_1, Y_1)$ and $P_2 = (X_2, Y_2)$, which are distinct and not opposite (for the group operations defined above). $P_3 = P_1 + P_2 = (X_3, Y_3)$ can be constructed in the following way:

Let $\mathcal{L}$ be the straight line which goes through P_1 and through P_2. $\mathcal{L}$ will intersect $\mathcal{E}(\mathbb{F}_{2^m})$ in a third point P_3'. We will obtain: $P_3 = -P_3'$ (P_3 will be the opposite of P_3' for the group law on $\mathcal{E}(\mathbb{F}_{2^m})$).

Let us show that this geometric construction indeed yields the foregoing expressions:

Let $Y = \lambda X + \gamma$ be the equation of $\mathcal{L}$: $\lambda = \frac{Y_1 + Y_2}{X_1 + X_2}$ and $\gamma = Y_1 + \lambda X_1$. A point $P = (X, \lambda X + \gamma)$ of $\mathcal{L}$ is at the same time onto $\mathcal{E}(\mathbb{F}_{2^m})$ if and only if $(\lambda X + \gamma)^2 + \lambda X^2 + \gamma X = X^3 + aX^2 + b$. Consider now the cubic equation $X^3 + (\lambda^2 + \lambda + a)X^2 + \gamma X + \gamma^2 + b = 0$.

We know: The coefficient $\lambda^2 + \lambda + a$ is the sum of the three roots of this equation: $X_1 + X_2 + X_3 = \lambda^2 + \lambda + a$. This gives the expression for X_3. For the ordinate of the third point of intersection of $\mathcal{L}$ with $\mathcal{E}(\mathbb{F}_{2^m})$ we get

$$Y_3' = \lambda X_3 + \gamma = \lambda(X_1 + X_3) + Y_1.$$

Passing to the opposite point gives finally: $Y_3 = \lambda(X_1 + X_3) + X_3 + Y_1$.

As to the doubling operation ($P_1 = P_2$), let us remain in the logic of the foregoing arguments and replace the straight line $\mathcal{L}$ which connects P_1 and P_2 by the tangent to $\mathcal{E}(\mathbb{F}_{2^m})$, at P_1.

Recall Let $F(X, Y) = 0$ be the equation of a plane curve, and let $P_1 = (X_1, Y_1)$ be a *simple point* of the curve ($\frac{\partial F}{\partial X}(P_1)$ and $\frac{\partial F}{\partial Y}(P_1)$ don't vanish simultaneously).

Then the equation of the tangent $\mathcal{L}$ to the curve defined by $F(X, Y) = 0$ at $P_1 = (X_1, Y_1)$ is given by

$$\frac{\partial F}{\partial X}(P_1)(X - X_1) + \frac{\partial F}{\partial Y}(P_1)(Y - Y_1) = 0.$$

Specialize down to our particular curve $\mathcal{E}(\mathbb{F}_{2^m}) : Y^2 + XY + X^3 + aX^2 + b = 0$.

We have: $\frac{\partial F}{\partial X} = X^2 + Y$, $\frac{\partial F}{\partial Y} = X$. This yields the equation of $\mathcal{L}$: $X_1 Y = (X_1^2 + Y_1)X + X_1^3$. In case $X_1 \neq 0$, we obtain $Y = \lambda X + \gamma$, with $\lambda = X_1 + \frac{Y_1}{X_1}$, $\gamma = X_1^2$. With precisely the same argument as before we arrive at:

$$X_3 = \lambda^2 + \lambda + a, \quad Y_3' = \lambda X_3 + \gamma = \lambda X_3 + X_1^2,$$

$$Y_3 = X_3 + Y_3' = X_1^2 + (\lambda + 1)X_3,$$

which was our claim ...

Example An elliptic curve over $\mathbb{F}_{16} = \mathbb{F}_2[x]/(x^4 + x + 1)$.
Recall: $\omega = x$ is a generator of the multiplicative group $\mathbb{F}_{16}^*$:

$\omega = x,$	$\omega^8 = x^2 + 1,$
$\omega^2 = x^2,$	$\omega^9 = x^3 + x,$
$\omega^3 = x^3,$	$\omega^{10} = x^2 + x + 1,$
$\omega^4 = x + 1,$	$\omega^{11} = x^3 + x^2 + x,$
$\omega^5 = x^2 + x,$	$\omega^{12} = x^3 + x^2 + x + 1,$
$\omega^6 = x^3 + x^2,$	$\omega^{13} = x^3 + x^2 + 1,$
$\omega^7 = x^3 + x + 1,$	$\omega^{14} = x^3 + 1,$
	$\omega^{15} = 1,$

Our elliptic curve $\mathcal{E}(\mathbb{F}_{16})$ will be defined by the equation $Y^2 + XY = X^3 + \omega^4 X^2 + 1$. And these are the 16 points on the curve:

$\mathbb{O}$	$(0, 1)$	$(1, \omega^6)$	$(1, \omega^{13})$
(ω^3, ω^8)	(ω^3, ω^{13})	(ω^5, ω^3)	(ω^5, ω^{11})
(ω^6, ω^8)	(ω^6, ω^{14})	(ω^9, ω^{10})	(ω^9, ω^{13})
(ω^{10}, ω^1)	(ω^{10}, ω^8)	$(\omega^{12}, 0)$	$(\omega^{12}, \omega^{12})$

Let us carry out, as an example, an addition and a doubling of points:
$$P_1 = (X_1, Y_1) = (\omega^6, \omega^8) \qquad P_2 = (X_2, Y_2) = (\omega^3, \omega^{13})$$

1. $P_3 = (X_3, Y_3) = P_1 + P_2$:
$\lambda = \frac{Y_1 + Y_2}{X_1 + X_2} = \frac{\omega^8 + \omega^{13}}{\omega^6 + \omega^3} = \omega,$
$X_3 = \lambda^2 + \lambda + X_1 + X_2 + a = \omega^2 + \omega + \omega^6 + \omega^3 + \omega^4 = 1,$
$Y_3 = \lambda(X_1 + X_3) + X_3 + Y_1 = \omega(\omega^6 + 1) + 1 + \omega^8 = \omega^{13}.$

2. $P_3 = (X_3, Y_3) = 2P_1$:
$\lambda = X_1 + \frac{Y_1}{X_1} = \omega^6 + \frac{\omega^8}{\omega^6} = \omega^3,$
$X_3 = \lambda^2 + \lambda + a = \omega^6 + \omega^3 + \omega^4 = \omega^{10},$
$Y_3 = X_1^2 + (\lambda + 1)X_3 = \omega^{12} + (\omega^3 + 1)\omega^{10} = \omega^8.$

Exercises

(1) Show that every elliptic curve $\mathcal{E}(\mathbb{F}_{2^m})$ admits a single (finite) point P of order 2 (i.e. such that $2P = \mathbb{O}$).

(2) Consider the elliptic curve $\mathcal{E}(\mathbb{F}_{16})$ of the foregoing example.
Since $\#\mathcal{E}(\mathbb{F}_{16}) = 16$, the order of a point P on $E(\mathbb{F}_{16})$ is 1, 2, 4, 8 or 16. Determine the order of each of the points on the curve. Which are the subgroups of $\mathcal{E}(\mathbb{F}_{16})$?

(3) Another elliptic curve over $\mathbb{F}_{16}$: $\mathcal{E}(\mathbb{F}_{16}) \equiv Y^2 + XY = X^3 + \omega^3$
  Let $G = (\omega^3, \omega^5)$
  (a) Show that $G \in E(\mathbb{F}_{16})$.
  (b) Compute $2G$, $3G$, $4G$, $5G$.
  (c) Show that $10G = (0, \omega^9)$.
  (d) Show that $E(\mathbb{F}_{16})$ is a *cyclic* group of 20 elements.
(4) An elliptic curve over $\mathbb{F}_{256} = \mathbb{F}_2[x]/(x^8 + x^4 + x^3 + x^2 + 1)$.

Notation: $\alpha = x \bmod (x^8 + x^4 + x^3 + x^2 + 1)$.
  We know already: α is a generator for the multiplicative group $\mathbb{F}_{256}^*$.

(a) Put $\omega = \alpha^{17} \equiv 10011000$.

Verify that $\quad \omega^4 + \omega + 1 = 0$.
  Deduce that the field $\mathbb{F}_{16} = \mathbb{F}_2[x]/(x^4 + x + 1)$ can be identified with the subfield $\mathbb{F}_2[\omega] = \{0, \omega, \omega^2, \ldots, \omega^{15} = 1\}$ of $\mathbb{F}_{256}$ (this means: the *multiplicative* group of 15th roots of unity in $\mathbb{F}_{256}^*$, augmented by 0, is *stable under addition*).
  We are interested in the elliptic curve defined by the same equation as in exercise (3), but now over $\mathbb{F}_{256}$: $\mathcal{E}(\mathbb{F}_{256}) \equiv Y^2 + XY = X^3 + \omega^3$
  Our goal will be to find $\#\mathcal{E}(\mathbb{F}_{256})$, i.e. the order of the group of $\mathbb{F}_{256}$-rational points on the curve (note that we are allowed to solve our given elliptic equation over *any* field k which is an extension of $\mathbb{F}_{16}$).

(b) Let $P = (\alpha, \alpha^{88}) = (00000010, 11111110) \in \mathbb{F}_{256}^2$.
  Show that $P \in \mathcal{E}(\mathbb{F}_{256})$.
(c) Compute $2P$, $4P$, $8P$, $16P$, $32P$, $64P$, $128P$, $256P$.
(d) Show that $\#\mathcal{E}(\mathbb{F}_{256}) = 280$.

(Hint: $\mathcal{E}(\mathbb{F}_{16})$ is a subgroup of $\mathcal{E}(\mathbb{F}_{256})$; hence $\#\mathcal{E}(\mathbb{F}_{256})$ is divisible by 20. On the other hand, $226 \leq \#\mathcal{E}(\mathbb{F}_{256}) \leq 288$, by the Hasse estimation. Conclude now by means of (c)).

Finding Elliptic Curves with Prescribed Order: The Complex Multiplication Elliptic Curve Generation Method (*Lay–Zimmer method*)

We want to discuss (and to solve) the following highly non-trivial problem:
The *input data*:

(1) A *field of scalars* $\mathbb{F}_{2^m} = \mathbb{F}_2[x]/(p(x))$, where $p(x) = x^m + \cdots + 1$ is an irreducible binary pentanomial (or even a trinomial).
(2) A large *prime number n* – which will determine the magnitude of the cyclic range for our signature arithmetic, i.e. which will be the *order* of a (cyclic) subgroup of some elliptic curve $E(\mathbb{F}_{2^m})$.

The *output data*:

The *equation* $Y^2 + XY = X^3 + b$, $b \in \mathbb{F}_{2^m}^*$ of an elliptic curve $\mathcal{E}(\mathbb{F}_{2^m})$ over $\mathbb{F}_{2^m}$ such that $N = \#\mathcal{E}(\mathbb{F}_{256})$ is *nearly prime*[13] in the following sense: $N = 2^r \cdot n$, $r \geq 2$ (actually, we only are interested in $N = 4n$).

In the sequel, we shall describe the algorithmic steps which finally realize our programme. It should be pointed out that the underlying mathematical structures are rich and rather complicated (we are in the heart of Algebraic Number Theory) – so, an exhaustive presentation would be largely beyond the scope of this book.

Our goal is to make the algorithms as transparent as possible. Nevertheless, at the end of this section, we shall try to give a brief sketch of the theoretical background. But let us resume:

Starting with a scalar field of 2^m elements and a prescribed group order $N = 4n$, we are in search of the non-zero constant term b of the elliptic equation $Y^2 + XY = X^3 + b$ which will have the desired number of solutions.

We will have to solve three algorithmic problems:

(1) Find a CM discriminant D for 2^m.
    [D is the arithmetical link between the scalar-field degree and the desired group order N]
(2) Compute the reduced class polynomial $w_D(t)$ of D.
(3) Find a particular root β (in $\mathbb{F}_{2^m}$) of $w(t) = w_D(t) \bmod 2$; then the constant term b of our elliptic equation will be either equal to β or to β^3 (depending on $D \equiv 0 \bmod 3$ or not).

So, our first goal is to introduce the parameter D which controls the transit from our input data to our output data.

CM Discriminants

Let $\mathcal{E} = \mathcal{E}(\mathbb{F}_{2^m})$ be an elliptic curve over $\mathbb{F}_{2^m}$, of order N. Then $Z = 2^{m+2} - (2^m + 1 - N)^2$ is positive, by the Hasse Theorem. There is a unique factorization $Z = DV^2$ with D (odd and) square-free. We can write

$$(*) \quad 2^{m+2} = W^2 + DV^2$$

$$(**) \quad N = 2^m + 1 \pm W$$

We shall say that $\mathcal{E} = \mathcal{E}(\mathbb{F}_{2^m})$ has *complex multiplication* by D (better: by $\sqrt{-D}$).

D will be called a *CM discriminant* for 2^m.

---

[13] We define near primality as simply as needed; whenever $N \equiv 0 \bmod 4$, we are allowed to search for elliptic equations without quadratic term in X, i.e. satisfying $a = 0$.

Exercise Show that we have necessarily: $D \equiv 7 \bmod 8$.

In order to well understand the following algorithmic steps (CM discriminant test and class number computation), we should briefly recall some elementary facts from Algebraic Number Theory.

Ideal Classes in Imaginary Quadratic Fields – a Summary

We shall restrict to the case which we actually need. So, let D be a positive square-free integer, $D \equiv 7 \bmod 8$, and put $d = -D$. Note that $d \equiv 1 \bmod 4$.[14]

$K = \mathbb{Q}(\sqrt{d}) = \{r + s\sqrt{d} : r, s \in \mathbb{Q}\} \subset \mathbb{C}$ is the imaginary quadratic field defined by adjunction of $\sqrt{d}$ to $\mathbb{Q}$.

Put $\omega = \frac{1+\sqrt{d}}{2} = \frac{1+i\sqrt{D}}{2}$.

$R = \mathbb{Z}[\omega] = \{n + m\omega : n, m \in \mathbb{Z}\}$ is the ring of (algebraic) integers in K (consisting of the elements of K which satisfy an integer quadratic equation with leading coefficient equal to 1).

R is a *lattice* in the complex plane, freely $\mathbb{Z}$-generated by 1 and $\omega = \frac{1+\sqrt{d}}{2}$.

Notation $R = (1, \omega)_{\mathbb{Z}}$.

We are interested in the *sublattices* of R which are *ideals* of R (i.e. which are stable under multiplication with complex numbers of the form $n + m\omega$, $n, m \in \mathbb{Z}$).

Proposition *A sublattice $I \subset R$ is an ideal of R $\Longleftrightarrow$ there exist $a, b, c, \gamma \in \mathbb{Z}$ such that $b^2 - 4ac = d, a > 0, \gamma > 0$, and $I = \gamma(a, \frac{b+i\sqrt{D}}{2})_{\mathbb{Z}} = \{n\gamma a + m\gamma\frac{b+i\sqrt{D}}{2}, n, m \in \mathbb{Z}\}$.*

The arithmetical complexity of K (equivalently: of R) is roughly encoded in the *classgroup* $\mathrm{Cl}(K)$ of $K = \mathbb{Q}(\sqrt{d})$. So we have to speak about fractional ideals.

A *fractional ideal* $J \subset K$ is an R-submodule of K of the form $J = \frac{1}{n}I$, with some fixed $n \in \mathbb{N}$, and some ideal I of R (you should consider J as a refinement to the order n of the lattice I). Hence, as a consequence of proposition 1, a fractional ideal J is of the form $J = \gamma(a, \frac{b+i\sqrt{D}}{2})_{\mathbb{Z}}$, with a, b, c as in Proposition 2.4.5, and some fixed positive $\gamma \in \mathbb{Q}$.

Fractional ideals multiply in a straightforward way:

$IJ = \{$ all finite sums of products with one factor in I and one factor in $J\}$. In particular, consider, for a fractional ideal J, its conjugate $\sigma(J) = \{\bar{\xi} : \xi \in J\}$. Then $\sigma(J)J = J\sigma(J) = (N(J)) \equiv$ the principal (fractional) ideal generated by the norm of J (which is, for $J = \gamma(a, \frac{b+i\sqrt{D}}{2})_{\mathbb{Z}}$, equal to the rational number $\gamma^2 a$).

---

[14] This singles out one of two types of imaginary quadratic fields, but will not substantially touch upon the generality of most of our results.

As a consequence, we get: $\frac{1}{N(J)}\sigma(J)J = (1)$, i.e. every fractional ideal J admits a multiplicative inverse $J^{-1} = \frac{1}{N(J)}\sigma(J)$. Thus the set of all non-zero fractional ideals of $K = \mathbb{Q}(\sqrt{d})$ is an abelian group under multiplication (of fractional ideals). This group is much too large in order to appropriately describe the *type* of the particular number field $K = \mathbb{Q}(\sqrt{d})$.

Finally, the good object is the quotient group modulo the subgroup of principal (fractional) ideals – the *classgroup* $\mathrm{Cl}(K)$ of $K = \mathbb{Q}(\sqrt{d})$. We shall see in a moment that $\mathrm{Cl}(K)$ is a *finite* group. Its order, the *classnumber* $h(K)$ of K, is an important invariant of the number field K.

So far, we did not substantially make use of the imaginary type of our quadratic number field. But now we shall proceed to give a "minimal parametrization" for the classgroup $\mathrm{Cl}(K)$, where it will play an essential role.

Proposition *Every ideal class of an imaginary quadratic field with discriminant[15] $d = -D$ admits a representative of the form $I = (1, \frac{b+\mathrm{i}\sqrt{D}}{2a})_{\mathbb{Z}}$, with $a, b \in \mathbb{Z}, a > 0, 4ac - b^2 = D$ (and appropriate $c \in \mathbb{Z}$).*

The ideal class $[I]$ is completely determined by $\tau = \frac{b+\mathrm{i}\sqrt{D}}{2a}$, which lies in the upper half-plane $\mathbf{H} = \{z \in \mathbb{C} : \mathrm{Im}(z) > 0\}$.

Two points $\tau = \frac{b+\mathrm{i}\sqrt{D}}{2a}$ and $\tau' = \frac{b'+\mathrm{i}\sqrt{D}}{2a'}$ of $\mathbf{H}$ represent the same ideal class $\Longleftrightarrow$ there exists $A = \begin{pmatrix} \alpha & \beta \\ \gamma & \delta \end{pmatrix} \in SL_2(\mathbb{Z}) \left[\begin{array}{l} \text{the group of integer } 2 \times 2 \\ \equiv \text{ matrices with determinant} \\ \text{equal to 1} \end{array} \right]$

with $\tau' = \frac{\alpha\tau+\beta}{\gamma\tau+\delta}$. Now consider the set of all maps

$$A : \mathbf{H} \ni z \longmapsto A(z) = \frac{\alpha z+\beta}{\gamma z+\delta} \in \mathbf{H} \text{ with } A = \begin{pmatrix} \alpha & \beta \\ \gamma & \delta \end{pmatrix} \in SL_2(\mathbb{Z}).^{16}$$

We get the *modular group* $\boldsymbol{\Gamma}$. Since the matrix A determines the transformation A up to a sign, we may identify:

$$\boldsymbol{\Gamma} = PSL_2(\mathbb{Z}) = SL_2(\mathbb{Z})/\{\pm 1\}.$$

Note that $\boldsymbol{\Gamma}$ is generated by the two "natural" transformations T and S induced by the matrices

$$T = \begin{pmatrix} 1 & 1 \\ 0 & 1 \end{pmatrix} \text{ and } S = \begin{pmatrix} 0 & -1 \\ 1 & 0 \end{pmatrix} \text{ in } SL_2(\mathbb{Z}).$$

Geometrically, we have $T(z) = z + 1$ and $S(z) = -\frac{1}{z}$, i.e. T is a horizontal translation by a unit, and S is a reflection on the unit-circle, followed by a reflection on the imaginary axis. Observe that the order of S as a matrix is 4, whereas the order of S as a mapping is 2.

According to proposition 2, an ideal class of $K = \mathbb{Q}(\sqrt{d})$ determines an orbit for the action of $\boldsymbol{\Gamma}$ onto $\mathbf{H}$. Thus we are interested in *canonical representatives* for orbits.

---

[15] Recall: We focus on the case $d \equiv 1 \bmod 4$.

[16] We do not notationally distinguish the matrix from the induced transformation.

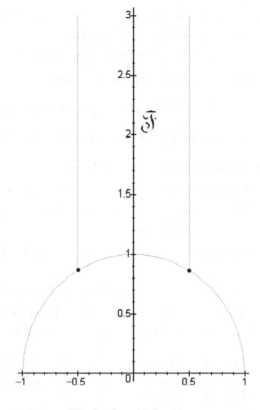

The fundamental region.

The *fundamental region.* **F** of the modular group Γ is defined as follows:
$$\mathbf{F} = \{z \in \mathbb{C} :\mid Re(z) \mid \leq \tfrac{1}{2}, \mid z \mid \geq 1\}.$$

Proposition F *is a fundamental region for the (action of the) modular group* Γ *in the following sense:*

(a) *Every* $z \in \mathbf{H}$ *is* modΓ *equivalent to a point* $z' \in \mathbf{H}$*: there is* $A \in \Gamma$*:* $z' = A(z) \in \mathbf{F}$.

(b) *Let* z *and* z' *be two distinct points of* $\mathbf{F}$*. Then* $z \equiv z' \bmod \Gamma \quad \Longleftrightarrow \quad z$ *and* z' *lie on the boundary of* $\mathbf{F}$*; more precisely: one of the following alternatives hods:*

(i) $\mid Re(z) \mid = \mid Re(z') \mid = \tfrac{1}{2}$, *and* $z' = z \pm 1$.
(ii) $\mid z \mid = \mid z' \mid = 1$, *and* $z' = -\bar{z}$.

Hence: Every $z \in \mathbf{H}$ *admits modulo* Γ *a unique representative* $\tau \in \mathbf{F}$ – *with* $Re(\tau) \geq 0$ *if* $\mid Re(\tau) \mid = \tfrac{1}{2}$ *or* $\mid \tau \mid = 1$.

Consequence (Reduced ideals) In every ideal class of $K = \mathbb{Q}(\sqrt{-D})$ there is a *unique* ideal $I = \left(a, \frac{b+\mathrm{i}\sqrt{D}}{2}\right)_{\mathbb{Z}}$ such that

(1) $a, b \in \mathbb{Z}, a > 0, b^2 + D = 4ac$ with appropriate $c \in \mathbb{Z}$,
(2) $\mid b \mid \le a \le c$, and $b \ge 0$ whenever either $\mid b \mid = a$ or $a = c$.

$\left[\text{Note that for } \tau = \frac{b+\mathrm{i}\sqrt{D}}{2a} \text{ we have: } Re(\tau) = \frac{b}{2a} \text{ and } \mid \tau \mid^2 = \frac{c}{a}\right]$

$I = \left(a, \frac{b+\mathrm{i}\sqrt{D}}{2}\right)_{\mathbb{Z}}$ is said to be *reduced* (whenever (1) and (2) hold).

Now it is easy to compute the classgroup $\mathrm{Cl}(K)$:

You only have to determine all triples (a, b, c) satisfying (1) and (2).

Note that $b^2 + D = 4ac$ and $\mid b \mid \le a \le c$ yield $\mid b \mid \le a \le \sqrt{\frac{D}{3}}$; thus there is only a *finite* number of possibilities.

Example $D = 163$ (a little bit non-thematic: $D \not\equiv 7 \bmod 8$, but still $-D \equiv 1 \bmod 4$).

$\sqrt{\frac{163}{3}} \approx 7.37$, this gives $1, 3, 5, 7$ as positive candidates for b (which has to be odd: $b^2 + 163 \equiv 0 \bmod 4$). Now, $\frac{b^2+163}{4} = 41, 43, 47, 53$ for $b = 1, 3, 5, 7$.
Thus only $(a, b, c) = (1, 1, 41)$ is possible: $\mathrm{Cl}(\mathbb{Q}(\sqrt{-163})) = \{1\}$.

$\Big[$**Remark** What can we say about the second type of imaginary quadratic fields $K = \mathbb{Q}(\sqrt{d})$, with square-free $d < 0$ and $d \equiv 2, 3 \bmod 4$?
Here, the field discriminant is equal to $4d = -D$.
We insist: For $\mathbb{Q}(\sqrt{-1})$, the field discriminant is -4, for $\mathbb{Q}(\sqrt{-2})$, the field discriminant is -8, for $\mathbb{Q}(\sqrt{-5})$, the field discriminant is -20, etc.
The classgroup computation is relative to $D \equiv 0 \bmod 4$ (since normal forms for fractional ideals are relative to D).

Example $D = 20$ (corresponding to $K = \mathbb{Q}(\sqrt{-5})$).
We search for (a, b, c) with $b^2 + 20 = 4ac$ (hence b will be even), $\mid b \mid \le a \le c$, and $b \ge 0$ whenever either $\mid b \mid = a$ or $a = c$.
Now, $\mid b \mid \le a \le \sqrt{\frac{20}{3}}$ implies $b = 0$ or $b = 2$; this gives

$$(a, b, c) = \begin{cases} (1, 0, 5), \\ (2, 2, 3), \end{cases}$$

Thus $h(\mathbb{Q}(\sqrt{-5})) = 2.\Big]$

Exercises

(1) Recall the two standard generators $T = \begin{pmatrix} 1 & 1, \\ 0 & 1 \end{pmatrix}$ and $S = \begin{pmatrix} 0 & -1, \\ 1 & 0 \end{pmatrix}$

of $SL_2(\mathbb{Z})$. Put $R = \begin{pmatrix} 0 & -1, \\ 1 & 1 \end{pmatrix}$, $R_1 = -R$.

(a) Check that $R = ST$ and that $R(z) = \frac{-1}{z+1}$ (as a mapping).

(b) Check that $T = SR_1$.

(c) Show that $S^4 = R_1^3 = \begin{pmatrix} 1 & 0 \\ 0 & 1 \end{pmatrix}$. Hence: $S^2 = R^3 = Id$ as mappings.

(2) Show that $h(\mathbb{Q}(\sqrt{-7})) = 1$.

(3) Find the seven reduced representatives in the classgroup of $\mathbb{Q}(\sqrt{-71})$.

$$\left[\mathbf{Answer}\ (a,b,c) = \begin{cases} (1,1,18), \\ (2,\pm 1,9), \\ (3,\pm 1,6), \\ (4,\pm 3,5). \end{cases} \right]$$

(4) Find, for positive square-free $D \equiv 7 \bmod 8$, the minimal values such that $h(\mathbb{Q}(\sqrt{-D})) = 2,3,4,5,6,7$.

$$\left[\mathbf{Answer}\ h(\mathbb{Q}(\sqrt{-D})) = \begin{cases} 2 & \text{for}\quad D = 15, \\ 3 & \text{for}\quad D = 23, \\ 4 & \text{for}\quad D = 39, \\ 5 & \text{for}\quad D = 47, \\ 6 & \text{for}\quad D = 63, \\ 7 & \text{for}\quad D = 71. \end{cases} \right]$$

Remark Let $h_D = h(\mathbb{Q}(\sqrt{-D}))$, where $-D$ is the discriminant of the imaginary quadratic field. Then

$$\lim_{D \to \infty} \frac{\log(h_D)}{\log(\sqrt{D})} = 1.$$

Thus, for large D, $h = h_D$ is roughly of the magnitude of $\sqrt{D}$.

Historically, the classification of fractional ideals in imaginary quadratic number fields has been imbedded in the classification of certain *binary quadratic forms*.

So, consider (positive definite) homogeneous quadratic forms

$$q(X,Y) = aX^2 + bXY + cY^2 = (X \quad Y) \begin{pmatrix} a & \frac{b}{2} \\ \frac{b}{2} & c \end{pmatrix} \begin{pmatrix} X \\ Y \end{pmatrix}$$

with integer coefficients $a,b,c \in \mathbb{Z}, a > 0$, and such that $D = 4ac - b^2$ is positive and square-free. We shall say that $q(X,Y)$ *represents* the integer $u \in \mathbb{Z} \iff u = q(x,y)$ for appropriate *integers* $x,y \in \mathbb{Z}$. Note that $q(X,Y)$ always represents the coefficient a (for $(x,y) = (1,0)$).

The quadratic form $q'(X,Y) = a'X^2 + b'XY + c'Y^2$ is said to be *equivalent* to

$$q(X,Y) = aX^2 + bXY + cY^2 \text{ whenever there is } U = \begin{pmatrix} \alpha & \beta \\ \gamma & \delta \end{pmatrix} \in SL_2(\mathbb{Z})$$

with $q'(X,Y) = q(\alpha X + \beta Y, \gamma X + \delta Y)$ (invertible integer change of coordinates).

This simply means that $\begin{pmatrix} a' & \frac{b'}{2} \\ \frac{b'}{2} & c' \end{pmatrix} = U^t \begin{pmatrix} a & \frac{b}{2} \\ \frac{b}{2} & c \end{pmatrix} U.$

We get an equivalence relation for (positive definite) quadratic forms (with negative square-free discriminant $-D$.[17])

Equivalent quadratic forms $q(X, Y)$ and $q'(X, Y)$ represent the *same* integers:

$$u = q(x, y) \Longleftrightarrow u = q'(x', y') \text{ with } \begin{pmatrix} x' \\ y' \end{pmatrix} = U^{-1} \begin{pmatrix} x \\ y \end{pmatrix} \begin{bmatrix} U \text{ has the same} \\ \text{meaning as above.} \end{bmatrix}$$

Now, the equivalence relation for quadratic forms with discriminant $-D$ is "the same" as the equivalence relation for (normalized) fractional ideals of $K = \mathbb{Q}(\sqrt{-D})$. More precisely:

Exercise

Consider $\begin{pmatrix} a' & \frac{b'}{2} \\ \frac{b'}{2} & c' \end{pmatrix} = U^t \begin{pmatrix} a & \frac{b}{2} \\ \frac{b}{2} & c \end{pmatrix} U,$ with $U = \begin{pmatrix} \alpha & \beta \\ \gamma & \delta \end{pmatrix} \in SL_2(\mathbb{Z})$ and positive square-free $D = 4ac - b^2$, $a > 0$.

Put $\quad \tau = \frac{b+i\sqrt{D}}{2a} \quad \tau' = \frac{b'+i\sqrt{D}}{2a'}$. Then: $\tau' = \frac{\delta\tau+\beta}{\gamma\tau+\alpha}$.

Thus a *reduced triple* (a, b, c) (with $a, b, c \in \mathbb{Z}, a > 0$, $b^2 + D = 4ac$ and such that $|b| \le a \le c$ – with $b \ge 0$ whenever either $|b| = a$ or $a = c$) has two meanings:

It (minimally) represents

- The ideal class $[I] = [(a, \frac{b+i\sqrt{D}}{2})_{\mathbb{Z}}]$ of $K = \mathbb{Q}(\sqrt{-D})$
- The (class of the) quadratic form $q(X, Y) = aX^2 + bXY + cY^2$.

In the sequel, it will be convenient to work with *integer* symmetric matrices $\begin{pmatrix} A & B \\ B & C \end{pmatrix}$ such that $A > 0$; the determinant $D = AC - B^2$ is supposed to be positive and square-free (we shall always have $D \equiv 7 \bmod 8$).

Thus we consider (positive definite) quadratic forms $Q(X, Y) = AX^2 + 2BXY + CY^2$ with discriminant equal to $-4D$.

Note that this "arithmetical zoom to the order 4" is not completely harmless.

First, trivially, $\mathbb{Q}(\sqrt{-4D}) = \mathbb{Q}(\sqrt{-D})$.

But how can we get a faithful parametrization of the classgroup $Cl(\mathbb{Q}(\sqrt{-D})))$ in terms of (appropriately) *reduced matrices* $[A, B, C]^{18}$ = $\begin{pmatrix} A & B \\ B & C \end{pmatrix}$?

Let us explain our problem by a simple

---

[17] This definition of the quadratic-form discriminant aims to harmonize with field theory.

[18] This is a *definition* of the bracket-triple.

Example D = 7

We already know that $\text{Cl}(\mathbb{Q}(\sqrt{-7}))) = \{1\}$ (there is a single reduced $(a, b, c) = (1, 1, 2)$). But there are *two* non-$SL_2(\mathbb{Z})$-equivalent reduced quadratic forms with discriminant $-4 \cdot 7$:

$$[A, B, C] = \begin{cases} [1, 0, 7] \\ [2, 1, 4] \end{cases}$$

$\Big[$ Reduction via the fundamental-region argument: We have to search for all $[A, B, C]$ such that $D + B^2 = AC$, $|\, 2B \,| \leq A \leq C$ and $B \geq 0$ if $|\, 2B \,| = A$ or $A = C.\Big]$

We note that we will have to sift out "redundant" representatives $[A, B, C]$ by an extra-condition. This finally gives the following.

Proposition *Let* $-D \equiv 1 \bmod 4$.

Then $\text{Cl}(\mathbb{Q}(\sqrt{-D})))$ *can be identified with the set of symmetric matrices* $[A, B, C]$ *such that* $D + B^2 = AC$, *and which are strongly reduced in the following sense:*

(1) $|\, 2B \,| \leq A \leq C$ *and* $B \geq 0$ *if* $|\, 2B \,| = A$ *or* $A = C$.
(2) $\gcd[A, 2B, C] = 1$ *(i.e. either* A *or* C *has to be odd).*

The ambitious reader is invited to prove the statement as an exercise.

But let us look at an

Example D = 71. We get for $\text{Cl}(\mathbb{Q}(\sqrt{-71})))$ the following alternative parametrizations:

$$[A, B, C] = \begin{cases} [1, 0, 71], \\ [3, \pm 1, 24], \\ [8, \pm 1, 9], \\ [5, \pm 2, 15] \end{cases} \qquad (a, b, c) = \begin{cases} (1, 1, 18) \\ (2, \pm 1, 9), \\ (3, \pm 1, 6), \\ (4, \pm 3, 5). \end{cases}$$

For odd $B = b$, the transcription from $[A, B, C]$ to (a, b, c) is easy to understand. But for even B? How can we get $(4, \pm 3, 5)$ from $[5, \pm 2, 15]$?

Start with $Q(X, Y) = 5X^2 + 4XY + 15Y^2$ and substitute $X \longmapsto \frac{X+Y}{2}$, $Y \longmapsto \frac{Y-X}{2}$.

This gives $q(X, Y) = 4X^2 - 5XY + 6Y^2$. $(a, b, c) = (4, -5, 6)$ is not yet reduced.

$(4, -5, 6) \longmapsto (4, 3, 5)$ is the desired reduction: $b \longmapsto b \bmod 2a$ corresponds to

$$\begin{pmatrix} 1 & 0 \\ k & 1 \end{pmatrix} \begin{pmatrix} a & \frac{b}{2} \\ \frac{b}{2} & c \end{pmatrix} \begin{pmatrix} 1 & k \\ 0 & 1 \end{pmatrix} = \begin{pmatrix} a & \frac{2ka+b}{2} \\ \frac{2ka+b}{2} & ak^2 + bk + c \end{pmatrix}$$

for appropriate $k \in \mathbb{Z}$. This reads, in terms of the $\mathbf{\Gamma}$-action on $\mathbf{H}$: $T^k(\frac{b+i\sqrt{D}}{2a}) = \frac{b+i\sqrt{D}}{2a} + k = \frac{b+2ka+i\sqrt{D}}{2a}$. By the way, $S(\frac{b+i\sqrt{D}}{2a}) = \frac{-b+i\sqrt{D}}{2c}$ – which yields the reduction step $(a, b, c) \longmapsto (c, -b, a)$.

Finally note that our substitution $X \longmapsto \frac{X+Y}{2}, Y \longmapsto \frac{Y-X}{2}$ corresponds to

$$\tau = \frac{B + i\sqrt{D}}{A} \longmapsto \tau' = \frac{\tau + 1}{-\tau + 1} = \frac{b + i\sqrt{D}}{2a},$$

with

$$a = \frac{(A - 2B + C)}{4}, \quad b = \frac{(A - C)}{2}.$$

This finishes our summary on ideal classes of (certain) imaginary quadratic fields.

Now we are ready to treat our first problem: Let $m \in \mathbb{N}$ be a fixed field degree.

How can we decide whether a given positive square-free integer D (with $D \equiv 7 \bmod 8$) is a CM discriminant for 2^m?

The question is: Does the diophantine equation $2^{m+2} = X^2 + DY^2$ admit a *primitive* solution $(x, y) \in \mathbb{Z}^2$ (i.e. with coprime coordinates – note that x then must be odd)? (in another terminology: does the quadratic form $Q(X, Y) = X^2 + DY^2$ *properly* represent the integer 2^{m+2}?)

Before presenting the compliance test for D, we have to resolve some minor technical (and conceptual) problems.

Exercises

(1) (Finding square roots modulo a power of 2).
Let $d \equiv 1 \bmod 8$.
Consider the congruence $X^2 \equiv d \bmod 2^n$, $n \geq 3$.
Show that it admits exactly four solutions; more precisely: The four solutions are of the form $\{t, 2^{n-1} - t, 2^{n-1} + t, 2^n - t\}$ with $1 \leq t < 2^{n-2}$.

$\Big[$ Hint You should proceed by induction on n. Note that the four solutions on level n give eight candidates on level $n+1$, which split into two families: $\{t, 2^n - t, 2^n + t, 2^{n+1} - t\}$ and $\{2^{n-1} - t, 2^{n-1} + t, 2^n + 2^{n-1} - t, 2^{n+1} - 2^{n-1} + t\}$. Only one of them is acceptable. $\Big]$

(2) Find, following the "exclusion method" proposed by exercise (1), the four solutions of
$$X^2 + 71 \equiv 0 \bmod 1024$$

$\Big[$ Answer: 235, 277, 747, 789. $\Big]$

(3) (Square roots modulo a power of 2: The algorithm.)

Let $d \equiv 1 \bmod 8$.
Consider the congruence $X^2 \equiv d \bmod 2^n$, $n \geq 3$.
We search b, $1 \leq b < 2^{n-2}$, such that $b^2 \equiv d \bmod 2^n$ (note that b is *unique!*).

Show that the following algorithm yields b.

$d = \delta_{n-1}\delta_{n-2}\cdots\delta_1\delta_0$ (binary notation)

$t = \tau_{n-1}\tau_{n-2}\cdots\tau_1\tau_0$ (binary notation – b will be constructed via t step-by-step, i.e. from the tail to the head)

$u = u_{n-1}u_{n-2}\cdots u_1 u_0$ (binary notation – u means t^2, the approximation of $d = b^2$ according to the widening of the arithmetical window in the spirit of the argument of exercise (1))

1. Initialize: $t = 1$, $u = 1$.
   $\Big[$ Note that $\delta_2\delta_1\delta_0 = 001$ and $\tau_1\tau_0 = 01$ $\Big]$
2. $t \longmapsto b$:   For $2 \le j \le n-2$ do:
   If $u_{j+1} \ne \delta_{j+1}$, set $\tau_j = 1$.
   If $j < \frac{n}{2}$, set $u = (u + 2^{j+1}t - 2^{2j}) \bmod 2^n$.
   else   set $u = (u + 2^{j+1}t) \bmod 2^n$.
3. Normalization: If $\tau_{n-2} = 0$, set $b = t$, else set $b = 2^{n-1} - t$.

$\Big[$ Recall the hint to exercise (1): Whenever $X^2 \equiv d \bmod 2^{j+1}$ has exactly the solutions $x \equiv \pm t \bmod 2^j$, then $X^2 \equiv d \bmod 2^{j+2}$ has either $x \equiv \pm t \bmod 2^{j+1}$, or $x \equiv \pm(2^j + t) \bmod 2^{j+1}$ as its only solutions.
Now, $t^2 - d = 2^{j+1} \cdot k$, $k \in \mathbb{Z}$, means $\delta_j = u_j$, $\delta_{j-1} = u_{j-1}$, $\cdots$
k even $\Longleftrightarrow \delta_{j+1} = u_{j+1}$ k odd $\Longleftrightarrow \delta_{j+1} \ne u_{j+1}$.
But this parity alternative on k precisely corresponds to the alternative on the x-parametrization $\bmod 2^{j+1}$.$\Big]$

(4) Following the preceding algorithm, find the minimal positive solution b of:
(a) $X^2 \equiv 73 \bmod 256$
(b) $X^2 \equiv 73 \bmod 65536$
$\Big[$ $b = 29$ in the first case, and $b = 4253$ in the second case.$\Big]$
(5) Find, by means of the algorithm of exercise (3), the minimal positive integer b such that $b^2 + 71 \equiv 0 \bmod 2^{18}$.
$\Big[$ Answer $b = 17173$.$\Big]$
(6) Let $Q(X,Y) = AX^2 + 2BXY + CY^2$ be reduced: $|2B| \le A \le C$.

(a) Show that $Q(x,y) \ge (A - 2|B| + C) \cdot \min(x^2, y^2)$, and hence

$$Q(x,y) \ge A - 2|B| + C \quad \text{for } xy \ne 0.$$

(b) Deduce from (a) that the outer coefficients A and C of $Q(X,Y)$ give the minimum values properly represented by any equivalent form.

(7) Show the following assertion: If $[A,B,C]$ is equivalent to the reduced form $[A',B',C']$, then $[A,-B,C]$ is equivalent to the reduced form $[A',-B',C']$.
Consequence: If $B' = 0$ or necessarily positive ($B \geq 0$ if $| 2B |= A$ or $A = C$), then $[A,B,C]$ is equivalent to $[A,-B,C]$.

$\Big[$ Hint Every reduction is a chain of alternating actions of $S = \begin{pmatrix} 0 & -1 \\ 1 & 0 \end{pmatrix}$

and $T^k = \begin{pmatrix} 1 & k \\ 0 & 1 \end{pmatrix}$, for appropriate $k \in \mathbb{Z}$. Show that the assertion is

true for every step. $\Big]$

(8) Suppose that $Q(X,Y) = X^2 + DY^2$ properly represents 2^{m+2} (i.e. with relatively prime coordinates x and y).
Find an equivalent quadratic form $Q'(X,Y) = 2^{m+2}X^2 + 2BXY + CY^2$ (which trivially represents 2^{m+2}).

$\Big[$ Hint Write $2^{m+2} = x^2 + Dy^2$, $(x,y) \in \mathbb{Z}^2$ – where x and y are coprime. So, $rx + sy = 1$ for appropriate $r,s \in \mathbb{Z}$. Then compute $\begin{pmatrix} x & y \\ -s & r \end{pmatrix}\begin{pmatrix} 1 & 0 \\ 0 & D \end{pmatrix}\begin{pmatrix} x & -s \\ y & r \end{pmatrix}.\Big]$

(9) Show the following equivalence:

$$X^2 + DY^2 \text{ properly represents } 2^{m+2} \Longleftrightarrow \begin{array}{c} 4X^2 + 2BXY + CY^2 \\ (D + B^2 = 4C) \text{ properly} \\ \text{represents } 2^{m+2} \text{ by } (x,y), \\ \text{with } y \equiv 0 \bmod 2. \end{array}$$

$\Big[$ *Sketch of proof.* First note that $AX^2 + 2XY + CY^2 = \frac{1}{A}[(AX+BY)^2 + DY^2]$.
Thus $4X^2 + 2BXY + CY^2 = 2^{m+2} \Longleftrightarrow (2X + B\frac{Y}{2})^2 + D(\frac{Y}{2})^2 = 2^{m+2}$.
Then, you have to find the argument for the correctness of the transformations
$\begin{array}{ll} \xi = 2x + B\frac{y}{2} & x = \frac{1}{2}(\xi - B\eta) \\ \eta = \frac{y}{2} & y = 2\eta \end{array}\Big]$

(10) $D \geq 23,$[19] $D \equiv 7 \bmod 8$, $m \geq 1$.

Let $Q(X,Y) = 2^{m+2}X^2 + 2BXY + CY^2$ be a quadratic form with "leading coefficient" $A = 2^{m+2}$. So, $D + B^2 = 2^{m+2}C$.

(a) Show that there are – up to equivalence – only two couples of possibilities for $Q(X,Y)$: $[2^{m+2}, \pm B_1, C_1]$ and $[2^{m+2}, \pm B_2, C_2]$.
(b) If $X^2 + DY^2$ properly represents 2^{m+2}, then one of the couples reduces to $[1,0,D]$, whereas the other couple reduces to $[4, \pm B, C]$.

---

[19] $D = 7, 15$ present some particularities...

$\Big[$ Help

(a) $\begin{pmatrix} 1 & 0 \\ k & 1 \end{pmatrix} \begin{pmatrix} 2^{m+2} & B \\ B & C \end{pmatrix} \begin{pmatrix} 1 & k \\ 0 & 1 \end{pmatrix} = \begin{pmatrix} 2^{m+2} & B' \\ B' & C' \end{pmatrix}$ with $B' =$
$k \cdot 2^{m+2} + B$, i.e. $B' \equiv B \bmod 2^{m+2}$. Hence, appropriately choosing k, we get $0 < B' < 2^{m+2}$. But there are precisely four distinct roots of $X^2 + D \equiv 0 \bmod 2^{m+2}$ (cf exercise (1)).

(b) First note that reduction respects sign-alternatives in the middle coefficient B; so, whenever $X^2 + DY^2$ properly represents 2^{m+2}, one of the couples of (a) must reduce to $[1, 0, D]$. Now use exercise (9) – also look at the hints for the next group of exercises...$\Big]$

The **following algorithm** will

(1) Decide, whether the diophantine equation $2^{m+2} = X^2 + DY^2$ admits a primitive solution;
(2) Provide, in the affirmative case, the (odd) x-value of a solution.

The *idea* of the algorithm is simple: Start with the quadratic form $Q(X, Y) = 2^{m+2}X^2 + 2BXY + CY^2$ (such that $D + B^2 = 2^{m+2}C$) and find the *equivalent reduced form*. The answer of the algorithm is positive, whenever we arrive at the situation described in exercise (10); else, it is negative.

Let us first explain the basic *reduction step* of the algorithm:
It transforms a (positive definite) matrix

$$S = \begin{pmatrix} A & B \\ B & C \end{pmatrix}$$

(with $A > 0$) into an equivalent matrix

$$S' = \begin{pmatrix} A' & B' \\ B' & C' \end{pmatrix}$$

which will be "closer to reduced form". More precisely, let

$$S = \begin{pmatrix} A & B \\ B & C \end{pmatrix},$$

and put $\delta = \lfloor \frac{B}{C} + \frac{1}{2} \rfloor^{20}$,

$$U = \begin{pmatrix} 0 & -1 \\ 1 & \delta \end{pmatrix} \in SL_2(\mathbb{Z}).$$

Then

$$S' = \begin{pmatrix} A' & B' \\ B' & C' \end{pmatrix} = U^t S U.$$

---

[20] This is an asymmetrical round.

We have: $A' = C$, $B' = -B + \delta C$, $C' = A - 2\delta B + \delta^2 C$.
Since $\delta \leq \frac{B}{C} + \frac{1}{2} < \delta + 1$, i.e. $\delta C \leq B + \frac{1}{2}C < \delta C + C$, we get $\mid 2B' \mid \leq A'$.
For $A' \leq C'$, reduced form is achieved. Otherwise, we have to continue, etc.
How do we approach reducedness?

Since $D = A'C' - B'^2 = CC' - B'^2$, we have $D \leq CC'$, i.e. $\frac{D}{C^2} \leq \frac{C'}{A'}$. Hence:
$A' \leq C'$ whenever $C \leq \sqrt{D}$. Now, at every iteration of the reduction step, the
value of A decreases (we only iterate the reduction step in case $C < A$ – but
then $A' = C$). So, after a finite number of steps, $A' = C$ will be sufficiently
small; this means that we dispose of the desired inequality $C \leq \sqrt{D}$, and we
can stop.

Proposition (Testing for CM discriminants)
   *The following algorithm decides, for a square-free positive integer $D \geq 23$,
$D \equiv 7 \bmod 8$ and a field degree $m \geq 1$, whether D is a CM discriminant
for 2^m.*

   *In the affirmative case, it provides an (odd) integer x such that $2^{m+2} =
x^2 + Dy^2$ (for some $y \in \mathbb{Z}$).*

   Otherwise, the message will be "Not a CM discriminant".

Algorithm   *(1) Compute B such that $B^2 \equiv -D \bmod 2^{m+2}$ (see exercises
   (1) to (5)).*

(2) Initialize $A = 2^{m+2}$, $C = \frac{B^2+D}{2^{m+2}}$, $S = \begin{pmatrix} A & B \\ B & C \end{pmatrix}$, and $(x,y) = (1,0)$.

(3) Iterate the reduction step until $\mid 2B \mid \leq A \leq C$:
$$S = \begin{pmatrix} A & B \\ B & C \end{pmatrix} \longmapsto S' = \begin{pmatrix} A' & B' \\ B' & C' \end{pmatrix} = U^t SU,$$

with $\delta = \lfloor \frac{B}{C} + \frac{1}{2} \rfloor$, $U = \begin{pmatrix} 0 & -1 \\ 1 & \delta \end{pmatrix} \in SL_2(\mathbb{Z})$.

Compute $(x', y') = (\delta x + y, -x)$.
Reset the target-variables to source-variables.
(4) The final decision:
   If $A = 1$, then output x and stop.
   If $A = 4$ and $y = 2\eta$ is even, then output $2x + B\eta$ and stop.
   Otherwise, output "Not a CM discriminant".

Proof Everything has been treated in previous exercises. The exclusion of
$D = 7, 15$ is explained in the next set of exercises.

Note that we can get rid of this exclusion by a modification of the algorithm
which stems from the following observation (cf the hints to the next exercises):

Whenever we initialize with a middle-term B that corresponds to an *odd*
outer coefficient C – and this is one alternative for the minimal positive choice
of B – then the alternative in the final decision disappears, and we can treat
all $D \equiv 7 \bmod 8$ – without any exclusion.

Exercises

(1) Show that our reduction algorithm always provides a *primitive* solution (x, y) of the final reduced equation $Q(X, Y) = 2^{m+2}$ (primitive: x and y are relatively prime).

(2) $D = 7$.
Note that $X^2 + 7Y^2 = 64$ admits $(1, 3)$ as a primitive solution.
Now consider $Q_1(X, Y) = 64X^2 + 22XY + 2Y^2$ and $Q_2(X, Y) = 64X^2 + 42XY + 7Y^2$.
(a) Show that $Q_1(X, Y)$ reduces to $2X^2 + 2XY + 4Y^2$.
(b) Show that $Q_2(X, Y)$ reduces to $X^2 + 7Y^2$.
Thus we have to be cautious: (a) will provoke a negative (hence false) output of our test algorithm.

(3) $D = 7$. $m \geq 1$.
Show that $X^2 + 7Y^2 = 2^{m+2}$ always admits a primitive solution (x, y).
$\left[\right.$ Hint $2^{m+2}X^2 + 2BXY + CY^2$, with $7 + B^2 = 2^{m+2} \cdot C$, reduces to either $[1, 0, 7]$ or $[2, 1, 4]$.
The alternative depends on the parity of the outer coefficient C.
Show: Replacing B by $2^{m+1} - B$ changes the parity of C. $\left.\right]$

(4) $D = 15$.
(a) Show that we get four reduced quadratic forms $AX^2 + 2BXY + CY^2$ and that the classnumber $h(\mathbb{Q}(\sqrt{-15})) = 2$.
(b) Show that $X^2 + 15Y^2$ properly represents $2^4 = 16$ and $2^6 = 64$, but not $2^5 = 32$ and $2^7 = 128$.
(c) Now consider the case $2^{m+2} = 2^8 = 256$.
Start with $Q_1(X, Y) = 256X^2 + 78XY + 6Y^2$ and $Q_2(X, Y) = 256X^2 + 178XY + 31Y^2$.
Show that $Q_1(X, Y)$ is equivalent to $4X^2 + 2XY + 4Y^2$ and that $Q_2(X, Y)$ is equivalent to $X^2 + 15Y^2$.
Caution *The reduction of $Q_1(X, Y)$ to $4X^2 + 2XY + 4Y^2$ transforms $(1, 0)$ to the primitive solution $(6, -7)$, which will cause a negative answer of our compliance test. Clearly, a transposition $X \longleftrightarrow Y$ would resolve our problem ...*
(d) Finally, consider the case $2^{m+2} = 2^9 = 512$.
Start with $Q_1(X, Y) = 512X^2 + 78XY + 3Y^2$ and $Q_2(X, Y) = 512X^2 + 434XY + 92Y^2$.
Show that $Q_1(X, Y)$ is equivalent to $3X^2 + 5Y^2$ and that $Q_2(X, Y)$ is equivalent to $2X^2 + 2XY + 8Y^2$.
Thus $D = 15$ is not a CM discriminant for $2^7 = 128$.

(5) $D = 15$. $m \geq 1$.
(a) Show that $X^2 + 15Y^2$ properly represents 2^{m+2} $\Longleftrightarrow$ m is even.
(b) Show that $2X^2 + 2XY + 8Y^2$ properly represents 2^{m+2} $\Longleftrightarrow$ m is odd.

$\Big[$ Hint First note that starting our test with $Q(X,Y) = 2^{m+2}X^2+2BXY+ CY^2$ (such that $D + B^2 = 2^{m+2} \cdot C$), we always may assume $B = t$ or $B = 2^{m+1} - t$, with $0 < t < 2^m$. For one of these cases, C is even, for the other C is odd. Only the *odd case* can reduce to $X^2 + DY^2$.

For $D = 15$ we get: The odd case reduces to either $[1, 0, 15]$ or $[3, 0, 5]$. The even case reduces to either $[2, 1, 8]$ or $[4, 1, 4]$.

(a) You should eliminate undesirable constellations by arguments mod 3.

(b) $2X^2 + 2XY + 8Y^2 = 2^{m+2} \iff (2X + Y)^2 + 15Y^2 = 2^{m+3}.\Big]$

(6) $D = 23$.

   (a) Show that $D = 23$ is a CM discriminant for $2^3 = 8$, but not a CM discriminant for $2^4 = 16$.

   (b) Is $D = 23$ a CM discriminant for (i) $2^5 = 32$?
                                          (ii) $2^6 = 64$?

   $\Big[$ Answer (b)(i) no   (ii) yes.

   Note that (i) gives an interesting negative answer: $X^2 + 23Y^2 = 128$ admits the solution $(x, y) = (6, 2)$. But $128X^2 + 38XY + 3Y^2$ reduces to $3X^2 - 2XY + 8Y^2$ (and $128X^2 + 90XY + 16Y^2$ reduces to $2X^2 + 2XY + 12Y^2$). Thus there is *no primitive* solution of $X^2 + 23Y^2 = 128$.

   (ii) $256X^2 + 90XY + 8Y^2$ reduces to $4X^2 + 2XY + 6Y^2$, which admits the solution $(x, y) = (5, -6)$. This gives the solution $(x, y) = (7, -3)$ for $X^2 + 23Y^2 = 256.\Big]$

(7) $D = 71$.

   Show, following the reduction method of our compliance test, that $D = 71$ is *not* a CM discriminant for $2^8 = 256$.

   $\Big[$ Solution: Let us be generous: We shall reduce both $Q_1(X,Y) = 1024X^2 + 470XY + 54Y^2$ and $Q_2(X,Y) = 1024X^2 + 554XY + 75Y^2$.
   In the first case, we obtain $8X^2 + 6XY + 10Y^2$.

   In the second case, we obtain $8X^2 + 2XY + 9Y^2.\Big]$

(8) Decide whether $D = 71$ is a CM discriminant for 2^{16}.

   $\Big[$ Solution: The reduction steps of our algorithm:
   $[262144, 17173, 1125] \longmapsto [1125, -298, 79] \longmapsto [79, -18, 5] \longmapsto [5, -2, 15]$

   Hence $D = 71$ is not a CM discriminant for $2^{16}.\Big]$

Let us sum up:

The essential mathematical content of this paragraph is the following:

First, the couple $(N, 2^m)$ which describes the desired field degree m as well as the desired number of points N on the elliptic curve to be designed, introduces the integer D, which is interpreted as the (opposite of the) discriminant of an imaginary quadratic number field $K = \mathbb{Q}(\sqrt{-D})$.

Then, the $h = h(K)$ ideal classes of $K = \mathbb{Q}(\sqrt{-D})$ define h distinct *lattices* of the complex plane, parameterised by h numbers $\tau_1 = \frac{b_1 + i\sqrt{D}}{2a_1}, \ldots,$ $\tau_h = \frac{b_h + i\sqrt{D}}{2a_h}$ in the fundamental region (of the modular group).

It is precisely the appearance of this "lattice-shaped" geometry which will help us to realize our programme.

Note that the quadratic forms are only "technical slaves", without conceptual importance.

Finding a Nearly Prime Order

Having fixed the desired field degree $m \geq 1$ and a lower bound for the prime order n of the cyclic subgroup to support our signature arithmetic, we have to find

> a square-free positive integer $D \equiv 7 \bmod 8$ and a sufficiently large prime n, such that $N = 4n$ is the order of an elliptic curve over $\mathbb{F}_{2^m}$ with complex multiplication by D.

We have to proceed as follows:

1. Choose a square-free positive $D \equiv 7 \bmod 8$.
2. Compute $h = h(\mathbb{Q}(\sqrt{-D}))$.
3. If m does not divide h, return to 1.
4. Test whether D is a CM discriminant for 2^m. If not, return to 1. Otherwise the result of the compliance test is $x \in \mathbb{Z}$.
5. The possibilities for N are $2^m + 1 \pm x$. Now the question is whether $N = 4n$, with n prime (and sufficiently large). If the corresponding order is o.k., output $(D, 4n)$. Else return to 1.

Example Fix $m = 155$.

For $D = 942679$ we get $h = 620$. So, m correctly divides h.

Our CM discriminant test yields: $2^{157} = x^2 + Dy^2$, with

$x = 229529878683046820398181,$

$y = -371360755031779037497.$

Finally, we get

$$2^{155} + 1 - x = 4n,$$

where n is the prime $n = 1141798154164767904846623037312629032935 6873447.$

Thus there is an elliptic curve over $\mathbb{F}_{2^{155}}$ of order $N = 4n$ having complex multiplication by D.

Reduced Class Polynomials and their Roots Modulo 2

After these rather lengthy preliminaries on CM discriminants, our next goal is a shameless and direct algorithmic attack of our main problem: to find quickly the constant term b of the desired elliptic equation.

First let us have a look at the tip of the modular function iceberg: Consider

$$F(z) = 1 + \sum_{j=1}^{\infty} (-1)^j \left(z^{(3j^2-j)/2} + z^{(3j^2+j)/2} \right) = 1 - z - z^2 + z^5 + z^7 - z^{12} - z^{15} + \cdots$$

Let D be a positive square-free integer, $D \equiv 7 \bmod 8$, and let $[A, B, C]$ be an integer symmetric matrix with determinant $D = AC - B^2$.
Put $\theta = \exp(\frac{-\sqrt{D}+B i}{A}\pi)$

Let
$$\mathbf{f_0}(A, B, C) = \theta^{-\frac{1}{24}} \cdot \frac{F(-\theta)}{F(\theta^2)}$$

$$\mathbf{f_1}(A, B, C) = \theta^{-\frac{1}{24}} \cdot \frac{F(\theta)}{F(\theta^2)}$$

$$\mathbf{f_2}(A, B, C) = \sqrt{2} \cdot \theta^{\frac{1}{12}} \cdot \frac{F(\theta^4)}{F(\theta^2)}$$

Observation $|\theta| < e^{-\pi\sqrt{3}/2} = 0.0658287$.

Thus we control perfectly the speed of convergence of the power series $F(z)$ used in computing the numbers $\mathbf{f}_k(A, B, C)$, $k = 0, 1, 2$.

If D and $[A, B, C]$ are as above, then we define the *class invariant* $c[A, B, C]$ of $[A, B, C]$ as follows:

$$c[A, B, C]_0 = \begin{cases} \frac{1}{\sqrt{2}} \cdot (-1)^{\frac{A^2-1}{8}} \cdot e^{\frac{\pi i}{24} B(A-C+A^2C)} \cdot \mathbf{f_0}(A, B, C) & AC \text{ odd} \\ \frac{1}{\sqrt{2}} \cdot e^{\frac{\pi i}{24} B(A+2C-AC^2)} \cdot \mathbf{f_1}(A, B, C) & C \text{ even} \\ \frac{1}{\sqrt{2}} \cdot e^{\frac{\pi i}{24} B(A-C-AC^2)} \cdot \mathbf{f_2}(A, B, C) & A \text{ even} \end{cases}$$

$$c[A, B, C] = \begin{cases} c[A, B, C]_0 & \text{for} \quad D \not\equiv 0 \bmod 3; \\ (c[A, B, C]_0)^3 & \text{for} \quad D \equiv 0 \bmod 3. \end{cases}$$

Let $h = h(\mathbb{Q}(\sqrt{-D}))$ be the classnumber of $K = \mathbb{Q}(\sqrt{-D})$, and let $[A_1, B_1, C_1]$ $\ldots [A_h, B_h, C_h]$ be the h reduced symmetric matrices which represent faithfully the ideal classes.

Definition *The reduced class polynomial for D*

$$w_D(t) = \prod_{j=1}^{h} (t - c[A_j, B_j, C_j]).$$

Proposition $w_D(t) \in \mathbb{Z}[t]$ (*i.e. $w_D(t)$ has integer coefficients*).

Note that this result facilitates the computation of $w_D(t)$: You only need a rough approximation of $w_D(t)$, which uniquely rounds to an integer-coefficient result.

Example D = 71.

Recall the seven reduced symmetric matrices with determinant $D = 71$:
$[1, 0, 7]$, $[3, \pm 1, 24]$, $[8, \pm 1, 9]$, $[5, \pm 2, 15]$,

$$\mathbf{c}[1, 0, 7] = \frac{1}{\sqrt{2}}\mathbf{f}_0(1, 0, 7),$$

$$\mathbf{c}[3, 1, 24] = \frac{1}{\sqrt{2}}e^{\frac{\pi i}{8}}\mathbf{f}_1(3, 1, 24),$$

$$\mathbf{c}[3, -1, 24] = \frac{1}{\sqrt{2}}e^{-\frac{\pi i}{8}}\mathbf{f}_1(3, -1, 24),$$

$$\mathbf{c}[8, 1, 9] = \frac{1}{\sqrt{2}}e^{\frac{23\pi i}{24}}\mathbf{f}_2(8, 1, 9),$$

$$\mathbf{c}[8, -1, 9] = \frac{1}{\sqrt{2}}e^{\frac{-23\pi i}{24}}\mathbf{f}_2(8, -1, 9)$$

$$\mathbf{c}[5, 2, 15] = -\frac{1}{\sqrt{2}}e^{\frac{5\pi i}{12}}\mathbf{f}_0(5, 2, 15),$$

$$\mathbf{c}[5, -2, 15] = -\frac{1}{\sqrt{2}}e^{\frac{-5\pi i}{12}}\mathbf{f}_0(5, -2, 15).$$

We obtain:

$$\mathbf{c}[1, 0, 7] = 2.1306068298388953300559146868894 2503\ldots$$
$$\mathbf{c}[3, 1, 24] = 0.9596917853056702525079704764550 7504\ldots$$
$$+0.3491607100126965479985531629392 6907\ldots i$$
$$\mathbf{c}[3, -1, 24] = 0.9596917853056702525079704764550 7504\ldots$$
$$-0.3491607100126965479985531629392 6907\ldots i$$
$$\mathbf{c}[8, 1, 9] = -0.7561356880400178905356401098531 772\ldots$$
$$+0.0737508631630889005240764944567 675\ldots i$$
$$\mathbf{c}[8, -1, 9] = -0.7561356880400178905356401098531 772\ldots$$
$$-0.0737508631630889005240764944567 675\ldots i$$
$$\mathbf{c}[5, 2, 15] = -0.2688595121851000270002877100466 102\ldots$$
$$-0.8410857740132980010364863422490 5292\ldots i$$
$$\mathbf{c}[5, -2, 15] = -0.2688595121851000270002877100466 102\ldots$$
$$+0.8410857740132980010364863422490 5292\ldots i$$

This yields: $w_{71}(t) = t^7 - 2t^6 - t^5 + t^4 + t^3 + t^2 - t - 1$.

Now we are able to attack our problem:

We start with: A field $\mathbb{F}_{2^m}$, a CM discriminant D for 2^m, and the desired curve order $N = 4n$.

We search: $b \in \mathbb{F}_{2^m}$ such that the elliptic curve $\quad Y^2 + XY = X^3 + b$ over $\mathbb{F}_{2^m}$ has order $N = 4n$.[21]

Recall n is a large prime number...

---

1. Compute $w_D(t)$ and $w(t) = w_D(t) \bmod 2$.

2. Find the smallest divisor d of m greater than $(\log_2 D) - 2$ such that D is a CM discriminant for 2^d.

3. Find $p(t) = $ a degree d factor of $w(t)$ (cf exercise (1)).

4. Compute $\beta = $ a root in $\mathbb{F}_{2^m}$ of $p(t) = 0$ (cf exercises (2), (3)).

5. $b = \begin{cases} \beta & \text{if } D \equiv 0 \bmod 3 \\ \beta^3 & \text{else} \end{cases}$

This is a somehow enigmatic protocol. We shall try to give the necessary explanations in the section "Elliptic Curves with Complex Multiplication".

Exercises

(1) Let the binary polynomial $f(t) = f_1(t) \cdots f_r(t)$, $r \geq 2$, be a product of r irreducible factors of the same degree d.

The following algorithm will produce a random degree-d factor of $f(t)$:
First the participants:

$g(t) \equiv $ a divisor of $f(t)$ whose degree will become – along several reduction rounds – smaller and smaller until we get the final random irreducible factor of $f(t)$.

$h(t) \equiv $ the end-of-round reduction of $g(t)$.

$c(t) \equiv $ a step-by-step approximation of $h(t)$.

The algorithm:

1. $g(t) = f(t)$.
2. While $\deg g(t) > d$:

---

[21] An elliptic curve with the considered one-parameter equation – i.e. without the X^2-term – *necessarily* has order $N \equiv 0 \bmod 4$!

2.1 Choose $u(t) \equiv$ a random polynomial of degree $2d - 1$.

2.2 Set $c(t) = u(t)$.

2.3 The (trial of) reduction round:
Compute $d - 1$ times $c(t) = c(t)^2 + u(t) \bmod g(t)$

2.4 Compute $h(t) = gcd(c(t), g(t))$
$h(t)$ is expected to be a shorter product of irreducible factors of $f(t)$ than $g(t)$. If this is not true or if $h(t) = 1$, return to 2.1.

2.5 Else, choose between $h(t)$ and $g(t)/h(t)$: the shorter expression will become the new $g(t)$.

3. Output $g(t)$.

Now the *exercise*; show the following:
Let $u(t)$ be a multiple of one of the irreducible factors of $g(t)$: $u(t) = m(t)f_j(t)$. Then $f_j(t)$ will be a factor of $h(t)$.

$\Big[$ **Recall:** $d(t) = gcd(a(t), b(t))$ via iterated Euclidean division:
1. $\alpha(t) = a(t)$, $\beta(t) = b(t)$.
2. While $\beta(t) \neq 0$: $\rho(t) = \alpha(t) \bmod \beta(t)$
   Reset $\alpha(t) = \beta(t)$, $\beta(t) = \rho(t)$.
3. $d(t) = \alpha(t).\Big]$

(2) Finding a root in $\mathbb{F}_{2^m}$ of an irreducible binary polynomial.

If $f(t)$ is an irreducible binary polynomial of degree m, then $f(t)$ has m distinct roots in the field $\mathbb{F}_{2^m}$.

(a) Try an (almost word-by-word) copy of the foregoing algorithm in order to obtain a random root of $f(t)$ in $\mathbb{F}_{2^m}$.
(b) Apply the algorithm to $f(t) = t^4 + t^3 + t^2 + t + 1$ and $\mathbb{F}_{16} = \mathbb{F}_2[x]/(x^4 + x + 1)$.

$\Big[$ Help (a) First consider the adaptation of the previous algorithm to our needs:

1. Set $g(t) = f(t)$.
2. While $\deg g(t) > 1$:

   2.1 Choose a random $u \in \mathbb{F}_{2^m}$.
   2.2 Set $c(t) = ut$.

   2.3 The (trial of) reduction round:

   Compute $m - 1$ times $c(t) = (c(t)^2 + ut) \bmod g(t)$

    2.4 Compute $h(t) = gcd(c(t), g(t))$

        $h(t)$ is expected to be of smaller degree than $g(t)$. If this is not true or if $h(t)$ is constant, return to 2.1.

    2.5 Else, choose between $h(t)$ and $g(t)/h(t)$: the shorter expression will become the new $g(t)$.

3. Output $g(0)$.

(b) Put $\alpha = x \bmod (x^4 + x + 1)$.

    We know: $f(t) = t^4 + t^3 + t^2 + t + 1 = (t + \alpha^3)(t + \alpha^6)(t + \alpha^9)(t + \alpha^{12})$.

    Let us try to find one of these four roots by means of our algorithm.

(i) First, we choose $u = \alpha^2$. The reduction round yields $c(t) = \alpha^{10}t^3 + \alpha^5t^2 + t + \alpha^8$. $h(t) = \alpha^5t^2 + \alpha^7t + \alpha^{14}$. This is our new $g(t)$.

(ii) Now, we choose $u = \alpha^3$ (one of the roots...) and get finally $c(t) = 1$. You are surprised? Note that we start with ut and not with $t - u - \mathrm{so}$ there is no reason for an analogy to the claim of exercise (1).

(iii) Finally, we take $u = \alpha$. Now, the reduction round yields $c(t) = \alpha^{13}t + \alpha$. $h(t) = t + \alpha^3$ allows to conclude.$\Big]$

(3) Let $f(t) = t^m + \cdots + 1$ be an (irreducible) binary polynomial (which admits m distinct roots in $\mathbb{F}_{2^m}$). Let b_1, b_2 be two roots of $f(t)$.

$$E_1 \equiv Y^2 + XY = X^3 + b_1$$

$$E_2 \equiv Y^2 + XY = X^3 + b_2$$

Show that $E_1 \simeq E_2$ (as abstract groups).

$\Big[$Hint First, observe that the set of roots of $f(t)$ is of the form $\{b, b^2, b^4, \ldots b^{2^{m-1}}\}$.

It suffices to show the claim for b and b^2.

Now show that the map $E_1 \ni (x, y) \longmapsto (x^2, y^2) \in E_2$ is well-defined, compatible with the group operations, and bijective.$\Big]$

(4) $D = 23$.

    (a) Compute $w_{23}(t) = t^3 - t - 1$.

    (b) $D = 23$ is a CM discriminant for $8 = 2^3$. We get over $\mathbb{F}_8 = \mathbb{F}_2[x]/(x^3 + x + 1)$: $w_3(t) \bmod 2 = t^3 + t + 1 = (t + \alpha)(t + \alpha^2)(t + \alpha^4)$,   with $\alpha = x \bmod (x^3 + x + 1)$.

    This yields three elliptic curves over $\mathbb{F}_8$:

$$E_1 \equiv Y^2 + XY = X^3 + \alpha^3,$$

$$E_2 \equiv Y^2 + XY = X^3 + \alpha^6,$$

$$E_3 \equiv Y^2 + XY = X^3 + \alpha^5.$$

Show that each curve is isomorphic to $\mathbb{Z}/12\mathbb{Z}$ (as an abstract group), and find a base-point (a generator) in each case.

(c) $D = 23$ is also a CM discriminant for $64 = 2^6$.

Consider $\mathbb{F}_8$ as a subfield of $\mathbb{F}_{64} = \mathbb{F}_2[x]/(x^6 + x + 1)$ in the following way: Put $\beta = x \bmod (x^6 + x + 1)$ and $\alpha = \beta^{27} = \beta^3 + \beta^2 + \beta$. Then $\alpha^3 + \alpha + 1 = 0$.

Thus $\mathbb{F}_8 = \mathbb{F}_2[\alpha] \subset \mathbb{F}_{64}$. Furthermore, $w_3(t) \bmod 2$ factors now as follows: $t^3 + t + 1 = (t + \beta^{27})(t + \beta^{54})(t + \beta^{45})$. We get over $\mathbb{F}_{64}$:

$$E_1 \equiv Y^2 + XY = X^3 + \beta^{18},$$

$$E_2 \equiv Y^2 + XY = X^3 + \beta^{36},$$

$$E_3 \equiv Y^2 + XY = X^3 + \beta^9.$$

The order of each of the three curves is 72 (why?). Find points of order 24 on each curve. Are there points of order 72 (i.e. are the curves cyclic groups)?

$\Big[$ Solution: (a) Choose $f(z) = 1 - z - z^2 + z^5 + z^7 - z^{12} - z^{15}$ as an optimistic approximation of $F(z)$. With $\quad \theta_0 = \exp(-\sqrt{23} \cdot \pi)$, $\theta_1 = \exp(\frac{-\sqrt{23}+i}{3}\pi)$, $\theta_2 = \exp(\frac{-\sqrt{23}-i}{3}\pi)$
we get

$\mathbf{f}_0(1,0,23) \approx \theta_0^{-\frac{1}{24}} \cdot \frac{f(-\theta_0)}{f(\theta_0^2)}$ $\qquad$ $\mathbf{c}[1,0,23] = \frac{1}{\sqrt{2}} \cdot \mathbf{f}_0(1,0,23)$

$\mathbf{f}_1(3,1,8) \approx \theta_1^{-\frac{1}{24}} \cdot \frac{f(\theta_1)}{f(\theta_1^2)}$ $\quad$ and $\quad$ $\mathbf{c}[3,1,8] = \frac{1}{\sqrt{2}} \cdot e^{\frac{19\pi i}{24}} \cdot \mathbf{f}_1(3,1,8)$

$\mathbf{f}_1(3,-1,8) \approx \theta_2^{-\frac{1}{24}} \cdot \frac{f(\theta_2)}{f(\theta_2^2)}$ $\qquad$ $\mathbf{c}[3,-1,8] = \frac{1}{\sqrt{2}} \cdot e^{\frac{-19\pi i}{24}} \cdot \mathbf{f}_1(3,-1,8)$

We approximately obtain:
$\mathbf{c}[1,0,23] = 1.324717957$,
$\mathbf{c}[3,1,8] = -0.6623589785 + 0.5622795120i$,
$\mathbf{c}[3,-1,8] = -0.6623589785 - 0.5622795120i$,

(b) $X^2 + 23Y^2 = 32$ admits the primitive solution $(x,y) = (3,1)$. This gives the order 9 ± 3. Since we have chosen the option of an equation without X^2-term, the order must be divisible by 4. Three base-points: $E_1 : G_1 = (\alpha^5, \alpha)$, $E_2 : G_2 = (\alpha^3, \alpha^2)$, $E_3 : G_3 = (\alpha^6, \alpha^3)$. (c) $X^2 + 23Y^2 = 256$ admits the primitive solution $(x,y) = (7,3)$. This gives the order 65 ± 7. Now we can argue as before, or use (b): Since the group of $\mathbb{F}_8$-rational points is a subgroup of the group of $\mathbb{F}_{64}$-rational points, the order of the latter must be divisible by 12. Three samples for points of order 24 on $E_1 \equiv Y^2 + XY = X^3 + \beta^{18}$: $P_1 = (\beta^{10}, \beta^{31})$, $P_2 = (\beta^{17}, \beta^{38})$, $P_3 = (\beta^{30}, \beta^{19})$. The underlying abstract group is *not cyclic.*$\Big]$

(5) $D = 39$.

    (a) Compute $w_{39}(t) = t^4 - 3t^3 - 4t^2 - 2t - 1$.

    (b) Show that $D = 39$ is a CM discriminant for $16 = 2^4$ and for $256 = 2^8$.

    (c) First consider the scalar field $\mathbb{F}_{16} = \mathbb{F}_2[x]/(x^4 + x + 1)$. Put $\alpha = x \bmod (x^4 + x + 1)$.

    $w_{39}(t) \bmod 2 = t^4 + t^3 + 1 = (t + \alpha^7)(t + \alpha^{11})(t + \alpha^{13})(t + \alpha^{14})$. This yields four elliptic curves over $\mathbb{F}_{16}$:

$$E_1 \equiv Y^2 + XY = X^3 + \alpha^7,$$

$$E_2 \equiv Y^2 + XY = X^3 + \alpha^{11},$$

$$E_3 \equiv Y^2 + XY = X^3 + \alpha^{13},$$

$$E_4 \equiv Y^2 + XY = X^3 + \alpha^{14}.$$

Show that E_1 is a cyclic group of order 12, generated by $G_1 = (\alpha^{12}, \alpha^4)$. Find base-points (i.e. generators) for E_2, E_3, E_4.

    (d) Now extend the scalar field $\mathbb{F}_{16}$ to $\mathbb{F}_{256} = \mathbb{F}_2[x]/(x^8 + x^4 + x^3 + x^2 + 1)$. Put $\beta = x \bmod (x^8 + x^4 + x^3 + x^2 + 1)$.[22]
$\mathbb{F}_{16} = \mathbb{F}_2[\alpha] \subset \mathbb{F}_{256}$, with $\alpha = \beta^{17} \equiv 10011000$.
Thus, over $\mathbb{F}_{256}$, we get
$w_{39}(t) \bmod 2 = t^4 + t^3 + 1 = (t + \beta^{119})(t + \beta^{187})(t + \beta^{221})(t + \beta^{238})$.
Our four elliptic curves now read as follows:

$$E_1 \equiv Y^2 + XY = X^3 + 10010011,$$

$$E_2 \equiv Y^2 + XY = X^3 + 11011100,$$

$$E_3 \equiv Y^2 + XY = X^3 + 01000101,$$

$$E_4 \equiv Y^2 + XY = X^3 + 00001011.$$

Show that each curve has order 264. What about cyclicity?

----

[22] Recall the log table for $\mathbb{F}_{256}^*$ in the AES-section.

$\Big[$ Help (a) We replace $F(z)$ by $f(z) = 1 - z - z^2 + z^5 + z^7 - z^{12} - z^{15}$.

With $\quad \theta_0 = \exp(-\sqrt{39} \cdot \pi)$, $\theta_1 = \exp(-\sqrt{39} \cdot \frac{\pi}{3})$, $\theta_2 = \exp(\frac{-\sqrt{39}+\mathrm{i}}{5}\pi)$,

$\theta_3 = \exp(\frac{-\sqrt{39}-\mathrm{i}}{5}\pi)$

we get

$$\mathbf{f}_0(1,0,39) \approx \theta_0^{-\frac{1}{24}} \cdot \frac{f(-\theta_0)}{f(\theta_0^2)} \qquad \mathbf{c}[1,0,39] = \left\{ \frac{1}{\sqrt{2}} \cdot \mathbf{f}_0(1,0,39) \right\}^3$$

$$\mathbf{f}_0(3,0,13) \approx \theta_1^{-\frac{1}{24}} \cdot \frac{f(-\theta_1)}{f(\theta_1^2)} \qquad \mathbf{c}[3,0,13] = \left\{ -\frac{1}{\sqrt{2}} \cdot \mathbf{f}_0(3,0,13) \right\}^3$$

and

$$\mathbf{f}_1(5,1,8) \approx \theta_2^{-\frac{1}{24}} \cdot \frac{f(\theta_2)}{f(\theta_2^2)} \qquad \mathbf{c}[5,1,8] = \left\{ \frac{1}{\sqrt{2}} \cdot e^{-\frac{11\pi i}{24}} \cdot \mathbf{f}_1(5,1,8) \right\}^3$$

$$\mathbf{f}_1(5,-1,8) \approx \theta_3^{-\frac{1}{24}} \cdot \frac{f(\theta_3)}{f(\theta_3^2)} \qquad \mathbf{c}[5,-1,8] = \left\{ \frac{1}{\sqrt{2}} \cdot e^{\frac{11\pi i}{24}} \cdot \mathbf{f}_1(5,-1,8) \right\}^3$$

You should obtain:
$\mathbf{c}[1,0,39] = 4.106964608$
$\mathbf{c}[3,0,13] = -0.8041889695$
$\mathbf{c}[5,1,8] = -0.1513878187 + 0.5290154688i$
$\mathbf{c}[5,-1,8] = -0.1513878187 - 0.5290154688i$

(b) $X^2 + 39Y^2 = 64$

admits the primitive solution $(x,y) = (5,1)$. $1024X^2 + 822XY + 165Y^2$ reduces to $X^2 + 39Y^2$; the initial solution $(x,y) = (1,0)$ finally becomes $(x,y) = (-7,5)$. (c) (d) follow the strategy of exercise (4). $\Big]$

(6) $D = 95$.

(a) Find the eight reduced matrices $[A,B,C]$ such that $AC - B^2 = 95$.

(b) Compute $w_{95}(t) = t^8 - 2t^7 - 2t^6 + t^5 + 2t^4 - t^3 + t - 1$.

$\Big[$ Answer (b) We get $w_{95}(t) = (t - \frac{1}{\sqrt{2}} \cdot \mathbf{f}_0(1,0,95))(t + \frac{1}{\sqrt{2}} \cdot \mathbf{f}_0(5,0,19))(t - \frac{1}{\sqrt{2}} \cdot e^{\frac{19\pi i}{24}} \cdot \mathbf{f}_1(3,1,32))(t - \frac{1}{\sqrt{2}} \cdot e^{-\frac{19\pi i}{24}} \cdot \mathbf{f}_1(3,-1,32))(t - \frac{1}{\sqrt{2}} \cdot e^{\frac{\pi i}{12}} \cdot \mathbf{f}_0(9,2,11))(t - \frac{1}{\sqrt{2}} \cdot e^{-\frac{\pi i}{12}} \cdot \mathbf{f}_0(9,-2,11))(t - \frac{1}{\sqrt{2}} \cdot e^{\frac{3\pi i}{8}} \cdot \mathbf{f}_2(8,3,13))(t - \frac{1}{\sqrt{2}} \cdot e^{-\frac{3\pi i}{8}} \cdot \mathbf{f}_2(8,-3,13))$.

The approximate values of the class invariants:
$\mathbf{c}[1,0,95] \approx 2.532681963$
$\mathbf{c}[5,0,19] \approx -0.9146479740$
$\mathbf{c}[3,\pm 1,32] \approx -0.8287391440 \pm 0.6954398630i$
$\mathbf{c}[9,\pm 2,11] \approx 0.8090169935 \pm 0.2096901122i$
$\mathbf{c}[8,\pm 3,13] \approx 0.2107051554 \pm 0.6954398639i$ $\Big]$

	Reduced class polynomials for small discriminants
D	$w_D(t)$
7	$t - 1$
15	$t^2 - t - 1$
23	$t^3 - t - 1$
31	$t^3 - t^2 - 1$
39	$t^4 - 3t^3 - 4t^2 - 2t - 1$
47	$t^5 - t^3 - 2t^2 - 2t - 1$
55	$t^4 - 2t^3 + t - 1$
71	$t^7 - 2t^6 - t^5 + t^4 + t^3 + t^2 - t - 1$
79	$t^5 - 3t^4 + 2t^3 - t^2 + t - 1$
87	$t^6 - 13t^5 - 11t^4 + 4t^3 - 4t^2 - t - 1$
95	$t^8 - 2t^7 - 2t^6 + t^5 + 2t^4 - t^3 + t - 1$
103	$t^5 - t^4 - 3t^3 - 3t^2 - 2t - 1$
111	$t^8 - 21t^7 - 26t^6 + 14t^5 + 4t^4 - 11t^3 - 6t^2 - t - 1$
119	$t^{10} - 4t^9 + 5t^8 - 8t^7 + 9t^6 - 7t^5 + 5t^4 - 4t^3 + 2t^2 - t + 1$
127	$t^5 - 3t^4 - t^3 + 2t^2 + t - 1$
143	$t^{10} - 6t^9 + 12t^8 - 13t^7 + 9t^6 - 3t^5 - 3t^4 + 6t^3 - 6t^2 + 3t - 1$
151	$t^7 - 3t^6 - t^5 - 3t^4 - t^2 - t - 1$
159	$t^{10} - 47t^9 - 146t^8 - 196t^7 - 219t^6 - 121t^5 - 63t^4 - 7t^3 - t^2 + t - 1$
167	$t^{11} - 2t^{10} - 4t^9 - 9t^8 - 7t^7 - 11t^6 - 6t^5 - 10t^4 - 4t^3 - 5t^2 - t - 1$
183	$t^8 - 71t^7 - 53t^6 + 157t^5 - 98t^4 + 10t^3 + 11t^2 - t - 1$
191	$t^{13} - 6t^{12} + 10t^{11} - 16t^{10} + 22t^9 - 19t^8 + 11t^7 - 5t^6 - t^5 + 5t^4 - 4t^3 + 2t - 1$
199	$t^9 - 5t^8 + 3t^7 - 3t^6 - 3t^3 - t - 1$
215	$t^{14} - 6t^{13} + 4t^{12} + 11t^{11} - 13t^{10} - 7t^9 + 16t^8 - 4t^7 - 13t^6 + 8t^5 + 3t^4 - 6t^3 + 2t - 1$
223	$t^8 - 5t^7 + t^5 + t^4 - 4t^3 - t^2 - 1$
231	$t^{12} - 138t^{11} - 33t^{10} + 517t^9 - 724t^8 + 349t^7 - 118t^6 - 32t^5 + 114t^4 - 45t^3 + 29t^2$ $-4t + 1$
239	$t^{15} - 6t^{14} + 2t^{13} + 8t^{12} + 4t^{11} - 27t^{10} + 13t^9 + 15t^8 - 4t^7 - 20t^6 + 13t^5 + 5t^4 - 4t^3$ $-4t^2 + 4t - 1$
247	$t^6 - 4t^5 - 7t^4 - 7t^3 - 6t^2 - 3t - 1$
255	$t^{12} - 186t^{11} - 194t^{10} + 839t^9 - 702t^8 - 287t^7 + 1012t^6 - 912t^5 + 513t^4 - 221t^3$ $+66t^2 - 11t + 1$
263	$t^{13} - 8t^{12} + 16t^{11} - 27t^{10} + 38t^9 - 36t^8 + 22t^7 - 12t^6 + 13t^5 - 19t^4 + 21t^3 - 15t^2$ $+6t - 1$
271	$t^{11} - 5t^{10} - 6t^9 - 5t^8 + 3t^7 + 6t^6 + 3t^5 - 3t^4 - t^3 - t^2 - 1$
287	$t^{14} - 8t^{13} + 9t^{12} + 6t^{11} - 5t^{10} - 7t^9 - 8t^8 + 6t^7 + 2t^6 - t^4 - t^3 + 3t^2 + t + 1$
295	$t^8 - 8t^7 + 9t^6 - t^5 - 7t^4 + 10t^3 - 7t^2 + 3t - 1$
303	$t^{10} - 325t^9 - 1302t^8 - 756t^7 - 720t^6 - 447t^5 - 173t^4 - 46t^3 - 36t^2 - 2t - 1$
311	$t^{19} - 4t^{18} - 16t^{17} - 37t^{16} - 42t^{15} - 38t^{14} - 4t^{13} + 10t^{12} + 25t^{11} + 18t^{10} + 9t^9 + t^8$ $-10t^7 - 13t^6 - 14t^5 - 8t^4 - 5t^3 - 2t^2 - t - 1$

Example For $D = 942679$ we have $h = h(\mathbb{Q}(\sqrt{-D})) = 620$, and

$$
\begin{aligned}
w(t) = w_D(t) \bmod 2 = \;& t^{620} + t^{616} + t^{610} + t^{606} + t^{604} + t^{602} + t^{597} + t^{596} \\
&+t^{593} + t^{592} + t^{591} + t^{589} + t^{588} + t^{585} + t^{580} + t^{568} \\
&+t^{566} + t^{565} + t^{563} + t^{560} + t^{559} + t^{558} + t^{557} + t^{556} \\
&+t^{555} + t^{552} + t^{550} + t^{548} + t^{547} + t^{546} + t^{545} + t^{543} \\
&+t^{541} + t^{540} + t^{539} + t^{536} + t^{533} + t^{531} + t^{529} + t^{519} \\
&+t^{518} + t^{512} + t^{510} + t^{507} + t^{505} + t^{503} + t^{501} + t^{498} \\
&+t^{496} + t^{495} + t^{492} + t^{491} + t^{488} + t^{487} + t^{486} + t^{484} \\
&+t^{483} + t^{482} + t^{481} + t^{476} + t^{475} + t^{473} + t^{471} + t^{467} \\
&+t^{466} + t^{465} + t^{464} + t^{462} + t^{460} + t^{459} + t^{458} + t^{456} \\
&+t^{454} + t^{453} + t^{452} + t^{451} + t^{450} + t^{448} + t^{447} + t^{443} \\
&+t^{442} + t^{441} + t^{440} + t^{439} + t^{438} + t^{430} + t^{429} + t^{426} \\
&+t^{425} + t^{424} + t^{423} + t^{422} + t^{421} + t^{417} + t^{416} + t^{411} \\
&+t^{410} + t^{409} + t^{408} + t^{407} + t^{406} + t^{403} + t^{401} + t^{398} \\
&+t^{395} + t^{392} + t^{389} + t^{388} + t^{387} + t^{383} + t^{382} + t^{379} \\
&+t^{377} + t^{376} + t^{373} + t^{372} + t^{371} + t^{368} + t^{366} + t^{365} \\
&+t^{360} + t^{359} + t^{358} + t^{357} + t^{355} + t^{354} + t^{353} + t^{351} \\
&+t^{346} + t^{345} + t^{344} + t^{341} + t^{340} + t^{335} + t^{333} + t^{329} \\
&+t^{328} + t^{327} + t^{325} + t^{323} + t^{322} + t^{320} + t^{318} + t^{317} \\
&+t^{314} + t^{313} + t^{312} + t^{311} + t^{309} + t^{306} + t^{305} + t^{304} \\
&+t^{302} + t^{301} + t^{300} + t^{299} + t^{297} + t^{292} + t^{290} + t^{289} \\
&+t^{288} + t^{287} + t^{284} + t^{281} + t^{276} + t^{274} + t^{273} + t^{272} \\
&+t^{264} + t^{263} + t^{262} + t^{260} + t^{257} + t^{255} + t^{253} + t^{252} \\
&+t^{250} + t^{249} + t^{248} + t^{245} + t^{244} + t^{242} + t^{241} + t^{239} \\
&+t^{238} + t^{237} + t^{235} + t^{234} + t^{233} + t^{228} + t^{225} + t^{223} \\
&+t^{221} + t^{219} + t^{211} + t^{210} + t^{209} + t^{208} + t^{207} + t^{201} \\
&+t^{200} + t^{195} + t^{192} + t^{191} + t^{190} + t^{189} + t^{186} + t^{185} \\
&+t^{180} + t^{179} + t^{176} + t^{175} + t^{172} + t^{171} + t^{167} + t^{166} \\
&+t^{161} + t^{160} + t^{158} + t^{157} + t^{155} + t^{154} + t^{152} + t^{150} \\
&+t^{148} + t^{147} + t^{146} + t^{145} + t^{141} + t^{140} + t^{134} + t^{133} \\
&+t^{130} + t^{129} + t^{128} + t^{126} + t^{125} + t^{123} + t^{121} + t^{119} \\
&+t^{117} + t^{114} + t^{110} + t^{109} + t^{108} + t^{106} + t^{103} + t^{102} \\
&+t^{101} + t^{100} + t^{99} + t^{96} + t^{93} + t^{91} + t^{87} + t^{83} + t^{82} \\
&+t^{76} + t^{75} + t^{74} + t^{72} + t^{71} + t^{70} + t^{69} + t^{68} + t^{67} \\
&+t^{65} + t^{61} + t^{59} + t^{57} + t^{56} + t^{54} + t^{51} + t^{49} + t^{45}
\end{aligned}
$$

$$+t^{43} + t^{42} + t^{41} + t^{37} + t^{36} + t^{35} + t^{33} + t^{30} + t^{27}$$
$$+t^{24} + t^{22} + t^{20} + t^{17} + t^{16} + t^{13} + t^{12} + t^{10} + t^6$$
$$+t^2 + 1.$$

This polynomial factors into four irreducibles over $\mathbb{F}_2$, each of degree 155. One of these is:

$$t^{155} + t^{154} + t^{145} + t^{143} + t^{140} + t^{139} + t^{138} + t^{137} + t^{136} + t^{134} + t^{132} + t^{131}$$
$$+t^{130} + t^{129} + t^{128} + t^{127} + t^{126} + t^{124} + t^{123} + t^{121} + t^{120} + t^{118} + t^{117} + t^{116}$$
$$+t^{115} + t^{111} + t^{104} + t^{103} + t^{101} + t^{97} + t^{93} + t^{91} + t^{89} + t^{88} + t^{86} + t^{85} + t^{82}$$
$$+t^{80} + t^{75} + t^{74} + t^{72} + t^{70} + t^{66} + t^{64} + t^{62} + t^{61} + t^{60} + t^{55} + t^{54} + t^{51} + t^{50}$$
$$+t^{49} + t^{31} + t^{29} + t^{27} + t^{26} + t^{23} + t^{22} + t^{19} + t^{18} + t^{16} + t^{15} + t^{14} + t^{13} + t^{11}$$
$$+t^{10} + t^9 + t^6 + t^2 + t + 1.$$

Call it $p(t)$. If β is a root of $p(t)$, then the curve

$$Y^2 + XY = X^3 + \beta^3$$

over $\mathbb{F}_{2^{155}}$ has order $4n$, where n is the prime

$$n = 11417981541647679048466230373126290329356873447.$$

Elliptic Curves with Complex Multiplication

Up to now, we have tried to explain the essential algorithmic steps of the Lay–Zimmer method relative to a minimum of mathematical prerequisites of the reader.

But there is some robust Number Theory interacting with Algebraic-Analytic Geometry which control our step-by-step arguments.

So, let's briefly sketch the underlying Mathematics.

1. *Complex multiplication.*

The pivotal structure is that of a *lattice* $\Lambda = (\omega_1, \omega_2)_\mathbb{Z}$ in the complex plane (note that ω_1 and ω_2 have to be $\mathbb{R}$-linearly independent). Two lattices Λ_1 and Λ_2 are equivalent (homothetic), whenever $\Lambda_2 = \alpha\Lambda_1$ for some $\alpha \in \mathbb{C}^*$. Thus every lattice is – up to equivalence and reindexing of the $\mathbb{Z}$-basis – of the form $\Lambda = (1, \tau)_\mathbb{Z}$, with $\tau \in \mathbf{H}$.

A lattice Λ (which is trivially stable under multiplication by integers) is said to admit *complex multiplication* whenever $\alpha\Lambda \subset \Lambda$ for some non-integer $\alpha \in \mathbb{C}$. In this case, the admissible multiplicators α constitute an *order R* in some imaginary quadratic number field $K = \mathbb{Q}(\sqrt{-D})$ [23], and Λ has to be (homothetic to) an invertible fractional R-ideal (inside K).

---

[23] i.e. R is a unitary subring of K, and a lattice – a free $\mathbb{Z}$-module of rank 2.

Example Recall our summary on imaginary quadratic fields $K = \mathbb{Q}(\sqrt{-D})$ – with $D \equiv 3 \bmod 4$.

We only considered fractional R-ideals for $R = R_K = \mathbb{Z}[\omega]$, with $\omega = \frac{1+\sqrt{-D}}{2}$, i.e. for the *maximal order* (the ring of integers) of K.

A typical non-maximal order would be $R = \mathbb{Z}[\sqrt{-D}]$ (we dealt with it implicitly when considering quadratic forms with discriminant $-4D$). Note that non-maximal orders R will admit non-invertible fractional R-ideals.

Every lattice $\Lambda = (\omega_1, \omega_2)_{\mathbb{Z}} \subset \mathbb{C}$ gives rise to a function field: the field $\mathbb{C}(\Lambda)$ of *elliptic functions for* Λ: $f(z)$ is elliptic for $\Lambda \Longleftrightarrow f(z)$ is defined on $\mathbb{C}$, except for isolated singularities, is meromorphic on $\mathbb{C}$, and doubly periodic in the following sense: $f(z+\omega_1) = f(z+\omega_2) = f(z)$ (Λ is the period lattice for $f(z)$).

The most important elliptic function for Λ is its *Weierstrass $\wp$-function*:

$$\wp(z; \Lambda) = \frac{1}{z^2} + \sum_{\omega \in \Lambda'} \left(\frac{1}{(z-\omega)^2} - \frac{1}{\omega^2} \right), \quad \Lambda' = \Lambda - \{0\}, \quad z \notin \Lambda$$

The singularities of $\wp(z) = \wp(z; \Lambda)$ consist of double poles at the points of the lattice Λ. $\wp(z) = \wp(z; \Lambda)$ satisfies the differential equation

$$\wp'(z)^2 = 4\wp(z)^3 - g_2\wp(z) - g_3 \text{ with } \begin{matrix} g_2 = g_2(\Lambda) = 60 \sum_{\omega \in \Lambda'} \frac{1}{\omega^4} \\ g_3 = g_3(\Lambda) = 140 \sum_{\omega \in \Lambda'} \frac{1}{\omega^6} \end{matrix}$$

$\Delta(\Lambda) = g_2(\Lambda)^3 - 27g_3(\Lambda)^2$ is always non-zero. Thus $j(\Lambda) = 1,728 \cdot \frac{g_2(\Lambda)^3}{\Delta(\Lambda)}$ is well defined. $j(\Lambda)$ is an important lattice invariant: $j(\Lambda) = j(\Lambda') \Longleftrightarrow \Lambda$ and Λ' are homothetic.

Now consider the mapping $\mathbb{C} - \Lambda \ni z \longmapsto (\wp(z), \wp'(z)) \in \mathbb{C}^2$.

We get an (analytic) parametrization of the affine algebraic curve $Y^2 = X^3 - g_2X - g_3$ (an elliptic curve, given by its *Weierstrass equation*).

Let $E(\mathbb{C})$ be the projective curve defined by $Y^2Z = X^3 - g_2XZ^2 - g_3Z^3$, together with its group structure given by the *addition-of-points* and *doubling-of-points* formulas (similar to those introduced at the beginning of this section). Then our $(\wp, \wp')$-parametrization induces a (biholomorphic) *isomorphism* $\mathbb{C}/\Lambda \simeq E(\mathbb{C})$.

Caution We are dealing with two group objects in slightly different categories: $\mathbb{C}/\Lambda$ is a complex torus, with its elliptic function field; $E(\mathbb{C})$ is an abelian variety (a group object in the category of projective algebraic varieties), with its field of rational functions. But $E(\mathbb{C})$ bears an analytic structure that makes it comparable to $\mathbb{C}/\Lambda$.

Uniformization Theorem: Let $E = E(\mathbb{C})$ be an elliptic curve, given by the Weierstrass equation $Y^2 = X^3 - g_2X - g_3$, $g_2, g_3 \in \mathbb{C}$, $g_3^3 - 27g_3^2 \neq 0$, then there is a unique lattice $\Lambda \subset \mathbb{C}$ such that $g_2 = g_2(\Lambda)$, $g_3 = g_3(\Lambda)$.

Thus we are able to speak of *elliptic curves with complex multiplication*, but have to refer to the lattice model. Now, let R be the ring of complex multiplicators for Λ, then R identifies naturally with $\mathrm{End}(\mathbb{C}/\Lambda) \equiv$ the endomorphism ring of the torus $\mathbb{C}/\Lambda$. But this endomorphism ring has its counterpart in $End(E(\mathbb{C}))$, the endomorphism ring of the abelian variety $E = E(\mathbb{C})$ (note that $\alpha \in End(E(\mathbb{C})) \Longleftrightarrow \alpha$ is a rational mapping that is a group homomorphism). This permits an *intrinsic definition* of an elliptic curve with complex multiplication: its endomorphism ring has to be non-trivial.

Example The lattice $\Lambda = (1, \sqrt{-2})_{\mathbb{Z}}$ admits complex multiplication by $\alpha = \sqrt{-2}$.
The corresponding Weierstrass equation is $E \equiv Y^2 = X^3 - 30X - 28$.
$End(E) = R = \mathbb{Z}[\sqrt{-2}]$.

In geometric language, the elements of $R = End(E)$ are *isogenies* (i.e. surjective group homomorphisms with finite kernel). For $\alpha \in R$ we have: $deg(\alpha) := \#\mathrm{Ker}\alpha = \#(\Lambda/\alpha\Lambda) = N(\alpha) \equiv$ the norm of α. In our case: $deg(\sqrt{-2}) = 2$.

The complex multiplication by $\alpha = \sqrt{-2}$ reads in affine coordinates as follows:

$$\sqrt{-2} \cdot (x, y) = \left(-\frac{2x^2 + 4x + 9}{4(x + 2)}, -\frac{1}{\sqrt{-2}} \cdot \frac{2x^2 + 8x - 1}{4(x + 2)^2} \cdot y \right)$$

(Note that x and y are coordinates relative to E – which explains the linearity in y.)

2. *Reduction of elliptic curves with given complex multiplication.* A first observation: Two elliptic curves E_1 and E_2 over $\mathbb{C}$ are isomorphic (as abelian varieties) if and only if the lattices Λ_1 and Λ_2 of their torus models are equivalent (homothetic).

Now return to our particular situation: Let D be a positive, square-free integer with $D \equiv 7 \bmod 8$, $R = (1, \frac{1+i\sqrt{D}}{2})_{\mathbb{Z}} \equiv$ the ring of integers in $K = \mathbb{Q}(\sqrt{-D})$.
$h = h_D \equiv$ the classnumber of $K = \mathbb{Q}(\sqrt{-D})$.

There are h non-isomorphic elliptic curves E over $\mathbb{C}$ having complex multiplication by R (i.e. such that $R = End(E)$): Namely, let $\tau_1, \ldots, \tau_h \in \mathbf{H}$ be the h canonical representatives for the ideal classes of $K = \mathbb{Q}(\sqrt{-D})$, and let $\Lambda_1 = (1, \tau_1)_{\mathbb{Z}}, \ldots \Lambda_h = (1, \tau_h)_{\mathbb{Z}}$ be the corresponding lattices. Then the associated elliptic curves are non-isomorphic, and admit complex multiplication by $R = (1, \frac{1+i\sqrt{D}}{2})_{\mathbb{Z}}$.

The j-invariants of the elliptic curves are given by $j(\tau_k) = j(\Lambda_k)$ $1 \leq k \leq h$.

They are *algebraic integers* and generate the *Hilbert class field* K_D of K.

K_D is the Galois extension of K whose Galois group is precisely the classgroup $\mathrm{Cl}(K)$ of K. The minimal polynomial of the $j(\tau_k), 1 \leq k \leq h$ is the *class polynomial*

$$h_D(t) = \prod_{k=1}^{h} (t - j(\tau_k)) \in \mathbb{Z}[t].$$

We insist: $h_D(t)$ is a polynomial of degree h, with *integer coefficients*.

Now consider $\boxed{q = 2^m, \text{ and } \mathbb{F}_q = \mathbb{F}_{2^m}.}$

If $4q = u^2 + Dv^2$ (we know this theme...), then 2 splits in $K = \mathbb{Q}(\sqrt{-D})$ (i.e. $2R = P_1P_2$ is a product of two distinct prime ideals of R; and $R/P_1 = R/P_2 = \mathbb{F}_2$) and is *unramified* in K_D:

There are r distinct prime ideals $\mathbf{p_1}, \dots \mathbf{p_r}$ in the ring R_D of integers of K_D such that $2R_D = \mathbf{p_1} \cdots \mathbf{p_r}$ and $R_D/\mathbf{p_k} = \mathbb{F}_{2^m}$ $1 \leq k \leq r$.

– Note that the field degree m must necessarily divide the classnumber $h = h_D$.

The reduction $\bmod \mathbf{p}$ – where $\mathbf{p}$ can be any of the $\mathbf{p_k}$ above – gives

$$\overline{h}_D(t) = \overline{h}_1(t) \cdots \overline{h}_r(t),$$

i.e. we get a product of r irreducible binary polynomials $\overline{h}_k(t)$ of the *same degree* m, each admitting m distinct roots in $\mathbb{F}_{2^m}$.

Important observation: The situation will remain unchanged if we replace $h_D(t)$ by some other minimal polynomial for any primitive element of K_D.

Now accept the non-trivial fact that our complex elliptic curves with complex multiplication $R \subset K = \mathbb{Q}(\sqrt{-D})$ admit defining equations with coefficients in $R_D \subset K_D$.

This makes them reducible mod $\mathbf{p}$ (where $\mathbf{p}$ is any of the prime ideals lying over 2). Thus we get $h = h_D$ elliptic curves defined over $\mathbb{F}_{2^m}$. The *fundamental observation* comes from Deuring's reduction theorem. Let $\overline{E}_\Omega$ be any of these reductions, considered as an elliptic curve over $\Omega = \overline{\mathbb{F}}_{2^m} \equiv$ the algebraic closure of $\mathbb{F}_2$. Then $End(\overline{E}_\Omega) = R$.

(Reduction is conservative with respect to complex multiplication!).[24]

The $h = h_D$ elliptic curves over $\mathbb{F}_{2^m}$, obtained by reduction mod $\mathbf{p}$, will be precisely (up to isomorphism) the curves over $\mathbb{F}_{2^m}$ with complex multiplication by R.

This leads us to adopt the following strategy:

Compute the class polynomial $h_D(t) \in \mathbb{Z}[t]$ and reduce it modulo 2.

It splits completely over $\mathbb{F}_{2^m}$, and each of its roots is (more or less) the j-invariant of a looked for elliptic curve (we never shall rigorously define the j-invariant for our characteristic 2 constellation; only note that whenever we deal with an equation of the form $Y^2 + XY = X^3 + b$, $b \neq 0$, then we have $b = j^{-1}$).

Unfortunately, $h_D(t)$ has very large coefficients. In order to remedy to this situation, it is possible to use a different polynomial $w_D(t) \in \mathbb{Z}[t]$,which also defines the Hilbert class field K_D. We only have to replace the function j by another well-chosen modular function.[25]

---

[24] Do not be afraid of complex numbers acting on curves defined over $\mathbb{F}_{2^m}$! Why should the mapping $P \longmapsto -7P$ not be the square of a "simpler" mapping $P \longmapsto \sqrt{-7} \cdot P$?

[25] $f(\tau)$ is *modular* whenever it is meromorphic on $\mathbf{H}$ and invariant under the action of $SL_2(\mathbb{Z})$.

Recall $w_D(t) = \prod_{k=1}^{h}(t - c(\tau_k)) \in \mathbb{Z}[t]$, where the *class invariants* $c(\tau_k)$, $1 \le k \le h$, are defined by means of the classical Weber functions. We would not get lost in technical details; let us altogether indicate the ingredients: Start with Dedekind's η-function:

For $\tau \in \mathbf{H}$, we put $q = e^{2\pi i \tau}$. Let $\zeta_n = e^{\frac{2\pi i}{n}}$. The Dedekind η-function is defined by

$$\eta(\tau) = q^{\frac{1}{24}} \prod_{n=1}^{\infty}(1 - q^n) = q^{\frac{1}{24}} \sum_{n=-\infty}^{\infty}(-1)^n q^{\frac{3n^2+n}{2}}.$$

$\eta(\tau)$ is defined (converges) for $\tau \in \mathbf{H}$. The classical Weber functions $\mathbf{f}$, $\mathbf{f}_1$, and $\mathbf{f}_2$ now are defined in terms of η as follows:

$\mathbf{f}(\tau) = \zeta_{48}^{-1} \frac{\eta((\tau+1)/2)}{\eta(\tau)}$,

$\mathbf{f}_1(\tau) = \frac{\eta(\tau/2)}{\eta(\tau)}$,

$\mathbf{f}_2(\tau) = \sqrt{2} \cdot \frac{\eta(2\tau)}{\eta(\tau)}$.

Now, every modular function is a rational function of $j(\tau)$ (considered as a function on the upper half-plane). The identity $j = \left(\frac{\mathbf{f}_1^{24}+16}{\mathbf{f}_1^8}\right)^3$ shows that the Weber functions are sufficiently general to support all modular arithmetic.

For example, the reader will note their massive intervention in the definition of our class invariants – up to normalizations which partially stem from our 4D-approach to the classgroup representatives.

What is the advantage to replace $h_D(t)$ by $w_D(t)$?

Example $D = 71$.

$h_{71}(t) = \mathbf{t}^7 + 5 \cdot 7 \cdot 31 \cdot 127 \cdot 233 \cdot 9769 t^6 - 2 \cdot 5 \cdot 7 \cdot 44171287694351 t^5$

$+ 2 \cdot 3 \cdot 7 \cdot 23427152097630431440031 t^4$

$- 3 \cdot 7 \cdot 31 \cdot 1265029590537220860166039 t^3$

$+ 2 \cdot 7 \cdot 11^3 \cdot 67 \cdot 229 \cdot 179740261924717851926 33 t^2$

$- 7 \cdot 11^6 \cdot 17^6 \cdot 14209133309 79618293 t$

$+ (11^3 \cdot 17^2 \cdot 23 \cdot 41 \cdot 47 \cdot 53)^3$

(all primes $<1{,}000$ are factored out of the coefficients).

$w_{71}(t) = t^7 - 2t^6 - t^5 + t^4 + t^3 + t^2 - t - 1$.

Our final point is to briefly explain the background for the system of the two equations establishing D as a CM discriminant for 2^m. Consider an elliptic curve $E = E(\mathbb{C})$ over the complex numbers, with complex multiplication by $R \subset K = \mathbb{Q}(\sqrt{-D})$. Suppose that $E = E(\mathbb{C})$ is reducible to $\overline{E} = E(\mathbb{F}_{2^m})$. Let $\Omega = \overline{\mathbb{F}}_2 \equiv$ the algebraic closure of $\mathbb{F}_{2^m}$.

By the Deuring reduction theorem, $R = \mathrm{End}(E) = \mathrm{End}_\Omega(\overline{E}_\Omega)$. (Reduction does not affect complex multiplication.) Now consider $\pi_q \in \mathrm{End}_\Omega(\overline{E}_\Omega)$, given by $\pi_q(x,y) = (x^q, y^q)$, $q = 2^m$.

This is an isogeny of degree[26] $q = 2^m$, and its fixed point set is precisely $E(\mathbb{F}_{2^m})$ (the points with coordinates in $\mathbb{F}_{2^m}$). Thus: $\text{Ker}(1 - \pi_q) = E(\mathbb{F}_{2^m})$ and $\deg(1 - \pi_q) = N = \#E(\mathbb{F}_{2^m})$.

When "lifting" π_q to $\pi \in R$, we dispose of a number theoretic interpretation for the degree of an isogeny (which is a complex mutiplicator): $\deg(\pi) = N(\pi) \equiv$ the norm of π.

Since reduction preserves the degree (of separable isogenies), we arrive at two norm equations

$N(\pi) = q = 2^m$,

$N(1 - \pi) = N = \#E(\mathbb{F}_{2^m})$.

Written down more explicitely, they become our familiar system for CM discriminants.

The Algorithm ECDSA

We shall present the algorithm ECDSA without going too much into detail. Nevertheless, we shall insist on its structure as an *additive twin* of DSA.

First the *elliptic data*, which will be *public*:

Let us start with a field $\mathbb{F}_{2^m} = \mathbb{F}_2[x]/(p(x))$, where $p(x)$ is an irreducible trinomial or pentanomial – so, the arithmetic of the field of constants will be public...

Choose an elliptic curve defined over $\mathbb{F}_{2^m}$ (note that there are approximately 2^m mutually non-isomorphic elliptic curves over $\mathbb{F}_{2^m}$ – we can safely choose...): $\mathcal{E}(\mathbb{F}_{2^m}) : Y^2 + XY = X^3 + aX^2 + b$ with $a, b \in \mathbb{F}_{2^m}$, $b \neq 0$. Then, we need a *base point*, generator of a $\boxed{\text{(cyclic) subgroup of prime order}}$ in $\mathcal{E}(\mathbb{F}_{2^m})$:

$G = (X_G, Y_G) \in E(\mathbb{F}_{2^m})$ such that $nG = \mathbb{O}$, where n is a (large) prime number: $n > max\{2^{160}, 4\sqrt{2^m}\}$.

Note that it is the prime number n which is the *main security parameter* for ECDSA[27].

– Generation of the keys:

$\boxed{d \equiv \text{The private key of the signatory}}$: a "sufficiently arbitrary" *integer* between 1 and $n - 1$.

The pseudo-random generation of d is guaranteed by auxiliary algorithms in the spirit of those that we have seen serving the DSA.

---

[26] The *degree* of an isogeny is the cardinality of its kernel.

[27] In order to guard our curve against existing attacks on ECDLP – the Elliptic Curve Discrete Logarithm Problem (to be explained in a moment) – the condition on the size of the prime n has to be complemented by two further conditions – the Menezes–Okamoto–Vanstone Condition and the Anomalous Condition – which will guarantee that no transfer to an easy Discrete Logarithm Problem in some scalar field $\mathbb{F}_{2^M}$ is possible.

$\boxed{Q \equiv \text{the public key of the signatory}}$: a distinguished *point* on the elliptic curve;
more precisely: $Q = (X_Q, Y_Q) = dG.$

-   Generation of the signature $\sigma = (r, s)$.
    Let M be the message (the document) to be signed.
    (a) Computation of $e = \text{SHA-1}(M)$.
    (b) Selection of $k \equiv$ the secret key of the signature act. k will be an integer between 1 and $n - 1$, "sufficiently arbitrary", exactly like d.
    (c) $(X_1, Y_1) = kG.$
    (d) X_1 is considered (by means of its binary notation) as an *integer*.
        Put then $\boxed{r = X_1 \bmod n.}$ If $r = 0$, return to b.
        $\boxed{s = k^{-1}(e + dr) \bmod n.}$ If $s = 0$, return to b.
    (e) $\sigma = (r, s)$ $\equiv$the signature (a couple of two integers).

-   Signature verification.
    -   The scenario is precisely the same as for DSA: The verifier of the signature receives the transmitted message M' along with the transmitted signature $\sigma' = (r', s')$. He will carry out essentially the same operations as in DSA.
        (a) Computation of $e' = \text{SHA-1}(M')$.
        (b) Verification that $1 < r', s' < n - 1$.
        (c) Computation of $w = (s')^{-1} \bmod n$.
        (d) Computation of
        $$u_1 = e'w \bmod n$$
        $$u_2 = r'w \bmod n$$
        (e) Computation of $(X_1, Y_1) = u_1 G + u_2 Q.$
            (If $u_1 G = -u_2 Q$, he rejects...)
        (f) If $r' = X_1$ (as *integers*), he accepts; otherwise, he rejects.

Summary of the Analogies Between DSA and ECDSA

1. *The relevant group information*

Group	$(\mathbb{Z}/p\mathbb{Z})^*$	$E(\mathbb{F}_{2^m})$
Group elements	The integers $\{1, 2, \ldots, p - 1\}$	The points (X, Y) which satisfy the equation of the elliptic curve, with the point at infinity $\mathbb{O}$
Group operation	Multiplication modulo p	Addition of points
Notation	The elements: g, h The multiplication: $g \cdot h$ The exponentiation: g^a	The elements: P, Q The addition: $P + Q$ Scalar multiple of a point: aP
Discrete logarithm problem	Given $g \in (\mathbb{Z}/p\mathbb{Z})^*$ and $h = g^a \bmod p$, find the integer a.	Given $P \in E(\mathbb{F}_{2^m})$ and $Q = aP$, find the integer a.

2. *The basic notation*

DSA Notation	ECDSA Notation
q	n
g	G
x	d
y	Q

3. *The algorithmic skeleton*

DSA Setup	ECDSA Setup
1. p and q are prime numbers, q divides $p - 1$.	1. E is an elliptic curve, defined over the field $\mathbb{F}_{2^m}$.
2. g is an element of order q in $(\mathbb{Z}/p\mathbb{Z})^*$.	2. G is a point of prime order n in $E(\mathbb{F}_{2^m})$.
3. We work in the group $\{g^0, g^1, g^2, \ldots, g^{q-1}\}$.	3. We work in the group $\{\mathbb{O}, G, 2G, \ldots, (n-1)G\}$.

4. *The key generation*

DSA Key Generation	ECDSA Key Generation
1. Select a random integer x between 1 and $q - 1$.	1. Select a unpredictable integer d between 1 and $n - 1$.
2. Compute $y = g^x \bmod p$.	2. Compute $Q = dG$.
3. The private key is x.	3. The private key is d.
4. The public key is y.	4. The public key is Q.

5. *The signature generation*

DSA Signature Generation	ECDSA Signature Generation
1. Select a random integer k between 1 and $q-1$.	1. Select a unpredictable integer k between 1 and $n-1$.
2. Compute $g^k \bmod p$.	2. Compute $kG = (X_1, Y_1)$.
3. Compute $r = (g^k \bmod p) \bmod q$.	3. Compute $r = X_1 \bmod n$.
4. Compute $m = $ SHA-1(M).	4. Compute $e = $ SHA-1(M).
5. Compute $s = k^{-1}(m + xr) \bmod q$.	5. Compute $s = k^{-1}(e + dr) \bmod n$.
6. The signature for M is (r, s).	6. The signature for M is (r, s).

6. *The signature verification*

DSA Verification	ECDSA Verification
1. Compute $m = $ SHA-1(M).	1. Compute $e = $ SHA-1(M).
2. Compute $s^{-1} \bmod q$.	2. Compute $s^{-1} \bmod n$.
3. Compute $u_1 = ms^{-1} \bmod q$.	3. Compute $u_1 = es^{-1} \bmod n$.
4. Compute $u_2 = rs^{-1} \bmod q$.	4. Compute $u_2 = rs^{-1} \bmod n$.
5. Compute $v' = g^{u_1} y^{u_2} \bmod p$.	5. Compute $u_1 G + u_2 Q = (X_1, Y_1)$.
6. Compute $v = v' \bmod q$.	6. Compute $v = X_1 \bmod n$.
7. Accept the signature if $v = r$.	7. Accept the signature if $v = r$.

Exercises

(1) Assume a correct transmission of the message M and of the signature $\sigma = (r, s)$. Adopt the standardized notation above. Show that you will get *necessarily*:
$r = X_1 \bmod n$, where X_1 is the abscissa of the point $u_1 G + u_2 Q$ on the given elliptic curve.

(2) Let $\mathbb{F}_{16} = \mathbb{F}_2[x]/(x^4 + x + 1)$, and let $\omega = x \bmod (x^4 + x + 1)$. Consider the elliptic curve $\mathcal{E}(\mathbb{F}_{16}) : Y^2 + XY = X^3 + \omega^3$.

(a) Let $G = (\omega, 0)$. Verify that $5G = \mathbb{O}$.

(b) The message digest of the received message M' is $e' = \mathrm{H}(M') = 1100$ (four bits – we shall be modest...). The transmitted signature $\sigma' = (r', s') = (2,2)$.

The public key of the signatory: $Q = (\omega^5, \omega^3)$.

Show that $Q \in \mathcal{E}(\mathbb{F}_{16})$, and verify the signature.

The Security of ECDSA

The security of the system ECDSA depends on the *algorithmic complexity* of the elliptic curve discrete logarithm problem (ECDLP):

Given $G, Q \in \mathrm{E}(\mathbb{F}_{2^m})$, $nG = \mathbb{O}$, find

$0 \leq l \leq n - 1$ such that $Q = lG$ (if l exists)

The best general algorithms known to date for ECDLP are the POLLARD-rho method and the POLLARD-lambda method, which take about $\sqrt{\pi n/2}$ and $3.28\sqrt{n}$ steps (one step means one elliptic curve addition), i.e. – according to the requirements of the standard concerning the size of n – at least 2^{81} steps (recall that we have to exclude certain "vulnerable" elliptic curves...).

Which is the power of the actual technology? Concerning software attacks, assume that a 1 million instructions per second (MIPS) machine can perform 4×10^4 elliptic curve additions per second. Then, the number of elliptic curve additions that can be performed by a 1 MIPS machine in one year is 2^{40}. As a consequence, if $m = 163$ (that is the degree of the scalar field) and if $n \approx 2^{160}$, then 10,000 computers, each rated at 1,000 MIPS, will find (with POLLARD-rho) an elliptic logarithm in 85000 years. Concerning hardware attacks, the reference is a study of van Oorschot and Wiener (1994) which estimates that if one uses a special-purpose hardware for parallel search, and that if $n \approx 10^{36} \approx 2^{120}$, then a machine with 32,5000 processors (that could be built for about 10 million dollars) would compute a single discrete elliptic logarithm in about 35 days. This is the reason for the standard to require that $n > 2^{160}$; so, hardware attacks of this kind would be infeasible.

An Example of ECDSA

Our scalar field will be $\mathbb{F}_{2^{191}} = \mathbb{F}_2[x]/(x^{191} + x^9 + 1)$. The elements of this field – which are binary words of length 191, i.e. the coefficient-sequences of the remainders of division by $p(x) = x^{191} + x^9 + 1$ – will be presented in blocks of six 32-bit words, in hexadecimal notation.

The elliptic curve:

$$\mathcal{E}(\mathbb{F}_{2^{191}}) : Y^2 + XY = X^3 + aX^2 + b$$

with

$a = $ 2866537b 67675263 6a68f565 54e12640 276b649e f7526267

$b = $ 2e45ef57 1f00786f 67b0081b 9495a3d9 5462f5de 0aa185ec

The base point $G = (X_G, Y_G)$:

$X_G = $ 36b3daf8 a23206f9 c4f299d7 b21a9c36 9137f2c8 4ae1aa0d,

$Y_G = $ 765be734 33b3f95e 332932e7 0ea245ca 2418ea0e f98018fb,

$nG = \mathbb{O}$,

with

$n = $ 1569275433846670190958947355803350458831205595451630533029.

$\#E(\mathbb{F}_{2^{191}}) = 2n$.

Key Generation

$d = $ 1275552191113212300012030439187146164646146646466749494799.

$Q = dG = (X_Q, Y_Q)$ with

$X_Q = $ 5de37e756bd55d72e3768cb396ffeb962614dea4ce28a2e7,

$Y_G = $ 55c0e0e02f5fb132caf416ef85b229bbb8e1352003125ba1.

Signature Generation

The message $M \equiv$ "abc" (familiar to us...)

1. The message digest:
   $e = \text{SHA-1}(M) = $ 968236873715988614170569073515315707566766479517.
2. Elliptic computations:
   2.1 Selection of k:
   $k = $ 1542725565216523985789236956265265265235675811949404040041.
   2.2 Computation of $R = kG = (X_1, Y_1)$:
   $X_1 = $ 438e5a11fb55e4c65471dcd49e266142a3bdf2bf9d5772d5,
   $Y_1 = $ 2ad603a05bd1d177649f9167e6f475b7e2ff590c85af15da,
   2.3 X_1 becomes an integer:
   $X_1 = $ 1656469817011541734314669640730254878828443186986697061077.
   2.4 $r = X_1 \bmod n$:
   $r = $ 8719438316487154335572228492690441999723759153506652048.
3. Congruence computations:
   Computation of $s = k^{-1}(e + dr) \bmod n$:

   $s = $ 3089926919658049473615416645490858952921537770257720636598.

4. The signature $\sigma = (r, s)$:

   $r = $ 8719438316487154335572228492690441999723759153506652048,

   $s = $ 3089926919658049473615416645490858952921537770257720636598.

Signature Verification

We suppose a correct transmission.

1. The message digest of M':
   $e' = \text{SHA-1}(M') = 968236873715988614170569073515315707566766479517.$
2. Elliptic computations:
   2.1 First $w = (s')^{-1} \bmod n$:
   $w = 952933666850866331568782284754801289889992082635386177703.$
   2.2 Now $u_1 = e'w \bmod n$ and $u_2 = r'w \bmod n$:
   $u_1 = 124888640715470785402243451608406250330179237436099440066,$
   $u_2 = 527017380977534012168222466016199849611971141652753464154.$
   2.3 Then $(X_1, Y_1) = u_1 G + u_2 Q$:
   $(u_1 G)_X = \text{1a045b0c 26af1735 9163e9b2 bf1aa57c 5475c320 78abe159}$
   $(u_1 G)_Y = \text{53eca58f ae7a4958 783e8173 cf1ca173 eac47049 dca02345}$
   $(u_2 Q)_X = \text{015cf19f e8485bed 8520ca06 bd7fa967 a2ce0b30 4ffcf0f5}$
   $(u_2 Q)_Y = \text{314770fa 4484962a ec673905 4a6652bc 07607d93 cac79921}$
   $X_1 = \text{438e5a11 fb55e4c6 5471dcd4 9e266142 a3bdf2bf 9d5772d5}$
   $Y_1 = \text{2ad603a0 5bd1d177 649f9167 e6f475b7 e2ff590c 85af15da}$
3. Validation of the signature:
   3.1 X_1 becomes an integer:
   $X_1 = 1656469817011541734314669640730254878828443186986697061077.$
   3.2 $v = X_1 \bmod n$:
   $v = 871943831648715433557222849269044199972375915350665 28048.$
   3.3 $v = r'$.   o.k.

3

Information Theory and Signal Theory: Sampling and Reconstruction

In this chapter, we shall treat the following question: what is the *digital skeleton* of a (traditionally acoustic) signal, which describes it faithfully and exhaustively? In a more technical language: we shall deal with conversion (without loss of information)

$$\text{analog} \longrightarrow \text{digital} \longrightarrow \text{analog}$$

in signal theory.

This means appropriate sampling, followed by reconstruction of the initial signal from the time series of its samples. The Sampling Theorems are always interpolation theorems (in a rather general sense: there will also be convergence questions – with sometimes delicate nuances...).

In principle, we could also adopt the following viewpoint: the reality is discrete (perhaps even finite); the "continuous world"[1] is the bridge between perception and modelling (simple forms described by simple laws). In this sense, a Sampling Theorem will control the uniqueness of the law (of the form) behind the empirical data – permitting to refine arbitrarily their "digital density".

But there are also classical down-to-earth arguments around digitization, i.e. conversion of a continuous-time waveform into a bitstream. First, a sampler will convert the continuous-time waveform into a discrete-time sequence of numbers (a time series). This process is a mathematical abstraction, and we will stop precisely here. Then, the resulting time series has to be quantized into a bit stream (how many bits per sample?) – a non-mathematical step that we shall neglect. At any rate, digitizing means loss of information (at least for those who believe in a perfect control of the continuous situation). But, miraculously, it is exactly at this point that the discrete world wins a strange battle against the analog world: analog channels have a finite capacity and often inflict considerable distortion due to noise; properly digitizing the source and transmitting a digital signal through the channel often results in

---

[1] The continuous event "film" is nothing but a time series of digital images.

less distortion than if the source output were left in analog form. So, perversely, digitization can even be considered as an information preserving process.

We shall proceed in the following way:

At the beginning, we shall introduce one of the main tools for subsequent arguments: the Discrete Fourier Transform. Then, we shall improvise on the Sampling Theorem in the periodic case. Finally, we shall attack the non-periodic case, where the Sampling Theorem (due to Whittaker, who proved it in 1935) has always been associated with the names of Shannon and of Nyquist.

Let us repeat: we shall always remain in an ideal mathematical world: quantization as well as distortion will be excluded.

3.1 The Discrete Fourier Transform

This fundamental section treats a chapter of Linear Algebra which is more or less the mathematical theory of *periodic* time series (i.e. sampled trigonometric polynomials).

So, everything is dominated by the reassuring image of the unit circle and its polygonal approximations. The great coherence and elegance of this theory go hand in hand with a remarkable robustness towards a lot of interesting applications of non-periodic character, in Discrete Mathematics and in Engineering.

3.1.1 Basic Properties

We shall choose an ad hoc approach and treat the Discrete Fourier Transform as an object of lectures in elementary Linear Algebra.

(a) Prelude: The nth Roots of Unity

Consider $S^1 = \{z \in \mathbb{C} : |z| = 1\}$, the unit circle in the complex plane. There is a "natural" 2π-periodic parametrization given by
$$e^{i\theta} = \cos\theta + i \cdot \sin\theta, \quad \theta \in \mathbb{R}.$$

Fundamental Identity

$e^{i(\theta_1 + \theta_2)} = e^{i\theta_1} e^{i\theta_2}, \quad \theta_1, \theta_2 \in \mathbb{R}.$

Let us verify:

$$e^{i(\theta_1 + \theta_2)} = \cos(\theta_1 + \theta_2) + i \cdot \sin(\theta_1 + \theta_2)$$
$$= \cos\theta_1 \cdot \cos\theta_2 - \sin\theta_1 \cdot \sin\theta_2 + i(\sin\theta_1 \cdot \cos\theta_2 + \cos\theta_1 \cdot \sin\theta_2)$$
$$= (\cos\theta_1 + i \cdot \sin\theta_1)(\cos\theta_2 + i \cdot \sin\theta_2) = e^{i\theta_1} e^{i\theta_2}.$$

A well-known consequence of our fundamental identity is the

De Moivre Formula

$(\cos\theta + i\cdot\sin\theta)^k = \cos(k\theta) + i\cdot\sin(k\theta)$, i.e. $(e^{i\theta})^k = e^{i(k\theta)}$ for every $k \in \mathbb{Z}$.

The argument:

For $k \geq 1$ one obtains the De Moivre Formula by recursion on k (since $e^{i(k\theta)} = e^{i\theta}\cdot e^{i(k-1)\theta}$). As to negative exponents, observe only that $\frac{1}{z} = \bar{z}$ (inverse = conjugate) for $z \in S^1$. Let us insist: $e^{-i\theta} = \overline{e^{i\theta}} = \frac{1}{e^{i\theta}}$.

Definition For $n \geq 1$, we put $\omega_n = e^{-\frac{2\pi i}{n}} = \cos\frac{2\pi}{n} - i\cdot\sin\frac{2\pi}{n}$.

This gives:

$$\omega_2 = -1, \qquad\qquad \omega_4 = -i,$$
$$\omega_3 = -\tfrac{1}{2} - \tfrac{i}{2}\sqrt{3}, \qquad \omega_8 = \tfrac{1}{2}\sqrt{2} - \tfrac{i}{2}\sqrt{2}.$$

Consider now, for fixed $n \geq 2$, the n (distinct) complex numbers $\omega_n, \omega_n^2, \omega_n^3, \ldots, \omega_n^n = 1$ on the unit circle. Geometrically, we get (for $n \geq 3$) the n vertices of a regular n-polygon inscribed in the unit circle. Algebraically, we obtain the n distinct roots of the polynomial $X^n - 1$.

This is the reason for their name: the nth roots of unity. Note that with increasing powers of ω_n we turn around the unit circle clockwise (this is a consequence of the "conjugate" definition of ω_n).

Examples

(1) $X^4 - 1 = (X-\omega_4)(X-\omega_4^2)(X-\omega_4^3)(X-\omega_4^4) = (X+i)(X+1)(X-i)(X-1)$.
(2) The eight 8th roots of unity:

$$\omega_8 = \tfrac{1}{2}\sqrt{2} - \tfrac{i}{2}\sqrt{2}, \quad \omega_8^2 = \omega_4 = -i, \quad \omega_8^3 = -\tfrac{1}{2}\sqrt{2} - \tfrac{i}{2}\sqrt{2}, \quad \omega_8^4 = \omega_4^2 = -1,$$
$$\omega_8^5 = -\tfrac{1}{2}\sqrt{2} + \tfrac{i}{2}\sqrt{2}, \quad \omega_8^6 = \omega_4^3 = i, \quad \omega_8^7 = \tfrac{1}{2}\sqrt{2} + \tfrac{i}{2}\sqrt{2}, \quad \omega_8^8 = 1.$$

We note that

$$\omega_8^{-1} = \bar{\omega}_8, \quad \omega_8^5 = -\omega_8, \quad \omega_8^3 = -\bar{\omega}_8.$$

Back to the general case: we have, for $n \geq 2 : \omega_n^k = \omega_n^{k'} \Leftrightarrow k \equiv k' \bmod n \Leftrightarrow k - k'$ is divisible by n.

Example $\omega_8^{21} = \omega_8^{-11} = \omega_8^5 = -\omega_8$.

Exercises

(1) Compute (the real number) $\alpha = (2 - 2i)\cdot\omega_8^{15}$.
(2) Compute $z = \omega_8^{11}\cdot i + \omega_8^{-3}$.
(3) Compute the sum $1 + \omega_8^3 + \omega_8^6 + \omega_8^9 + \omega_8^{12} + \omega_8^{15} + \omega_8^{18} + \omega_8^{21}$.
(4) Find the $k \in \mathbb{Z}$ such that $\omega_3^2 \cdot \omega_8^k = \omega_{24}$.

The Discrete Fourier Transform of order n

Fix $n \geq 1$, and consider the $\mathbb{C}$-linear transformation

$$F_n : \mathbb{C}^n \longrightarrow \mathbb{C}^n,$$

$$\begin{pmatrix} z_0 \\ z_1 \\ \cdot \\ \cdot \\ z_{n-1} \end{pmatrix} \longmapsto \begin{pmatrix} Z_0 \\ Z_1 \\ \cdot \\ \cdot \\ Z_{n-1} \end{pmatrix},$$

given by $Z_j = \sum_{k=0}^{n-1} \omega_n^{jk} z_k$, $0 \leq j \leq n-1$.

In matrix notation:

$$\begin{pmatrix} Z_0 \\ Z_1 \\ Z_2 \\ \cdot \\ \cdot \\ Z_{n-1} \end{pmatrix} = \begin{pmatrix} 1 & 1 & 1 & \cdot \cdot & 1 \\ 1 & \omega_n & \omega_n^2 & \cdot \cdot & \omega_n^{n-1} \\ 1 & \omega_n^2 & \omega_n^4 & \cdot \cdot & \omega_n^{2(n-1)} \\ \cdot & \cdot & \cdot & \cdot \cdot & \cdot \\ \cdot & \cdot & \cdot & \cdot \cdot & \cdot \\ 1 & \omega_n^{n-1} & \omega_n^{2(n-1)} & \cdot \cdot & \omega_n^{(n-1)(n-1)} \end{pmatrix} \begin{pmatrix} z_0 \\ z_1 \\ z_2 \\ \cdot \\ \cdot \\ z_{n-1} \end{pmatrix}.$$

We shall design by F_n as well the particular matrix as the linear transformation defined by it.

Hence, $F_n = (\omega_n^{jk})_{0 \leq j,k \leq n-1}$.

Note that $F_1 : \mathbb{C} \to \mathbb{C}$ is the identity.

Example

$$F_4 = \begin{pmatrix} 1 & 1 & 1 & 1 \\ 1 & -i & -1 & i \\ 1 & -1 & 1 & -1 \\ 1 & i & -1 & -i \end{pmatrix}.$$

The equations for the Discrete Fourier Transform of order 4 are then:

$$Z_0 = z_0 + z_1 + z_2 + z_3,$$

$$Z_1 = z_0 - iz_1 - z_2 + iz_3,$$

$$Z_2 = z_0 - z_1 + z_2 - z_3,$$

$$Z_3 = z_0 + iz_1 - z_2 - iz_3.$$

Exercise

Write the matrix F_8 in function of ± 1, $\pm i$, $\pm \omega_8$, $\pm \overline{\omega}_8$.

One obtains

$$
F_8 = \begin{pmatrix}
1 & 1 & 1 & 1 & 1 & 1 & 1 & 1 \\
1 & \omega_8 & -i & -\overline{\omega}_8 & -1 & -\omega_8 & i & \overline{\omega}_8 \\
1 & -i & -1 & i & 1 & -i & -1 & i \\
1 & -\overline{\omega}_8 & i & \omega_8 & -1 & \overline{\omega}_8 & -i & -\omega_8 \\
1 & -1 & 1 & -1 & 1 & -1 & 1 & -1 \\
1 & -\omega_8 & -i & \overline{\omega}_8 & -1 & \omega_8 & i & -\overline{\omega}_8 \\
1 & i & -1 & -i & 1 & i & -1 & -i \\
1 & \overline{\omega}_8 & i & -\omega_8 & -1 & -\overline{\omega}_8 & -i & \omega_8
\end{pmatrix}.
$$

Let us return to the general situation. Our first goal is to show the following result: For every $n \geq 1$, F_n is an invertible linear transformation, and the inverse transformation F_n^{-1} is "very similar to F_n".

Fundamental Observation

Let $z \in \mathbb{C}$, $z \neq 1$ and $z^n = 1$. Then $1 + z + z^2 + \cdots + z^{n-1} = 0$.
The argument is simple: $1 - z^n = (1 - z)(1 + z + z^2 + \cdots + z^{n-1})$.

Now we are able to be specific.

Proposition $n \geq 1$. Then F_n is invertible, and $F_n^{-1} = \frac{1}{n}\overline{F}_n$ (componentwise conjugation), i.e.

$$
F_n^{-1} = \frac{1}{n}\begin{pmatrix}
1 & 1 & 1 & \cdots & 1 \\
1 & \omega_n^{-1} & \omega_n^{-2} & \cdots & \omega_n^{-(n-1)} \\
1 & \omega_n^{-2} & \omega_n^{-4} & \cdots & \omega_n^{-2(n-1)} \\
\vdots & \vdots & \vdots & \ddots & \vdots \\
1 & \omega_n^{-(n-1)} & \omega_n^{-2(n-1)} & \cdots & \omega_n^{-(n-1)(n-1)}
\end{pmatrix} = \frac{1}{n}(\omega_n^{-jk})_{0 \leq j,\,k \leq n-1}.
$$

Proof We have to show that $\frac{1}{n}F_n \cdot \overline{F}_n = I_n \equiv$ the identity matrix of order n. Let $\mathbf{l}_j$ be the jth row of F_n, and let $\mathbf{c}_k$ be the kth column of $\overline{F}_n$, $0 \leq j,\,k \leq n-1$.

Let us show that $\frac{1}{n}\mathbf{l}_j \cdot \mathbf{c}_k = \begin{cases} 1 \text{ for } j = k, \\ 0 \text{ for } h \neq k. \end{cases}$

But

$$
\frac{1}{n}\mathbf{l}_j \cdot \mathbf{c}_k = \frac{1}{n}\left(1\ \omega_n^j\ \omega_n^{2j}\ \cdots\ \omega_n^{(n-1)j}\right)\begin{pmatrix} 1 \\ \omega_n^{-k} \\ \omega_n^{-2k} \\ \vdots \\ \omega_n^{-(n-1)k} \end{pmatrix}
$$

$$
= \frac{1}{n}\left(1 + \omega_n^{j-k} + \omega_n^{2(j-k)} + \cdots + \omega_n^{(n-1)(j-k)}\right).
$$

For $j = k$ we obtain $\frac{n}{n} = 1$, as needed. For $j \neq k$ we get the sum $\frac{1}{n}(1 + z + z^2 + \cdots + z^{n-1})$, with $z = \omega_n^{j-k} \neq 1, z^n = 1$. Thus our foregoing observation permits to conclude. $\square$

Example

$$F_4^{-1} = \frac{1}{4} \begin{pmatrix} 1 & 1 & 1 & 1 \\ 1 & i & -1 & -i \\ 1 & -1 & 1 & -1 \\ 1 & -i & -1 & i \end{pmatrix},$$

i.e. we have the backward equations
$$z_0 = \tfrac{1}{4}(Z_0 + Z_1 + Z_2 + Z_3),$$
$$z_1 = \tfrac{1}{4}(Z_0 + iZ_1 - Z_2 - iZ_3),$$
$$z_2 = \tfrac{1}{4}(Z_0 - Z_1 + Z_2 - Z_3),$$
$$z_3 = \tfrac{1}{4}(Z_0 - iZ_1 - Z_2 + iZ_3).$$

Note that the particular form of F_n^{-1}, very similar to F_n, allows the "algorithmic elimination" of F_n^{-1}. (The algorithms which compute $y = F_n(x)$ can be used for the computation of $x = F_n^{-1}(y)$).

More precisely, we have the following result:

$$F_n^{-1}(y) = \frac{1}{n}\overline{F_n(\bar{y})},$$

i.e. let

$$\begin{pmatrix} z_0 \\ z_1 \\ \cdot \\ \cdot \\ z_{n-1} \end{pmatrix} = F_n \begin{pmatrix} \overline{y_0} \\ \overline{y_1} \\ \cdot \\ \cdot \\ \overline{y_{n-1}} \end{pmatrix} \quad \text{then} \quad F_n^{-1} \begin{pmatrix} y_0 \\ y_1 \\ \cdot \\ \cdot \\ y_{n-1} \end{pmatrix} = \frac{1}{n} \begin{pmatrix} \overline{z_0} \\ \overline{z_1} \\ \cdot \\ \cdot \\ \overline{z_{n-1}} \end{pmatrix}.$$

We insist: The *inverse* Discrete Fourier Transform is – up to two componentwise conjugations – reducible to the Discrete Fourier Transform.
The argument is simple:
In generalizing the identity $\overline{az} = \bar{a}\cdot\bar{z}$ (for $a, z \in \mathbb{C}$) we obtain, for $A \in \mathbb{C}^{n\times n}$ and

$$z = \begin{pmatrix} z_0 \\ \cdot \\ \cdot \\ z_{n-1} \end{pmatrix} \in \mathbb{C}^n, \text{ the following identity: } \overline{Az} = \overline{A}\,\overline{z}$$

(the conjugation of a vector (of a matrix) is done componentwise).

Let us show our claim for $n = 2$:

$$\overline{\begin{pmatrix} a_{00} & a_{01} \\ a_{10} & a_{11} \end{pmatrix} \begin{pmatrix} z_0 \\ z_1 \end{pmatrix}} = \overline{\begin{pmatrix} a_{00}z_0 + a_{01}z_1 \\ a_{10}z_0 + a_{11}z_1 \end{pmatrix}} = \begin{pmatrix} \overline{a_{00}} \cdot \overline{z_0} + \overline{a_{01}} \cdot \overline{z_1} \\ \overline{a_{10}} \cdot \overline{z_0} + \overline{a_{11}} \cdot \overline{z_1} \end{pmatrix} = \overline{\begin{pmatrix} a_{00} & a_{01} \\ a_{10} & a_{11} \end{pmatrix}} \cdot \overline{\begin{pmatrix} z_0 \\ z_1 \end{pmatrix}}.$$

In our particular situation, we get
$$F_n^{-1}(y) = \tfrac{1}{n}\overline{F_n(\bar{y})} = \tfrac{1}{n}\overline{F_n}(\bar{y}), \text{ as promised.}$$

Exercises

(1) Write the matrix F_8^{-1} in function of ±1, $\pm i$, $\pm w_8$, $\pm\bar{w}_8$.
(2) Compute the matrix $J_4 = \frac{1}{4}F_4^2$.
(3) Find the matrix $J_n = \frac{1}{n}F_n^2$ (follow the proof of the foregoing proposition).
Answer:

$$J_n = \begin{pmatrix} 1\,0\,0\cdots0\,0\,0 \\ 0\,0\,0\cdots0\,0\,1 \\ 0\,0\,0\cdots0\,1\,0 \\ \cdots\cdots\cdots \\ 0\,0\,1\cdots0\,0\,0 \\ 0\,1\,0\cdots0\,0\,0 \end{pmatrix},$$

i.e. $\frac{1}{n}F_n\left(F_n\begin{pmatrix} z_0 \\ z_1 \\ \cdot \\ z_{n-1} \end{pmatrix}\right) = \begin{pmatrix} z_0 \\ z_{n-1} \\ \cdot \\ z_1 \end{pmatrix}.$

(4) Show that $F_n^{-1} = \frac{1}{n^2}F_n^3$.
(5) $n \geq 2$. Let l_j be the jth row of the matrix F_n, $0 \leq j \leq n-1$.
Verify, for the matrix F_8, that $l_7 = \bar{l}_1$, $l_6 = \bar{l}_2$, $l_5 = \bar{l}_3$, $l_4 = \bar{l}_4$.
Generalize: show that $l_{n-j} = \bar{l}_j$ for $1 \leq j \leq n-1$.

The Convolution Theorem

The Convolution Theorem describes the *characteristic* property of the Discrete Fourier Transform. Stated differently: the very definition of the Fourier transform encapsulates this result.

In the beginning, we shall adopt a purely formal viewpoint. We shall remain in the context of Linear Algebra, without reference to the language and the arguments of signal theory and of digital filtering.

Recall The circular convolution.

For $x = \begin{pmatrix} x_0 \\ x_1 \\ x_2 \\ \cdot \\ x_{n-1} \end{pmatrix} \in \mathbb{C}^n$, the associated circular matrix $C(x)$ is defined by

$$C(x) = \begin{pmatrix} x_0 & x_{n-1} & x_{n-2} & \cdots & x_1 \\ x_1 & x_0 & x_{n-1} & \cdots & x_2 \\ x_2 & x_1 & x_0 & \cdots & x_3 \\ \cdot & \cdot & \cdot & \cdots & \cdot \\ x_{n-1} & x_{n-2} & x_{n-3} & \cdots & x_0 \end{pmatrix}.$$

The multiplication of a circular matrix with a vector presents important particularities:

(1) *Commutativity.* $C(x) \cdot y = C(y) \cdot x$.
(2) *Invariance under cyclic permutation.* $C(x) \cdot \sigma(y) = \sigma(C(x) \cdot y)$, where

$$\sigma \begin{pmatrix} z_0 \\ z_1 \\ \vdots \\ z_{n-1} \end{pmatrix} = \begin{pmatrix} z_{n-1} \\ z_0 \\ \vdots \\ z_{n-2} \end{pmatrix}.$$

Combining (1) and (2), one obtains
(3) If $z = C(x) \cdot y$, then $C(z) = C(x)C(y) = C(y)C(x)$ (multiplication of circular matrices).

Definition *The (circular) convolution product* $* : \mathbb{C}^n \times \mathbb{C}^n \longrightarrow \mathbb{C}^n$.
Let us define, for

$$x = \begin{pmatrix} x_0 \\ x_1 \\ \cdot \\ x_{n-1} \end{pmatrix}, \quad y = \begin{pmatrix} y_0 \\ y_1 \\ \cdot \\ y_{n-1} \end{pmatrix} \in \mathbb{C}^n$$

their *(circular) convolution product* $z = x * y$ by $z = C(x) \cdot y = C(y) \cdot x$.
We immediately get:

(1) $x * y = y * x$,
(2) $\sigma(x * y) = (\sigma(x)) * y = x * (\sigma(y))$,
(3) $C(x * y) = C(x)C(y)$.

Note that all this is already well known: recall the operations on the words (the columns of the state array) in the cipher AES–Rijndael.

Observation Polynomial multiplication via circular convolution.
$p(T) = a_0 + a_1 T + \cdots + a_{n-1}T^{n-1}$,
$q(T) = b_0 + b_1 T + \cdots + b_{n-1}T^{n-1}$.
Put

$$A = \begin{pmatrix} a_0 \\ a_1 \\ \cdot \\ a_{n-1} \\ 0 \\ \cdot \\ 0 \end{pmatrix}, \quad B = \begin{pmatrix} b_0 \\ b_1 \\ \cdot \\ b_{n-1} \\ 0 \\ \cdot \\ 0 \end{pmatrix} \in \mathbb{C}^{2n}.$$

Compute

$$C = A * B = \begin{pmatrix} c_0 \\ c_1 \\ \cdot \\ c_{2n-2} \\ 0 \end{pmatrix} \in \mathbb{C}^{2n}.$$

Then: $p(T)q(T) = c_0 + c_1 T + \cdots + c_{2n-2} T^{2n-2}$.

In this way, we obtain the coefficients of the product polynomial by means of circular convolution. The practical interest of this "formal pirouette" comes from the *fast algorithms* for the computation of circular convolutions. This will be our next main theme.

Exercises

(1) Compute $\begin{pmatrix} 3 \\ -2 \\ 0 \\ 4 \end{pmatrix} * \begin{pmatrix} -1 \\ 2 \\ 5 \\ 1 \end{pmatrix}$.

(2) Compute $\begin{pmatrix} 2 \\ 3 \\ -2 \\ 0 \\ 1 \\ 5 \\ -7 \\ 4 \end{pmatrix} * \begin{pmatrix} 1 \\ 0 \\ -2 \\ 0 \\ 1 \\ 0 \\ -2 \\ 0 \end{pmatrix}$.

(3) Solve the equation $a * x = b$

with $a = \begin{pmatrix} 0 \\ 1 \\ 2 \\ 3 \end{pmatrix}$, $x = \begin{pmatrix} x_0 \\ x_1 \\ x_2 \\ x_3 \end{pmatrix}$, $b = \begin{pmatrix} -2 \\ -2 \\ 2 \\ 2 \end{pmatrix}$.

Notation   Let $f_0, f_1, \ldots, f_{n-1}$ be the n columns of the matrix F_n. (F_n is invertible; thus the n columns of F_n are a basis of the vector space $\mathbb{C}^n$).

(4) Let

$$A = \begin{pmatrix} 0\,3\,2\,1 \\ 1\,0\,3\,2 \\ 2\,1\,0\,3 \\ 3\,2\,1\,0 \end{pmatrix}.$$

Show that $\{f_0, f_1, f_2, f_3\}$ is a basis of eigenvectors for A, and compute the corresponding eigenvalues.

(Recall: v is an eigenvector for a matrix A whenever v is non-zero and $Av = \lambda v$, where the scalar λ is the associated eigenvalue: "A acts by multiplication with the scalar λ in direction v").

(5) Let

$$\sigma = \begin{pmatrix} 0 & 0 & \ldots & 0 & 1 \\ 1 & 0 & \ldots & 0 & 0 \\ 0 & 1 & \ldots & 0 & 0 \\ & & \cdots \cdots & & \\ 0 & 0 & \ldots & 1 & 0 \end{pmatrix} \in \mathbb{C}^{n \times n}.$$

Show that $\{\mathbf{f}_0, \mathbf{f}_1, \ldots, \mathbf{f}_{n-1}\}$ is a basis of eigenvectors for σ in $\mathbb{C}^n$, and find the corresponding eigenvalues.

(6) Let $A = C \begin{pmatrix} a_0 \\ a_1 \\ \cdot \\ a_{n-1} \end{pmatrix}$ be a circular matrix.

(a) Show that $A = a_0 I + a_1 \sigma + a_2 \sigma^2 + \cdots + a_{n-1} \sigma^{n-1}$.
(b) Show that $\{\mathbf{f}_0, \mathbf{f}_1, \ldots, \mathbf{f}_{n-1}\}$ is a basis of eigenvectors for A in $\mathbb{C}^n$, and find the corresponding eigenvalues.

Now we are ready to present the result which describes the *characteristic property* of the Discrete Fourier Transform.

Let us first fix the notation:

$$x = \begin{pmatrix} x_0 \\ x_1 \\ \cdot \\ \cdot \\ x_{n-1} \end{pmatrix}, \quad y = \begin{pmatrix} y_0 \\ y_1 \\ \cdot \\ \cdot \\ y_{n-1} \end{pmatrix}, \quad z = x * y = \begin{pmatrix} z_0 \\ z_1 \\ \cdot \\ \cdot \\ z_{n-1} \end{pmatrix} \in \mathbb{C},$$

$$X = \begin{pmatrix} X_0 \\ X_1 \\ \cdot \\ \cdot \\ X_{n-1} \end{pmatrix} = F_n \begin{pmatrix} x_0 \\ x_1 \\ \cdot \\ \cdot \\ x_{n-1} \end{pmatrix}, \quad Y = \begin{pmatrix} Y_0 \\ Y_1 \\ \cdot \\ \cdot \\ Y_{n-1} \end{pmatrix} = F_n \begin{pmatrix} y_0 \\ y_1 \\ \cdot \\ \cdot \\ y_{n-1} \end{pmatrix},$$

$$Z = \begin{pmatrix} Z_0 \\ Z_1 \\ \cdot \\ \cdot \\ Z_{n-1} \end{pmatrix} = F_n \begin{pmatrix} z_0 \\ z_1 \\ \cdot \\ \cdot \\ z_{n-1} \end{pmatrix}.$$

Then, we are able to state and prove the **Convolution Theorem**

$$F_n(x * y) = F_n(x) \cdot F_n(y),$$

i.e.

$$Z = \begin{pmatrix} Z_0 \\ Z_1 \\ \cdot \\ \cdot \\ Z_{n-1} \end{pmatrix} = \begin{pmatrix} X_0 Y_0 \\ X_1 Y_1 \\ \cdot \\ \cdot \\ X_{n-1} Y_{n-1} \end{pmatrix}.$$

Proof Convention: all indices have to be interpreted modulo n: we shall replace a negative index by its (non-negative) remainder of division by n.

We have, for $0 \leq j \leq n - 1$:

$$
Z_j = \sum_{0 \leq k \leq n-1} z_k \omega_n^{jk} = \sum_{0 \leq k \leq n-1} \left(\sum_{0 \leq m \leq n-1} x_m y_{k-m} \right) \omega_n^{jk}
$$
$$
= \sum_{0 \leq m \leq n-1} x_m \left(\sum_{0 \leq k \leq n-1} y_{k-m} \omega^{jk} \right).
$$

Let us make now vary $p = k - m$ between 0 and n; then we get
$Z_j = \sum_{0 \leq m \leq n-1} x_m \omega_n^{jm} \cdot \sum_{0 \leq p \leq n-1} y_p \omega_n^{jp} = X_j Y_j$, as claimed. $\square$

If we write now – not without ulterior motives – our theorem differently, then it may be read like this.

Convolution Theorem bis: The (circular) convolution product of two vectors x, $y \in \mathbb{C}^n$ can be computed in the following way:
$x * y = F_n^{-1}(F_n(x) \cdot F_n(y)) = \frac{1}{n} \overline{F}_n (F_n(x) \cdot F_n(y))$.

Remark The computation of a circular convolution according to the initial definition needs n^2 complex multiplications (where n is the length of the vectors to be multiplied). Using the foregoing identity, a zero-cost Fourier transform would reduce all this to n parallel complex multiplications. On the other hand, by traditional matrix algebra, we shall need *a priori* $3n^2 + n$ complex multiplications.

But admit for the moment the following result.

If $n = 2^m$ – n is a power of 2 – then there exists an algorithm (the *Fast Fourier Transform*) which reduces the computation of $y = F_n(x)$ to $m \cdot 2^{m-1} = \frac{1}{2} n \cdot \text{Log}_2 n$ complex multiplications. The computation of a convolution product of two vectors of length $n = 2^m$ will then need only $(3m + 2) \cdot 2^{m-1}$ complex multiplications (recall the "algorithmic elimination" of F_n^{-1}). For example, for $n = 2^8 = 256$, the direct method needs 65,536 complex multiplications, whereas the fast method will need only 3,328 complex multiplications.

There remains nevertheless the question: **why** are we interested in computing convolution products? The "natural" answer is given by signal theory (digital filters). We shall present it in a moment. But we dispose already of a rather convincing example, which comes from an absolutely non-periodic world.

Fast Multiplication of Large Numbers

(You may think of digital signature arithmetic, but the magnitudes are not really big there.)

The multiplication of two (large) integers p and q – in decimal notation, say – can be done in two steps:

(i) First, multiply p and q as polynomial expressions:
$$p = a_n a_{n-1} \cdots a_1 a_0 = a_n 10^n + a_{n-1} 10^{n-1} + \cdots + a_1 10 + a_0,$$
$$q = b_n b_{n-1} \cdots b_1 b_0 = b_n 10^n + b_{n-1} 10^{n-1} + \cdots + b_1 10 + b_0,$$
$$pq = c_{2n} 10^{2n} + c_{2n-1} 10^{2n-1} + \cdots + c_1 10 + c_0.$$

(ii) Then, renormalize pq in correct decimal notation:
$$pq = \bar{c}_{2n+1} \bar{c}_{2n} \cdots \bar{c}_1 \bar{c}_0, \text{ with } 0 \le \bar{c}_i \le 9,\ 0 \le i \le 2n+1.$$

Observe that only the first step is costly. But we have already seen how polynomial multiplication becomes circular convolution.

Exercises

(1) With the help of the Convolution Theorem, solve the equation $a * x = b$ in the two following cases:

(a) $a = \begin{pmatrix} 0 \\ 1 \\ 2 \\ 3 \end{pmatrix}$, $b = \begin{pmatrix} -2 \\ -2 \\ 2 \\ 2 \end{pmatrix}$.

(b) $a = \begin{pmatrix} 1 \\ 2 \\ 3 \\ 4 \end{pmatrix}$, $b = \begin{pmatrix} 2 \\ -2 \\ -2 \\ 2 \end{pmatrix}$.

(2) Decide, with the aid of the Convolution Theorem, if the circular matrix $C(x)$ with

$$x = \frac{1}{2} \begin{pmatrix} 1 + \sqrt{2} \\ \sqrt{2} \\ 1 - \sqrt{2} \\ -\sqrt{2} \end{pmatrix}$$

is invertible. In the affirmative case, find the inverse matrix.

(3) Show that $\frac{1}{n} F_n^{-1}(u * v) = F_n^{-1}(u) F_n^{-1}(v)$ (use the identity $F_n^{-1}(z) = \frac{1}{n} \overline{F_n(\bar{z})}$).

(4) Define the inner product $\langle , \rangle : \mathbb{C}^n \times \mathbb{C}^n \longrightarrow \mathbb{C}^n$ by $\langle x, y \rangle = \frac{1}{n} x^t \cdot \bar{y}$.
(*Attention:* $\langle y, x \rangle = \overline{\langle x, y \rangle}$ and $\langle x, \alpha y \rangle = \bar{\alpha} \langle x, y \rangle$ for $\alpha \in \mathbb{C}$).
(a) The n columns $\mathbf{f}_0, \mathbf{f}_1, \ldots, \mathbf{f}_{n-1}$ of the matrix F_n are an orthonormal basis for this inner product:

$$\langle \mathbf{f}_i, \mathbf{f}_j \rangle = \begin{cases} 1 \text{ for } i = j, \\ 0 \text{ otherwise.} \end{cases}$$

(b) $\frac{1}{n} \langle F_n(x), F_n(y) \rangle = \langle x, y \rangle$, i.e. $\langle F_n(x), y \rangle = \langle x, \overline{F_n}(y) \rangle$.
(c) $\langle F_n(x * y), z \rangle = \langle F_n(x), F_n^{-1}(\bar{y} * F_n(z)) \rangle$.

3.1.2 The Fast Fourier Transform Algorithm

The Fast Fourier Transform (FFT) is not an autonomous transformation. It is only a recursive algorithm which allows the fast computation of the vector $y = F_n x$ for $n = 2^m$ – i.e. whenever n *is a power of 2* – by iterated half-length reduction of the current vectors (a cascade visualizing in a binary tree).

The Splitting Theorem

Recall *The recursive structure of the Pascal triangle for the binomial coefficients is given by the relation* $\binom{n}{k} = \binom{n-1}{k-1} + \binom{n-1}{k}$ *(in older notation:* $C_n^k = C_{n-1}^{k-1} + C_{n-1}^k$*).*

We search a similar relation which allows, for $n = 2m$, to compute the vector $y = F_n x$ by means of the two vectors $y' = F_m x'$ and $y'' = F_m x''$, where x' and x'' are extracted from the vector x in a simple and direct manner.

Fundamental observation: $\boxed{\omega_m = \omega_n^2 \quad \text{for } n = 2m.}$

Verify: every row (column) of even index of the matrix F_8 depends only on ω_4.

Whence the **pivotal idea** for a recursive algorithm.

Reduce the computation of

$$\begin{pmatrix} y_0 \\ y_1 \\ \cdot \\ y_{n-1} \end{pmatrix} = F_n \begin{pmatrix} x_0 \\ x_1 \\ \cdot \\ x_{n-1} \end{pmatrix}$$

to the computation of

$$\begin{pmatrix} y'_0 \\ y'_1 \\ \cdot \\ y'_{m-1} \end{pmatrix} = F_m \begin{pmatrix} x_0 \\ x_2 \\ \cdot \\ x_{2m-2} \end{pmatrix}$$

and of

$$\begin{pmatrix} y''_0 \\ y''_1 \\ \cdot \\ y''_{m-1} \end{pmatrix} = F_m \begin{pmatrix} x_1 \\ x_3 \\ \cdot \\ x_{2m-1} \end{pmatrix}.$$

Hypothesis Suppose $\begin{pmatrix} y'_0 \\ y'_1 \\ \cdot \\ y'_{m-1} \end{pmatrix}$ and $\begin{pmatrix} y''_0 \\ y''_1 \\ \cdot \\ y''_{m-1} \end{pmatrix}$ to be already computed.

We would like to compute

$$\begin{pmatrix} y_0 \\ y_1 \\ \cdot \\ y_{n-1} \end{pmatrix}$$

in function of these two vectors.

Let us write explicitly:

$$y_j' = \sum_{0 \le k \le m-1} x_{2k}\omega_m^{kj}, \quad 0 \le j \le m-1,$$

$$y_j'' = \sum_{0 \le k \le m-1} x_{2k+1}\omega_m^{kj}, \quad 0 \le j \le m-1$$

$$y_j = \sum_{0 \le k \le n-1} x_k\omega_n^{kj}, \quad 0 \le j \le n-1.$$

Now we write once more y_j, separating the even and the odd indices:

$$y_j = \sum_{0 \le k \le m-1} x_{2k}\omega_n^{2kj} + \sum_{0 \le k \le m-1} x_{2k+1}\omega_n^{(2k+1)j}, \quad 0 \le j \le n-1.$$

But $\omega_n^2 = \omega_m$, i.e. $\omega_n^{2kj} = \omega_m^{kj}$, hence

$$y_j = \sum_{0 \le k \le m-1} x_{2k}\omega_m^{kj} + \omega_n^j \cdot \sum_{0 \le k \le m-1} x_{2k+1}\omega_m^{kj}, \quad 0 \le j \le m-1.$$

Finally:

$$y_j = y_j' + \omega_n^j \cdot y_j'', \quad 0 \le j \le m-1.$$

This gives formulas for the first half of the y-coordinates. We have still to find the formulas for $m \le j \le n-1$.

We decide to use the relations $\omega_n^{m+j} = -\omega_n^j$, and we obtain immediately

$$y_{m+j} = y_j' - \omega_n^j \cdot y_j'', \quad 0 \le j \le m-1.$$

Splitting Theorem

In the given situation, the components of $y = F_n x$ are represented in function of the components of $y' = F_m x'$ and of $y'' = F_m x''$ by the formulas

$$\boxed{\begin{aligned} y_j &= y_j' + \omega_n^j \cdot y_j'' \quad 0 \le j \le m-1, \\ y_{m+j} &= y_j' - \omega_n^j \cdot y_j'' \quad 0 \le j \le m-1. \end{aligned}}$$

The Algorithm for n = 8

Computation of $\begin{pmatrix} y_0 \\ y_1 \\ . \\ y_7 \end{pmatrix} = F_8 \begin{pmatrix} x_0 \\ x_1 \\ . \\ x_7 \end{pmatrix}.$

(1) If $\begin{pmatrix} y_0' \\ y_1' \\ y_2' \\ y_3' \end{pmatrix} = F_4 \begin{pmatrix} x_0 \\ x_2 \\ x_4 \\ x_6 \end{pmatrix}, \quad \begin{pmatrix} y_0'' \\ y_1'' \\ y_2'' \\ y_3'' \end{pmatrix} = F_4 \begin{pmatrix} x_1 \\ x_3 \\ x_5 \\ x_7 \end{pmatrix},$

then we get:

$$\begin{aligned}
y_0 &= y_0' + y_0'', & y_4 &= y_0' - y_0'', \\
y_1 &= y_1' + \omega_8 \cdot y_1'', & y_5 &= y_1' - \omega_8 \cdot y_1'', \\
y_2 &= y_2' - \mathrm{i} \cdot y_2'', & y_6 &= y_2' + \mathrm{i} \cdot y_2'', \\
y_3 &= y_3' + \omega_8^3 \cdot y_3'', & y_7 &= y_3' - \omega_8^3 \cdot y_3''.
\end{aligned}$$

(2) In order to compute

$$\begin{pmatrix} y_0 \\ y_1 \\ y_2 \\ y_3 \end{pmatrix} = F_4 \begin{pmatrix} x_0 \\ x_2 \\ x_4 \\ x_6 \end{pmatrix},$$

we need

$$\begin{pmatrix} y_0' \\ y_1' \end{pmatrix} = F_2 \begin{pmatrix} x_0 \\ x_4 \end{pmatrix}$$

and

$$\begin{pmatrix} y_0'' \\ y_1'' \end{pmatrix} = F_2 \begin{pmatrix} x_2 \\ x_6 \end{pmatrix}.$$

We shall get

$$\begin{aligned}
y_0 &= y_0' + y_0'', & y_2 &= y_0' - y_0'', \\
y_1 &= y_1' - \mathrm{i} \cdot y_1'', & y_3 &= y_1' + \mathrm{i} \cdot y_1''.
\end{aligned}$$

In order to compute

$$\begin{pmatrix} y_0 \\ y_1 \\ y_2 \\ y_3 \end{pmatrix} = F_4 \begin{pmatrix} x_1 \\ x_3 \\ x_5 \\ x_7 \end{pmatrix},$$

we need

$$\begin{pmatrix} y_0' \\ y_1' \end{pmatrix} = F_2 \begin{pmatrix} x_1 \\ x_5 \end{pmatrix}$$

and

$$\begin{pmatrix} y_0'' \\ y_1'' \end{pmatrix} = F_2 \begin{pmatrix} x_3 \\ x_7 \end{pmatrix}, \dots$$

(3) Finally, in order to compute $\begin{pmatrix} y_0 \\ y_1 \end{pmatrix} = F_2 \begin{pmatrix} x_0 \\ x_4 \end{pmatrix}$, we need $y_0' = F_1 x_0 = x_0$ and $y_0'' = F_1 x_4 = x_4$, and we obtain

$$y_0 = y_0' + y_0'', \quad y_1 = y_0' - y_0''.$$

Similar for

$$\begin{pmatrix} y_0 \\ y_1 \end{pmatrix} = F_2 \begin{pmatrix} x_2 \\ x_6 \end{pmatrix}, \begin{pmatrix} y_0 \\ y_1 \end{pmatrix} = F_2 \begin{pmatrix} x_1 \\ x_5 \end{pmatrix}$$

and

$$\begin{pmatrix} y_0 \\ y_1 \end{pmatrix} = F_2 \begin{pmatrix} x_3 \\ x_7 \end{pmatrix} \dots$$

$\left[\text{Note that } F_2 = \begin{pmatrix} 1 & 1 \\ 1 & -1 \end{pmatrix}, \text{ in perfect harmony with our formulas.} \right]$

Scheme for the computation of

$$\begin{pmatrix} y_0 \\ y_1 \\ \cdot \\ y_7 \end{pmatrix} = F_8 \begin{pmatrix} x_0 \\ x_1 \\ \cdot \\ x_7 \end{pmatrix}.$$

y_0	y_1	y_2	y_3	y_4	y_5	y_6	y_7
y_0'	y_1'	y_2'	y_3'	y_0''	y_1''	y_2''	y_3''
y_0	y_1	y_2	y_3	y_0	y_1	y_2	y_3
y_0'	y_1'	y_0''	y_1''	y_0'	y_1'	y_0''	y_1''
y_0	y_1	y_0	y_1	y_0	y_1	y_0	y_1
y_0'	y_0''	y_0'	y_0''	y_0'	y_0''	y_0'	y_0''
x_0	x_4	x_2	x_6	x_1	x_5	x_3	x_7

Example Computation of $y = F_8 x$ for

$$x = \frac{1}{8} \begin{pmatrix} 3 \\ -i \\ 1 \\ i \\ 3 \\ -i \\ 1 \\ i \end{pmatrix}.$$

3.1 The Discrete Fourier Transform

1	0	0	0	1	0	1	0
1	0	$\frac{1}{2}$	0	0	0	$-\frac{i}{2}$	0
$\frac{3}{4}$	0	$\frac{1}{4}$	0	$-\frac{i}{4}$	0	$\frac{i}{4}$	0
$\frac{3}{8}$	$\frac{3}{8}$	$\frac{1}{8}$	$\frac{1}{8}$	$-\frac{i}{8}$	$-\frac{i}{8}$	$\frac{i}{8}$	$\frac{i}{8}$

Exercises

(1) Compute, via the FFT scheme, $y = F_8 x$ for

$$x = \begin{pmatrix} 0 \\ 1 \\ \frac{1}{2} \\ 0 \\ 0 \\ 0 \\ \frac{1}{2} \\ 1 \end{pmatrix}, \quad x = \frac{1}{8}\begin{pmatrix} 1 \\ 2 \\ 3 \\ 4 \\ 5 \\ 4 \\ 3 \\ 2 \end{pmatrix}.$$

(2) Let $A = C(a)$ be the circular matrix the first column a of which is given by $a_0 = \frac{1}{2}$, $a_1 = 0$, $a_2 = \frac{1}{2}$, $a_3 = 0$, $a_4 = \frac{1}{2}$, $a_5 = 0$, $a_6 = -\frac{1}{2}$, $a_7 = 0$. Using the Convolution Theorem, show that the matrix $C(a)$ is invertible, and compute $C(a)^{-1}$.

(3) You dispose of a program FFT-256 that computes $y = F_{256} x$. How to use this program in order to compute $x = F_{128}^{-1} y$?

(4) Let $A = C(a)$ be the circular matrix the first column a of which is given by

$a_0 = \frac{1}{2}$, $a_1 = \frac{1}{4}$, $a_2 = \frac{1}{2}$, $a_3 = -\frac{1}{4} - \frac{1}{4}\sqrt{2}$, $a_4 = 0$, $a_5 = \frac{1}{4}$, $a_6 = 0$, $a_7 = -\frac{1}{4} + \frac{1}{4}\sqrt{2}$.

Using the Convolution Theorem, show that the matrix $C(a)$ is invertible, and compute $C(x) = C(a)^{-1}$.

(5) Compute $\frac{1}{16} F_{16}(x)$ for $x \in \mathbb{C}^{16}$ with

$$
\begin{array}{llll}
x_0 = 2, & x_1 = 1 + \sqrt{2} + i, & x_2 = 2\sqrt{2} + 2i, & x_3 = 1 - \sqrt{2} - i, \\
x_4 = 2 + 4i, & x_5 = 1 - \sqrt{2} + i, & x_6 = -2\sqrt{2} - 2i, & x_7 = 1 + \sqrt{2} - i, \\
x_8 = 2, & x_9 = 1 + \sqrt{2} + i, & x_{10} = -2\sqrt{2} + 2i, & x_{11} = 1 - \sqrt{2} - i, \\
x_{12} = 2 - 4i, & x_{13} = 1 - \sqrt{2} + i, & x_{14} = 2\sqrt{2} - 2i, & x_{15} = 1 + \sqrt{2} - i.
\end{array}
$$

(6) Compute $\frac{1}{16} F_{16}(x)$ for $x \in \mathbb{C}^{16}$ with

$$
\begin{array}{llll}
x_0 = 10, & x_1 = 1 + 3(1 + \sqrt{2})i, & x_2 = -2, & x_3 = 1 - 3(1 - \sqrt{2})i, \\
x_4 = -2, & x_5 = 1 + 3(1 - \sqrt{2})i, & x_6 = -2, & x_7 = 1 - 3(1 + \sqrt{2})i, \\
x_8 = -6, & x_9 = 1 + 3(1 + \sqrt{2})i, & x_{10} = -2, & x_{11} = 1 - 3(1 - \sqrt{2})i, \\
x_{12} = -2, & x_{13} = 1 + 3(1 - \sqrt{2})i, & x_{14} = -2, & x_{15} = 1 - 3(1 + \sqrt{2})i.
\end{array}
$$

Execution of the Algorithm FFT

We shall adopt a row-notation. Let $n = 2^l$. Consider first the binary tree of the input vectors: $\mathbf{x} = (x_k) = (x_0, x_1, \ldots, x_{n-1})$ will be sorted as follows:

$$\mathbf{x}^{(0,0)} = (x_k),$$
$$\mathbf{x}^{(1,0)} = (x_{2k}) \quad \mathbf{x}^{(1,1)} = (x_{2k+1}),$$
$$\mathbf{x}^{(2,0)} = (x_{4k}) \; \mathbf{x}^{(2,1)} = (x_{4k+2}) \; \mathbf{x}^{(2,2)} = (x_{4k+1}) \; \mathbf{x}^{(2,3)} = (x_{4k+3}),$$

$$\begin{array}{cccccccc} \mathbf{x}^{(3,0)}= & \mathbf{x}^{(3,1)}= & \mathbf{x}^{(3,2)}= & \mathbf{x}^{(3,3)}= & \mathbf{x}^{(3,4)}= & \mathbf{x}^{(3,5)}= & \mathbf{x}^{(3,6)}= & \mathbf{x}^{(3,7)}= \\ (x_{8k}) & (x_{8k+4}) & (x_{8k+2}) & (x_{8k+6}) & (x_{8k+1}) & (x_{8k+5}) & (x_{8k+3}) & (x_{8k+7}). \end{array}$$

The jth row, $0 \le j \le l$, consists of 2^j vectors, of length 2^{l-j} each. We will have, in general: $\mathbf{x}^{(j,m)} = (x_k^{(j,m)})_{0 \le k \le 2^{l-j}-1}$ with $0 \le j \le l$ (j is the index of the current row) and $0 \le m \le 2^j - 1$ (m is the index of the block k in the current row). Note the recursion formulas:

$$x_k^{(0,0)} = x_k,$$
$$x_k^{(j+1,2m)} = x_{2k}^{(j,m)},$$
$$x_k^{(j+1,2m+1)} = x_{2k+1}^{(j,m)}.$$

Notation: $x[q] := x_q^{(0,0)}$.

We shall try to find the index $q = q(j, m, k)$ such that $x_k^{(j,m)} = x[q(j, m, k)]$ ("the kth component of the mth block in the jth row is $x[?]$").

Example Distribution of the indices for $l = 3$.

$$000 \; 001 \; 010 \; 011 \; 100 \; 101 \; 110 \; 111,$$

$$000 \; 010 \; 100 \; 110|001 \; 011 \; 101 \; 111,$$

$$000 \; 100|010 \; 110|001 \; 101|011 \; 111,$$

$$000|100|010|110|001|101|011|111.$$

On the jth row, the $l-j$ first digits enumerate (in binary notation) the components, the j last digits enumerate the blocks. But the enumeration of the blocks is not monotone.

Definition The function "word inversion":

$$p_j(\beta_{j-1}2^{j-1} + \beta_{j-2}2^{j-2} + \cdots + \beta_1 2 + \beta_0) = (\beta_0 2^{j-1} + \beta_1 2^{j-2} + \cdots + \beta_{j-2}2 + \beta_{j-1}).$$

Example $p_5(11001) = 10011$.

Recursion Formulas

$$p_{j+1}(2m) = p_j(m), \quad p_{j+1}(2m+1) = p_j(m) + 2^j, \quad 0 \le m \le 2^j - 1.$$

Lemma $q(j, m, k) = 2^j \cdot k + p_j(m)$,
i.e. $x_k^{(j,m)} = x[2^j \cdot k + p_j(m)]$, $0 \le j \le l$, $0 \le m \le 2^j - 1$, $0 \le k \le 2^{l-j} - 1$.
In particular, we get for the singletons of the last row:
$x_0^{(l,m)} = x[p_l(m)]$.

Proof Recursion on j (Exercise).   □

Example Let $\mathbf{x} = (x_0, x_1, \ldots, x_{1,023})$ be a vector of 2^{10} components. Which component of $\mathbf{x}$ will occupy the 313th place of the last row of the input-tree for the algorithm FFT?
    Answer: $x_0^{(10,313)} = x[p_{10}(0100111001)] = x[1001110010] = x_{626}$.

Consider now the binary tree of the Fourier transforms (with $\mathbf{y}^{(j,m)} = F_{2^{l-j}} \mathbf{x}^{(j,m)}$):

$$\mathbf{y}^{(0,0)}$$

$$\mathbf{y}^{(1,0)} \qquad\qquad\qquad \mathbf{y}^{(1,1)}$$

$$\mathbf{y}^{(2,0)} \qquad\quad \mathbf{y}^{(2,1)} \qquad\quad \mathbf{y}^{(2,2)} \qquad\quad \mathbf{y}^{(2,3)}$$

$$\mathbf{y}^{(3,0)} \quad \mathbf{y}^{(3,1)} \quad \mathbf{y}^{(3,2)} \quad \mathbf{y}^{(3,3)} \quad \mathbf{y}^{(3,4)} \quad \mathbf{y}^{(3,5)} \quad \mathbf{y}^{(3,6)} \quad \mathbf{y}^{(3,7)}$$

$\mathbf{y}^{(0,0)} = F_n \mathbf{x}^{(0,0)}$ is the final result.
We **know** the row of the basis-singletons:
$\mathbf{y}^{(l,m)} = y_0^{(l,m)} = x_0^{(l,m)} = x[p_l(m)]$.
The vectors of the intermediate lines are computed by the Splitting Theorem.
    Put $r_j = \omega_{2^j} = e^{-2\pi i/2^j}$. Then

$$y_k^{(j-1,m)} = y_k^{(j,2m)} + r_{l-j+1}^k \cdot y_k^{(j,2m+1)},$$

$$y_{k+2^{l-j}}^{(j-1,m)} = y_k^{(j,2m)} - r_{l-j+1}^k \cdot y_k^{(j,2m+1)}.$$

Consider every row of our tree as a *single* vector of length $n = 2^l$.
    Now write $\mathbf{y}_j = (y_j[k])$, i.e. $y_j[2^{l-j} \cdot m + k] = y_k^{(j,m)}$, $0 \le m \le 2^j - 1$, $0 \le k \le 2^{l-j} - 1$ (we proceed horizontally, block after block).
Recall that we have, on the level of the singletons:

$$y_l[m] = x[p_l(m)], \quad 0 \le m \le 2^l - 1.$$

With this new notation, the splitting equations become

$$y_{j-1}[2^{l-j+1} \cdot m + k] = y_j[2^{l-j+1} \cdot m + k] + r_{l-j+1}^k \cdot y_j[2^{l-j+1} \cdot m + k + 2^{l-j}],$$

$$y_{j-1}[2^{l-j+1} \cdot m + k + 2^{l-j}] = y_j[2^{l-j+1} \cdot m + k]$$
$$- r_{l-j+1}^k \cdot y_j[2^{l-j+1} \cdot m + k + 2^{l-j}].$$

Put $h := 2^{l-j+1} \cdot m + k$, $s := 2^{l-j}$.

Auxiliary observation: $r_{l-j+1}^k = r_{l-j+1}^h$ ($h \equiv k \mod 2^{l-j+1}$). Whence our *recursion formulas* in final form:

$$\boxed{\begin{aligned} &y_{j-1}[h] = y_j[h] + r_{l-j+1}^h \cdot y_j[h+s] \\ &y_{j-1}[h+s] = y_j[h] - r_{l-j+1}^h \cdot y_j[h+s] \\ &\text{for } j = l, l-1, \dots, 1, \ s = 2^{l-j}, \ h = 2^{l-j+1} \cdot m + k, \\ &\text{with } m = 0, 1, \dots, 2^{j-1} - 1 \text{ and } k = 0, 1, \dots, s-1. \end{aligned}}$$

Conclusion If the vectors $\mathbf{y}_j = (y_j[h])$ are computed according to the recursion formulas above, then the $y_0[m]$, $0 \le m \le 2^l - 1$, are the components of $\mathbf{y} = F_n\mathbf{x}$.

Exercises

(1) Let $\mathbf{y} = F_8\mathbf{x}$.

True or false: the FFT scheme, applied to an input vector $\mathbf{x}$, the components of which are ranged in *natural order*, produces the transformed vector $\mathbf{y}$ in p_3-permuted order:

y_0	y_4	y_2	y_6	y_1	y_5	y_3	y_7
x_0	x_1	x_2	x_3	x_4	x_5	x_6	x_7

(2) Let $\mathbf{x} = \begin{pmatrix} x_0 \\ \cdot \\ x_7 \end{pmatrix}$ be a vector with *real* components, $\mathbf{y} = F_8\mathbf{x}$. Show that y_0 and y_4 will be real, and that $y_1 = \overline{y_7}$, $y_2 = \overline{y_6}$ and $y_3 = \overline{y_5}$.

3.2 Trigonometric Interpolation

In this section, we shall proceed to give a more conceptual interpretation of the Discrete Fourier Transform, in the language of signal theory.

More precisely, we shall try to clarify the mathematical meaning of the statement: the Fourier transform passes from the time domain to the frequency domain.

Despite a thunderous horizon of practical questions, our presentation will always remain naïvely mathematical; for example, we shall never touch upon the problem of digitization, i.e. of effective sampling.

3.2.1 Trigonometric Polynomials

Consider (complex-valued) *trigonometric polynomials*, in *real notation*, i.e. of the form

$$f(t) = \frac{a_0}{2} + a_1 \cdot cos2\pi t + b_1 \cdot sin2\pi t + a_2 \cdot cos4\pi t + b_2 \cdot sin4\pi t + \ldots a_m \cdot cos2\pi mt + b_m \cdot sin2\pi mt$$

with $a_0, a_1, \ldots, a_m, b_1, \ldots, b_m \in \mathbb{C}$.

Remarks

(1) $f(t)$ is a function of the real variable t ($t \equiv$ the time), taking complex values.
(2) The time t is *normalized*: $f(t+1) = f(t)$ for every $t \in \mathbb{R}$.
    $f(t)$ is thus a 1-periodic function; in another language: $f(t)$ is an elementary 1-periodic signal.
(3) *Example*:
    $$f(t) = 3 + (2+i)\cos 2\pi t - 3\sin 2\pi t - i\cos 4\pi t - (1-i)\sin 4\pi t = x(t) + iy(t)$$
    with
    $$x(t) = 3 + 2\cos 2\pi t - 3\sin 2\pi t - \sin 4\pi t,$$
    $$y(t) = \cos 2\pi t - \cos 4\pi t + \sin 4\pi t.$$

We observe: our formal generosity to admit complex coefficients means conceptually that we consider two periodic *real* signals $x(t)$ and $y(t)$ (the real part and the imaginary part of $f(t)$) *simultaneously*. You should imagine their simultaneous progression in two orthogonal planes.

Let us insist. The factor i serves merely as a *separator*: our 3D coordinate system is composed of the t-axis, together with the x-axis and the y-axis of the usual complex plane (orthogonal to the t-axis). Multiplication by i means a rotation of 90° around the t-axis (note that it has absolutely *no* phase shift signification in the current formal context).

Concerning the language. The notation of $f(t)$ is called *real*, since $f(t) = x(t) + iy(t)$ is given explicitly by two *real* functions, linear combinations of the elementary functions $\cos 2\pi \nu t$ and $\sin 2\pi \nu t$, $0 \leq \nu \leq m$.

Let us now pass to a purely *complex notation*.

Recall

$$e^{i\theta} = \cos\theta + i\sin\theta$$

hence $\cos\theta = \frac{1}{2}(e^{i\theta} + e^{-i\theta})$ and $\sin\theta = -\frac{i}{2}(e^{i\theta} - e^{-i\theta})$.

We obtain

$$f(t) = \frac{a_0}{2} + a_1 \cdot cos2\pi t + b_1 \cdot sin2\pi t + a_2 \cdot cos4\pi t + b_2 \cdot sin4\pi t$$

$$+ \ldots a_m \cdot cos2\pi mt + b_m \cdot sin2\pi mt = \frac{a_0}{2} + \sum_{1 \le \nu \le m} \frac{a_\nu}{2}(e^{2\pi i\nu t} + e^{-2\pi i\nu t})$$

$$- \sum_{1 \le \nu \le m} \frac{ib_\nu}{2}(e^{2\pi i\nu t} - e^{-2\pi i\nu t}) = \frac{a_0}{2} + \sum_{1 \le \nu \le m} \frac{1}{2}(a_\nu - ib_\nu)e^{2\pi i\nu t}$$

$$- \sum_{1 \le \nu \le m} \frac{1}{2}(a_\nu + ib_\nu)e^{-2\pi i\nu t} = \sum_{-m \le \nu \le m} c_\nu e^{2\pi i\nu t}$$

with the **transfer formulas** between the real notation and the complex notation:

$$c_\nu = \tfrac{1}{2}(a_\nu - ib_\nu), \; c_{-\nu} = \tfrac{1}{2}(a_\nu + ib_\nu), \quad 0 \le \nu \le m,$$
$$a_\nu = c_\nu + c_{-\nu}, \quad b_\nu = i(c_\nu - c_{-\nu}), \quad 0 \le \nu \le m.$$

(Note that the negative frequencies occurring in the complex notation have no conceptual interpretation and disappear when passing to the real notation.)

We shall call $f(t) = \sum_{-m \le \nu \le m} c_\nu e^{2\pi i\nu t}$ a **trigonometric polynomial** (in complex notation) of degree m.

And indeed: $f(t) = c_{-m}X^{-m} + c_{-m+1}X^{-m+1} + \cdots c_{-1}X^{-1} + c_0 + c_1 X + \cdots + c_{m-1}X^{m-1} + c_m X^m$ with $X = e^{2\pi it}$.

Exercises

(1) Let $f(t) = \frac{1}{2}e^{-8\pi it} + e^{-4\pi it} + 1 + \frac{1}{2}e^{8\pi it}$.
Find the real notation of $f(t)$ and compute the values of $f(t)$ for $t = 0, \frac{1}{8}, \frac{1}{4}, \frac{3}{8}, \frac{1}{2}, \frac{5}{8}, \frac{3}{4}, \frac{7}{8}$.

(2) Let $f(t) = 4 - 2\cos 2\pi t + 3\sin 4\pi t - \cos 6\pi t + 2\sin 6\pi t$.
Write $f(t)$ in complex notation $f(t) = \sum_{-3 \le \nu \le 3} c_\nu e^{2\pi i\nu t}$.

(3) Let $f(t) = \sum_{-m \le \nu \le m} c_\nu e^{2\pi i\nu t}$ be a trigonometric polynomial of degree m, in complex notation. How do you recognize in this notation that $f(t)$ is actually a real function?

(4) Let $f(t) = 2 - \cos 2\pi t + 3\sin 2\pi t + 4\cos 4\pi t - 2\sin 6\pi t + \cos 8\pi t$.

Find the complex notation of $\tilde{f}(t) = f(t - \frac{1}{8}) = \tilde{c}_{-4}e^{-8\pi it} + \tilde{c}_{-3}e^{-6\pi it} + \tilde{c}_{-2}e^{-4\pi it} + \tilde{c}_{-1}e^{-2\pi it} + \tilde{c}_0 + \tilde{c}_1 e^{2\pi it} + \tilde{c}_2 e^{4\pi it} + \tilde{c}_3 e^{6\pi it} + \tilde{c}_4 e^{8\pi it}$, i.e. find the coefficients $\tilde{c}_\nu$ for $-4 \le \nu \le 4$.

3.2.2 Sampling and Reconstruction

In the following pages we shall promote a Sampling Theorem which is the purely discrete periodic version of a celebrated theorem (associated with the name of Shannon), which we shall encounter in a moment. From an unpretentious mathematical viewpoint, we are simply in a situation of (trigonometric) interpolation.

Consider, as a first step, for $\nu \in \mathbb{Z}$, the elementary 1-periodic signal

$$e^{[\nu]}(t) = e^{2\pi i \nu t} = \cos 2\pi \nu t + i \sin 2\pi \nu t$$

(geometrically, this is one of the possible parametrizations of the unit circle).

Now, for $n \geq 2$, carry out a *sampling of order* n, i.e. consider the n values $e_0^{[\nu]}, e_1^{[\nu]}, \ldots, e_{n-1}^{[\nu]}$, where $e_k^{[\nu]} = e^{[\nu]}(\frac{k}{n}) = e^{(2\pi i/n)\cdot\nu k} = w_n^{-\nu k}, 0 \leq k \leq n-1$.

We observe: the vector

$$\frac{1}{n}\begin{pmatrix} e_0^{[\nu]} \\ e_1^{[\nu]} \\ \cdot \\ \cdot \\ e_{n-1}^{[\nu]} \end{pmatrix}$$

of these n samples (for $t = 0, \frac{1}{n}, \frac{2}{n}, \ldots, \frac{n-1}{n}$) is one of the columns of the matrix F_n^{-1}.

More precisely: let ν_0 be the (non-negative) remainder of division of ν by n ($\nu_0 \equiv \nu \bmod n$), then we are confronted with the ν_0th column of F_n^{-1} ($0 \leq \nu_0 \leq n-1$).

Conclusion

$$\frac{1}{n}F_n\begin{pmatrix} e_0^{[\nu]} \\ e_1^{[\nu]} \\ \cdot \\ \cdot \\ e_{n-1}^{[\nu]} \end{pmatrix} = e_{\nu_0} = \begin{pmatrix} 0 \\ \cdot \\ 1 \\ \cdot \\ 0 \end{pmatrix} \leftarrow \text{position } \nu_0.$$

Let us make this explicit for $n = 8$:

$$\frac{1}{8}F_8\begin{pmatrix} e_0^{[0]} \\ e_1^{[0]} \\ e_2^{[0]} \\ e_3^{[0]} \\ e_4^{[0]} \\ e_5^{[0]} \\ e_6^{[0]} \\ e_7^{[0]} \end{pmatrix} = e_0 = \begin{pmatrix} 1 \\ 0 \\ 0 \\ 0 \\ 0 \\ 0 \\ 0 \\ 0 \end{pmatrix}, \quad \frac{1}{8}F_8\begin{pmatrix} e_0^{[1]} \\ e_1^{[1]} \\ e_2^{[1]} \\ e_3^{[1]} \\ e_4^{[1]} \\ e_5^{[1]} \\ e_6^{[1]} \\ e_7^{[1]} \end{pmatrix} = e_1 = \begin{pmatrix} 0 \\ 1 \\ 0 \\ 0 \\ 0 \\ 0 \\ 0 \\ 0 \end{pmatrix},$$

$$\frac{1}{8}F_8\begin{pmatrix} e_0^{[2]} \\ e_1^{[2]} \\ e_2^{[2]} \\ e_3^{[2]} \\ e_4^{[2]} \\ e_5^{[2]} \\ e_6^{[2]} \\ e_7^{[2]} \end{pmatrix} = e_2 = \begin{pmatrix} 0 \\ 0 \\ 1 \\ 0 \\ 0 \\ 0 \\ 0 \\ 0 \end{pmatrix}, \quad \frac{1}{8}F_8\begin{pmatrix} e_0^{[3]} \\ e_1^{[3]} \\ e_2^{[3]} \\ e_3^{[3]} \\ e_4^{[3]} \\ e_5^{[3]} \\ e_6^{[3]} \\ e_7^{[3]} \end{pmatrix} = e_3 = \begin{pmatrix} 0 \\ 0 \\ 0 \\ 1 \\ 0 \\ 0 \\ 0 \\ 0 \end{pmatrix},$$

$$\frac{1}{8}F_8\begin{pmatrix} e_0^{[4]} \\ e_1^{[4]} \\ e_2^{[4]} \\ e_3^{[4]} \\ e_4^{[4]} \\ e_5^{[4]} \\ e_6^{[4]} \\ e_7^{[4]} \end{pmatrix} = \frac{1}{8}F_8\begin{pmatrix} e_0^{[-4]} \\ e_1^{[-4]} \\ e_2^{[-4]} \\ e_3^{[-4]} \\ e_4^{[-4]} \\ e_5^{[-4]} \\ e_6^{[-4]} \\ e_7^{[-4]} \end{pmatrix} = e_4 = \begin{pmatrix} 0 \\ 0 \\ 0 \\ 0 \\ 1 \\ 0 \\ 0 \\ 0 \end{pmatrix}, \quad \frac{1}{8}F_8\begin{pmatrix} e_0^{[-3]} \\ e_1^{[-3]} \\ e_2^{[-3]} \\ e_3^{[-3]} \\ e_4^{[-3]} \\ e_5^{[-3]} \\ e_6^{[-3]} \\ e_7^{[-3]} \end{pmatrix} = e_5 = \begin{pmatrix} 0 \\ 0 \\ 0 \\ 0 \\ 0 \\ 1 \\ 0 \\ 0 \end{pmatrix},$$

$$\frac{1}{8}F_8\begin{pmatrix} e_0^{[-2]} \\ e_1^{[-2]} \\ e_2^{[-2]} \\ e_3^{[-2]} \\ e_4^{[-2]} \\ e_5^{[-2]} \\ e_6^{[-2]} \\ e_7^{[-2]} \end{pmatrix} = e_6 = \begin{pmatrix} 0 \\ 0 \\ 0 \\ 0 \\ 0 \\ 0 \\ 1 \\ 0 \end{pmatrix}, \quad \frac{1}{8}F_8\begin{pmatrix} e_0^{[-1]} \\ e_1^{[-1]} \\ e_2^{[-1]} \\ e_3^{[-1]} \\ e_4^{[-1]} \\ e_5^{[-1]} \\ e_6^{[-1]} \\ e_7^{[-1]} \end{pmatrix} = e_7 = \begin{pmatrix} 0 \\ 0 \\ 0 \\ 0 \\ 0 \\ 0 \\ 0 \\ 1 \end{pmatrix}.$$

Let now $f(t) = \sum_{-m \le \nu \le m} c_\nu e^{2\pi i \nu t}$ be a trigonometric polynomial of degree m, in complex notation.

We sample to the order $n = 2m$:

$$f_0 = f(0), f_1 = f\left(\frac{1}{n}\right), f_2 = f\left(\frac{2}{n}\right), \ldots, f_{n-1} = f\left(\frac{n-1}{n}\right).$$

Now,

$$f(t) = \sum_{-m \le \nu \le m} c_\nu e^{[\nu]}(t),$$

hence we have

$$f_k = \sum_{-m \le \nu \le m} c_\nu e_k^{[\nu]}, \qquad 0 \le k \le n-1.$$

As a consequence

$$
\frac{1}{n}F_n
\begin{pmatrix}
f_0 \\
f_1 \\
\cdot \\
\cdot \\
\cdot \\
\cdot \\
\cdot \\
f_{n-1}
\end{pmatrix}
=
\sum_{-m \le \nu \le m} c_\nu \cdot \frac{1}{n}F_n
\begin{pmatrix}
e_0^{[\nu]} \\
e_1^{[\nu]} \\
\cdot \\
\cdot \\
\cdot \\
\cdot \\
\cdot \\
e_{n-1}^{[\nu]}
\end{pmatrix}
=
\begin{pmatrix}
c_0 \\
c_1 \\
\cdot \\
c_{m-1} \\
c_m + c_{-m} \\
c_{-m+1} \\
\cdot \\
c_{-1}
\end{pmatrix}.
$$

In order to get rid of the nasty and ambiguous central term $c_m + c_{-m}$, observe the following simple fact.

A sampling to the order $n = 2m$ of
$$f(t) = \frac{a_0}{2} + a_1 \cdot cos2\pi t + b_1 \cdot sin2\pi t + a_2 \cdot cos4\pi t + b_2 \cdot sin4\pi t + \ldots a_m \cdot$$
$cos2\pi mt + b_m \cdot sin2\pi mt$

is perfectly **insensible** to the value of b_m: $\sin 2\pi \frac{mk}{n} = \sin k\pi = 0$ for $0 \le k \le n-1$.

In other words, it is impossible to recover the contribution of the term $b_m \cdot \sin 2\pi mt$ after a sampling to the order $n = 2m$.

With unicity of reconstruction in mind, we shall limit our attention to **balanced** trigonometric polynomials, i.e. to trigonometric polynomials of the form

$$f(t) = \frac{a_0}{2} + a_1 \cdot cos2\pi t + b_1 \cdot sin2\pi t + a_2 \cdot cos4\pi t + b_2 \cdot sin4\pi t + \ldots \frac{a_m}{2} \cdot cos2\pi mt$$

(hence with $b_m = 0$). This is equivalent to a complex notation of the form

$$f(t) = \frac{c_{-m}}{2} e^{-2\pi imt} + \sum_{-m+1 \le \nu \le m-1} c_\nu e^{2\pi i\nu t} + \frac{c_m}{2} e^{2\pi imt}$$

with $c_m = c_{-m}$. Note that we have still $a_m = c_m + c_{-m}$ (this is the reason of the factor $\frac{a_m}{2}$).

Why is the factor $\frac{1}{2}$ appended to the extremal terms of the complex notation?

The reason is simply the pleasure of a formal harmony: we want the central term of the transformed vector to be called c_m ($= c_{-m}$) and not $2c_m (= c_m + c_{-m})$.

Remark Note that we already have convincing arguments for the Discrete Fourier Transform to pass from the time domain to the frequency domain. The input vectors are time-indexed, i.e. they are (naturally extensible to) periodic time series, the output vectors are frequency-indexed lists of (eventually cumulative) amplitudes for interpolating trigonometric polynomials. Let us sum up:

Sampling Theorem

$$f(t) = \frac{c_{-m}}{2}e^{-2\pi i m t} + \sum_{-m+1\leq\nu\leq m-1} c_\nu e^{2\pi i \nu t} + \frac{c_m}{2}e^{2\pi i m t}$$

a balanced trigonometric polynomial of degree m.
We sample to the order $n = 2m$ [2]:

$$f_0 = f(0), f_1 = f\left(\frac{1}{n}\right), f_2 = f\left(\frac{2}{n}\right), \dots, f_{n-1} = f\left(\frac{n-1}{n}\right).$$

Then

$$\frac{1}{n}F_n \begin{pmatrix} f_0 \\ f_1 \\ \cdot \\ \cdot \\ \cdot \\ \cdot \\ f_{n-1} \end{pmatrix} = \begin{pmatrix} c_0 \\ c_1 \\ \cdot \\ c_{m-1} \\ c_m = c_{-m} \\ c_{-m+1} \\ \cdot \\ c_{-1} \end{pmatrix}.$$

Note that our Sampling Theorem can also be read as a
Trigonometric Interpolation Theorem in the following manner:

Let $f_0, f_1, \dots, f_{n-1}$ be $n = 2m$ complex values.
Then there exists a *unique* balanced trigonometric polynomial of degree $\leq m$
$$f(t) = \frac{a_0}{2} + a_1 \cdot \cos 2\pi t + b_1 \cdot \sin 2\pi t + a_2 \cdot \cos 4\pi t + b_2 \cdot \sin 4\pi t + \dots \frac{a_m}{2} \cdot \cos 2\pi m t$$
with $f(\frac{k}{n}) = f_k$, $0 \leq k \leq n-1$.

FFT scheme for trigonometric interpolation in case $n = 8$:

	c_0	c_1	c_2	c_3	c_4	c_{-3}	c_{-2}	c_{-1}
$\frac{1}{8} \times$								
	$f(0)$	$f(\frac{1}{8})$	$f(\frac{1}{4})$	$f(\frac{3}{8})$	$f(\frac{1}{2})$	$f(\frac{5}{8})$	$f(\frac{3}{4})$	$f(\frac{7}{8})$

---

[2] Note that our reconstruction works under the assumption that the sampling frequency n is *twice* the maximal frequency m in the (naïve) spectrum of $f(t)$. This will be Ariadne's thread for Sampling Theorems.

Exercises

(1) Find the balanced trigonometric polynomial $f(t) = \frac{a_0}{2} + a_1 \cdot cos2\pi t + b_1 \cdot$
$sin2\pi t + a_2 \cdot cos4\pi t + b_2 \cdot sin4\pi t + a_3 \cdot cos6\pi t + b_3 \cdot sin6\pi t + \frac{a_4}{2} \cdot cos8\pi t$,
such that $f(0) = 1$, $f(\frac{1}{8}) = 2$, $f(\frac{1}{4}) = 3$, $f(\frac{3}{8}) = 4$, $f(\frac{1}{2}) = 5$, $f(\frac{5}{8}) = 4$,
$f(\frac{3}{4}) = 3$, $f(\frac{7}{8}) = 2$.

(2) Which is the balanced trigonometric polynomial $f(t) = \frac{a_0}{2} + a_1 \cdot cos2\pi t +$
$b_1 \cdot sin2\pi t + a_2 \cdot cos4\pi t + b_2 \cdot sin4\pi t + a_3 \cdot cos6\pi t + b_3 \cdot sin6\pi t + \frac{a_4}{2} \cdot cos8\pi t$
with $f(0) = f(\frac{1}{8}) = 1$, $f(\frac{1}{4}) = f(\frac{3}{8}) = -1$, $f(\frac{1}{2}) = f(\frac{5}{8}) = 1$, $f(\frac{3}{4}) =$
$f(\frac{7}{8}) = -1$?

(3) Find the eight balanced trigonometric polynomials of degree 4 $f_0(t), f_1(t)$,
$\ldots, f_7(t)$, which sample (to the order 8) into the eight unit vectors
$e_0, e_1, \ldots, e_7$.

(4) Transmission of a block of 16 bits (two 8-bit bytes) by frequency modulation.

$XY = x_0 x_1 \cdots x_7 y_0 y_1 \cdots y_7$ will be transmitted by means of

$$f(t) = \frac{a_0}{2} + a_1 \cdot cos2\pi t + b_1 \cdot sin2\pi t + a_2 \cdot cos4\pi t + b_2 \cdot sin4\pi t$$

$$+ a_3 \cdot cos6\pi t + b_3 \cdot sin6\pi t + \frac{a_4}{2} \cdot cos8\pi t$$

with

$$a_0 = x_0 + iy_0,$$
$$a_1 = x_1 + iy_1, \qquad b_1 = x_4 + iy_4,$$
$$a_2 = x_2 + iy_2, \qquad b_2 = x_5 + iy_5,$$
$$a_3 = x_3 + iy_3, \qquad b_3 = x_6 + iy_6.$$
$$\frac{a_4}{2} = x_7 + iy_7,$$

After sampling $f(t)$ to the order 8, it is possible to reconstruct XY with
the help of the Discrete Fourier Transform.

You sample:

$$f(0) = 2 + 4i, \; f\left(\tfrac{1}{8}\right) = 1 + \sqrt{2} + i, \; f\left(\tfrac{1}{4}\right) = 0, \; f\left(\tfrac{3}{8}\right) = 1 + \sqrt{2} + i,$$
$$f\left(\tfrac{1}{2}\right) = 2, \quad f\left(\tfrac{5}{8}\right) = 1 - \sqrt{2} + i, \; f\left(\tfrac{3}{4}\right) = 0, \; f\left(\tfrac{7}{8}\right) = 1 - \sqrt{2} + i.$$

Find XY, i.e. the 16 transmitted bits.

3.3 The Whittaker–Shannon Theorem

In order to better understand (and perhaps to appreciate) our modest Sampling Theorem in the elementary periodic case, let us now face the sacrosanct result for all arguments concerning the transition analog $\to$ digital $\to$ analog: the theorem of Nyquist–Shannon (as it is ordinarily called in engineering textbooks; let us nevertheless point out that the priority is Whittaker's, who proved it in 1935, 20 years before Shannon).

3.3.1 Fourier Series

In the core of this section, we will be confronted with the following interpolation problem: what type of function (of signal) is **determined** by an appropriate sequence of its samples?

There is a similar problem, that of developing a periodic (piecewise continuous) function into a Fourier series. In this case, the sequence of the Fourier coefficients determines (the form of) the periodic function.

Actually, these two problems are intimately linked – we shall see this in a moment.

Let us begin with a brief summary of Fourier series.

The Situation – Temporal Version[3]:

Let $f(\theta)$ be a function of the real variable θ, with (real or) complex values, and periodic, of period $P = 2W$.

Put $T = \frac{1}{2W}$.

($2\pi T$ will be the pulsation associated with the period $P = 2W$).

In order to begin decently, we shall suppose $f(\theta)$ to be **continuous** (eventually with a finite number of simple discontinuities in the period interval).

Consider the sequence of the (complex-valued) **Fourier coefficients** of $f(\theta)$:

$$c_k = \frac{1}{2W} \int_{-W}^{W} f(\theta) e^{-2\pi i k T\theta} d\theta, \quad k \in \mathbb{Z}.$$

These coefficients – arguably somehow enigmatic for the novice – are naturally obtained as the solutions of an *extremal problem*:

The trigonometric polynomials

$\sigma_N f(\theta) = \sum_{-N \le k \le N} c_k e^{2\pi i k T\theta}$

are, for every value of $N \ge 0$, the *best approximation* of $f(\theta)$ in the following sense:

Consider, for N fixed, all P-periodic trigonometric polynomials

$$p(\theta) = \sum_{-N \le k \le N} \gamma_k e^{2\pi i k T\theta}$$

the degree of which does not exceed N.

---

[3] Fourier series occupy two sensibly different places in mathematical signal theory. First, they formalize correctly the time domain analysis of periodic phenomena; then, they introduce "spectral periodicity" as Fourier transforms of time series. It is the first aspect which they were invented for – but it is the second aspect which guarantees their mathematical interest.

Measure the distance between $f(\theta)$ and each $p(\theta)$ by the "energy-norm" of their difference:

$$\|f(\theta) - p(\theta)\| = \sqrt{\frac{1}{2W} \int_{-W}^{W} |f(\theta) - p(\theta)|^2 d\theta}.$$

Then each $\sigma_N f(\theta)$ is *the* trigonometric polynomial of degree $\leq N$ which *minimizes* this "energy-distance".

Let us insist:

$$\sigma_N f(\theta) = \sum_{-N \leq k \leq N} c_k e^{2\pi i k T\theta}$$

best approximates (replaces) the function $f(\theta)$ **to the order** N for a specific notion of distance between functions (one computes a certain separating area). Note that a priori $\sigma_N f(\theta)$ is *not* attached to any interpolation problem. Pass now to the trigonometric series

$$\sigma f(\theta) = \lim_{N \to \infty} \sigma_N f(\theta) = \sum_{k \in \mathbb{Z}} c_k e^{2\pi i k T\theta}.$$

The fundamental question is then the following:

> What is the relation between the periodic function $f(\theta)$ and its *Fourier series* $\sigma f(\theta)$?

The answer is given by the *Theorem of Dirichlet*, which reads in small talk version like this.

If $f(\theta)$ is sufficiently regular, then the series $\sigma f(\theta)$ will converge for every $\theta \in \mathbb{R}$, and will represent the function $f(\theta)$:

$$f(\theta) = \sigma f(\theta) = \sum_{k \in \mathbb{Z}} c_k e^{2\pi i k T\theta}.$$

Let us underline: In this case, $f(\theta)$ is entirely *determined* by the sequence $(c_k)_{k \in \mathbb{Z}}$ of its Fourier coefficients. Remains to clarify the condition of *regularity* on $f(\theta)$, which forces the equality $f(\theta) = \sigma f(\theta)$.

(A) *Weak regularity.* $f(\theta)$ is of *bounded variation* in the period interval.
   Consider, for every finite partition $\mathbb{P} = (-W = \theta_0 < \theta_1 < \cdots < \theta_n = W)$ of the period interval, the sum

$$V(\mathbb{P}) = \sum_{0 \leq k \leq n-1} |f(\theta_{k+1}) - f(\theta_k)|.$$

   If the set $\{V(\mathbb{P})\}$ of all these values admits a finite upper bound, then $f(\theta)$ is said to be of bounded variation in $[-W, W]$.
   In particular, this is the case for differentiable $f(\theta)$, admitting a bounded derivation $f'(\theta)$.

(B) *Strong regularity.* If $f(\theta)$ is twice continuously differentiable, or if the series $\sum_{k \in \mathbb{Z}} c_k$ is absolutely convergent, then $\sigma f(\theta)$ converges absolutely and uniformly to $f(\theta)$.

Exercises

We shall consider only real 2π – periodic functions (i.e. with $W = \pi$), which will be specified on an interval of length 2π. It is clear that the transit formulas between complex notation and real notation introduced for trigonometric polynomials are the same for trigonometrical series:

(1) $f(\theta) = \begin{cases} \theta & \text{for } 0 \le \theta \le \pi, \\ 2\pi - \theta & \text{for } \pi \le \theta \le 2\pi. \end{cases}$

Show that

$$\sigma f(\theta) = \frac{\pi}{2} - \frac{4}{\pi}\left(\cos\theta + \frac{\cos 3\theta}{3^2} + \frac{\cos 5\theta}{5^2} + \frac{\cos 7\theta}{7^2} + \cdots\right).$$

(2) $f(\theta) = \begin{cases} \theta & \text{for } -\pi < \theta \le 0, \\ 0 & \text{for } 0 \le \theta \le \frac{\pi}{2}, \\ \theta - \frac{\pi}{2} & \text{for } \frac{\pi}{2} \le \theta \le \pi. \end{cases}$

Find $\sigma f(\theta)$.

(3) $f(\theta) = \theta$ for $-\pi < \theta \le \pi$.

Show that

$$\sigma f(\theta) = 2\left(\sin\theta - \frac{\sin 2\theta}{2} + \frac{\sin 3\theta}{3} - \frac{\sin 4\theta}{4} + \cdots\right).$$

(4) $f(\theta) = \theta^2$ for $-\pi < \theta \le \pi$.

Show that

$$\sigma f(\theta) = \frac{\pi^2}{3} - 4\left(\cos\theta - \frac{\cos 2\theta}{2^2} + \frac{\cos 3\theta}{3^2} - \frac{\cos 4\theta}{4^2} + \cdots\right).$$

Fourier Series – Spectral Version

The development of a periodic function into a Fourier series is a standard modelling method in engineering (of acoustic, electroacoustic and other wave-dominated phenomena).

On the other hand, mathematical signal theory uses Fourier series almost exclusively as support for formal information, i.e. in the frequency domain: they appear as the Fourier transforms of certain time series (i.e. of discrete-time waveforms). We shall see this when following the arguments of the proof of the Nyquist–Shannon theorem.

From a formal viewpoint, the most natural presentation of the subject emerges (as often in Mathematics) from a **geometrization** of the situation. This makes the Hilbert spaces of square summable sequences and of (classes of) square integrable functions appear.

Recall

(1) $l^2(\mathbb{Z})$ is the space of (complex-valued) sequences $\mathbf{c} = (c_k)_{k \in \mathbb{Z}}$ such that

$$\sum_{k \in \mathbb{Z}} |c_k|^2 < \infty.$$

The inner product. $\langle \mathbf{c}, \mathbf{c}' \rangle = \sum_{k \in \mathbb{Z}} c_k \overline{c'_k}$ (conjugation of the second factor)
The norm. $\|\mathbf{c}\|^2 = \langle \mathbf{c}, \mathbf{c} \rangle = \sum_{k \in \mathbb{Z}} |c_k|^2$
The Cauchy–Schwarz inequality. $\langle \mathbf{c}, \mathbf{c}' \rangle \leq \|\mathbf{c}\| \cdot \|\mathbf{c}'\|$
$l^2(\mathbb{Z})$ is complete for the associated uniform structure: every Cauchy sequence in $l^2(\mathbb{Z})$ converges (to an element of $l^2(\mathbb{Z})$).

(2) $\mathbf{L}^2[-\pi, \pi]$ is the space of (classes of) functions which are square integrable over the interval $] -\pi, \pi]$.

More precisely, a (complex-valued) function $\hat{f}(\omega)$ of the real variable ω, $-\pi < \omega \leq \pi$, is square integrable over $] -\pi, \pi]$ if

$$\int_{-\pi}^{\pi} |\hat{f}(\omega)|^2 \mathrm{d}\omega < \infty$$

(i.e. the relevant Lebesgue integral is defined and takes a finite value).
Consider the equivalence relation.
$\hat{f}(\omega) \sim \hat{g}(\omega)$:

$\Longleftrightarrow \hat{f}$ and $\hat{g}$ are equal *almost everywhere*
$\Longleftrightarrow \{\omega : \hat{f}(\omega) \neq \hat{g}(\omega)\}$ is a set of zero measure.

$\mathbf{L}^2[-\pi, \pi]$ will then be the space of square integrable functions on $] -\pi, \pi]$, identified according to the relation of almost everywhere (a.e.) equality.

In the sequel, we shall always write *functions* – but think in terms of *classes of functions*.
Hence:

The inner product. $\langle \hat{f}, \hat{g} \rangle = \frac{1}{2\pi} \int_{-\pi}^{\pi} \hat{f}(\omega) \overline{\hat{g}(\omega)} \mathrm{d}\omega$
The associated norm. $\|\hat{f}\|^2 = \langle \hat{f}, \hat{f} \rangle = \frac{1}{2\pi} \int_{-\pi}^{\pi} |\hat{f}(\omega)|^2 \mathrm{d}\omega$
The Cauchy–Schwarz inequality. $\langle \hat{f}, \hat{g} \rangle \leq \|\hat{f}\| \cdot \|\hat{g}\|$

$\mathbf{L}^2[-\pi, \pi]$ is complete, exactly like $l^2(\mathbb{Z})$.

But an *indispensable remark* concerning the convergence in $\mathbf{L}^2[-\pi, \pi]$:

$$\hat{f} = \lim_{n \to \infty} \hat{f}_n \Longleftrightarrow \lim_{n \to \infty} \sqrt{\frac{1}{2\pi} \int_{-\pi}^{\pi} |\hat{f}(\omega) - \hat{f}_n(\omega)|^2 \mathrm{d}\omega} = 0$$

(convergence in quadratic mean).
Conceptually, we deal more or less with a *convergence in shape*: a separating area becomes smaller and smaller, hence the line of $\hat{f}(\omega)$ is "generically" (almost everywhere) approached. But attention: a priori, convergence

in quadratic mean does *not* involve (almost everywhere) simple convergence (an element of $\mathbf{L}^2[-\pi, \pi]$ has no values – it has a shape).

Note, moreover, that $\mathbf{L}^2[-\pi, \pi]$ is, from another viewpoint, the space of (classes of) 2π-periodic, locally square integrable functions: $]-\pi, \pi]$ will figure as the standard period interval of length 2π.

Let us now pass to the formalism which identifies $\mathrm{l}^2(\mathbb{Z})$ as *coordinate space* with respect to the classical orthonormal basis of $\mathbf{L}^2[-\pi, \pi]$.

For $\hat{f}(\omega)$, square integrable on $]-\pi, \pi]$, we define the sequence of its Fourier coefficients as follows:

$$f[k] = \langle \hat{f}(\omega), \mathrm{e}^{-\mathrm{i}\omega k} \rangle = \frac{1}{2\pi} \int_{-\pi}^{\pi} \hat{f}(\omega) \mathrm{e}^{\mathrm{i}\omega k} \mathrm{d}\omega, \quad k \in \mathbb{Z}.$$

[We note a friction with the time domain formalism: a natural change of variable $\omega = 2\pi T\theta$ yields a sequence of Fourier coefficients enumerated in the opposite sense – our frequency domain definition will finally be justified by its harmonizing with the formulary of the *inverse* Fourier transform.]

Then we get the following **fundamental result**:

The correspondence $\hat{f}(\omega) \longmapsto (f[k])_{k \in \mathbb{Z}}$
induces an *isomorphism* of Hilbert spaces between $\mathbf{L}^2[-\pi, \pi]$ and $\mathrm{l}^2(\mathbb{Z})$.

More precisely:

(1) If $\hat{f} \in \mathbf{L}^2[-\pi, \pi]$, then the series of its Fourier coefficients is square summable, and

$$\sum_{k \in \mathbb{Z}} |f[k]|^2 = \frac{1}{2\pi} \int_{-\pi}^{\pi} |\hat{f}(\omega)|^2 \mathrm{d}\omega$$

(Plancherel Formula). More generally, if $\hat{f}, \hat{g} \in \mathbf{L}^2[-\pi, \pi]$, then

$$\sum_{k \in \mathbb{Z}} f[k]\overline{g[k]} = \frac{1}{2\pi} \int_{-\pi}^{\pi} \hat{f}(\omega)\overline{\hat{g}(\omega)} \mathrm{d}\omega$$

(Parseval Identity). The Fourier series of $\hat{f}$ converges in quadratic mean to $\hat{f}$:

$$\lim_{N \to \infty} \frac{1}{2\pi} \int_{-\pi}^{\pi} \left| \sum_{k=-N}^{N} f[k]\mathrm{e}^{-\mathrm{i}\omega k} - \hat{f}(\omega) \right|^2 \mathrm{d}\omega = 0.$$

In other words, $\{\mathrm{e}^{-\mathrm{i}\omega k}, k \in \mathbb{Z}\}$ is an *orthonormal basis* of $\mathbf{L}^2[-\pi, \pi]$.[4]

(2) Conversely, if $(f[k])_{k \in \mathbb{Z}}$ is such that $\sum_{k \in \mathbb{Z}} |f[k]|^2 < \infty$, then the series $\sum_{k \in \mathbb{Z}} f[k]\mathrm{e}^{-\mathrm{i}\omega k}$ converges in quadratic mean to an element $\hat{f} \in \mathbf{L}^2[-\pi, \pi]$, and $\hat{f}$ is the only element of $\mathbf{L}^2[-\pi, \pi]$ which admits the sequence $f[k]$, $k \in \mathbb{Z}$, as the sequence of its Fourier coefficients.

---

[4] I.e. the linear span of the given orthonormal system is dense in $\mathbf{L}^2[-\pi, \pi]$.

There remains nevertheless the question: which is the result that replaces the Dirichlet theorem in this context?

The best imaginable statement should be the following:

If $\hat{f} \in \mathbf{L}^2[-\pi, \pi]$, then the Fourier series of $\hat{f}$ converges almost everywhere.

This is in fact a theorem which has been proved by *Carleson* in 1966 (!: the memoir of Dirichlet on the convergence of Fourier series dates from 1829).

Remark Do not forget that non-constructive results which are true "almost everywhere" may be *empty* from the algorithmic viewpoint: the field of the real numbers which are effectively computable is of measure zero!

3.3.2 The Whittaker–Shannon Theorem for Elementary Periodic Functions

We shall try to reconstruct a simple periodic function
$$x(t) = \frac{a_0}{2} + a_1 \cdot cos2\pi t + b_1 \cdot sin2\pi t + a_2 \cdot cos4\pi t + b_2 \cdot sin4\pi t + \ldots a_m \cdot cos2\pi mt + b_m \cdot sin2\pi mt$$
(i.e. a trigonometric polynomial of degree m) with the help of an equidistant sampling to the order $n > 2m$ ($n = 2m$ in the balanced case), *without* passing by the frequency domain. The solution will be a little mathematical perversity – we shall have approached something simple by something complicated – but at the same time we will have constructed a bridge between the Discrete Fourier Transform landscape and the highlands of Whittaker–Shannon.

Let us begin with an (elementary, but non-trivial) result of Calculus:

Recall Development of certain trigonometric functions into series of simple fractions.

(1) $\frac{\pi}{\sin(\pi\theta)} = \frac{1}{\theta} + \sum_{k \geq 1}(-1)^k \left(\frac{1}{\theta+k} + \frac{1}{\theta-k} \right),$

(2) $\pi \cdot \cot(\pi\theta) = \frac{1}{\theta} + \sum_{k \geq 1} \left(\frac{1}{\theta+k} + \frac{1}{\theta-k} \right)$ for every $\theta \notin \mathbb{Z}$.

$\Big[$Note that both series are absolutely convergent: for example, the second satisfies, for $k \geq 2|\theta|$:

$$\left| \frac{1}{\theta+k} + \frac{1}{\theta-k} \right| = \frac{2|\theta|}{k^2 - \theta^2} \leq \frac{8|\theta|}{3k^2},$$

which permits us to obtain an estimation of the significant remainder by $\frac{8|\theta|}{3} \sum \frac{1}{k^2}.\Big]$

This yields the following identities, valid for every $\theta \in \mathbb{R}$:

(1) $1 = \sum_{k \in \mathbb{Z}}(-1)^k \frac{\sin(\pi\theta)}{\pi(\theta-k)},$

(2) $\cos(\pi\theta) = \sum_{k \in \mathbb{Z}} \frac{\sin(\pi\theta)}{\pi(\theta-k)}.$

But, $\cos(k\pi) = (-1)^k$, which gives finally:

(1) $1 = \sum_{k\in\mathbb{Z}} \frac{\sin \pi(\theta-k)}{\pi(\theta-k)}$,

(2) $\cos(\pi\theta) = \sum_{k\in\mathbb{Z}} \cos(k\pi) \cdot \frac{\sin \pi(\theta-k)}{\pi(\theta-k)}$.

In order to obtain a formally well-established expression, let us replace θ by $2Wt$, and put $T = \frac{1}{2W}$.

This yields:

(1) $1 = \sum_{k\in\mathbb{Z}} \frac{\sin 2\pi W(t-kT)}{2\pi W(t-kT)}$,

(2) $\cos 2\pi Wt = \sum_{k\in\mathbb{Z}} \cos(2\pi W(kT)) \cdot \frac{\sin 2\pi W(t-kT)}{2\pi W(t-kT)}$.

We insist:

Let $x(t) = 1$ or $x(t) = \cos 2\pi Wt$, then

$$x(t) = \sum_{k\in\mathbb{Z}} x_k \cdot \frac{\sin 2\pi W(t-kT)}{2\pi W(t-kT)},$$

where $x_k = x(kT)$, $k \in \mathbb{Z}$ with $T = \frac{1}{2W}$.

An elementary signal of the form $x(t) = \frac{a_0}{2} + a_1 \cdot \cos 2\pi Wt$ can thus be reconstructed, after equidistant (infinite) sampling, by means of the sequence $(x_k = x(kT))_{k\in\mathbb{Z}}$ of its samples, in form of the series

$$x(t) = \sum_{k\in\mathbb{Z}} x_k \cdot \frac{\sin 2\pi W(t-kT)}{2\pi W(t-kT)},$$

where $T = \frac{1}{2W}$ (W is the frequency of $x(t)$).

This is precisely the theorem of Whittaker–Shannon for a signal of the form

$$x(t) = \frac{a_0}{2} + a_1 \cdot \cos 2\pi Wt.$$

Note that at this rather primitive stage, the sequence of the samples is 2-periodic.

Clearly, this innocent result can be generalized. We will show the

Nyquist–Shannon theorem – the elementary periodic case.[5]

Let

$$x(t) = \frac{a_0}{2} + a_1 \cdot \cos 2\pi t + b_1 \cdot \sin 2\pi t + a_2 \cdot \cos 4\pi t + b_2 \cdot \sin 4\pi t + \dots \frac{a_m}{2} \cdot \cos 2\pi mt$$

be a balanced trigonometric polynomial of degree m.

We sample to the order $n = 2m$: $x_k = x(kT)$ with $T = \frac{1}{n} = \frac{1}{2m}$ (the sequence $(x_k)_{k\in\mathbb{Z}}$ is n-periodic). Then

$$x(t) = \sum_{k\in\mathbb{Z}} x_k \cdot \frac{\sin 2\pi m(t-kT)}{2\pi m(t-kT)}.$$

---

[5] Despite Whittaker's priority, let us occasionally respect the traditional labelling.

Proof First, note that we have normalized the time, as we did often in the preceding sections.

We shall begin with the case of a trigonometric polynomial which is an even function (i.e. $b_1 = b_2 = \cdots = b_m = 0$).

Auxiliary Result

The function $f(\theta) = \cos(\alpha\theta)$, $\alpha \notin \mathbb{Z}$, admits for $-\pi \leq \theta \leq \pi$ the following development in a Fourier series:

$$\cos(\alpha\theta) = \frac{\sin(\alpha\pi)}{\pi} \left[\frac{1}{\alpha} - \frac{2\alpha}{\alpha^2 - 1^2} \cdot \cos\theta + \frac{2\alpha}{\alpha^2 - 2^2} \cdot \cos 2\theta - \cdots \right].$$

Exchange now α and θ:

$$\cos(\alpha\theta) = \frac{\sin(\pi\theta)}{\pi} \left[\frac{1}{\theta} - \left(\frac{1}{\theta+1} + \frac{1}{\theta-1} \right) \cos\alpha \right.$$
$$\left. + \left(\frac{1}{\theta+2} + \frac{1}{\theta-2} \right) \cos 2\alpha - \cdots \right]$$
$$= \sum_{k\in\mathbb{Z}} (-1)^k \cos(k\alpha) \cdot \frac{\sin(\pi\theta)}{\pi(\theta-k)} = \sum_{k\in\mathbb{Z}} \cos(k\alpha) \cdot \frac{\sin\pi(\theta-k)}{\pi(\theta-k)}.$$

Replace θ by $2Wt$ and α by $\frac{p}{q}\pi$, $0 < \frac{p}{q} \leq 1$; we get, with $T = \frac{1}{2W}$:

$$(\sharp) \quad \cos 2\pi \left(\frac{p}{q}W \right) t = \sum_{k\in\mathbb{Z}} \cos \left(2\pi \left(\frac{p}{q}W \right)(kT) \right) \cdot \frac{\sin 2\pi W(t-kT)}{2\pi W(t-kT)}.$$

Let now
$x(t) = \frac{a_0}{2} + a_1 \cos 2\pi t + a_2 \cos 4\pi t + \cdots + a_m \cos 2\pi mt$
be an even trigonometric polynomial of degree m. Put $W = m$, $n = 2m$, i.e. $T = \frac{1}{n}$. Then we immediately obtain from $(\sharp)$ and by linearity of the sampling operation:

$$x(t) = \sum_{k\in\mathbb{Z}} x_k \cdot \frac{\sin 2\pi m \left(t - \frac{k}{n} \right)}{2\pi m \left(t - \frac{k}{n} \right)},$$

where $x_k = x(\frac{k}{n})$, $k \in \mathbb{Z}$. This is clearly the Nyquist–Shannon theorem for an even trigonometric polynomial.

In order to get our theorem in full generality, we have still to show the odd case.

Exercise

(a) Let $g(\theta) = \sin(\alpha\theta)$, $\alpha \notin \mathbb{Z}$, and let $\sigma g(\theta) = \lim_{N\to\infty} \sigma_N g(\theta)$ be its Fourier series. Show that

$$\sigma g(\theta) = \frac{\sin(\alpha\pi)}{\pi} \cdot \sum_{k \geq 1} (-1)^k \left(\frac{1}{\alpha - k} - \frac{1}{\alpha + k} \right) \cdot \sin(k\theta).$$

We shall admit that $g(\theta) = \sigma g(\theta)$ for $-\pi < \theta < \pi$.

(b) Deduce that we get, for $x(t) = \sin 2\pi \left(\frac{p}{q} W \right) t$ with $0 < \frac{p}{q} < 1$ and $T = \frac{1}{2W}$

$$x(t) = \sum_{k \in \mathbb{Z}} x_k \cdot \frac{\sin 2\pi W(t - kT)}{2\pi W(t - kT)},$$

where $x_k = x(kT)$, $k \in \mathbb{Z}$.

Combining the result of the exercise with our previous result in the even case, we finally obtain the theorem as stated. $\square$

Note that the reconstruction of the function $\sin 2\pi W t$ by a sampling with step size $T = \frac{1}{2W}$ is not possible: $\sin(k\pi) = 0$ for $k \in \mathbb{Z}$. This explains the appearance of the balanced form in the statement of the theorem.

Important Remark

The theorem of Nyquist–Shannon – elementary periodic case – as stated above, is an *equivalent* version of our Sampling Theorem of the preceding section.

More precisely: let $n = 2m$. Then, for every n-periodic sequence $(x_k)_{k \in \mathbb{Z}}$ of (real or) complex numbers, the series of functions

$$x(t) = \sum_{k \in \mathbb{Z}} x_k \cdot \frac{\sin 2\pi m(t - \frac{k}{n})}{2\pi m(t - \frac{k}{n})}$$

is *convergent* for every $t \in \mathbb{R}$, and it *equals* the balanced trigonometric polynomial

$$x(t) = \frac{c_{-m}}{2} e^{-2\pi i m t} + \sum_{-m+1 \leq \nu \leq m-1} c_\nu e^{2\pi i \nu t} + \frac{c_m}{2} e^{2\pi i m t},$$

where

$$\begin{pmatrix} c_0 \\ c_1 \\ \cdot \\ c_{m-1} \\ c_m = c_{-m} \\ c_{-m+1} \\ \cdot \\ c_{-1} \end{pmatrix} = \frac{1}{n} F_n \begin{pmatrix} x_0 \\ x_1 \\ \cdot \\ \cdot \\ \cdot \\ \cdot \\ x_{n-1} \end{pmatrix}$$

Question Is it reasonable to compute the values of a trigonometric polynomial with the aid of the Nyquist–Shannon interpolator?

Answer   Consult the following exercises.

Exercises

(1) Let $x(t) = \cos 2\pi t$. One knows: $x\left(\frac{1}{6}\right) = \cos\frac{\pi}{3} = \frac{1}{2}$.

(a) $x(t) = \sum_{k\in\mathbb{Z}}(-1)^k \cdot \frac{\sin 2\pi\left(t-\frac{k}{2}\right)}{2\pi\left(t-\frac{k}{2}\right)}$ according to our foregoing results.

Put

$$x_N(t) = \sum_{-N\leq k\leq N}(-1)^k \cdot \frac{\sin 2\pi\left(t-\frac{k}{2}\right)}{2\pi\left(t-\frac{k}{2}\right)}.$$

Show that

$$x_N\left(\frac{1}{6}\right) = \frac{3\sqrt{3}}{\pi} \cdot \left(\frac{1}{2} - \sum_{k=1}^{N}\frac{1}{9k^2-1}\right)$$

and compute $x_4\left(\frac{1}{6}\right)$, then $x_{10}(\frac{1}{6})$.

(b) How many terms of the series above must we compute in order to obtain $x\left(\frac{1}{6}\right) = \frac{1}{2}$ with a two digit decimal precision? In other words, find $N \geq 1$ such that $\left|\frac{1}{2} - x_N(\frac{1}{6})\right| < \frac{1}{100}$.

(2) Let still $x(t) = \cos 2\pi t$, but now sample to the order $n = 8$. We will have:

$$x(t) = \sum_{k\in\mathbb{Z}}\cos 2\pi\left(\frac{k}{8}\right) \cdot \frac{\sin 8\pi\left(t-\frac{k}{8}\right)}{8\pi\left(t-\frac{k}{8}\right)}.$$

Let $x_N(t)$ be the "truncated development of order N" of $x(t)$, as in the preceding exercise. Compute $x_4(\frac{1}{6})$, and compare with the result above.

Summary   Restricting our attention to elementary periodic signals, we observe that the problems of sampling and of reconstruction (i.e. of interpolation) are sensibly better resolved by the methods of Linear Algebra – via the Discrete Fourier Transform (and the associated algorithm FFT) – than by Analysis: the Nyquist–Shannon interpolators converge very slowly, and are thus rather imprecise (clearly, you can always improve by oversampling – as in exercise (2) above).

At this point, we have to face two natural objections:

(1) The minor objection: our treatment of periodic signals is too simplistic. The restriction to trigonometric polynomials simulates idyllic situations. What about *arbitrary* periodic signals?
(2) The major objection: the everyday signal is *not periodic* (although often *locally* periodic, particularly in acoustics, for example). How can we sample and reconstruct in these cases?

The answer to (1) is intimately linked to the following observation.

Remark (Uniform trigonometrical approximation – the theorem of Fejér):
Let $f(\theta)$ be a **continuous** and P-periodic function (we shall put, as always, $P = 2W$ and $T = \frac{1}{2W}$).

Consider the sequence of the *arithmetical means* of the Fourier polynomials $\sigma_n f(\theta)$ associated with $f(\theta)$

$$\mu_N f(\theta) = \frac{1}{N} \sum_{n=0}^{N-1} \sigma_n f(\theta), \quad N \geq 1,$$

i.e.

$$\mu_1 f(\theta) = \sigma_0 f(\theta) = c_0,$$
$$\mu_2 f(\theta) = \tfrac{1}{2}(\sigma_0 f(\theta) + \sigma_1 f(\theta)) = \tfrac{1}{2}c_{-1}e^{-2\pi i T\theta} + c_0 + \tfrac{1}{2}c_1 e^{2\pi i T\theta},$$
$$\mu_3 f(\theta) = \tfrac{1}{3}(\sigma_0 f(\theta) + \sigma_1 f(\theta) + \sigma_2 f(\theta)) = \tfrac{1}{3}c_{-2}e^{-4\pi i T\theta} + \tfrac{2}{3}c_{-1}e^{-2\pi i T\theta}$$
$$+ c_0 + \tfrac{2}{3}c_1 e^{2\pi i T\theta} + \tfrac{1}{3}c_2 e^{4\pi i T\theta},$$
$$\vdots$$
$$\mu_N f(\theta) = \sum_{-N+1 \leq k \leq N-1} \left(1 - \frac{|k|}{N}\right) c_k e^{2\pi i k T\theta}.$$

(One weakens the influence of the high frequencies.)

Then the sequence $(\mu_N f(\theta))_{N \geq 1}$ converges *uniformly* to $f(\theta)$:
For every $\varepsilon > 0$ there exists $N \geq 1$ such that $|f(\theta) - \mu_N f(\theta)| < \varepsilon$ for *every* $\theta \in \mathbb{R}$. ($f(\theta)$ and $\mu_N f(\theta)$ take the *same values*, up to an ε-deviation).

In this sense, every continuous periodic function "is" a trigonometric polynomial.

Exercise

Let $f(\theta)$ be the continuous 2π-periodic function

$$f(\theta) = \begin{cases} \theta & \text{for } 0 \leq \theta \leq \pi, \\ 2\pi - \theta & \text{for } \pi \leq \theta \leq 2\pi \end{cases}$$

(a sawtooth curve).

Recall $\sigma f(\theta) = \dfrac{\pi}{2} - \dfrac{4}{\pi}\left[\cos\theta + \dfrac{\cos 3\theta}{3^2} + \dfrac{\cos 5\theta}{5^2} + \dfrac{\cos 7\theta}{7^2} + \cdots\right].$

Find $N \geq 1$ such that $|f(\theta) - \mu_N f(\theta)| < \frac{1}{10}$ for every $\theta \in \mathbb{R}$.

The answer to the objection (2) is given by the usual Whittaker–Shannon theorem (in an essentially *non-periodic* spirit).

Let us begin with the mathematical basement.

3.3.3 The (Continuous) Fourier Transform: A Sketch

Let $f(t) = x(t) + iy(t)$ be a complex-valued function of the real variable t ("the time").

Suppose first that $f = f(t) \in L^1(\mathbb{R})$ ("$f = f(t)$ is summable over $\mathbb{R}$"), i.e. that the (Lebesgue) integral $\int_{-\infty}^{+\infty} |f(t)|dt$ (is defined and) converges.

Associate with $f(t)$ the function of the variable ω ("the normalized frequency")

$$\hat{f}(\omega) = \int_{-\infty}^{+\infty} f(t)e^{-i\omega t}dt.$$

More explicitly:

$$\hat{f}(\omega) = \int_{-\infty}^{+\infty} (x(t)\cos\omega t + y(t)\sin\omega t)dt + i\int_{-\infty}^{+\infty} (y(t)\cos\omega t - x(t)\sin\omega t)dt.$$

$\boxed{\hat{f}(\omega) = \mathbb{F}[f(t)] \text{ is called } \textit{the Fourier transform} \text{ of } f(t).}$

First Observation

$\hat{f}(\omega)$ is a *continuous* function of the variable ω (provided $f(t)$ is a summable function).

Basic example

$$e^{[0]}(t) = \begin{cases} 1 \text{ for } |t| \le \frac{1}{2}, \\ 0 \text{ for } |t| > \frac{1}{2} \end{cases}$$

(this is a function which defines a "window of length 1", according to common language in signal theory).

Its Fourier transform: $\hat{e}^{[0]}(\omega) = 2\frac{\sin\frac{1}{2}\omega}{\omega} = \sin c\left(\frac{\omega}{2\pi}\right)$. This introduces definitely the **convolutional sine**[6]

$$\sin c(x) = \begin{cases} 1 & \text{for } x = 0, \\ \frac{\sin \pi x}{\pi x} & \text{for } x \ne 0 \end{cases}$$

– we already became (implicitly) accustomed to it.

We observe: $\hat{e}^{[0]}(\omega)$ is infinitely differentiable, and each of its derivatives is bounded. On the other hand, $\hat{e}^{[0]}(\omega)$ is **not** summable: $\int_{-\infty}^{+\infty} |\hat{e}^{[0]}(\omega)|d\omega$ is **divergent** (although $\int_{-\infty}^{+\infty} \hat{e}^{[0]}(\omega)d\omega = 2\pi$).

Before going further, we should briefly speak about the frequency variable.

Remark A priori, the variable ω describes a **formal change of viewpoint** (which proves to be extremely fruitful), but which should not be overloaded by obstinate conceptual interpretations.[7] Down-to-earth frequencies – in the

---

[6] We try to give a meaning to the fourth letter "c" – referring to the fundamental impulse response property of the function.

[7] Do not forget that frequency without repetition is non-sense.

physical sense of the term – appear abundantly in the *time domain description* of periodic phenomena, and are traditionally designated by the variable ν. These (physical) frequencies will propagate into the (formal) frequency domain according to the identity:

$$\boxed{\omega = 2\pi\nu.}$$

But there is an important nuance in this seemingly flat change of variable: the variable ν "hashes" the time, the variable ω "forgets about" the time.

Let us go back to our time–window example, which can be easily generalized. Consider

$$e^{[\nu_0]}(t) = \begin{cases} e^{2\pi i \nu_0 t} & \text{for } |t| \leq \frac{1}{2}, \\ 0 & \text{for } |t| > \frac{1}{2}. \end{cases}$$

Then $\hat{e}^{[\nu_0]}(\omega) = \operatorname{sin} c\left(\frac{\omega-\omega_0}{2\pi}\right)$, $\omega_0 = 2\pi\nu_0$.

$\left[\text{Observe that we have, in general: if } \hat{f}(\omega) \text{ is the Fourier transform of } f(t), \right.$
$\left. \text{then } \hat{f}(\omega - \omega_0) \text{ is the Fourier transform of } e^{i\omega_0 t} f(t).\right]$

We note: the function $\hat{e}^{[\nu_0]}(\omega) = \operatorname{sin} c\left(\frac{\omega-\omega_0}{2\pi}\right)$ is *the characteristic function* of the elementary periodic function $e^{[\nu_0]}(t) = e^{2\pi i \nu_0 t}$, "seen through the window of length 1, symmetric to the origin".

It attains its maximum for $\omega = \omega_0$, corresponding to the fixed characteristic frequency, and tends symmetrically (with respect to the vertical line $\omega = \omega_0$) to 0, for $\omega \to \pm\infty$, in smaller and smaller becoming undulations.[8]

Exercise

Show the following elementary result.
    Let $\hat{f}(\omega) = \mathbb{F}[f(t)]$ and let $a \in \mathbb{R}^*$.
    Then $\mathbb{F}[f(at)] = \frac{1}{|a|}\hat{f}(\frac{\omega}{a})$.

Consequence $\hat{e}_n^{[\nu_0]}(\omega) = n \cdot \operatorname{sin} c(n \cdot \frac{\omega-\omega_0}{2\pi})$ (the index n indicates the length of the time domain window), i.e. if we "open" the truncation window for $f(t) = e^{2\pi i \nu_0 t}$ indefinitely, the corresponding characteristic functions will tend to the "function"

$$\hat{e}_\infty^{[\nu_0]}(\omega) = \begin{cases} \infty & \text{for } \omega = \omega_0, \\ 0 & \text{otherwise} \end{cases}$$

(in a more advanced formalism, $\hat{e}_\infty^{[\nu_0]}(\omega)$ is simply the Dirac $\delta_{(\omega_0)}$).

---

[8] A wavelet – we shall see more of this specimen in the last chapter of this book.

In a first approach, this example will justify the denomination of ω as *the frequency variable*. (We note, in the foregoing example, the correct appearance of the (time domain) frequency value ν_0 in our ω – formalism: passing to the limit creates a "point density" which indicates precisely this frequency – in normalized notation.)

Remark Although the (continuous) Fourier transform makes the transit to the "frequency domain", it is structurally hostile to periodicity: a non-zero periodic and continuous function $f(t)$ is *not* summable over all of $\mathbb{R}$, hence its (continuous) Fourier transform *does not exist* – at least as a **function** of the variable ω.

But if we extend the theory of the (continuous) Fourier transform to the *tempered distributions*, then we are in the correct situation:

$$\mathbb{F}\left[\sum_{-m \leq \nu_0 \leq m} c_{\nu_0} e^{2\pi i \nu_0 t}\right] = \sum_{-m \leq \nu_0 \leq m} c_{\nu_0} \delta_{(2\pi\nu_0)}.$$

The density defined by a trigonometric polynomial $f(t)$ will be transformed into the characteristic distribution of the frequencies "occurring in $f(t)$" – every $\delta_{(2\pi\nu_0)}$ is a Dirac distribution (i.e. a point density "centred in $\omega_0 = 2\pi\nu_0$").

Let us now move on to the **inverse** Fourier transform. We define, for a function $G(\omega) \in L^1(\mathbb{R})$, i.e. a summable function of the variable ω,

$$g(t) = \overline{\mathbb{F}}[G(\omega)] = \frac{1}{2\pi} \int_{-\infty}^{+\infty} G(\omega) e^{i\omega t} d\omega.$$

We obtain thus a continuous function of the variable t.

For a **continuous** and summable function $f(t)$, such that $\hat{f}(\omega) = \mathbb{F}[f(t)]$ is (continuous and) **summable**, we have

$$f(t) = \overline{\mathbb{F}}\mathbb{F}[f(t)], \quad \hat{f}(\omega) = \mathbb{F}\overline{\mathbb{F}}[\hat{f}(\omega)],$$

i.e. $\mathbb{F}$ and $\overline{\mathbb{F}}$ are inverse transformations, respectively.

Attention If $f_1(t) = f(t)$ almost everywhere (the two functions are distinct only on a set of measure zero – for example on a countable set), then $\hat{f}_1(\omega) = \hat{f}(\omega)$ for *every* $\omega \in \mathbb{R}$ (the integration "does not remark" the change of values of a summable function on a set of measure zero). It is only possible to recover the initial function, by inverse Fourier transform, up to a (a priori unknown) set of measure zero. But every class of summable functions (equivalence = equality almost everywhere) admits at most one single *continuous* representative. In this sense, our reciprocity statement needs the coupling of summabilty and of continuity.

Finally, let us state the *Fundamental property* of the Fourier transform.

The Convolution Theorem

Define, for $f = f(t)$, $h = h(t) \in L^1(\mathbb{R})$, their convolution product $g = f * h \in L^1(\mathbb{R})$ by

$$g(t) = \int_{-\infty}^{+\infty} f(t - \theta) h(\theta) d\theta = \int_{-\infty}^{+\infty} f(\theta) h(t - \theta) d\theta.$$

Then, $\mathbb{F}[g(t)] = \mathbb{F}[f(t)] \cdot \mathbb{F}[h(t)]$, i.e. $\hat{g}(\omega) = \hat{f}(\omega) \cdot \hat{h}(\omega)$.

This is the **characteristic** and **essential** property of the Fourier transform.

Towards a Hilbert Space Formalism

Notation $\mathbb{I}_E$ is the characteristic function of the subset $E \in \mathbb{R}$.

Example $\mathbb{I}_{[-1/2,1/2]}(t) = e^{[0]}(t)$.

Recall   The Fourier transform of the function $e^{[\nu_0]}(t) = e^{2\pi i \nu_0 t} \cdot \mathbb{I}_{[-1/2,1/2]}$ is the convolutional sine $\hat{e}^{[\nu_0]}(\omega) = \sin c \left(\frac{\omega - \omega_0}{2\pi} \right)$, $\omega_0 = 2\pi\nu_0$.

Although the $\sin c$ function is not summable (over $\mathbb{R}$), let us show *altogether* that

$$e^{[\nu_0]}(t) = \frac{1}{2\pi} \int_{-\infty}^{+\infty} \hat{e}^{[\nu_0]}(\omega) e^{i\omega t} d\omega.$$

This will be an immediate consequence of the following exercise.

Exercise   (Fourier transform of the convolutional sine)
Let $h_0(t) = \sin c(t) = \frac{\sin \pi t}{\pi t}$ (with continuous prolongation for $t = 0$).
Show that $\hat{h}_0(\omega) = \mathbb{I}_{[-\pi,\pi]} = \begin{cases} 1 \text{ for } |\omega| \leq \pi, \\ 0 \text{ else.} \end{cases}$
(Help: use $\int_{-\infty}^{+\infty} \sin c(at) dt = \frac{1}{|a|}$ for $a \in \mathbb{R}$.)

Solution

$$\hat{h}_0(\omega) = \int_{-\infty}^{+\infty} \sin c(t) \cdot e^{-i\omega t} dt = \int_{-\infty}^{+\infty} \sin c(t) \cdot \cos(\omega t) dt$$

$$= \frac{1}{2} \int_{-\infty}^{+\infty} \left[\frac{\sin(\pi + \omega)t}{\pi t} + \frac{\sin(\pi - \omega)t}{\pi t} \right] dt.$$

Put $\nu = \frac{\omega}{2\pi}$. Then:

$$\hat{h}_0(\omega) = \frac{1}{2} \int_{-\infty}^{+\infty} \left[\frac{\sin \pi(1+2\nu)t}{\pi t} + \frac{\sin \pi(1-2\nu)t}{\pi t} \right] dt$$

$$= \frac{1}{2} \left[\frac{1+2\nu}{|1+2\nu|} + \frac{1-2\nu}{|1-2\nu|} \right].$$

Whence: $\hat{h}_0(\omega) = \begin{cases} 1 \text{ for } |\nu| < \frac{1}{2}, \\ 0 \text{ for } |\nu| > \frac{1}{2}, \end{cases} = \begin{cases} 1 \text{ for } |\omega| < \pi, \\ 0 \text{ for } |\omega| > \pi. \end{cases}$

Consequence[9]

Put $h_k(t) = h_0(t-k) = \sin c(t-k)$, $k \in \mathbb{Z}$. This is a family of convolutional sines, centred at the particular $k \in \mathbb{Z}$. The family of their Fourier transforms $(\hat{h}_k(\omega))_{k\in\mathbb{Z}}$ are the sinusoids (in complex notation)

$$\hat{h}_k(\omega) = e^{-i\omega k} \cdot \mathbb{I}_{[-\pi,\pi]}, \quad k \in \mathbb{Z}.$$

We have already seen (frequency domain formalism of Fourier series) that this family is an orthonormal basis of the Hilbert space $\mathbf{L}^2[-\pi, \pi]$.

Question What is the *characteristic* Hilbert space property of the family $(h_k(t))_{k\in\mathbb{Z}}$?

The answer will be given by the theorem of Whittaker–Shannon.

Consider the (Hilbert) space $\mathbf{L}^2(\mathbb{R})$ of (classes of) square integrable functions on $\mathbb{R}$. Our family of convolutional sines $(h_k)_{k\in\mathbb{Z}}$ yields an example of typical elements of this space.

Our first goal: to define an **extension of the Fourier transform**

$$\mathbb{F} : \mathbf{L}^2(\mathbb{R}) \longrightarrow \mathbf{L}^2(\mathbb{R}),$$

which will be an *isomorphism* of Hilbert spaces.

Principal lemma. Let $f, h \in \mathbf{L}^1(\mathbb{R}) \cap \mathbf{L}^2(\mathbb{R})$. Then

$$\int_{-\infty}^{+\infty} f(t)\overline{h(t)}dt = \frac{1}{2\pi} \int_{-\infty}^{+\infty} \hat{f}(\omega)\overline{\hat{h}(\omega)}d\omega \quad \text{(Parseval)}.$$

In particular:

$$\int_{-\infty}^{+\infty} |f(t)|^2 dt = \frac{1}{2\pi} \int_{-\infty}^{+\infty} |\hat{f}(\omega)|^2 d\omega \quad \text{(Plancherel)}.$$

Proof Exercise (you have to use the Convolution Theorem relative to $g = f * \tilde{h}$, with $\tilde{h}(t) = \overline{h(-t)}$). □

---

[9] We shall become convinced that the coupling: frequency domain window ↔ time domain sin c function offers a robust skeleton to Fourier analysis.

Extension of $\mathbb{F}$ to the $\mathbf{L}^2(\mathbb{R})$ by density

How to define $\hat{f}(\omega) = \mathbb{F}[f(t)]$ in case $f = f(t)$ is square integrable, but not necessarily summable?

$\mathbf{L}^1(\mathbb{R}) \cap \mathbf{L}^2(\mathbb{R})$ is *dense* in $\mathbf{L}^2(\mathbb{R})$ (for the Hilbert space norm). Consequently, there exists $(f_n)_{n \in \mathbb{N}}$ in $\mathbf{L}^1(\mathbb{R}) \cap \mathbf{L}^2(\mathbb{R})$ such that $\| f - f_n \| \longrightarrow 0$ in $\mathbf{L}^2(\mathbb{R})$. $(f_n)_{n \in \mathbb{N}}$ is a Cauchy sequence in $\mathbf{L}^2(\mathbb{R})$, hence $(\hat{f}_n)_{n \in \mathbb{N}}$ is also a Cauchy sequence in $\mathbf{L}^2(\mathbb{R})$, by the Plancherel formula. Let then $\hat{f} \in \mathbf{L}^2(\mathbb{R})$ be the limit. We shall put $\mathbb{F}[f(t)] = \hat{f}(\omega)$.

Consequence $\mathbb{F} : \mathbf{L}^2(\mathbb{R}) \longrightarrow \mathbf{L}^2(\mathbb{R})$ is an *isomorphism* of Hilbert spaces.

In particular, we have the identities

$$\langle f, g \rangle = \langle \hat{f}, \hat{g} \rangle \quad (Parseval),$$

$$\| f \| = \| \hat{f} \| \quad (Plancherel).$$

Thus we can state:

Our system $(h_k(t))_{k \in \mathbb{Z}}$ of shifted convolutional sines is an *orthonormal system* in $\mathbf{L}^2(\mathbb{R})$.

Remark The foregoing result is a remedy to a certain uneasiness in face of the $\mathbf{L}^1$ – formalism: although we got reciprocity of $\mathbb{F}$ and of $\overline{\mathbb{F}}$ on certain classes of continuous *and* summable functions, we were *not* able to speak of a bijection of the (time domain) $\mathbf{L}^1(\mathbb{R})$ with the (frequency domain) $\mathbf{L}^1(\mathbb{R})$: the Fourier transform of a summable function which is not continuous (modulo equivalence) *cannot* be summable (in this sense, the example of a "window function" is absolutely typical).

Let us insist upon the *structural particularity* of the foregoing $\mathbf{L}^2$ – isomorphism.

The $\mathbf{L}^2$ Fourier transform, as well as its inverse, **does not produce functions, but only classes of functions (i.e. distributions).**

In simpler words: | Every $\mathbf{L}^2$ – result which has been established via Fourier transform, is only valid "almost everywhere".

3.3.4 The Sampling Theorem

The Nyquist–Shannon theorem, in its popular version, reads like this:

let $f(t)$ be a signal such that $\hat{f}(\nu) = 0$ for $|\nu| \geq W$.

Put $T = \frac{1}{2W}$.

Carry out an equidistant (infinite) sampling for $t_k = k \cdot T$, $k \in \mathbb{Z}$.

Then $f(t)$ can be reconstructed by means of the samples $f_k = f(kT)$, $k \in \mathbb{Z}$.

This statement offers a kind of universal key for the transit between the analog world and the digital world. But at first glance there are already some question marks: the common sense is (quite rightly) contradicted by the

mathematically indispensable appearance of an **infinite** sampling. The theoretical spirit asks: how can the function $f(t)$, apparently rather arbitrary, be entirely determined by the sequence of values $f_k = f(kT)$, $k \in \mathbb{Z}$?

It is the second problem which will be our main concern.

We shall give an $\mathbf{L}^2$ – version of the Whittaker–Shannon theorem (the label "Nyquist–Shannon" has become universal tradition, so we shall not completely abandon it). First we look at a **normalized** situation.

Consider the space $\mathbf{L}^2[-\pi, \pi]$ as a Hilbert subspace of $\mathbf{L}^2(\mathbb{R})$:

$$\mathbf{L}^2[-\pi, \pi] = \{\hat{f} \in \mathbf{L}^2(\mathbb{R}) : \hat{f}(\omega) = 0 \quad \text{for } |\omega| > \pi\}.$$

Let $\mathbf{L}_1^2(\mathbb{R})$ be the time domain copy of $\mathbf{L}^2[-\pi, \pi]$:

$$\mathbf{L}_1^2(\mathbb{R}) = \{f \in \mathbf{L}^2(\mathbb{R}) : \hat{f} \in \mathbf{L}^2[-\pi, \pi]\}.$$

We obtain an induced isomorphism of Hilbert spaces

$$\boxed{\mathbb{F} : \mathbf{L}_1^2(\mathbb{R}) \longrightarrow \mathbf{L}^2[-\pi, \pi],}$$

which makes correspond $(h_k(t))_{k \in \mathbb{Z}}$ to $(e^{-i\omega k})_{k \in \mathbb{Z}}$.[10]

Consequence   Whittaker–Shannon Theorem – Austere Version

The system $(h_k(t))_{k \in \mathbb{Z}}$ is an *orthonormal basis* of $\mathbf{L}_1^2(\mathbb{R})$.

More explicitly: consider, for $f \in \mathbf{L}_1^2(\mathbb{R})$, the Fourier series of its Fourier transform[11]:

$$\hat{f}(\omega) = \sum_{k \in \mathbb{Z}} f[k] e^{-i\omega k}$$

with

$$f[k] = \langle \hat{f}(\omega), e^{-i\omega k} \rangle, \quad k \in \mathbb{Z}.$$

Attention   The convergence is a convergence in quadratic mean (i.e. with respect to the Hilbert norm) and the equality is an equality almost everywhere.

All these natural operations will give in the time domain:
$f(t) = \sum_{k \in \mathbb{Z}} f[k] h_k(t)$ with $f[k] = \langle f(t), h_k(t) \rangle$, $k \in \mathbb{Z}$.

Let us point out: the "coordinates" of $f = f(t)$ with respect to the orthonormal basis $(h_k(t))_{k \in \mathbb{Z}}$ of $\mathbf{L}_1^2(\mathbb{R})$ are the Fourier coefficients of $\hat{f}(\omega)$.

But the representation of the $f[k]$ as **time domain** inner products permits their identification as the values of a certain function:

$$\langle f(t), h_k(t) \rangle = \int_{-\infty}^{+\infty} f(t) h_0(t - k) dt = (f * h_0^-)(k)$$

---

[10] We do not change notation for restricted sinusoids with compact support $[-\pi, \pi]$.

[11] I.e. we do a 2π-periodic prolongation of the non-trivial branch of the Fourier transform, then develop it into a Fourier series, finally restrict everything back to $]-\pi, \pi]$.

with $h_0^-(t) = h_0(-t)$. But, $(f * h_0^-)(t) = \overline{\mathbb{F}}[\hat{f}(\omega)\hat{h}_0(-\omega)]$ and $\hat{f}(\omega)\hat{h}_0(-\omega) = (\hat{f} \cdot \mathbb{I}_{[-\pi,\pi]})(\omega) = \hat{f}(\omega)$. Hence, for $f \in \mathbf{L}_1^2(\mathbb{R})$ we have: $(f * h_0^-)(t) = \overline{\mathbb{F}}[\hat{f}(\omega)]$. Put $f_1(t) = \overline{\mathbb{F}}\mathbb{F}[f(t)]$. This yields finally:

$$f(t) = \sum_{k\in\mathbb{Z}} f_1(k)h_k(t).$$

We note: $f_1(t)$ is *infinitely differentiable* in virtue of the following result:

Lemma Let $G(\omega)$ be a function of the variable ω such that each of the functions $\omega^m G(\omega)$, $m \geq 0$, is summable.

Then $g(t) = \overline{\mathbb{F}}[G(\omega)] = \frac{1}{2\pi} \int_{-\infty}^{+\infty} G(\omega)e^{i\omega t}d\omega$ is an *infinitely differentiable* function (the differentiation is simply done under the integral sign).

This holds manifestly for $f_1(t)$ (since $\hat{f}(\omega)$ admits a bounded support).

Whittaker–Shannon Theorem – Normalized Version.

Let $f(t)$ be a square integrable function on $\mathbb{R}$ and such that $\hat{f}(\omega) = 0$ for $|\omega| > \pi$.

Then the function $f_1(t) = \overline{\mathbb{F}}[\hat{f}(\omega)]$ is infinitely differentiable, and

$$f(t) = \sum_{k\in\mathbb{Z}} f_1(k)h_k(t)$$

(Attention: convergence and equality are *in* $\mathbf{L}_1^2(\mathbb{R})$).

Commentary As stated above, the Whittaker–Shannon theorem is *not* a real interpolation theorem. The sequence $(f_1(k))_{k\in\mathbb{Z}} = (f[k])_{k\in\mathbb{Z}}$ is the sequence of the Fourier coefficients of $\hat{f}(\omega)$, hence it does *not* depend on the values of $f(t)$ on $\mathbb{Z}$.

On the other hand, the Shannon interpolator of (the series of convolutional sines) *is* the function $f_1(t)$, which differs from $f(t)$ only on a set of measure zero. In this sense, we obtain a "generic" reconstruction of the line of $f(t)$ by regularization (smoothing).

If $f(t)$ is *smooth*, i.e. infinitely differentiable, then $f(t) = f_1(t)$, and $f(t)$ is effectively *determined* by the sequence $(f(k))_{k\in\mathbb{Z}}$.

So, do not forget:

> The Whittaker–Shannon theorem controls only **smooth** phenomena.

This is by no means surprising, since an equidistant sampling in itself is a rather weak constraint. You have to create a rigid situation by a finiteness condition (concerning the spectrum) and the (ensuing) regularity of the interpolator.

Exercise

The orthogonal projection $\mathbf{L}^2(\mathbb{R}) \longrightarrow \mathbf{L}_1^2(\mathbb{R})$.

Let $h_0(t) = \sin c(t)$ with $\hat{h}_0(\omega) = \mathbb{I}_{[-\pi,\pi]}$.
Show:

(a) For every $f \in \mathbf{L}^2(\mathbb{R})$ we have: $f_1 = f * h_0 \in \mathbf{L}_1^2(\mathbb{R})$.
(b) $\|f - f_1\| = \min\{\|f - g\| : g \in \mathbf{L}_1^2(\mathbb{R})\}$.

Note: h_0 is the impulse response of the linear and time-invariant filter which associates with every (square integrable) signal f the nearest signal (for the Hilbert space distance) which has its spectrum supported in $[-\pi, \pi]$.

Remark We can also read the Whittaker–Shannon theorem as a true interpolation theorem – in direction digital $\longrightarrow$ analog exclusively.

Let then $(f[k])_{k\in\mathbb{Z}}$ be an arbitrary time series, which is square summable: $\sum_{k\in\mathbb{Z}} |f[k]|^2 < \infty$. Consider the Whittaker–Shannon interpolator

$$f(t) = \sum_{k\in\mathbb{Z}} f[k] h_k(t).$$

Then $f(t)$ is infinitely differentiable and square integrable, and we get for its Fourier transform:

$$\hat{f}(\omega) = 0 \quad \text{for } |\omega| > \pi$$

and

$$\hat{f}(\omega) = \sum_{k\in\mathbb{Z}} f[k] \mathrm{e}^{-\mathrm{i}\omega k} \quad \text{for } -\pi \le \omega \le \pi.$$

This observation leads to the *definition* of the Fourier transform of a digital (square summable) signal – which will always be tacitly considered as the sequence of the samples of its Whittaker–Shannon interpolator.

In order to finally arrive at the correct statement of the popular form of our theorem (recall the beginning of this paragraph), we only have to carry out the evident changes of variable.

Nyquist–Shannon Theorem ω-Version

Let $f = f(t)$ be a function which is square integrable on $\mathbb{R}$.
If $\hat{f}(\omega) = 0$ outside $[-\frac{\pi}{T}, \frac{\pi}{T}]$,
then we may suppose $f = f(t)$ infinitely differentiable, and we obtain

$$f(t) = \sum_{k\in\mathbb{Z}} f(k \cdot T) h_T(t - k \cdot T)$$

with $h_T(t) = \sin c(\frac{t}{T})$.

Nyquist–Shannon Theorem ν-Version

Let $f = f(t)$ be a function which is square integrable on $\mathbb{R}$.
Assume $\hat{f}(\nu) = 0$ outside $[-W, W]$, and put $T = \frac{1}{2W}$.

Then we can suppose $f = f(t)$ infinitely differentiable, and obtain

$$f(t) = \sum_{k \in \mathbb{Z}} f(k \cdot T) h_T(t - k \cdot T)$$

with $h_T(t) = \sin c(\frac{t}{T})$.

Remarks   How should we read the theorem of $\begin{cases} \text{Whittaker–Shannon} \\ \text{Nyquist–Shannon} \end{cases}$?

(1) Let us begin with the mathematically correct situation: the signal $f(t)$ is well known *in its analytical form*, hence $\hat{f}(\omega)$ is (hopefully) calculable, and one is able to verify that the hypotheses of the Nyquist–Shannon theorem are satisfied. In this case, the reconstruction of $f(t)$ by the Nyquist–Shannon formula may seem ridiculous, since $f(t)$ is known (recall the development of $\cos 2\pi t$ in a series of convolutional sines).

(2) Pass now to a more realistic situation, which is mathematically a little bit frustrating: the signal $f(t)$ is *not* explicitly known in its analytical form. The hypotheses of the interpolation theorem **cannot** be verified. The theorem will be applied by virtue of an extra-mathematical argument based on the intuitive sense of the notion of "frequency" (all acoustic phenomena). Then $f(t)$ **is nothing but** the sequence of its samples. "Reconstruction" will simply signify "interpolation": the Nyquist–Shannon interpolator will only give a **smooth interpretation** of the sequence of samples of $f(t)$. Thus one creates a posteriori an analog signal which satisfies the hypotheses of the interpolation theorem. This answers another objection of mathematical pedantry: a function with compact support can never satisfy the hypotheses of the Nyquist–Shannon theorem (see exercise (3) at the end). A finite sampling (and there are no other samplings in practice), which corresponds conceptually to the profile of a signal with compact support, will nevertheless give rise to a "mathematically sound" interpolator.

(3) Let us come back to the validity of the **periodic** Nyquist–Shannon theorem. We have seen that there is equivalence with trigonometric interpolation via Discrete Fourier Transform. This suggests a strategy of primitive, but logical reconstruction: consider our (acoustic) signal $f(t)$ as a trigonometric polynomial, with *variable* but *locally constant* coefficients $c_k = c_k(t)$. Then the periodic Nyquist–Shannon theorem may be (locally) applied. One will sample to the frequency imposed by the physical data, and one will shoulder off Nyquist–Shannon interpolation while practicing reconstruction via Discrete Fourier Transform. You will obtain this way a "piecewise trigonometric" adjustment (similar to the piecewise linear construction of curves in primitive computer graphics).

Exercises

(1) We know that $|\sin t| \geq \frac{1}{2}$ for $k\pi - \frac{5}{6}\pi \leq t \leq k\pi - \frac{\pi}{6}$ and $k \in \mathbb{Z}$.

    (a) Deduce that $\int_0^{k\pi} \frac{|\sin t|}{t} dt > \frac{1}{2} \sum_{\nu=1}^{k} \mathrm{Log} \frac{\nu - \frac{1}{6}}{\nu - \frac{5}{6}}$,

    (b) Write $\mathrm{Log} \frac{\nu - \frac{1}{6}}{\nu - \frac{5}{6}} = \mathrm{Log}\left(1 + \frac{2}{3} \cdot \frac{1}{\nu - \frac{5}{6}}\right)$.

    Observe that $\mathrm{Log}\left(1 + \frac{2}{3} \cdot \frac{1}{\nu - \frac{5}{6}}\right) > \frac{1}{3} \cdot \frac{1}{\nu - \frac{5}{6}}$ for almost all $\nu \geq 1$.

    Conclude: the integral $\int_{-\infty}^{+\infty} |\frac{\sin t}{t}| dt$ is divergent.

(2) Let $f(t) = \pi \cdot \left(\frac{\sin(\pi t)}{\pi t}\right)^2$.

    (a) Compute its Fourier transform $\hat{f}(\omega)$, and show that the hypotheses of the Nyquist–Shannon theorem are satisfied.

    (b) Represent $f(t)$ by $f_1(t)$, its Nyquist–Shannon interpolator.

(3) Show that the (non-periodic) Nyquist–Shannon theorem can *never* rigorously apply to a concrete situation.

Let $f(t)$ be a signal, observed during a finite time $(0 \leq t \leq L)$, and consider it as non-periodic: logically, you have to put $f(t) = 0$ for $t \leq 0$ and $t \geq L$.

Show that in this situation the principal hypothesis of the Nyquist–Shannon theorem can *never* be satisfied: $f(t)$ and $\hat{f}(\nu)$ cannot admit *simultaneously* a support of finite length.

$\Big[$*Help*: suppose $\hat{f}(\nu) = 0$ for $|\nu| \geq W$. Put $T = \frac{1}{2W}$. Then you get: $f(t) = f_1(t) = \sum_{0 \leq k \leq N} f_k \frac{\sin 2\pi W(t - kT)}{2\pi W(t - kT)}$, with $f_k = f(kT)$ and $NT \geq L$.

$f(t)$ must necessarily be infinitely differentiable on $\mathbb{R}$. But $f(t) = 0$ for $t \leq 0$.

Hence: $f^{(m)}(0) = 0$ for $m \geq 0$.

Deduce that $f_0 = f_1 = \cdots = f_N = 0.\Big]$

4

Error Control Codes

The first chapter (on lossless data compression) has turned around the question: How can we *eliminate* redundancies in the treatment of information? Now, we shall be concerned with the opposite question: Why should we feel obliged to *introduce* redundancies in the treatment of information, and how can we algorithmically control the situation? So, we have to speak about error correcting codes.

The elementary theory of the subject has been treated extensively in the literature. Consequently, we shall focus our attention on two more advanced topics, which are best understood via ideas and methods of signal theory: The Reed–Solomon codes and the convolutional codes. In the first case, the decoding algorithm allows a pretty appearance of the Discrete Fourier Transform in binary polynomial arithmetic, in the second case, the mostly acclaimed decoding algorithm (the Viterbi algorithm) is a kind of step-by-step deconvolution – in the spirit of digital filtering.

4.1 The Reed–Solomon Codes

In this first section, we shall show how the use of the Discrete Fourier Transform, with coefficients in certain finite fields of characteristic 2 ($1 + 1 = 0!$), makes the algorithmic control of a (very important) class of error correcting codes considerably easier: We shall speak about the Reed–Solomon codes.

4.1.1 Preliminaries: Polynomial Codes

Recall the fundamental facts, beginning with a simple Hamming code. Suppose that our information is encoded in the 16 binary words of length 4:

$$0000, 0001, \ldots, 1111.$$

A transmission error of the information bitstream *cannot* be detected (and, a fortiori, cannot be corrected), since it necessarily changes *meaningful* words into *meaningful* words.

In order to detect (and to correct) transmission errors, we have to introduce *redundancies,* i.e. we have to turn to an encoding where every code word (every *significant* word) has a "neighbourhood" of *non-significant* words – the appearance of which will reveal the transmission error (and, hopefully, the kind of the error). Let us encode our 16 binary words (of length 4) into 16 code words (of length 7) in the following way:

0000	$\longmapsto$	0000000	1000	$\longmapsto$	1011000,
0001	$\longmapsto$	0001011	1001	$\longmapsto$	1010011,
0010	$\longmapsto$	0010110	1010	$\longmapsto$	1001110,
0011	$\longmapsto$	0011101	1011	$\longmapsto$	1000101,
0100	$\longmapsto$	0101100	1100	$\longmapsto$	1110100,
0101	$\longmapsto$	0100111	1101	$\longmapsto$	1111111,
0110	$\longmapsto$	0111010	1110	$\longmapsto$	1100010,
0111	$\longmapsto$	0110001	1111	$\longmapsto$	1101001,

We note: There are 128 binary words of length 7. Only 16 among them are elements of our code $\mathcal{C}$ (i.e. encode some information, have a *meaning*).

We observe further: The *minimum distance* between the words of the code $\mathcal{C}$ is equal to 3 (the distance between two words is the number of positions where they are distinct): One cannot pass from a code word to another code word without changing at least three positions.

Consequence *Double errors (per block of length 7) can be detected. Single errors (per block of length 7) can be corrected.*

The reason: Every binary word of length 7 has a distance at most equal to 1 to some *unique* code word.

The argument is the following: Consider the 16 "balls" of binary words of length 7 centered each at one of the 16 code words and filled with the 7 derived words that are obtained if we change one of the 7 positions of the "central" code word. These 16 "balls" are mutually disjoint (since the minimum distance between the code words is equal to 3), hence their union has $16 \cdot 8 = 128$ elements, which is exhaustive.

How did we succeed in creating this pretty situation?

The answer comes from an arithmetical trick. Let us write down our information words in $\boxed{\text{polynomial notation}}$:

$0101 = T^2 + 1,$
$1100 = T^3 + T^2,$ etc.

Now choose a *generator polynomial* $g(T)$ for our code:

$g(T) = T^3 + T + 1.$

The code words – in polynomial notation – are obtained via multiplication of the information polynomials with the generator polynomial:

$$c(T)=i(T)g(T)$$

Let us verify:

We encoded: $1001 \longmapsto 1010011$

But, indeed: $(T^3 + 1)(T^3 + T + 1) = T^6 + T^4 + T + 1$.

The algorithm for the detection of transmission errors is simple: *Euclidian division by $g(T)$*:

One divides the transmitted word – in polynomial notation – by our generator polynomial. If the remainder of division is zero, then one will suppose an error free transmission (we note that we recover at the same time the information word). If the remainder of division – the syndrome $s(T)$ of the transmitted word – is non-zero, then one will suppose a single-error transmission.

How to determine the error pattern, i.e. the position of the bit in error?

Let $c(T)$ be the correct (polynomial) word. A single transmission error at the position k (enumeration of the positions: 6.5.4.3.2.1.0) will create the received (polynomial) word

$v(T) = c(T) + T^k$.

Hence: $s(T) = v(T) \bmod g(T) = T^k \bmod g(T)$. This permits to determine k.

Example $v = 1110011$ i.e. $v(T) = T^6 + T^5 + T^4 + T + 1$.

We have, in $\mathbb{F}_8 = \mathbb{F}_2[T]/(T^3 + T + 1)$:

$$
\begin{aligned}
T^3 &= T + 1, \\
T^4 &= T^2 + T, \\
T^5 &= T^2 + T + 1, \\
T^6 &= T^2 + 1,
\end{aligned}
$$

Hence: $T^6 + T^5 + T^4 + T + 1 = (T^2 + 1) + (T^2 + T + 1) + (T^2 + T) + T + 1 = T^2 + T + 1$ i.e. $s(T) = T^2 + T + 1 \equiv T^5 \bmod (T^3 + T + 1)$. We conclude: The error is at the position 5. We correct: $v(T) \longmapsto c(T) = T^6 + T^4 + T + 1$. We recover the information: $T^6 + T^4 + T + 1 : (T^3 + T + 1) = T^3 + 1$ i.e. $i(T) = T^3 + 1 \equiv 1001$.

Summary *A polynomial code (of binary words of length n) admits the following description:*

(1) *The encoder. One multiplies the (polynomial) information words $i(T)$ of length $n - r$ ($\deg i(T) \le n - r - 1$) by a generator polynomial $g(T)$ (of degree r):*

$$c(T) = i(T)g(T).$$

(2) The decoder. One divides the received (polynomial) word $v(T)$ by the generator polynomial:

$$v(T) = m(T)g(T) + s(T)$$

$s(T)$, the remainder of division, is the syndrome of $v(T)$.
If $s(T) = 0$, one supposes an error free transmission, and one recovers the information:
$i(T) = m(T)$.
If $s(T) \neq 0$, one has detected a transmission error.

The correction of the error needs a deeper algorithmic control of $\mathbb{F}_2[T]/(g(T))$: you have to decide if $s(T)$ is an error monomial, or binomial, or trinomial, etc., indicators of the current error pattern.

Attention

Error correction means: Find the nearest code word to the received word. One corrects always the most innocent error .

Exercises

(1) Let $\mathcal{C}$ be the Hamming code treated in the foregoing example (with generator polynomial $g(T) = T^3 + T + 1$).
Decode the following block of four words: $v_1 v_2 v_3 v_4 = 1110111.011001.$
$1010101.1110011.$

(2) Let us go a little bit further: Consider the Hamming code $\mathcal{C}$ with generator polynomial $g(T) = T^4 + T + 1$.
The information words will have length 11 (represented by the binary polynomials of degree ≤ 10), the code words will thus have 15 bits each.
(a) Show that the 15 non-zero syndromes correspond precisely to the 15 single-error patterns (recall the cyclic group $(\mathbb{F}_2[T]/(T^4 + T + 1))^*$).
(b) Deduce that every binary word of length 15 is either a code word or at distance 1 of a *unique* code word. The minimum distance between the code words is thus 3.
(c) Decode $v_1 = 111011101110111$, then $v_2 = 100101010001010$.

(3) Let $\mathcal{C}$ be a binary polynomial code, with generator polynomial $g(T)$, and let n be the length of the code words. Show: $\mathcal{C}$ is *cyclic* (invariant under cyclic permutation of the bit-positions of its words) $\Longleftrightarrow$ $g(T)$ divides $T^n + 1$.
[This is true for the two Hamming codes that we have treated before: $T^3 + T + 1$ divides $T^7 + 1$, and $T^4 + T + 1$ divides $T^{15} + 1$.
Help: Let σ be the cyclic-left-shift operator; then $\sigma v(T) \equiv T v(T)$ mod $(T^n + 1)$.]

(4) (Towards the Meggitt decoder). Let $\mathcal{C}$ be a (binary) cyclic code, $g(T)$ its generator polynomial, n the length of its words. $v(T)$ a binary word of length n, in polynomial notation, $s(T)$ its syndrome. $v'(T) = \sigma v(T)$ the cyclically left-shifted word, $s'(T)$ its syndrome. Show that

$$\boxed{s'(T) \equiv Ts(T) \bmod g(T)}$$

(5) Let $\mathcal{C}$ be the cyclic binary code defined by the generator polynomial $g(T) = T^8 + T^7 + T^6 + T^4 + 1 = (T^4 + T + 1)(T^4 + T^3 + T^2 + T + 1)$ (dividing $T^{15} + 1$). The length of the code words will then be equal to 15, and the length of the information words will be equal to 7.
Accept. The minimum distance between the code words is equal to 5. $\mathcal{C}$ will thus admit the correction of single or double transmission errors (per block of 15 bits). Decode $v_1 = 110101011001000$ and $v_2 = 001011100001110$.
[Help: The syndromes $s_0(T) = 1$ and $s_{k,0}(T) = T^k + 1$, $1 \le k \le 7$, correspond to error patterns of a single error at the last position, and of double errors, concerning the last position and the kth position before. Using the result of exercise (4), you can reduce – via iterated cyclic shift – the decoding to the identification of one of these standard syndromes, while counting the number of necessary cyclic shifts).]

(6) *The binary Golay code* The information words: The 4,096 binary words of length 12. The generator polynomial: $g(T) = T^{11} + T^{10} + T^6 + T^5 + T^4 + T^2 + 1$. The code words (in polynomial version) $c(T) = i(T)g(T)$ are then of length 23.
Cyclicity. $T^{23} + 1 = (T + 1)g(T)\tilde{g}(T)$, with $\tilde{g}(T) = T^{11} + T^9 + T^7 + T^6 + T^5 + T + 1$.
Accept. The minimum distance between the code words is 7.
Thus $\mathcal{C}$ admits the correction of error patterns of a single, a double or a triple error (per block of 23 bits).

(a) Let v be a binary word of length 23, derived from a word of $\mathcal{C}$ after a transmission error of three faults at positions 18, 13 and 2. Which will be its syndrome?

(b) You receive the word $v = 10111101100101111010011$. Decode and recover the information.

4.1.2 Reed–Solomon Codes

The *Reed–Solomon codes* are polynomial codes, but on a (slightly) superior level to that of our foregoing description.

The information units – the letters – are no longer the bits 0 and 1, but, for given $n \ge 2$, the 2^n binary words of length n. In other words, we will have two different levels of polynomial arithmetic: First, the arithmetic of the finite fields $\mathbb{F}_{2^n}$ – i.e. that of the remainders of division by an irreducible binary polynomial of degree n – which will be the arithmetic of the letters of the chosen alphabet.

Then, the arithmetic of the polynomials with coefficients in the chosen field $\mathbb{F}_{2^n}$, which places us in the algorithmic context of polynomial coding.

Let us fix the notation (and the arithmetic) for the sequel:

$$\mathbb{F}_2 = \mathbb{Z}/2\mathbb{Z} = \{0,1\},$$

$$\mathbb{F}_4 = \mathbb{F}_2[x]/(x^2 + x + 1),$$

$$\mathbb{F}_8 = \mathbb{F}_2[x]/(x^3 + x + 1),$$

$$\mathbb{F}_{16} = \mathbb{F}_2[x]/(x^4 + x + 1),$$

$$\mathbb{F}_{64} = \mathbb{F}_2[x]/(x^6 + x + 1),$$

$$\mathbb{F}_{256} = \mathbb{F}_2[x]/(x^8 + x^4 + x^3 + x^2 + 1).$$

Then, we have to indicate the parameters of our presentation: Let $\mathbb{F}_{2^n}$, $n \geq 2$, be chosen, and let N be a divisor of $2^n - 1$.

The Reed–Solomon code $\mathrm{RS}(N, N - 2t)$ over $\mathbb{F}_{2^n}$ will be a polynomial code, with generator polynomial $g(T) \in \mathbb{F}_{2^n}[T]$. The degree of $g(T)$: $2t$. A code word will have N letters in $\mathbb{F}_{2^n}$, hence it will be a block of $N \cdot n$ bits. An information word will have $N - 2t$ letters in $\mathbb{F}_{2^n}$, hence it will be a block of $(N - 2t) \cdot n$ bits. $\mathrm{RS}(N, N - 2t)$ guarantees the correction of error patterns of maximally t errors.

Attention *One error means one letter in error, i.e. a block of n bits in error.*

An illustration: Consider the code $\mathrm{RS}(15, 7)$ over $\mathbb{F}_{256}$.
The length of the code words: 15 8-bit bytes $= 120$ bits.
The length of the information words: 7 bytes $= 56$ bits.

$\mathrm{RS}(15, 7)$ admits the correction of 4-error patterns, i.e. of four bytes in error (in a 15-byte word). For example: A transmission error which affects 26 successive bits (inside a block of 32 bits) can be corrected.

This points at the great practical interest of the Reed–Solomon codes (a convincing application: the error control code used in the CD system employs two concatenated Reed–Solomon codes, which are interleaved cross-wise: the – Cross-Interleaved Reed–Solomon Code (*CIRC*) – which is able to correct error bursts up to 3,500 bits (2.4 mm in length) and compensates for error bursts up to 12,000 bits (8.5 mm) that may be caused by minor scratches).

The theory (and the practice) of the Reed–Solomon codes owes its elegance (and its efficiency) to the interaction of polynomial ideas with "spectral" techniques based on the use of the Discrete Fourier Transform over the finite fields $\mathbb{F}_{2^n}$, $n \geq 2$.

Definition and First Properties of Reed–Solomon Codes

Fix $n \geq 2$, and consider $\mathbb{F}_{2^n} = \mathbb{F}_2[x]/(p(x))$, where $p(x)$ is an irreducible (primitive) binary polynomial of degree n.

Let N be a divisor of $2^n - 1$, and let $\omega = \omega_N \in \mathbb{F}_{2^n}$ be a (primitive) Nth root of unity: $\omega^N = 1$, but $\omega^k \neq 1$ for $1 \leq k < N$. (We note that whenever $p(x)$ is primitive, we have $\omega = x^{N'}$, with $N' = \frac{2^n - 1}{N}$.)

Hence we dispose of the Discrete Fourier Transform of order N[1]

$$F_N : (\mathbb{F}_{2^n})^N \longrightarrow (\mathbb{F}_{2^n})^N.$$

Recall our notational convention. The components of the transformed vector of a vector with variable components will be noted in upper-case letters.

Fix now t: $1 \leq t < \frac{N}{2}$.

Definition *of the Reed–Solomon code* RS$(N, N-2t)$ *over* $\mathbb{F}_{2^n}$:

$$\mathrm{RS}(N, N-2t) = \left\{ \begin{pmatrix} z_0 \\ z_1 \\ \vdots \\ z_{N-1} \end{pmatrix} \in (\mathbb{F}_{2^n})^N : \begin{array}{l} Z_{N-2t} = 0 \\ Z_{N-2t+1} = 0 \\ \vdots \\ Z_{N-1} = 0 \end{array} \right\}.$$

RS$(N, N-2t)$ *is the linear subspace of* $(\mathbb{F}_{2^n})^N$ *which consists of the N-tuples with the property that the 2t last components of their Fourier transform are equal to zero.*

Example The code RS$(5,3)$ over $\mathbb{F}_{16} = \mathbb{F}_2[x]/(x^4 + x + 1)$.

$\omega = \omega_5 = x^3$

$$F_5 = \begin{pmatrix} 1 & 1 & 1 & 1 & 1 \\ 1 & x^3 & x^6 & x^9 & x^{12} \\ 1 & x^6 & x^{12} & x^3 & x^9 \\ 1 & x^9 & x^3 & x^{12} & x^6 \\ 1 & x^{12} & x^9 & x^6 & x^3 \end{pmatrix} = \begin{pmatrix} 0001\ 0001\ 0001\ 0001\ 0001 \\ 0001\ 1000\ 1100\ 1010\ 1111 \\ 0001\ 1100\ 1111\ 1000\ 1010 \\ 0001\ 1010\ 1000\ 1111\ 1100 \\ 0001\ 1111\ 1010\ 1100\ 1000 \end{pmatrix}.$$

$$\mathrm{RS}(5,3) = \left\{ \begin{pmatrix} z_0 \\ z_1 \\ z_2 \\ z_3 \\ z_4 \end{pmatrix} \in (\mathbb{F}_{16})^5 : \begin{array}{l} Z_3 = 0 \\ Z_4 = 0 \end{array} \right\}$$

---

[1] We only need a group of Nth roots of unity in $(\mathbb{F}_{2^n})^*$ – in formal analogy with the group of Nth roots of unity in $\mathbb{C}^*$. Note that the matrix F_N^{-1} is simply the "conjugate" of F_N: Replace every coefficient by its multiplicative inverse. Since N is odd, the factor $\frac{1}{N} = 1$.

$$= \left\{ \begin{pmatrix} z_0 \\ z_1 \\ z_2 \\ z_3 \\ z_4 \end{pmatrix} \in (\mathbb{F}_{16})^5 : \begin{array}{l} z_0 + (1010)z_1 + (1000)z_2 + (1111)z_3 + (1100)z_4 = 0 \\ z_0 + (1111)z_1 + (1010)z_2 + (1100)z_3 + (1000)z_4 = 0 \end{array} \right\}.$$

This gives the code $RS(5,3)$ over $\mathbb{F}_{16}$ by its *parity-check equations*.

Exercise

(1) What is the probability that a word of 20 bits (chosen randomly) is a code word of $RS(5,3)$ over $\mathbb{F}_{16}$?

(2) Still the Reed–Solomon code $RS(5,3)$ over $\mathbb{F}_{16}$.
    You receive the word 0001.0101.$****$.$****$.1010, where eight successive binary positions are illegible. Supposing an otherwise error free transmission, decode and recover the information.

(3) The code $RS(7,3)$ over $\mathbb{F}_{64} = \mathbb{F}_2[x]/(x^6 + x + 1)$. Recall the cyclic group of 63th roots of unity in $\mathbb{F}_{64}$:

$x^6 = x + 1,$	$x^{25} = x^5 + x,$	$x^{44} = x^5 + x^3 + x^2 + 1,$
$x^7 = x^2 + x,$	$x^{26} = x^2 + x + 1,$	$x^{45} = x^4 + x^3 + 1,$
$x^8 = x^3 + x^2,$	$x^{27} = x^3 + x^2 + x,$	$x^{46} = x^5 + x^4 + x,$
$x^9 = x^4 + x^3,$	$x^{28} = x^4 + x^3 + x^2,$	$x^{47} = x^5 + x^2 + x + 1,$
$x^{10} = x^5 + x^4,$	$x^{29} = x^5 + x^4 + x^3,$	$x^{48} = x^3 + x^2 + 1,$
$x^{11} = x^5 + x + 1,$	$x^{30} = x^5 + x^4 + x + 1,$	$x^{49} = x^4 + x^3 + x,$
$x^{12} = x^2 + 1,$	$x^{31} = x^5 + x^2 + 1,$	$x^{50} = x^5 + x^4 + x^2,$
$x^{13} = x^3 + x,$	$x^{32} = x^3 + 1,$	$x^{51} = x^5 + x^3 + x + 1,$
$x^{14} = x^4 + x^2,$	$x^{33} = x^4 + x,$	$x^{52} = x^4 + x^2 + 1,$
$x^{15} = x^5 + x^3,$	$x^{34} = x^5 + x^2,$	$x^{53} = x^5 + x^3 + x,$
$x^{16} = x^4 + x + 1,$	$x^{35} = x^3 + x + 1,$	$x^{54} = x^4 + x^2 + x + 1,$
$x^{17} = x^5 + x^2 + x,$	$x^{36} = x^4 + x^2 + x,$	$x^{55} = x^5 + x^3 + x^2 + x,$
$x^{18} = x^3 + x^2 + x + 1,$	$x^{37} = x^5 + x^3 + x^2,$	$x^{56} = x^4 + x^3 + x^2 + x + 1,$
$x^{19} = x^4 + x^3 + x^2 + x,$	$x^{38} = x^4 + x^3 + x + 1,$	$x^{57} = x^5 + x^4 + x^3 + x^2 + x,$
$x^{20} = x^5 + x^4 + x^3 + x^2,$	$x^{39} = x^5 + x^4 + x^2 + x,$	$x^{58} = x^5 + x^4 + x^3 + x^2$
		$\qquad\qquad + x + 1,$
$x^{21} = x^5 + x^4 + x^3 + x + 1,$	$x^{40} = x^5 + x^3 + x^2 + x + 1,$	$x^{59} = x^5 + x^4 + x^3 + x^2 + 1,$
$x^{22} = x^5 + x^4 + x^2 + 1,$	$x^{41} = x^4 + x^3 + x^2 + 1,$	$x^{60} = x^5 + x^4 + x^3 + 1,$
$x^{23} = x^5 + x^3 + 1,$	$x^{42} = x^5 + x^4 + x^3 + x,$	$x^{61} = x^5 + x^4 + 1,$
$x^{24} = x^4 + 1,$	$x^{43} = x^5 + x^4 + x^2 + x + 1,$	$x^{62} = x^5 + 1.$

"The" seventh (primitive) root of unity : $\omega = \omega_7 = x^9$

(a) Write the matrix F_7 of the Discrete Fourier Transform of order 7 over $\mathbb{F}_{64}$.

(b) Write the parity-check equations for $RS(7,3)$ over $\mathbb{F}_{64} = \mathbb{F}_2[x]/(x^6 + x + 1)$.

$$\left[F_7 = \begin{pmatrix} 1 & 1 & 1 & 1 & 1 & 1 & 1 \\ 1 & x^9 & x^{18} & x^{27} & x^{36} & x^{45} & x^{54} \\ 1 & x^{18} & x^{36} & x^{54} & x^9 & x^{27} & x^{45} \\ 1 & x^{27} & x^{54} & x^{18} & x^{45} & x^9 & x^{36} \\ 1 & x^{36} & x^9 & x^{45} & x^{18} & x^{54} & x^{27} \\ 1 & x^{45} & x^{27} & x^9 & x^{54} & x^{36} & x^{18} \\ 1 & x^{54} & x^{45} & x^{36} & x^{27} & x^{18} & x^9 \end{pmatrix}\right.$$

$$= \begin{pmatrix} 000001 & 000001 & 000001 & 000001 & 000001 & 000001 & 000001 \\ 000001 & 011000 & 001111 & 001110 & 010110 & 011001 & 010111 \\ 000001 & 001111 & 010110 & 010111 & 011000 & 001110 & 011001 \\ 000001 & 001110 & 010110 & 001111 & 011001 & 011000 & 010110 \\ 000001 & 010110 & 011000 & 011001 & 001111 & 010111 & 001110 \\ 000001 & 011001 & 001110 & 011000 & 010111 & 010110 & 001111 \\ 000001 & 010111 & 011001 & 010110 & 001110 & 001111 & 011000 \end{pmatrix}$$

$$\text{RS}(7,3) = \left\{ \begin{pmatrix} z_0 \\ z_1 \\ z_2 \\ z_3 \\ z_4 \\ z_5 \\ z_6 \end{pmatrix} : \begin{array}{l} z_0 + x^{27}z_1 + x^{54}z_2 + x^{18}z_3 + x^{45}z_4 + x^9 z_5 + x^{36}z_6 = 0 \\ z_0 + x^{36}z_1 + x^9 z_2 + x^{45}z_3 + x^{18}z_4 + x^{54}z_5 + x^{27}z_6 = 0 \\ z_0 + x^{45}z_1 + x^{27}z_2 + x^9 z_3 + x^{54}z_4 + x^{36}z_5 + x^{18}z_6 = 0 \\ z_0 + x^{54}z_1 + x^{45}z_2 + x^{36}z_3 + x^{27}z_4 + x^{18}z_5 + x^9 z_6 = 0 \end{array} \right\} \right]$$

(3) What is the probability that a word of 42 bits (chosen randomly) is a code word of RS(7,3) over $\mathbb{F}_{64}$?

(4) Some immediately recognizable code words.

  (a) Let $z_0.z_1.z_2.z_3.z_4$ be a "constant" word of five times four bits: $z_0 = z_1 = z_2 = z_3 = z_4$. Show that it is a code word of RS(5,3) over $\mathbb{F}_{16}$.

  (b) Let $z_0.z_1.z_2.z_3.z_4.z_5.z_6$ be a "constant" word of seven times six bits: $z_0 = z_1 = z_2 = z_3 = z_4 = z_5 = z_6$. Show that it is a code word of RS(7,3) over $\mathbb{F}_{64}$.

  (c) Generalize.

(5) Cyclicity of the Reed–Solomon codes.

  (a) Let $z_0.z_1.z_2.z_3.z_4$ be a code word of RS(5,3) over $\mathbb{F}_{16}$. Show that then $z_4.z_0.z_1.z_2.z_3$ is also a code word of RS(5,3) over $\mathbb{F}_{16}$.

  (b) Let $z_0.z_1.z_2.z_3.z_4.z_5.z_6$ be a code word of RS(7,3) over $\mathbb{F}_{64}$. Show that then $z_6.z_0.z_1.z_2.z_3.z_4.z_5$ is also a code word of RS(7,3) over $\mathbb{F}_{64}$.

We return to the general case:

Important Observation

Write for the vector (for the word)

$$\begin{pmatrix} z_0 \\ z_1 \\ \vdots \\ z_{N-1} \end{pmatrix}$$

the *polynomial* $p(T) = z_0 + z_1 T + \cdots + z_{N-1}T^{N-1}$. Then

$$F_N \begin{pmatrix} z_0 \\ z_1 \\ z_2 \\ \vdots \\ z_{N-1} \end{pmatrix} = \begin{pmatrix} p(1) \\ p(\omega) \\ p(\omega^2) \\ \vdots \\ p(\omega^{N-1}) \end{pmatrix}.$$

We insist: The Fourier transform of a word of length N is obtained as the sample vector of the polynomial of this word, evaluated "around the discrete unit circle of order N" in $\mathbb{F}_{2^n}$.

Put now

$$g(T) = (T - \omega^{N-2t})(T - \omega^{N-2t+1}) \ldots (T - \omega^{N-1}).$$

Then, $\mathrm{RS}(N, N - 2t) = \{p(T) : g(T) \text{ divides } p(T)\}$. The argument: $\mathrm{RS}(N, N - 2t)$ is the set of the words of length N which, in polynomial notation, take the value zero for the $2t$ "last" powers of ω, i.e. for ω^{N-2t}, $\omega^{N-2t+1}, \ldots, \omega^{N-1}$.

Consequence $\mathrm{RS}(N, N - 2t)$ is a polynomial code over $\mathbb{F}_{2^n}$ with generator polynomial $g(T) = (T - \omega^{N-2t})(T - \omega^{N-2t+1}) \ldots (T - \omega^{N-1}) \in \mathbb{F}_{2^n}[T]$.

Exercises

(1) Compute the generator polynomial $g(T)$ for $\mathrm{RS}(5, 3)$ over $\mathbb{F}_{16}$.
   Verify that the code word (of length 5) which corresponds to the coefficients of $g(T)$ satisfies the parity-check equations.
(2) Compute the generator polynomial $g(T)$ for $\mathrm{RS}(7, 3)$ over $\mathbb{F}_{64}$.
   Verify that the code word (of length 7) which corresponds to the coefficients of $g(T)$ satisfies the parity-check equations.
(3) Cyclicity of the Reed–Solomon codes. Consider the general case of a Reed–Solomon code $\mathrm{RS}(N, N - 2t)$ over $\mathbb{F}_{2^n}$.
   (a) Show that the generator polynomial $g(T)$ divides the polynomial $T^N + 1$.
   (b) Deduce the cyclicity of $\mathrm{RS}(N, N - 2t)$: If $z_0.z_1.\ldots.z_{N-2}.z_{N-1}$ is a code word, then $z_{N-1}.z_0.z_1.\ldots.z_{N-2}$ is also a code word.
(4) In the following list of words, how many are words of the code $\mathrm{RS}(5,3)$ over $\mathbb{F}_{16}$?

0000.0011.1110.0101.0111	0000.0110.1111.1010.1110
1110.0000.0110.1111.1010	1110.0101.0111.0000.0011
0111.0000.0011.1110.0101	1010.1110.0000.0110.1111
0000.0011.1111.0101.0101	1000.0111.1011.1010.0000

(5) In the following list of words, how many are words of the code RS(5,3) over $\mathbb{F}_{16}$?

0001.1100.0101.1101.0101   1000.0110.0111.0110.1111

0000.1100.0101.0001.0000   0010.1011.1010.1001.1010

0111.0110.1111.1000.0110   0000.0000.0000.0000.0000

1101.0101.0001.1100.1111   0000.0000.0001.0100.1010

(6) In the following list of words of the code RS(7,3) over $\mathbb{F}_{64}$, find the masked letters.

$z_0.z_1.x^{36}.x^{54}.1.x^{36}.1$ $z_0.0.z_2.0.z_4.0.z_6$

$x^{36}.x^{45}.z_2.z_3.z_4.z_5.x^{45}$ $x^9.x^{18}.z_2.z_3.z_4.z_5.x^{18}$

$x^{36}.z_1.1.x^{36}.z_4.z_5.z_6$ $z_0.z_1.x^{54}.x^{54}.x^{36}.z_5.z_6$

The encoding algorithm is simple:

Let $a(T) = a_0 + a_1 T + \cdots + a_{N-2t-1} T^{N-2t-1}$ be the polynomial of the information word

$$\begin{pmatrix} a_0 \\ a_1 \\ \vdots \\ a_{N-2t-1} \end{pmatrix} \in (\mathbb{F}_{2^n})^{N-2t}$$

$c(T) = a(T)g(T) = z_0 + z_1 T + \cdots + z_{N-1} T^{N-1}$ will be the polynomial of the code word

$$\begin{pmatrix} z_0 \\ z_1 \\ \vdots \\ z_{N-1} \end{pmatrix}$$

associated with

$$\begin{pmatrix} a_0 \\ a_1 \\ \vdots \\ a_{N-2t-1} \end{pmatrix}.$$

Exercises

(1) Still our code $RS(5, 3)$ over $\mathbb{F}_{16}$.
Verify that the information word $1010.1111.0101$ is coded in
$0001.1100.0101.1101.0101$.

(2) Change your viewpoint: Consider the code $RS(5, 3)$ over $\mathbb{F}_{16}$ as being
parametrized in the following way:

$$\begin{pmatrix} z_0 \\ z_1 \\ z_2 \\ z_3 \\ z_4 \end{pmatrix} = F_5^{-1} \begin{pmatrix} Z_0 \\ Z_1 \\ Z_2 \\ 0 \\ 0 \end{pmatrix}.$$

Find the information word $a_0.a_1.a_2$ of the polynomial encoding which
corresponds to $Z_0.Z_1.Z_2 = 0001.1010.1111$ of the matrix encoding.

(3) Consider now the code $RS(7, 3)$ over $\mathbb{F}_{64}$.
Find the code word associated with the information word $a_0.a_1.a_2 = 010111.011001.000001$. *Recall:* $g(T) = x^{36} + x^{54}T + T^2 + x^{36}T^3 + T^4$.

Decoding and Error Correction

Remark $RS(N, N - 2t)$ permits the correction of t-error patterns.

Proof Let us first show that every non-zero word of $RS(N, N - 2t)$ admits at
least $2t + 1$ non-zero components:

$$F_N \begin{pmatrix} z_0 \\ z_1 \\ \vdots \\ z_{N-1} \end{pmatrix} = \begin{pmatrix} Z_0 \\ \cdot \\ Z_{N-2t-1} \\ 0 \\ \cdot \\ 0 \end{pmatrix}.$$

Consider $q(T) = Z_0 + Z_1 T + \cdots + Z_{N-2t-1} T^{N-2t-1}$. $q(T)$ admits at most
$N - (2t + 1)$ zeros. But

$$\begin{pmatrix} z_0 \\ z_1 \\ \vdots \\ z_{N-1} \end{pmatrix} = \begin{pmatrix} q(1) \\ q(\omega^{-1}) \\ q(\omega^{-2}) \\ \vdots \\ q(\omega^{-N+1}) \end{pmatrix},$$

which yields the claim. $\square$

Hence: Two distinct words of the code are distinct in at least $2t + 1$ positions. (The argument: The word sum is a code word, and it is non-zero
precisely at the positions where the two words are distinct).

As a consequence, a transmission error of (maximally) t faults in a given code word will create a word in error which remains still at a distance $\geq t+1$ from *every other* code word.

Now let us pass on to *the error-correcting algorithm*. A received word $\mathbf{v}$ can be written as the sum of the correct word $\mathbf{z}$ and of the error word $\mathbf{e}$ (the components of $\mathbf{e}$ indicate the "error events"):

$$\begin{pmatrix} v_0 \\ v_1 \\ \vdots \\ v_{N-1} \end{pmatrix} = \begin{pmatrix} z_0 \\ z_1 \\ \vdots \\ z_{N-1} \end{pmatrix} + \begin{pmatrix} e_0 \\ e_1 \\ \vdots \\ e_{N-1} \end{pmatrix}.$$

This gives for the Fourier transforms:

$$\begin{pmatrix} V_0 \\ V_1 \\ \vdots \\ \vdots \\ V_{N-1} \end{pmatrix} = \begin{pmatrix} Z_0 \\ . \\ Z_{N-2t-1} \\ 0 \\ . \\ 0 \end{pmatrix} + \begin{pmatrix} E_0 \\ . \\ E_{N-2t-1} \\ \vdots \\ E_{N-1} \end{pmatrix}.$$

Hence: Starting with the received word $\mathbf{v}$, one *knows* the $2t$ last positions $E_{N-2t}, \ldots, E_{N-1}$ of

$$F_N \begin{pmatrix} e_0 \\ e_1 \\ \vdots \\ e_{N-1} \end{pmatrix}.$$

We have to compute the components $E_0, \ldots E_{N-2t-1}$ in function of the $2t$ last components
under the hypothesis that there are *at most t non-zero* components e_i, $0 \leq i \leq N-1$.

$\square$

Lemma *There exist* $\Lambda_1, \Lambda_2, \ldots, \Lambda_t$ *with*

$$\begin{pmatrix} 1 \\ \Lambda_1 \\ . \\ \Lambda_t \\ 0 \\ . \\ 0 \end{pmatrix} * \begin{pmatrix} E_0 \\ E_1 \\ \vdots \\ \vdots \\ E_{N-1} \end{pmatrix} = \begin{pmatrix} 0 \\ 0 \\ \vdots \\ \vdots \\ 0 \end{pmatrix} \quad \Longleftrightarrow \quad \begin{pmatrix} e_0 \\ e_1 \\ \vdots \\ e_{N-1} \end{pmatrix} \quad \begin{array}{l} \textit{admits at most } t \\ \textit{non-zero components.} \end{array}$$

Proof $\begin{pmatrix} e_0 \\ \vdots \\ e_{N-1} \end{pmatrix}$ admits at most t non-zero components $\iff$ there exists

$\begin{pmatrix} \alpha_0 \\ \vdots \\ \alpha_{N-1} \end{pmatrix}$, admitting at most t zero components, and such that $\alpha_0 e_0 = \alpha_1 e_1 = \cdots = \alpha_{N-1} e_{N-1} = 0$.

Write

$$\begin{pmatrix} \alpha_0 \\ \vdots \\ \alpha_{N-1} \end{pmatrix} = F_N^{-1} \begin{pmatrix} \beta_0 \\ \vdots \\ \beta_{N-1} \end{pmatrix} = \begin{pmatrix} p(1) \\ p(\omega^{-1}) \\ \cdot \\ p(\omega^{-N+1}) \end{pmatrix}$$

with $p(T) = \beta_0 + \beta_1 T + \cdots + \beta_{N-1} T^{N-1}$.

Our hypothesis becomes: $p(\omega^{-j}) = 0$ for at most t exponents $j_1, j_2, \ldots, j_s$. Write $p(T) = p_0(T) q(T)$ with $p_0(T) = (T - \omega^{-j_1})(T - \omega^{-j_2}) \cdots (T - \omega^{-j_s})$. $q(\omega^j) \neq 0$ for $0 \le j \le N-1$. We thus can replace $p(T)$ by $p_0(T)$ and normalize to 1 the constant term of $p_0(T) : p_0(T) = 1 + \Lambda_1 T + \Lambda_2 T^2 + \cdots + \Lambda_t T^t$. But

$$F_N^{-1}\left(\begin{pmatrix} 1 \\ \Lambda_1 \\ \cdot \\ \Lambda_t \\ 0 \\ \cdot \\ 0 \end{pmatrix} * \begin{pmatrix} E_0 \\ E_1 \\ \vdots \\ \vdots \\ E_{N-1} \end{pmatrix} \right) = \begin{pmatrix} p_0(1) e_0 \\ p_0(\omega^{-1}) e_1 \\ \vdots \\ \vdots \\ p_0(\omega^{-N+1}) e_{N-1} \end{pmatrix}$$

according to the Convolution Theorem for F_N^{-1}. This provides our claim. $\square$

When applying the lemma to our situation, we obtain the following linear system:

$$\boxed{\begin{aligned} E_0 &= \Lambda_1 E_{N-1} + \Lambda_2 E_{N-2} + \cdots + \Lambda_t E_{N-t} \\ E_1 &= \Lambda_1 E_0 + \Lambda_2 E_{N-1} + \cdots + \Lambda_t E_{N-t+1} \\ &\vdots \\ E_{N-t} &= \Lambda_1 E_{N-t-1} + \Lambda_2 E_{N-t-2} + \cdots + \Lambda_t E_{N-2t} \\ &\vdots \\ E_{N-1} &= \Lambda_1 E_{N-2} + \Lambda_2 E_{N-3} + \cdots + \Lambda_t E_{N-t-1} \end{aligned}}$$

This gives a $t \times t$ linear system for $\Lambda_1, \Lambda_2, \ldots, \Lambda_t$:

$$\begin{pmatrix} E_{N-t-1} & E_{N-t-2} & \cdots & E_{N-2t} \\ \vdots & & & \vdots \\ E_{N-2} & E_{N-3} & \cdots & E_{N-t-1} \end{pmatrix} \begin{pmatrix} \Lambda_1 \\ \vdots \\ \Lambda_t \end{pmatrix} = \begin{pmatrix} E_{N-t} \\ \vdots \\ E_{N-1} \end{pmatrix}.$$

One resolves, i.e. one finds $\Lambda_1, \Lambda_2, \ldots, \Lambda_t$, then one computes $E_0, E_1, \ldots,$ E_{N-t-1}.

Let us sum up the decoding mechanism for our two favorite Reed–Solomon codes:

RS$(5, 3)$ over $\mathbb{F}_{16}$ RS$(7, 3)$ over $\mathbb{F}_{64}$

Correction of $t = 1$ letter in error   Correction of $t = 2$ letters in error
per transmitted word   per transmitted word

Let $\begin{pmatrix} E_3 \\ E_4 \end{pmatrix} = \begin{pmatrix} V_3 \\ V_4 \end{pmatrix}$

Let $\begin{pmatrix} E_3 \\ E_4 \\ E_5 \\ E_6 \end{pmatrix} = \begin{pmatrix} V_3 \\ V_4 \\ V_5 \\ V_6 \end{pmatrix}$

Compute Λ_1 :   Compute Λ_1, Λ_2 :

$E_4 = \Lambda_1 \cdot E_3$

$\begin{pmatrix} E_4 & E_3 \\ E_5 & E_4 \end{pmatrix} \begin{pmatrix} \Lambda_1 \\ \Lambda_2 \end{pmatrix} = \begin{pmatrix} E_5 \\ E_6 \end{pmatrix}$

then: then:

$E_0 = \Lambda_1 E_4$ $E_0 = \Lambda_1 E_6 + \Lambda_2 E_5$
$E_1 = \Lambda_1 E_0$ $E_1 = \Lambda_1 E_0 + \Lambda_2 E_6$
$E_2 = \Lambda_1 E_1$ $E_2 = \Lambda_1 E_1 + \Lambda_2 E_0$

□

Exercises

(1) Consider the code RS$(5, 3)$ over $\mathbb{F}_{16}$.
    You receive the word 1000.0100.0111.0110.1111.
    (a) Decode (verify, whether there is a transmission error; in the affirmative case, correct the error).
    (b) Recover the 12 information bits.

Solution (a) Apply first the Fourier transformation of order 5:

$$\begin{pmatrix} V_0 \\ V_1 \\ V_2 \\ V_3 \\ V_4 \end{pmatrix} = F_5 \begin{pmatrix} v_0 \\ v_1 \\ v_2 \\ v_3 \\ v_4 \end{pmatrix} = \begin{pmatrix} 1 & 1 & 1 & 1 & 1 \\ 1 & x^3 & x^6 & x^9 & x^{12} \\ 1 & x^6 & x^{12} & x^3 & x^9 \\ 1 & x^9 & x^3 & x^{12} & x^6 \\ 1 & x^{12} & x^9 & x^6 & x^3 \end{pmatrix} \begin{pmatrix} x^3 \\ x^2 \\ x^{10} \\ x^5 \\ x^{12} \end{pmatrix} = \begin{pmatrix} x \\ x^3 + x^2 + x + 1 \\ x^3 + x^2 + x + 1 \\ x^2 + x + 1 \\ x^3 + x^2 + 1 \end{pmatrix}.$$

Hence: $E_3 = x^2 + x + 1$, $E_4 = x^3 + x^2 + 1$. The single-error correcting algorithm starts with the equations

$E_0 = \Lambda_1 E_4,$
$E_1 = \Lambda_1 E_0,$
$E_2 = \Lambda_1 E_1,$
$E_3 = \Lambda_1 E_2,$
$E_4 = \Lambda_1 E_3.$

We thus get: $\Lambda_1 = \frac{x^3+x^2+1}{x^2+x+1} = \frac{x^{13}}{x^{10}} = x^3$

$$\begin{pmatrix} E_0 \\ E_1 \\ E_2 \\ E_3 \\ E_4 \end{pmatrix} = \begin{pmatrix} x \\ x+1 \\ x^3+x+1 \\ x^2+x+1 \\ x^3+x^2+1 \end{pmatrix}, \text{ hence } \begin{pmatrix} Z_0 \\ Z_1 \\ Z_2 \\ Z_3 \\ Z_4 \end{pmatrix} = \begin{pmatrix} 0 \\ x^3+x^2 \\ x^2 \\ 0 \\ 0 \end{pmatrix}.$$

The code word:

$$\begin{pmatrix} z_0 \\ z_1 \\ z_2 \\ z_3 \\ z_4 \end{pmatrix} = F^{-1} \begin{pmatrix} 0 \\ x^3+x^2 \\ x^2 \\ 0 \\ 0 \end{pmatrix} = \begin{pmatrix} x^3 \\ x^2+x \\ x^2+x+1 \\ x^2+x \\ x^3+x^2+x+1 \end{pmatrix} = \begin{pmatrix} 1000 \\ 0110 \\ 0111 \\ 0110 \\ 1111 \end{pmatrix}.$$

(b) Divide $(1111)T^4+(0110)T^3+(0111)T^2+(0110)T+(1000)$ by the generator polynomial $T^2 + (0101)T + (1100)$:

$(x^{12}T^4 + x^5T^3 + x^{10}T^2 + x^5T + x^3) : (T^2 + x^8T + x^6) = x^{12}T^2 + x^{12}$

$\underline{x^{12}T^4 + x^5T^3 + x^3T^2}$
$\qquad\qquad x^{12}T^2 + x^5T + x^3$
$\qquad\qquad \underline{x^{12}T^2 + x^5T + x^3}$
$\qquad\qquad\qquad\qquad 0$

The information word is: 1111.0000.1111.

(2) Once more, the Reed–Solomon code RS(5,3) over $GF(16) = \mathbb{F}_2[x]/(x^4 + x + 1)$.

You receive the word
1000.1111.0111.0110.1111.

(a) Decode (verify, whether there is a transmission error; in the affirmative case, correct the error).

(b) Recover the 12 information bits.

(3) Consider now the Reed–Solomon code   RS(7, 3)   over $\mathbb{F}_{64} = \mathbb{F}_2[x]/(x^6 + x + 1)$.

You receive the word   000000.000000.000000.000000.000111.101101. 101010.

(a) Decode (verify, whether there is a transmission error; in the affirmative case, correct the error).

(b) Recover the 18 information bits.

(4) Still the Reed–Solomon code RS(7,3) over $GF(64) = \mathbb{F}_2[x]/(x^6 + x + 1)$.

You receive the word
100001.100110.∗∗∗∗∗∗.∗∗∗∗∗∗.∗∗∗∗∗∗.∗∗∗∗∗∗.111000,   where 24 successive bits have become illegible. We suppose an otherwise error free

transmission. Find the relevant code word and recover the 18 information bits.

(5) Consider $\mathbb{F}_8 = \{0, 1, x^9, x^{18}, x^{27}, x^{36}, x^{45}, x^{54}\} \subset \mathbb{F}_{64}$. How many words of the code RS(7, 3) over $\mathbb{F}_{64}$ have all (seven) letters in $\mathbb{F}_8$?

(6) Define the Discrete Fourier Transform of order 7 over $\mathbb{F}_{64}$ with respect to $\tilde{\omega} = x^{54}$ (we exchange the matrices F_7 and F_7^{-1} : $\tilde{F}_7 = F_7^{-1}$ and $\tilde{F}_7^{-1} = F_7$). Which will be now the generator polynomial $\tilde{g}(T)$ of the new Reed–Solomon code $\tilde{RS}(3,3)$ over $\mathbb{F}_{64}$?

(7) How many words have the two Reed–Solomon codes RS(7, 3) and $\tilde{RS}(7,3)$ over $\mathbb{F}_{64}$ (cf the preceding question (6)) in common?

(8) Consider the code RS(7, 3)   over $\mathbb{F}_{64} = \mathbb{F}_2[x]/(x^6 + x + 1)$.
Decode the word $v = 000000.000000.000011.011101.111111.000100.111000$, i.e.
   – verify if there is a transmission error;
   – in the affirmative case, correct the error;
   – recover the 18 information bits.

(9) The same situation.
   (a) Decode the word $v = 010110.001111.001110.011001.011000.011001.010110$, i.e.
      – verify if there is a transmission error;
      – in the affirmative case, correct the error;
      – recover the 18 information bits.
   (b) Decode the word $v' = 011000.011001.010110.010110.001111.001110.011001$.
      Try to make use of (a)!

(10) Still the Reed–Solomon code RS(7,3) over $\mathbb{F}_{64} = \mathbb{F}_2[x]/(x^6 + x + 1)$.
You receive the word
$\mathbf{v} = 010110.000001.000000.000000.010110.010110.000000$.
   (a) Decode (verify if there is a transmission error; in the affirmative case, correct the error).
   (b) Recover the 18 information bits, by means of the polynomial method.

(11) The situation of the preceding exercise.
One receives the word
$\mathbf{v} = 000000.000000.000001.000001.001110.000001.001110$.
   (a) Decode.
   (b) Recover the 18 information bits.

(12) The Reed–Solomon code RS(15, 7) over $\mathbb{F}_{256} = \mathbb{F}_2[x]/(x^8+x^4+x^3+x^2+1)$.
Let us first sum up the situation: The letters will be 8-bit bytes, identified with remainders of division modulo $p(x) = x^8 + x^4 + x^3 + x^2 + 1$. The multiplicative group of the non-zero residues is cyclic, and generated by (the remainder of division) x: This gives the log table for the 255 non-zero bytes in the preliminary section to the cipher AES-Rijndael.

Notation $11111111 = ff = x^7 + x^6 + x^5 + x^4 + x^3 + x^2 + x + 1 = x^{175}$.

A code word will have 15 letters, hence 120 bits. We will be able to correct transmission errors of four bytes per code word.

(a) First compute the generator polynomial $g(T)$ of the code $RS(15,7)$ over $\mathbb{F}_{256}$.

(b) Decode the word

$\mathbf{v} = v_0 v_1 \cdots v_{14} =$ ca 5d a2 48 ca 7d 7d 58 38 f4 80 15 48 20 ff   (hex notation) $= x^{73}.x^{56}.x^{209}.x^{226}.x^{73}.x^{243}.x^{243}.x^{241}.x^{201}.x^{230}.x^{7}.x^{141}.x^{226}.x^{5}. x^{175}$ i.e.

– verify if there is a transmission error;

– in the affirmative case, correct the error;

– recover the 56 information bits, by the polynomial method.

[Help: $g(T) = x^{153}+x^{136}T+x^{187}T^2+x^{221}T^3+x^{187}T^4+x^{119}T^5+x^{238}T^6+ x^{85}T^7 + T^8$

The matrix of the Discrete Fourier Transform of order 15 over $\mathbb{F}_{256}$:

$$F_{15} = \begin{pmatrix}
1 & 1 & 1 & 1 & 1 & 1 & 1 & 1 & 1 & 1 & 1 & 1 & 1 & 1 & 1 \\
1 & x^{17} & x^{34} & x^{51} & x^{68} & x^{85} & x^{102} & x^{119} & x^{136} & x^{153} & x^{170} & x^{187} & x^{204} & x^{221} & x^{238} \\
1 & x^{34} & x^{68} & x^{102} & x^{136} & x^{170} & x^{204} & x^{238} & x^{17} & x^{51} & x^{85} & x^{119} & x^{153} & x^{187} & x^{221} \\
1 & x^{51} & x^{102} & x^{153} & x^{204} & 1 & x^{51} & x^{102} & x^{153} & x^{204} & 1 & x^{51} & x^{104} & x^{153} & x^{204} \\
1 & x^{68} & x^{136} & x^{204} & x^{17} & x^{85} & x^{153} & x^{221} & x^{34} & x^{102} & x^{170} & x^{238} & x^{51} & x^{119} & x^{187} \\
1 & x^{85} & x^{170} & 1 & x^{85} & x^{170} & 1 & x^{85} & x^{170} & 1 & x^{85} & x^{170} & 1 & x^{85} & x^{170} \\
1 & x^{102} & x^{204} & x^{51} & x^{153} & 1 & x^{102} & x^{204} & x^{51} & x^{153} & 1 & x^{102} & x^{204} & x^{51} & x^{153} \\
1 & x^{119} & x^{238} & x^{102} & x^{221} & x^{85} & x^{204} & x^{68} & x^{187} & x^{51} & x^{170} & x^{34} & x^{153} & x^{17} & x^{136} \\
1 & x^{136} & x^{17} & x^{153} & x^{34} & x^{170} & x^{51} & x^{187} & x^{68} & x^{204} & x^{85} & x^{221} & x^{102} & x^{238} & x^{119} \\
1 & x^{153} & x^{51} & x^{204} & x^{102} & 1 & x^{153} & x^{51} & x^{204} & x^{102} & 1 & x^{153} & x^{51} & x^{204} & x^{102} \\
1 & x^{170} & x^{85} & 1 & x^{170} & x^{85} & 1 & x^{170} & x^{85} & 1 & x^{170} & x^{85} & 1 & x^{170} & x^{85} \\
1 & x^{187} & x^{119} & x^{51} & x^{238} & x^{170} & x^{102} & x^{34} & x^{221} & x^{153} & x^{85} & x^{17} & x^{204} & x^{136} & x^{68} \\
1 & x^{204} & x^{153} & x^{102} & x^{51} & 1 & x^{204} & x^{153} & x^{102} & x^{51} & 1 & x^{204} & x^{153} & x^{102} & x^{51} \\
1 & x^{221} & x^{187} & x^{153} & x^{119} & x^{85} & x^{51} & x^{17} & x^{238} & x^{204} & x^{170} & x^{136} & x^{102} & x^{68} & x^{34} \\
1 & x^{238} & x^{221} & x^{204} & x^{187} & x^{170} & x^{153} & x^{136} & x^{119} & x^{102} & x^{85} & x^{68} & x^{51} & x^{34} & x^{17}
\end{pmatrix},$$

then,

$$\mathbf{V} = \begin{pmatrix} V_0 \\ V_1 \\ V_2 \\ V_3 \\ V_4 \\ V_5 \\ V_6 \\ V_7 \\ V_8 \\ V_9 \\ V_{10} \\ V_{11} \\ V_{12} \\ V_{13} \\ V_{14} \end{pmatrix} = \begin{pmatrix} V_0 \\ V_1 \\ V_2 \\ V_3 \\ V_4 \\ V_5 \\ V_6 \\ E_7 \\ E_8 \\ E_9 \\ E_{10} \\ E_{11} \\ E_{12} \\ E_{13} \\ E_{14} \end{pmatrix} = \begin{pmatrix} x^{138} \\ x^{73} \\ x^{160} \\ x^{100} \\ x^{150} \\ x^{247} \\ x^{96} \\ x^{166} \\ x^{182} \\ x^{52} \\ x^{89} \\ x^{180} \\ x^{73} \\ x^{146} \\ x^{158} \end{pmatrix}.$$

The linear system for $\Lambda_1, \Lambda_2, \Lambda_3, \Lambda_4$ becomes:

$$\begin{pmatrix} x^{89} & x^{52} & x^{182} & x^{166} \\ x^{180} & x^{89} & x^{52} & x^{182} \\ x^{73} & x^{180} & x^{89} & x^{52} \\ x^{146} & x^{73} & x^{180} & x^{89} \end{pmatrix} \begin{pmatrix} \Lambda_1 \\ \Lambda_2 \\ \Lambda_3 \\ \Lambda_4 \end{pmatrix} = \begin{pmatrix} x^{180} \\ x^{73} \\ x^{146} \\ x^{158} \end{pmatrix}.$$

By Gauss-elimination, we obtain: $\Lambda_1 = x^{68}$, $\Lambda_2 = x^{51}$, $\Lambda_3 = x^{102}$, $\Lambda_4 = x^{68}$.
$E_0 = x^{138}$ $E_1 = x^{175}$ $E_2 = x^{233}$ $E_3 = x^{171}$

Thus:

$E_4 = x^{46}$ $E_5 = x^{106}$ $E_6 = x^{122}$

The transmitted code word: $x^{73}.x^{56}.x^{209}.x^{226}.x^{73}.x^{243}.x^{243}.\underline{x^{56}}.\underline{x^{226}}.0.\underline{x^5}.x^{141}.$
$x^{226}.x^5.x^{175}$

The information word: $a_0a_1a_2a_3a_4a_5a_6 = $ ff 00 ff 00 ff 00 ff $= x^{175}(1 + x^2 + x^4 + x^6)$]

4.2 Convolutional Codes

The convolutional codes occupy a similar place in the domain of error correcting codes as arithmetical coding in the domain of data compaction: A bitstream, *in principle unlimited*, will be transformed into a binary code stream (in principle also unlimited) by means of a kind of *arithmetical filtering* which works in complete formal analogy to usual digital filtering – with the laws of *binary arithmetic*, of course (recall the cryptosystem AES and the transformation of the state array by circular convolution).

The decoding algorithm has to solve a convolution equation which will be, in general, perturbed by a transmission error. If the number of errors per "inspection window" does not exceed a certain threshold, then the decoding algorithm (usually the algorithm of *Viterbi*) will properly produce the initial information stream and, in this way, will correct the transmission error.

Our presentation avoids the usual geometrization of the theory of convolutional codes (on the one hand, as trellis codes, on the other hand, as finite automata). The reason: Most algorithmic problems about convolutional codes become rather inaccessible under the mask of "convincing" geometrical arguments.

4.2.1 Encoding: Digital Filtering in Binary Arithmetic

Generalities:

A convolutional encoder is defined – as already mentioned – in formal analogy with a digital filter. It will transform an information bitstream into a code bitstream .

The code bitstream has to be "thicker" than the information bitstream: There is no detection and no correction of transmission errors without redundancies! This means: More code bits than information bits per unit of time. More precisely: The information bitstream $a_0a_1a_2 \cdots a_t \cdots$ is presented in *frames* a_j, $j \geq 0$.

Every frame is a binary word of (fixed) length k, $k \geq 1$. We shall adopt a notation in columns, in order to underline the parallelism of arrival:

$$a_0 = \begin{pmatrix} a_{10} \\ a_{20} \\ \vdots \\ a_{k0} \end{pmatrix}, \quad a_1 = \begin{pmatrix} a_{11} \\ a_{21} \\ \vdots \\ a_{k1} \end{pmatrix}, \quad \cdots \quad a_t = \begin{pmatrix} a_{1t} \\ a_{2t} \\ \vdots \\ a_{kt} \end{pmatrix}, \quad \cdots$$

The code bitstream $c_0 c_1 c_2 \cdots c_t \cdots$ is equally presented in frames c_j, $j \geq 0$. Every frame is a binary word of (fixed) length n, $n \geq 1$:

$$c_0 = \begin{pmatrix} c_{10} \\ c_{20} \\ \vdots \\ c_{n0} \end{pmatrix}, \quad c_1 = \begin{pmatrix} c_{11} \\ c_{21} \\ \vdots \\ c_{n1} \end{pmatrix}, \quad \cdots \quad c_t = \begin{pmatrix} c_{1t} \\ c_{2t} \\ \vdots \\ c_{nt} \end{pmatrix}, \quad \cdots$$

In order to guarantee that the code bitstream is thicker than the information bitstream, one demands $n > k$. We will speak of an (n, k) *convolutional code*. We shall only treat the case $k = 1$: *Every information frame will be a single bit.* The transformation of the information bitstream into the code bitstream will be carried out by an encoder which functions like (the convolution with) the impulse response of a digital filter.

In order to obtain a neat formalism, let us introduce a variable x, which will be *discrete-time carrier*. Then x^t signifies "at the moment t". The information bitstream thus becomes a formal power series $a(x) = \sum_{t \geq 0} a_t x^t$ where the information frame a_t "at the moment t" is the coefficient of x^t.

The code bitstream will also be a formal power series $c(x) = \sum_{t \geq 0} c_t x^t$. The coefficients c_t are binary n-tuples.

A convolutional encoder thus becomes a transformation $T : \mathbb{F}_2^k[[x]] \longrightarrow \mathbb{F}_2^n[[x]]$ between two groups of formal power series with vector coefficients (of length k and n, respectively), where T will have a rather specific form:

$$\boxed{c(x) = T(a(x)) = G(x)a(x)}$$

$G(x)$ is a (fixed) polynomial with *matrix coefficients*: $G(x) = G_0 + G_1 x + G_2 x^2 + \cdots + G_m x^m$ with $G_j \in \mathbb{F}_2^{n \times k}[x]$, $0 \leq j \leq m$.

$\Bigg[$ We insist: The coefficients of $G(x)$ are matrices over $\mathbb{F}_2 = \{0, 1\}$, consisting of n lines and k columns. They can be multiplied with the information frames, thus producing vectors in the format of the code frames. The variable x commutes with everything in order to guarantee that the result of the operation becomes an element of $\mathbb{F}_2^n[[x]]$. $\Bigg]$

Formally, $G(x)$ occupies at the same time the place of the *generator polynomial* of a polynomial code and of the *impulse response* of a digital filter.

Remark *If $k = 1$ (the information frames are simply bits), then $G(x) = T(1) = T(1 + 0x + 0x^2 + 0x^3 + \cdots)$*

If $k > 1$, $G(x)$ does not allow an interpretation as a code bitstream; the notion of impulse response is then not really adapted.

Examples *(1)* $k = 1$, $n = 3$.
Choose

$$G(x) = \begin{pmatrix} 1 \\ 1 \\ 1 \end{pmatrix} + \begin{pmatrix} 1 \\ 0 \\ 1 \end{pmatrix} x + \begin{pmatrix} 0 \\ 1 \\ 1 \end{pmatrix} x^2.$$

The information bitstream $a_0 a_1 a_2 \cdots a_t \cdots$ *is transformed into the code bitstream*

$$c_{10} \; c_1 1 \; c_1 2 \; c_1 3 \; \ldots \; c_{1t} \; \ldots$$
$$c_{20} \; c_2 1 \; c_2 2 \; c_2 3 \; \ldots \; c_{2t} \; \ldots$$
$$c_{30} \; c_3 1 \; c_3 2 \; c_3 3 \; \ldots \; c_{3t} \; \ldots$$

in the following way: Write $a(x) = a_0 + a_1 x + a_2 x^2 + \cdots$
$c(x) = \mathbf{c}_0 + \mathbf{c}_1 x + \mathbf{c}_2 x^2 + \cdots$
then $\mathbf{c}(x) = G(x)a(x) = a(x)G(x)$ *is computed, frame by frame, in the following way (Cauchy formula for the coefficients of a product of formal power series):*

$$\mathbf{c}_t = a_t \mathbf{g}_0 + a_{t-1} \mathbf{g}_1 + a_{t-2} \mathbf{g}_2 = a_t \begin{pmatrix} 1 \\ 1 \\ 1 \end{pmatrix} + a_{t-1} \begin{pmatrix} 1 \\ 0 \\ 1 \end{pmatrix} + a_{t-2} \begin{pmatrix} 0 \\ 1 \\ 1 \end{pmatrix}$$

This gives in componentwise notation:

$$c_{1t} = a_t + a_{t-1}$$
$$c_{2t} = a_t + a_{t-2}$$
$$c_{3t} = a_t + a_{t-1} + a_{t-2}$$

Consider, for example, the information bitstream
$a_0 a_1 a_2 a_3 \cdots \equiv 10101010 \cdots$
We obtain the following code bitstream:
$$1 \; 1 \; 1 \; 1 \; 1 \; 1 \cdots$$
$\mathbf{c}_0 \mathbf{c}_1 \mathbf{c}_2 \mathbf{c}_3 \cdots \equiv 1 \; 0 \; 0 \; 0 \; 0 \; 0 \cdots$
$$1 \; 1 \; 0 \; 1 \; 0 \; 1 \cdots$$

(2) $k = 2$, $n = 3$.
Choose

$$G(x) = \begin{pmatrix} 1 & 0 \\ 0 & 1 \\ 1 & 1 \end{pmatrix} + \begin{pmatrix} 1 & 1 \\ 1 & 0 \\ 1 & 0 \end{pmatrix} x + \begin{pmatrix} 0 & 1 \\ 1 & 1 \\ 0 & 1 \end{pmatrix} x^2.$$

The beginning of our information bitstream will be

$$a_0 a_1 a_2 a_3 \cdots \equiv \begin{matrix} 1 \; 1 \; 0 \; 0 \; 1 \; 1 \cdots \\ 0 \; 1 \; 1 \; 0 \; 0 \; 0 \cdots \end{matrix}$$

i.e.

$$a(x) = \begin{pmatrix} 1 \\ 0 \end{pmatrix} + \begin{pmatrix} 1 \\ 1 \end{pmatrix} x + \begin{pmatrix} 0 \\ 1 \end{pmatrix} x^2 + \begin{pmatrix} 0 \\ 0 \end{pmatrix} x^3 + \begin{pmatrix} 1 \\ 0 \end{pmatrix} x^4 + \begin{pmatrix} 1 \\ 0 \end{pmatrix} x^5 + \cdots$$

The frames of the code bitstream $\mathbf{c}(x) = G(x)a(x) = \mathbf{c}_0 + \mathbf{c}_1 x + \mathbf{c}_2 x^2 + \mathbf{c}_3 x^3 + \cdots$
are computed by means of the convolution relation

$$\mathbf{c}_t = G_0 a_t + G_1 a_{t-1} + G_2 a_{t-2} = \begin{pmatrix} 1 & 0 \\ 0 & 1 \\ 1 & 1 \end{pmatrix} a_t + \begin{pmatrix} 1 & 1 \\ 1 & 0 \\ 1 & 0 \end{pmatrix} a_{t-1} + \begin{pmatrix} 0 & 1 \\ 1 & 1 \\ 0 & 1 \end{pmatrix} a_{t-2}.$$

So, we obtain

$$\mathbf{c}_0 = \begin{pmatrix} 1 & 0 \\ 0 & 1 \\ 1 & 1 \end{pmatrix}\begin{pmatrix} 1 \\ 0 \end{pmatrix} = \begin{pmatrix} 1 \\ 0 \\ 1 \end{pmatrix},$$

$$\mathbf{c}_1 = \begin{pmatrix} 1 & 0 \\ 0 & 1 \\ 1 & 1 \end{pmatrix}\begin{pmatrix} 1 \\ 1 \end{pmatrix} + \begin{pmatrix} 1 & 1 \\ 1 & 0 \\ 1 & 0 \end{pmatrix}\begin{pmatrix} 1 \\ 0 \end{pmatrix} = \begin{pmatrix} 0 \\ 0 \\ 1 \end{pmatrix},$$

$$\mathbf{c}_2 = \begin{pmatrix} 1 & 0 \\ 0 & 1 \\ 1 & 1 \end{pmatrix}\begin{pmatrix} 0 \\ 1 \end{pmatrix} + \begin{pmatrix} 1 & 1 \\ 1 & 0 \\ 1 & 0 \end{pmatrix}\begin{pmatrix} 1 \\ 1 \end{pmatrix} + \begin{pmatrix} 0 & 1 \\ 1 & 1 \\ 0 & 1 \end{pmatrix}\begin{pmatrix} 1 \\ 0 \end{pmatrix} = \begin{pmatrix} 0 \\ 1 \\ 0 \end{pmatrix}$$

and so on ...
Our code bitstream will then be:

$$\mathbf{c}_0\mathbf{c}_1\mathbf{c}_2\mathbf{c}_3\mathbf{c}_4 \cdots = \begin{matrix} 1\,0\,0\,0\,0 \cdots \\ 0\,0\,1\,0\,1 \cdots \\ 1\,1\,0\,1\,0 \cdots \end{matrix}$$

Attention When applying the formulas for the first values of t, we suppose tacitly that every a_t, for $t < 0$, is zero.

The $(n, 1)$ convolutional codes: Some polynomial algebra

Recall: A $(n, 1)$ convolutional code is obtained by means of a generator polynomial

$$G(x) = \mathbf{g}_0 + \mathbf{g}_1 x + \mathbf{g}_2 x^2 + \cdots + \mathbf{g}_m x^m,$$

which transforms (by polynomial multiplication) information bitstreams, presented as binary formal power series (a frame = one bit) into code bitstreams, presented as formal power series with vector coefficients (a frame = n bits).

Write $G(x)$ more explicitly:

$$G(x) = \begin{pmatrix} g_{10} \\ g_{20} \\ \vdots \\ g_{n0} \end{pmatrix} + \begin{pmatrix} g_{11} \\ g_{21} \\ \vdots \\ g_{n1} \end{pmatrix} x + \cdots + \begin{pmatrix} g_{1m} \\ g_{2m} \\ \vdots \\ g_{nm} \end{pmatrix} x^m = \begin{pmatrix} g_1(x) \\ g_2(x) \\ \vdots \\ g_n(x) \end{pmatrix}$$

with

$$g_1(x) = g_{10} + g_{11}x + g_{12}x^2 + \cdots + g_{1m}x^m,$$
$$g_2(x) = g_{20} + g_{21}x + g_{22}x^2 + \cdots + g_{2m}x^m,$$
$$\vdots$$
$$g_n(x) = g_{n0} + g_{n1}x + g_{n2}x^2 + \cdots + g_{nm}x^m.$$

Thus we have:

$$\mathbf{c}(x) = a(x)G(x) = \begin{pmatrix} a(x)g_1(x) \\ a(x)g_2(x) \\ \vdots \\ a(x)g_n(x) \end{pmatrix}.$$

A $(n, 1)$ convolutional code is defined by n usual generator polynomials which act *simultaneously* (by binary polynomial multiplication) on *the same* information bitstream (written as a binary formal power series).

Example

$$G(x) = \begin{pmatrix} 1 \\ 1 \\ 1 \end{pmatrix} + \begin{pmatrix} 1 \\ 0 \\ 1 \end{pmatrix} x + \begin{pmatrix} 0 \\ 1 \\ 1 \end{pmatrix} x^2 = \begin{pmatrix} g_1(x) \\ g_2(x) \\ g_3(x) \end{pmatrix} = \begin{pmatrix} 1 + x \\ 1 + x^2 \\ 1 + x + x^2 \end{pmatrix}.$$

Then let $a(x) = a_0 + a_1 x + a_2 x^2 + \cdots + a_t x^t + \cdots$ be the information bitstream, in formal power series notation. The first bit of every code frame is obtained by the convolution $a(x)(1 + x)$.

The second bit of every code frame is obtained by the convolution $a(x)(1 + x^2)$. The third bit of every code frame is obtained by the convolution $a(x)(1 + x + x^2)$.

The ensuing natural question is the following: Are there algebraic relations between the polynomials $g_1(x), g_2(x), \ldots, g_n(x)$ which have an influence on the quality of the convolutional coding? The answer is yes.

Before giving details, let us recall some simple facts:

A convolutional code (i.e. the collection of its words) is *linear* : The sum of two code words is a code word (addition of binary sequences: position by position – across the frames). This stems from the fact that the multiplication by $G(x)$ is a linear operation on the information sequences. A *transmission error* is formalized as in the case of polynomial codes:

Then let

$\mathbf{v} = \mathbf{v}_0 \mathbf{v}_1 \mathbf{v}_2 \mathbf{v}_3 \cdots \equiv$ the stream of received frames

$\mathbf{c} = \mathbf{c}_0 \mathbf{c}_1 \mathbf{c}_2 \mathbf{c}_3 \cdots \equiv$ the transmitted code bitstream (as produced by the encoder)

$\mathbf{e} = \mathbf{e}_0 \mathbf{e}_1 \mathbf{e}_2 \mathbf{e}_3 \cdots \equiv$ the error bitstream (written as a sequence of frames)

$$\mathbf{e}_t = \begin{pmatrix} e_{1t} \\ e_{2t} \\ \vdots \\ e_{nt} \end{pmatrix} \qquad e_{kt} = \begin{cases} 0 & \text{this position has not been affected} \\ 1 & \text{this position is in error} \end{cases}$$

the code frame $\mathbf{c}_t$ has undergone an error of the type $\mathbf{e}_t$

Notation in formal power series: $\boxed{\mathbf{v}(x) = \mathbf{c}(x) + \mathbf{e}(x)}$

Look now at interesting algebraic relations between the n binary polynomials $g_1(x), g_2(x), \ldots, g_n(x)$.

Example Consider

$$G(x) = \begin{pmatrix} g_1(x) \\ g_2(x) \end{pmatrix} = \begin{pmatrix} 1 + x^2 \\ x + x^2 \end{pmatrix},$$

$$pgcd(g_1(x), g_2(x)) = 1 + x.$$

Let $u = 11111111 \cdots$ be the constant sequence which gives the formal power series: $u(x) = \sum x^t = 1 + x + x^2 + x^3 + \cdots$ $u(x) = \frac{1}{1+x}$ as a formal power series over $\mathbb{F}_2 = \{0, 1\}$.[2]

Hence: $u(x)G(x) = \begin{pmatrix} u(x)g_1(x) \\ u(x)g_2(x) \end{pmatrix} = \begin{pmatrix} 1+x \\ x \end{pmatrix}$

In other words: Our convolutional encoder transforms $u = 11111111 \ldots$ into

$$\begin{pmatrix} 1\ 1\ 0\ 0\ 0 \ldots \\ 0\ 1\ 0\ 0\ 0 \ldots \end{pmatrix}$$

The consequence: If after a transmission only the first bit of $\mathbf{c}_0$ and the two bits of $\mathbf{c}_1$ are in error (i.e. only three bits are in error), then this corresponds altogether to a total bit-inversion affecting *all* positions of the information bitstream. Such a convolutional encoder is called *catastrophic*.

A non-catastrophic convolutional code will thus be characterized by the following coprimality condition: $pgcd(g_1(x), g_2(x), \ldots, g_n(x)) = 1$.

Example

$$G(x) = \begin{pmatrix} 1+x \\ 1+x^2 \\ 1+x+x^2 \end{pmatrix}$$

defines a non-catastrophic convolutional code (since $g_3(x) = 1+x+x^2$ is irreducible over $\mathbb{F}_2$).

How can we make a clever use of the foregoing coprimality condition?

The answer comes from an algebraic result which is already familiar to us: recall the question of multiplicative inversion in the rings of residues.

Lemma $pgcd(g_1(x), g_2(x), \ldots, g_n(x)) = 1$ $\iff$ *There exist polynomials* $b_1(x), b_2(x), \ldots, b_n(x) \in \mathbb{F}_2[x]$ *with* $b_1(x)g_1(x) + b_2(x)g_2(x) + \cdots + b_n(x)g_n(x) = 1$.

In the case $n = 2$ the two polynomials $b_1(x)$ and $b_2(x)$ are obtained by the Euclidean algorithm which yields the $gcd(g_1(x), g_2(x))$; the case $n > 2$ is easily reducible – by an evident recursion scheme – to the case $n = 2$.

Example

$$g_1(x) = 1+x, \quad g_2(x) = 1+x^2, \quad g_3(x) = 1+x+x^2.$$

Here, we have $g_1(x) + g_2(x) + g_3(x) = 1$. Thus, $b_1(x) = b_2(x) = b_3(x) = 1$.

All this allows an easy *reconstruction* of the information bitstream in function of the code bitstream: Write

$$\mathbf{c}(x) = \begin{pmatrix} c_1(x) \\ c_2(x) \\ \vdots \\ c_n(x) \end{pmatrix},$$

where

the formal power series which describes the
$c_j(x) \equiv$ bitstream of the jth position of the code
frames, across the frames $1 \leq j \leq n$

---

[2] i.e. $u(x)$ and $v(x) = 1 + x$ are multiplicative inverses.

$$c_1(x) = a(x)g_1(x),$$
$$c_2(x) = a(x)g_2(x),$$
$$\vdots$$
$$c_n(x) = a(x)g_n(x).$$

Then $a(x) = b_1(x)c_1(x) + b_2(x)c_2(x) + \cdots + b_n(x)c_n(x)$.

In our foregoing *example* we get immediately: The information bitstream is obtained from the code bitstream as follows: compute, for every code frame, the sum of the frame-bits (in XOR, of course); this gives the corresponding information bit!

Attention *Our arguments are "error free" – we assume a correct transmission.*

Exercises

(1) Let $\mathcal{C}$ be the convolutional code defined by

$$G(x) = \begin{pmatrix} 1+x \\ 1+x^2 \\ 1+x+x^2 \end{pmatrix}.$$

You receive the following stream of frames:

$$\mathbf{v} = \mathbf{v_0 v_1 v_2 v_3 v_4} \cdots = \begin{pmatrix} 1\,0\,0\,1\,0\ldots \\ 0\,1\,0\,1\,1\ldots \\ 1\,0\,1\,0\,1\ldots \end{pmatrix}.$$

There must be a transmission error. Why?

(2) Which of the following convolutional codes are catastrophic?

(a) $g_1(x) = x^2$, $\qquad\qquad g_2(x) = 1+x+x^3$,
(b) $g_1(x) = 1+x^2+x^4$, $\qquad g_2(x) = 1+x+x^3+x^4$,
(c) $g_1(x) = 1+x+x^2+x^4$, $\quad g_2(x) = 1+x^3+x^4$,
(d) $g_1(x) = 1+x^4+x^5+x^6$, $\quad g_2(x) = 1+x+x^3+x^5$,

(3) Let $\mathcal{C}$ be the convolutional code defined by

$$G(x) = \begin{pmatrix} g_1(x) \\ g_2(x) \end{pmatrix} = \begin{pmatrix} 1+x^2 \\ 1+x+x^2 \end{pmatrix}.$$

(a) Find $b_1(x), b_2(x)$ with $b_1(x)g_1(x) + b_2(x)g_2(x) = 1$.
(b) Recover the six first information bits $a_0a_1a_2a_3a_4a_5$ from

$$\mathbf{c} = \begin{pmatrix} 1\,1\,1\,1\,1\,1\ldots \\ 1\,0\,0\,1\,1\,0\ldots \end{pmatrix}.$$

(4) Let $\mathcal{C}$ be the convolutional code defined by

$$G(x) = \begin{pmatrix} g_1(x) \\ g_2(x) \\ g_3(x) \end{pmatrix} = \begin{pmatrix} 1+x^2+x^3 \\ 1+x+x^3 \\ 1+x+x^2+x^3 \end{pmatrix}.$$

(a) Find $b_1(x), b_2(x), b_3(x)$ with $b_1(x)g_1(x) + b_2(x)g_2(x) + b_3(x)g_3(x) = 1$.

(b) Recover the six first information bits $a_0a_1a_2a_3a_4a_5$ from

$$c = c_0c_1c_2c_3c_4c_5\cdots = \begin{pmatrix} 1\,0\,0\,0\,0\,0\ldots \\ 1\,1\,1\,1\,0\,0\ldots \\ 1\,1\,0\,1\,1\,1\ldots \end{pmatrix}.$$

Towards Error-Detection and Error-Correction: The Minimum Distance d^* Between the Words of a Convolutional Code

Formally, an $(n,1)$ convolutional code $\mathcal{C}$ is a set of sequences $c = c_0c_1c_2c_3\ldots$ over the alphabet of the 2^n binary frames of length n.

The generation of $\mathcal{C}$ by means of an encoder of the form $c(x) = a(x)G(x)$ (language of formal power series) has two consequences:

---

(1) $\mathcal{C}$ is linear, i.e. it is stable by addition (of infinite binary matrices with n lines: we act via Boolean XOR, binary position by binary position).

(2) $\mathcal{C}$ is stable under "annihilation of leading zero frames" :
if $c = c_0c_1c_2c_3\cdots \in \mathcal{C}$ and if c_0 is the zero vector, then $c' = c_1c_2c_3\cdots \in \mathcal{C}$.
(note that – conceptually – this is merely a time-unit shift ...)

The argument which gives (2) is the following:

$$G(x) = g_0 + g_1x + g_2x^2 + \cdots + g_mx^m$$

and g_0 is not the zero vector.
Hence, if $c(x) = a(x)G(x)$, then

$$c_0 = \begin{pmatrix} 0 \\ 0 \\ \vdots \\ 0 \end{pmatrix} \quad\Longleftrightarrow\quad a_0 = 0.$$

We get:
$c_1 = a_1g_0,$
$c_2 = a_2g_0 + a_1g_1,$
$c_3 = a_3g_0 + a_2g_1 + a_1g_2.$
Thus, $c'(x) = c_1 + c_2x + c_3x^2 + \cdots = (a_1 + a_2x + \ldots)G(x) \in \mathcal{C}.$ $\square$

It is because of property (2) that we will often assume, in primarily theoretical considerations, that the code words in question should begin with a non-zero frame.

As in the case of ordinary polynomial codes, the encoder – given by the multiplicative action of the generator polynomial $G(x)$ on the various information bit-streams – must create *distances* between the code words: In order to make sure that a (decent) transmission error can be *detected* (and maybe even corrected), it must above all be *impossible* that such a transmission error transforms a correct word (the transmitted code word) into a different *correct* word (the received word). The correct words (the code words) have to be "sufficiently distant" of one another. This explains the interest of the *minimum distance* between the code words.

In the situation of the convolutional codes, we are confronted with words of (theoretically) *infinite length*. This does not change our usual arguments: The distance between two words (i.e. the number of binary positions where they are distinct) is precisely the *weight* ($\equiv$ the number of non-zero binary positions) of the word which is their *sum*. A convolutional code being linear, the *minimum distance* $\mathbf{d}_\infty$ between its words is nothing but the *minimum weight* of its non-zero words (beginning with a non-zero frame).

Let now $G(x) = \mathbf{g}_0 + \mathbf{g}_1 x + \mathbf{g}_2 x^2 + \cdots + \mathbf{g}_m x^m$ be the generator polynomial of the considered convolutional code $\mathcal{C}$. $G(x) \in \mathcal{C}$, i.e. the word $\mathbf{G} = \mathbf{g}_0 \mathbf{g}_1 \mathbf{g}_2 \cdots \mathbf{g}_m 000 \cdots$ is a code word in $\mathcal{C}$.

The weight of $\mathbf{G}$: The number of non-zero positions of the (vector) coefficients of $G(x)$. Hence: $\mathbf{d}_\infty \le$ the weight of $\mathbf{G}$. For the decoding arguments (error-detection and error-correction) we need another notion of minimum distance. It is clear that a practicable decoding algorithm only can work by means of an *inspection window (of fixed finite length)* moving over the bitstream of received frames, which, itself, will be – at least in principle – unlimited.

We guess that we should be able to correct "dispersed" transmission errors of the following kind: The number of binary errors per *inspectable sector* does not exceed a certain threshold. We shall see in a moment that here the "natural" minimum distance between the code words will be the following:

$$\mathbf{d}^* \;=\; \min\{d_{m+1}(\mathbf{c},\mathbf{c}') \;:\; \mathbf{c},\mathbf{c}' \in \mathcal{C}, \mathbf{c}_0 \neq \mathbf{c}'_0\} \quad \boxed{m = \deg G(x)} \quad \text{where}$$

$d_{m+1}(\mathbf{c},\mathbf{c}') =$ the number of binary positions where the two initial segments $\mathbf{c}_0 \mathbf{c}_1 \mathbf{c}_2 \ldots \mathbf{c}_m$ and $\mathbf{c}'_0 \mathbf{c}'_1 \mathbf{c}'_2 \ldots \mathbf{c}'_m$ of $\mathbf{c}$ and $\mathbf{c}'$ are distinct

$\mathbf{d}^*$ is then the minimum weight of the $m+1$ first frames $\mathbf{c}_0 \mathbf{c}_1 \mathbf{c}_2 \ldots \mathbf{c}_m$ for all code words $\mathbf{c}$ beginning with a non-zero frame.

Since $\mathbf{G} = \mathbf{g}_0 \mathbf{g}_1 \mathbf{g}_2 \ldots \mathbf{g}_m$ (the coefficients of $G(x)$) is in the competition, we will have $\mathbf{d}^* \le$ the weight of $\mathbf{G} = \mathbf{g}_0 \mathbf{g}_1 \mathbf{g}_2 \ldots \mathbf{g}_m$.

Example

$$G(x) = \begin{pmatrix} 1+x \\ 1+x^2 \\ 1+x+x^2 \end{pmatrix} = \begin{pmatrix} 1 \\ 1 \\ 1 \end{pmatrix} + \begin{pmatrix} 1 \\ 0 \\ 1 \end{pmatrix} x + \begin{pmatrix} 0 \\ 1 \\ 1 \end{pmatrix} x^2.$$

The weight of $\mathbf{G} = \mathbf{g}_0 \mathbf{g}_1 \mathbf{g}_2 = \begin{pmatrix} 1\,1\,0 \\ 1\,0\,1 \\ 1\,1\,1 \end{pmatrix}$ is equal to 7.

We search for $\mathbf{d}^*$ and $\mathbf{d}_\infty$.

We know already: $\mathbf{d}^* \le \mathbf{d}_\infty \le 7$. In order to find $\mathbf{d}^*$, we need to consider all "prefixes" of three frames ($m = 2$) for code words which begin with a non-zero frame (i.e. with

$$\mathbf{g}_0 = \begin{pmatrix} 1 \\ 1, \\ 1, \end{pmatrix},$$

which is the only possible non-zero beginning for a code word).

So, let us compute $\mathbf{c} = (1 + a_1 x + a_2 x^2) G(x)$ for $a_1 a_2 = 00, 01, 10$ and 11.

(00) $\mathbf{c} = \begin{pmatrix} 1\,1\,0\,0\,0\,0\ldots \\ 1\,0\,1\,0\,0\,0\ldots \\ 1\,1\,1\,0\,0\,0\ldots \end{pmatrix}$ the weight of $\mathbf{c}$: 7 ;   the 3-weight of $\mathbf{c}$: 7 .

(01) $\mathbf{c} = \begin{pmatrix} 1\,1\,1\,1\,0\,0\,... \\ 1\,0\,0\,0\,1\,0\,... \\ 1\,1\,0\,1\,1\,0\,... \end{pmatrix}$ the weight of $\mathbf{c}$: 10; the 3-weight of $\mathbf{c}$: 6.

(10) $\mathbf{c} = \begin{pmatrix} 1\,0\,1\,0\,0\,0\,... \\ 1\,1\,1\,1\,0\,0\,... \\ 1\,0\,0\,1\,0\,0\,... \end{pmatrix}$ the weight of $\mathbf{c}$: 8; the 3-weight of $\mathbf{c}$: 6 .

(11) $\mathbf{c} = \begin{pmatrix} 1\,0\,0\,1\,0\,0\,... \\ 1\,1\,0\,1\,1\,0\,... \\ 1\,0\,1\,0\,1\,0\,... \end{pmatrix}$ the weight of $\mathbf{c}$: 9; the 3-weight of $\mathbf{c}$: 5.

We see: $\mathbf{d}^* = 5$ and $\mathbf{d}_\infty = 7$.

(Note that we have necessarily: $\mathbf{d}_\infty \geq \mathbf{d}^* +$ the weight of the highest coefficient of $G(\mathbf{x})$.)

Exercises

(1) Find $\mathbf{d}^*$ and $\mathbf{d}_\infty$ for the convolutional code given by the generator polynomial

$$G(x) = \begin{pmatrix} 1 + x + x^3 \\ 1 + x + x^2 + x^3 \end{pmatrix} .$$

(2) Find $\mathbf{d}^*$ and $\mathbf{d}_\infty$ for the convolutional code given by the generator polynomial

$$G(x) = \begin{pmatrix} 1 + x^2 \\ 1 + x + x^2 \\ 1 + x + x^2 \end{pmatrix}.$$

(Here, $g_2(x) = g_3(x)$; this is permitted and usual...).

(3) Find $\mathbf{d}^*$ and $\mathbf{d}_\infty$ for the convolutional code given by the generator polynomial

$$G(x) = \begin{pmatrix} 1 + x^2 + x^3 \\ 1 + x + x^3 \\ 1 + x + x^3 \\ 1 + x + x^2 + x^3 \end{pmatrix}.$$

(4) Once more the code of the preceding exercise:

$$G(x) = \begin{pmatrix} 1 \\ 1 \\ 1 \\ 1 \end{pmatrix} + \begin{pmatrix} 0 \\ 1 \\ 1 \\ 1 \end{pmatrix} x + \begin{pmatrix} 1 \\ 0 \\ 0 \\ 1 \end{pmatrix} x^2 + \begin{pmatrix} 1 \\ 1 \\ 1 \\ 1 \end{pmatrix} x^3.$$

Preliminary observation: The first frame $\mathbf{c}_0$ of a code word $c(x)$ is either the word zero or $\mathbf{g}_0$, the constant term of $G(x)$, according to the alternative $\begin{cases} 0 \\ 1 \end{cases}$ for the initial bit a_0 of the information word $a(x)$.

Moreover, $(1 + a_1 x + a_2 x^2 + ...)G(x) = G(x) + (a_1 x + a_2 x^2 + ...)G(x)$. Taking this observation into account, find $\quad a_0 a_1 a_2 a_3 a_4 a_5$ – the first six bits of the information bitstream – which give the following code bitstream:

$$\mathbf{c} = \mathbf{c}_0 \mathbf{c}_1 \mathbf{c}_2 \mathbf{c}_3 \mathbf{c}_4 \mathbf{c}_5 \mathbf{c}_6 \mathbf{c}_7 \mathbf{c}_8 \cdots = \begin{matrix} 1\,1\,1\,1\,0\,0\,1\,1\,0\,... \\ 1\,0\,1\,1\,1\,0\,0\,1\,1\,... \\ 1\,0\,1\,1\,1\,0\,0\,1\,1\,... \\ 1\,0\,0\,0\,0\,0\,0\,0\,0\,... \end{matrix}$$

(5) (Refrain from too optimistic decoding). Consider the convolutional code, given by

$$G(x) = \begin{pmatrix} 1+x \\ 1+x^2 \\ 1+x+x^2 \end{pmatrix} = \begin{pmatrix} 1 \\ 1 \\ 1 \end{pmatrix} + \begin{pmatrix} 1 \\ 0 \\ 1 \end{pmatrix} x + \begin{pmatrix} 0 \\ 1 \\ 1 \end{pmatrix} x^2.$$

[We know: The first frame c_0 of an error free code bitstream c is either $\begin{pmatrix} 0 \\ 0 \\ 0 \end{pmatrix}$ (for $a_0 = 0$) or $\begin{pmatrix} 1 \\ 1 \\ 1 \end{pmatrix}$ (for $a_0 = 1$).]

We receive:

$$\mathbf{v} = \begin{matrix} 0\,0\,0\,1\,0\,0\,0\ldots \\ 0\,1\,0\,1\,1\,0\,0\ldots \\ 0\,0\,1\,0\,1\,0\,0\ldots \end{matrix}$$

Are we allowed to conclude that the information bitstream begins with $a_0 = 0$?

(a) Compute $\mathbf{c}^{(1)}(x) = (1 + x + x^2)G(x)$ (corresponding to the information bitstream $a_0 a_1 a_2 \cdots = 11100 \cdots$) and determine $d(\mathbf{v}, \mathbf{c}^{(1)}) \equiv$ the number of binary positions where $\mathbf{v}$ and $\mathbf{c}^{(1)}$ are distinct.

(b) Show that $d(\mathbf{v}, \mathbf{c}) > d(\mathbf{v}, \mathbf{c}^{(1)})$ for every code word $\mathbf{c} \neq \mathbf{c}^{(1)}$.

The Down-to-Earth Decoding

Some Simple Formulas

Notation In the sequel, the *derivation sign* will be used to indicate a simple left shift (with annihilation of the constant term): $(a_0 + a_1 x + a_2 x^2 + a_3 x^3 + \ldots)' = a_1 + a_2 x + a_3 x^2 + \ldots$.

Settle in the following situation:

$$G(x) = \begin{pmatrix} g_1(x) \\ \vdots \\ g_n(x) \end{pmatrix} = \mathbf{g}_0 + \mathbf{g}_1 x + \mathbf{g}_2 x^2 + \ldots \mathbf{g}_m x^m$$

the generator polynomial of an $(n, 1)$ convolutional code.

The information bitstream $a_0 a_1 a_2 \ldots a_t \ldots$ is transformed into the code bitstream $\mathbf{c}_0 \mathbf{c}_1 \mathbf{c}_2 \ldots \mathbf{c}_t \ldots$ by "binary convolution": $\boxed{\mathbf{c}(x) = a(x)G(x)}$

where
$$a(x) = a_0 + a_1 x + a_2 x^2 + \cdots + a_t x^t + \ldots$$
$$\mathbf{c}(x) = \mathbf{c}_0 + \mathbf{c}_1 x + \mathbf{c}_2 x^2 + \cdots + \mathbf{c}_t x^t + \ldots$$

The code bitstream will pass through a channel (in the broad sense), which will give rise to a bitstream of frames eventually in error:

$$\mathbf{v}_0 \mathbf{v}_1 \mathbf{v}_2 \ldots \mathbf{v}_t \ldots$$

The transmission error is described by an *error bitstream*

$$\mathbf{e}_0 \mathbf{e}_1 \mathbf{e}_2 \ldots \mathbf{e}_t \ldots,$$

which simply adds (Boolean XOR, binary position by binary position) to the code bitstream.

With
$$\mathbf{v}(x) = \mathbf{v}_0 + \mathbf{v}_1 x + \mathbf{v}_2 x^2 + \cdots + \mathbf{v}_t x^t + \ldots$$

$$\mathbf{e}(x) = \mathbf{e}_0 + \mathbf{e}_1 x + \mathbf{e}_2 x^2 + \cdots + \mathbf{e}_t x^t + \ldots$$

thus we get:
$$\boxed{\mathbf{v}(x) = \mathbf{c}(x) + \mathbf{e}(x)}$$

Suppose that you have decoded the first bit a_0 by an algorithm which recovers the beginning of the information bitstream from (the beginning of) the stream of the received frames – under the condition that the number of transmission errors is decent, of course.

How to *reinitialize* the situation?

$$\mathbf{c}_1(x) = a'(x)G(x) = (a_1 + a_2 x + a_3 x^2 + \ldots)(\mathbf{g}_0 + \mathbf{g}_1 x + \mathbf{g}_2 x^2 + \cdots + \mathbf{g}_m x^m)$$

$$= \frac{1}{x}(\mathbf{c}_0(x) + a_0 \cdot G(x)) = \frac{1}{x}(\mathbf{v}(x) + \mathbf{e}(x) + a_0 \cdot G(x))$$

$$= \frac{1}{x}(\mathbf{v}_0 + x\mathbf{v}'(x) + \mathbf{e}_0 + x\mathbf{e}'(x) + a_0\mathbf{g}_0 + a_0 \cdot xG'(x)).$$

Now: $\mathbf{v}_0 = a_0\mathbf{g}_0 + \mathbf{e}_0$; hence $\mathbf{c}_1(x) = \mathbf{v}'(x) + \mathbf{e}'(x) + a_0 \cdot G'(x)$

Put
$$\boxed{\mathbf{v}_1(x) = \mathbf{v}'(x) + a_0 \cdot G'(x)}$$

then
$$\mathbf{c}_1(x) = \mathbf{v}_1(x) + \mathbf{e}'(x).$$

We insist:

$$a_0 = 0 : \mathbf{c}_1(x) = \mathbf{c}_0'(x) = \mathbf{v}'(x) + \mathbf{e}'(x),$$

$$a_0 = 1 : \mathbf{c}_1(x) = \{\mathbf{v}'(x) + G'(x)\} + \mathbf{e}'(x).$$

In the case where $a_0 = 1$, the bitstream of the received frames will thus be reinitialized to $|\mathbf{v}_1 + \mathbf{g}_1|\mathbf{v}_2 + \mathbf{g}_2| \ldots |\mathbf{v}_m + \mathbf{g}_m|\mathbf{v}_{m+1}|\mathbf{v}_{m+2}| \ldots$

Thus, in a step-by-step decoding, with iterated reinitialization, the sequence of the received frames will be always readjusted according to the "arithmetical radiation" of the first bit ($= 1$) over the block of the $m + 1$ leading frames. Of course, all this can easily be generalized.

Exercises

(1) Suppose that we have decoded – in a single step – the first four bits a_0, a_1, a_2, a_3 of the information bitstream. Consider the corresponding reinitialization. Let $c_4(x) = a^{(4)}(x)G(x) = (a_4 + a_5x + a_6x^2 + \ldots)(g_0 + g_1x + g_2x^2 + \cdots + g_mx^m)$. Show that $c_4(x) = v_4(x) + e^{(4)}(x)$, where $v_4(x) = \{v(x) + (a_0 + a_1x + a_2x^2 + a_3x^3)G(x)\}^{(4)}$.

(2) Still the situation of the preceding exercise.

Let thus $a_0a_1a_2a_3$ be the (decoded) prefix of an information bitstream such that the corresponding bitstream of received frames is $\mathbf{v} = v_0v_1v_2\ldots v_t\ldots$

Let $\mathbf{w} = w_0w_1w_2\ldots w_t\ldots$ be the reinitialized stream of received frames: $v_4(x) = w_0 + w_1x + w_2x^2 + w_3x^3 + \ldots$

These will be our fixed data. Consider now *all* information bitstreams with prefix $a_0a_1a_2a_3$. For such a sequence $a_0a_1a_2a_3\alpha_4\alpha_5\alpha_6\ldots$ and for $N \geq 5$ let $c_{(N-1)} = c_0c_1\ldots c_{N-1}$ be the segment of the N first frames of the code bitstream, and $\tilde{c}_{(N-5)} = \tilde{c}_0\tilde{c}_1\ldots\tilde{c}_{N-5}$ be the segment of the $N - 4$ first frames of the code bitstream associated with $\alpha_4\alpha_5\alpha_6\ldots: c_4(x) = (\alpha_4 + \alpha_5x + \alpha_6x^2 + \ldots)(g_0 + g_1x + g_2x^2 + \cdots + g_mx^m) = \tilde{c}_0 + \tilde{c}_1x + \tilde{c}_2x^2 + \ldots$

Show that then

$$d(\tilde{c}_{(N-5)}, \mathbf{w}_{(N-5)}) = d(c_{(N-1)}, \mathbf{v}_{(N-1)}) - - d(c_{(3)}, \mathbf{v}_{(3)})$$

(We note that the term $d(c_{(3)}, \mathbf{v}_{(3)})$ only depends on the common prefix $a_0a_1a_2a_3$, and is thus a constant of our problem ...)

The Decoding

Recall, once more, the formulary for our decoding situation:

So, let

$$G(x) = \begin{pmatrix} g_1(x) \\ \vdots \\ g_n(x) \end{pmatrix} = g_0 + g_1x + g_2x^2 + \cdots + g_mx^m$$

be the generator polynomial of an $(n, 1)$ convolutional code.

The information bitstream $a_0a_1a_2\ldots a_t\cdots$ will be transformed into the code bitstream $c_0c_1c_2\cdots c_t\cdots$ according to the familiar operation

$$c(x) = a(x)G(x),$$

$$a(x) = a_0 + a_1x + a_2x^2 + \cdots + a_tx^t + \cdots$$

where

$$c(x) = c_0 + c_1x + c_2x^2 + \cdots + c_tx^t + \ldots$$

The bitstream of received frames (eventually in error):

$$v_0v_1v_2\ldots v_t\cdots$$

The error bitstream $e_0e_1e_2\cdots e_t\cdots$ adds simply to the code bitstream. Put

$v(x) = v_0 + v_1x + v_2x^2 + \cdots + v_tx^t + \ldots$

$e(x) = e_0 + e_1x + e_2x^2 + \cdots + e_tx^t + \cdots$

which gives $v(x) = c(x) + e(x)$

The decoder has to reconstruct the correct information bitstream $a_0 a_1 a_2 \ldots a_t \ldots$ from the stream of received frames $\mathbf{v}_0 \mathbf{v}_1 \mathbf{v}_2 \ldots \mathbf{v}_t \ldots$ *under the condition* that the error bitstream $\mathbf{e}_0 \mathbf{e}_1 \mathbf{e}_2 \ldots \mathbf{e}_t \ldots$ is "locally light".

More precisely: Let $m \equiv$ the degree of $G(x)$, and let $\mathbf{d}^*$ be the minimum distance of the convolutional code. Make a window of $m + 1$ successive frames move along the error bitstream. If the number of binary errors over $m + 1$ successive frames never exceeds the threshold $\frac{1}{2}(\mathbf{d}^* - 1)$, then the decoding algorithm will correctly reconstruct the initial information bitstream. The *decoding* is done (in the worst case) *bit by bit*.

Compute, for every binary word $b_0 b_1 \ldots b_m$ of length $m + 1$, the $m + 1$ first frames of its code word: $\mathbf{c}(b_0 b_1 \ldots b_m)_{(m)} = \bar{\mathbf{c}}_0 \bar{\mathbf{c}}_1 \ldots \bar{\mathbf{c}}_m$ then compute, in each case, the distance towards the $m + 1$ first frames of the received word : $\bar{d} = d(\bar{\mathbf{c}}_0 \bar{\mathbf{c}}_1 \ldots \bar{\mathbf{c}}_m, \mathbf{v}_0 \mathbf{v}_1 \ldots \mathbf{v}_m)$.

According to our hypothesis (on the number of binary errors per inspection window of length $m + 1$), the correct prefix $a_0 a_1 \ldots a_m$ of the current information bitstream will give $d = d(\mathbf{c}_0 \mathbf{c}_1 \ldots \mathbf{c}_m, \mathbf{v}_0 \mathbf{v}_1 \ldots \mathbf{v}_m) \leq \frac{1}{2}(\mathbf{d}^* - 1)$. Let now $a_0' a_1' \ldots a_m'$ be another segment of length $m + 1$ such that $d' = d(\mathbf{c}_0' \mathbf{c}_1' \ldots \mathbf{c}_m', \mathbf{v}_0 \mathbf{v}_1 \ldots \mathbf{v}_m) \leq \frac{1}{2}(\mathbf{d}^* - 1)$. Then $d(\mathbf{c}_0' \mathbf{c}_1' \ldots \mathbf{c}_m', \mathbf{c}_0 \mathbf{c}_1 \ldots \mathbf{c}_m) \leq \mathbf{d}^* - 1$. This means $\mathbf{c}_0' = \mathbf{c}_0$, i.e. $a_0' = a_0$ (g_0 will always be the constant vector of n values 1).

Let us sum up: If the transmission errors are weak in the sense that the number of binary errors over $m+1$ successive received frames never exceeds the value $\frac{1}{2}(\mathbf{d}^* - 1)$, then every binary word $a_0' a_1' \ldots a_m'$ with $d' = d(\mathbf{c}_0' \mathbf{c}_1' \ldots \mathbf{c}_m', \mathbf{v}_0 \mathbf{v}_1 \ldots \mathbf{v}_m) \leq \frac{1}{2}(\mathbf{d}^* - 1)$ begins with the first bit of the correct information bitstream: $a_0' = a_0$. One saves the first bit which has been identified this way, and one reinitializes for the next step.

Remark Whenever there is only *one single* word $a_0 a_1 \ldots a_m$ such that $d = d(\mathbf{c}_0 \mathbf{c}_1 \ldots \mathbf{c}_m, \mathbf{v}_0 \mathbf{v}_1 \ldots \mathbf{v}_m) \leq \frac{1}{2}(\mathbf{d}^* - 1)$, then it is the right beginning of the information bitstream, and we can save the entire block of the $m + 1$ bits, then reinitialize according to the method discussed above.

We shall see examples in the following exercises.

Exercises

(1) Consider $G(x) = \begin{pmatrix} 1 + x^2 \\ 1 + x + x^2 \end{pmatrix} = \begin{pmatrix} 1 \\ 1 \end{pmatrix} + \begin{pmatrix} 0 \\ 1 \end{pmatrix} x + \begin{pmatrix} 1 \\ 1 \end{pmatrix} x^2$

   (a) Show that $\mathbf{d}^* = 3$. We can thus decode correctly provided there is maximally one single binary error over three successive received frames.

   (b) Consider the first 16 frames of the received bitstream :

$\mathbf{v}_0$	$\mathbf{v}_1$	$\mathbf{v}_2$	$\mathbf{v}_3$	$\mathbf{v}_4$	$\mathbf{v}_5$	$\mathbf{v}_6$	$\mathbf{v}_7$	$\mathbf{v}_8$	$\mathbf{v}_9$	$\mathbf{v}_{10}$	$\mathbf{v}_{11}$	$\mathbf{v}_{12}$	$\mathbf{v}_{13}$	$\mathbf{v}_{14}$	$\mathbf{v}_{15}$
0	1	1	1	1	1	1	0	1	1	1	1	1	0	1	1
1	0	0	1	0	0	0	1	1	0	1	1	1	0	0	1

   Find the first 12 information bits $a_0 a_1 a_2 \ldots a_{11}$.

(2)   Let $G(x) = \begin{pmatrix} 1 + x^2 + x^3 \\ 1 + x + x^3 \\ 1 + x + x^3 \\ 1 + x + x^2 + x^3 \end{pmatrix} = \begin{pmatrix} 1 \\ 1 \\ 1 \\ 1 \end{pmatrix} + \begin{pmatrix} 0 \\ 1 \\ 1 \\ 1 \end{pmatrix} x + \begin{pmatrix} 1 \\ 0 \\ 0 \\ 1 \end{pmatrix} x^2 + \begin{pmatrix} 1 \\ 1 \\ 1 \\ 1 \end{pmatrix} x^3$

(a) Show that $\mathbf{d}^* = 7$.

We will decode correctly if there are not more than three binary errors over four successive received frames.

(b) Look at the first 20 frames of the received bitstream :

$$\mathbf{v}_0\ \mathbf{v}_1\ \mathbf{v}_2\ \mathbf{v}_3\ \mathbf{v}_4\ \mathbf{v}_5\ \mathbf{v}_6\ \mathbf{v}_7\ \mathbf{v}_8\ \mathbf{v}_9\ \mathbf{v}_{10}\ \mathbf{v}_{11}\ \mathbf{v}_{12}\ \mathbf{v}_{13}\ \mathbf{v}_{14}\ \mathbf{v}_{15}\ \mathbf{v}_{16}\ \mathbf{v}_{17}\ \mathbf{v}_{18}\ \mathbf{v}_{19}$$

```
1 0 0 0 1 0 0 1 1 0 0 1 1 1 0 1 0 0 1 0
0 1 1 1 1 1 0 0 0 1 1 1 1 0 0 1 0 1 1 0
0 1 1 1 1 1 0 0 0 0 0 1 1 1 0 1 0 1 1 0
0 1 0 1 1 1 1 0 1 1 0 1 1 1 0 0 1 0 1 0
```

Find the first 12 information bits $a_0a_1a_2\ldots a_{11}$.

(3) In the two preceding exercises, the decoding can be done in blocks of $m+1$ bits (i.e. in blocks of 3 and 4 bits, respectively). This is always possible for these two codes. The reason?

(4) Consider, once more, the convolutional code of exercise (2). Let now the stream of the received frames be the following:

$$\mathbf{v}_0\ \mathbf{v}_1\ \mathbf{v}_2\ \mathbf{v}_3\ \mathbf{v}_4\ \mathbf{v}_5\ \mathbf{v}_6\ \mathbf{v}_7\ \mathbf{v}_8\ \mathbf{v}_9\ \mathbf{v}_{10}\ \mathbf{v}_{11}\ \mathbf{v}_{12}\ \mathbf{v}_{13}\ \mathbf{v}_{14}\ \mathbf{v}_{15}\ \mathbf{v}_{16}$$

```
0 1 0 1 0 1 1 1 0 1 1 1 0 1 1 1 0
0 0 0 1 0 1 1 1 0 1 1 1 0 1 1 1 0
0 0 0 1 0 1 1 1 0 1 1 1 0 1 1 1 0
1 0 1 0 0 0 0 0 0 0 0 0 0 0 0 0 0
```

Find the first 12 information bits $a_0a_1a_2\ldots a_{11}$.

4.2.2 Decoding: The Viterbi Method

The Viterbi decoder is an ambitious (and universally acclaimed) attempt to replace the dull simplicity of the down-to-earth decoding by a more flamboyant algorithmic approach.

What is the gain? In sufficiently decent situations, we will decode – in a single step of the algorithm – rather long prefixes of the information bitstream (but we will have to invest time and space).

What is the loss? The mathematical control of the situation needs slightly stronger hypotheses than the down-to-earth decoding.

Recall, once more, the decoding situation:

$\mathbf{v}_0\mathbf{v}_1\mathbf{v}_2\ldots\mathbf{v}_t\ldots\ \equiv$ the bitstream of the received frames – eventually in error,

$\mathbf{e}_0\mathbf{e}_1\mathbf{e}_2\ldots\mathbf{e}_t\ldots\ \equiv$ the error bitstream,

$\mathbf{c}_0\mathbf{c}_1\mathbf{c}_2\ldots\mathbf{c}_t\ldots\ \equiv$ the code bitstream.

We have: $\mathbf{v}(x) = \mathbf{c}(x) + \mathbf{e}(x)$ (formal power series notation). The code bitstream is generated from the information bitstream $a_0a_1a_2\ldots a_t\ldots$ by "arithmetical filtering": $\mathbf{c}(x) = a(x)G(x)$ (formal power series notation), where

$$G(x) = \begin{pmatrix} g_1(x) \\ \vdots \\ g_n(x) \end{pmatrix} = \mathbf{g}_0 + \mathbf{g}_1 x + \mathbf{g}_2 x^2 + \cdots \mathbf{g}_m x^m$$

is the generator polynomial of the $(n, 1)$ convolutional code.

The Viterbi decoder should reconstruct the correct information bitstream $a_0 a_1 a_2 \ldots a_t \ldots$ from the bitstream of received frames $\mathbf{v}_0 \mathbf{v}_1 \mathbf{v}_2 \cdots \mathbf{v}_t \cdots$ *under the condition* that the error bitstream $\mathbf{e}_0 \mathbf{e}_1 \mathbf{e}_2 \ldots \mathbf{e}_t \ldots$ is "locally thin".

Competing with the down-to-earth decoder, we want the familiar condition:

Let $m \equiv$ the degree of $G(x)$, and let $\mathbf{d}^*$ be the minimum distance of the considered convolutional code. If the number of binary errors over $m + 1$ successive frames does not exceed the threshold $\frac{1}{2}(\mathbf{d}^* - 1)$, then the Viterbi algorithm will correctly reconstruct the initial information bitstream.

But we shall encounter some (minor) problems. In order to understand the complications, let us look first at the structure of the algorithm.

- *The interior algorithm* establishes – in widening step-by-step the horizon of evaluation – a list of binary words which are the *candidates* for the beginning of the information bitstream.
- *The exterior algorithm* chooses the *common prefix* of the candidates as the beginning of the decoded information bitstream; then, it reinitializes according to the well-known procedure.

The correct functioning of the algorithm is guaranteed whenever:

(1) The beginning of the correct information bitstream *is* actually in the list of the candidates of the interior algorithm;
(2) *There is* always a common prefix of the candidates.

As we have already underlined, we would like that the hypotheses which allow the decoding by the down-to-earth algorithm guarantee similarly the decoding by the Viterbi algorithm. Alas, this is not exactly true.[3]

The Interior Viterbi Algorithm

Fix first the length N of an "inspection window". N will be an appropriate multiple of n, i.e. of the number of components of the code bitstream. Often $N = 2n$ will be sufficient.[4]

Recall: $m = \deg G(x)$.

The *interior algorithm* associates with $\mathbf{v}_0 \mathbf{v}_1 \mathbf{v}_2 \ldots \mathbf{v}_{N-1}$ (the N first received frames) a list of (at least) 2^m initial segments $a_0 a_1 a_2 \ldots a_{N-1}$ of the possible information bitstream, which are the *candidates* (*the most probable* segments) for a decoding of $\mathbf{v}_0 \mathbf{v}_1 \mathbf{v}_2 \ldots \mathbf{v}_{N-1}$. What is the *logic* of this list?

---

[3] For the practitioner, these problems don't exist: A deficiency of the decoding algorithm for him always means that there were too many transmission errors – and he is not wrong.

[4] This strange appearance of n seems to come from a visualization which concatenates horizontal frames.

One would like to find the *correct beginning* of the information bitstream. Hence: The m *last* binary positions $a_{N-m} \ldots a_{N-1}$ of the candidates will be *free* and thus *without significance* for the decoding. The $N - m$ *first* binary positions $a_0 a_1 \ldots a_{N-m-1}$ are more and more *significant* towards the head, and are determined by an "iterated bifurcation" algorithm. Step by step – always in function of the m last (parametrizing) positions – one chooses one of the two alternatives for a test segment, in comparing the distances of the associated code segments towards the current segment of the received bitstream .

So, we shall obtain a final list of (at least) 2^m candidates, which comes up this way: For every intermediate length L ($m < L \leq N$) we establish a list of $2^m + \epsilon_L$ preliminary candidates $a_0 a_1 a_2 \ldots a_{L-1}$ the m last positions of which will be free (non-significant) and the $L - m$ first positions of which will be significant for the probable beginning of the original information bitstream. The term ϵ_L stems from the fact that the algorithm is sometimes forced to accept the *two* possibilities of an alternative where it should make a choice – an event which hopefully will become rare towards the end of a round.

Description of the Interior Algorithm

Step no 1:

Every binary word $a_1 a_2 \ldots a_m$ is a terminal segment of two information words: $0 a_1 a_2 \ldots a_m$, $1 a_1 a_2 \ldots a_m$. Let us determine, in each case, the most probable segment (with respect to our decoding situation):

(1) Compute
$\mathbf{c}^0 \equiv$ the $m + 1$ first frames of the code word of $0 + a_1 x + a_2 x^2 + \cdots + a_m x^m$
$\mathbf{c}^1 \equiv$ the $m + 1$ first frames of the code word of $1 + a_1 x + a_2 x^2 + \cdots + a_m x^m$.

(2) Let $\mathbf{v}_{(m)} = \mathbf{v}_0 \mathbf{v}_1 \mathbf{v}_2 \ldots \mathbf{v}_m$. Compute
$d^0 = d(\mathbf{v}_{(m)}, \mathbf{c}^0) \equiv$ the number of binary positions where $\mathbf{v}_{(m)}$ and $\mathbf{c}^0$ are distinct
$d^1 = d(\mathbf{v}_{(m)}, \mathbf{c}^1)$

(3) Save
$0 a_1 a_2 \ldots a_m$ if $d^0 < d^1$
$1 a_1 a_2 \ldots a_m$ if $d^0 > d^1$
both   if $d^0 = d^1$

We obtain thus $2^m + \epsilon_1$ segments $a_0 a_1 \ldots a_m$ (parametrized by the last m bits). The term ϵ_1 comes from the case without exclusion.

Step no 2:

Every binary word $a_2 a_3 \ldots a_m a_{m+1}$ (of length m) gives rise to two types of possible prolongations, in function of the choices of step no 1:
$a_0 0 a_2 a_3 \ldots a_m a_{m+1}$ $a_0 1 a_2 a_3 \ldots a_m a_{m+1}$
Note that a_0 is determined by $0 a_2 a_3 \ldots a_m$ (and by $1 a_2 a_3 \ldots a_m$) according to the preceding step (and will *not* necessarily have the same value in these two cases – you can also have "multiplicities", due to non-exclusions in the preceding step). Let us find, for each of the m-tuples $a_2 a_3 \cdots a_m a_{m+1}$ the most probable prolongation:

(1) Compute
$\mathbf{c}^0 \equiv$ the $m + 2$ first frames of the code word of
$a_0 + a_2 x^2 + \cdots + a_m x^m + a_{m+1} x^{m+1}$,
$\mathbf{c}^1 \equiv$ the $m + 2$ first frames of the code word of
$a_0 + x + a_2 x^2 + \cdots + a_m x^m + a_{m+1} x^{m+1}$.

(2) Let $\quad \mathbf{v}_{(m+1)} \;=\; \mathbf{v}_0\mathbf{v}_1\mathbf{v}_2\ldots\mathbf{v}_{m+1}$. Compute $d^0 \;=\; d(\mathbf{v}_{(m+1)},\mathbf{c}^0)$ $d^1 \;=\;$ $d(\mathbf{v}_{(m+1)},\mathbf{c}^1)$.

(3) Save $\begin{array}{ll} a_0 0 a_2\ldots a_{m+1} & \text{if } d^0 < d^1 \\ a_0 1 a_2\ldots a_{m+1} & \text{if } d^0 > d^1 \\ \text{both} & \text{if } d^0 = d^1. \end{array}$

We obtain $2^m + \epsilon_2$ segments $a_0a_1a_2\ldots a_{m+1}$ (parametrized by the last m bits).
a_0a_1 is *determined* by $a_2a_3\ldots a_m a_{m+1}$.

Step no t:

Consider the 2^m binary words $a_ta_{t+1}\ldots a_{m+t-2}a_{m+t-1}$ (of m bits each). According to the choices of the previous steps, each of these words gives rise to two types of information words which extend the preliminary candidates on the list of the preceding step:

$$a_0\ldots a_{t-2}0a_t\ldots a_{m+t-2}a_{m+t-1},\; a_0\ldots a_{t-2}1a_t\ldots a_{m+t-2}a_{m+t-1}$$

Note that the prefix $a_0a_1\ldots a_{t-2}$ is a function of the segment $\begin{smallmatrix}0\\1\end{smallmatrix}a_t\ldots a_{m+t-2}$ according to the choices of the preceding step (attention to possible "multiplicities", due to former non-exclusions).

Let us find, for each of the m-tuples $a_ta_{t+1}\cdots a_{m+t-2}a_{m+t-1}$ the most probable prolongation (in the sense of a logical decoding):

(1) Compute
$$\mathbf{c}^0 \equiv \begin{array}{l}\text{the } m+t \text{ first frames of the code word of}\\ a_0+\cdots+a_{t-2}x^{t-2}+a_tx^t+\cdots+a_{m+t-1}x^{m+t-1}\end{array}$$

$$\mathbf{c}^1 \equiv \begin{array}{l}\text{the } m+t \text{ first frames of the code word of}\\ a_0+\cdots+a_{t-2}x^{t-2}+x^{t-1}+a_tx^t+\cdots+a_{m+t-1}x^{m+t-1}\end{array}$$

(2) Let $\quad \mathbf{v}_{(m+t-1)} = \mathbf{v}_0\mathbf{v}_1\mathbf{v}_2\ldots\mathbf{v}_{m+t-1}$. Compute

$$d^0 = d(\mathbf{v}_{(m+t-1)},\mathbf{c}^0)$$
$$d^1 = d(\mathbf{v}_{(m+t-1)},\mathbf{c}^1)$$

(3) Save

$$\begin{array}{ll} a_0\ldots a_{t-2}0a_t\ldots a_{m+t-1} & \text{if } d^0 < d^1 \\ a_0\ldots a_{t-2}1a_t\ldots a_{m+t-1} & \text{if } d^0 > d^1 \\ \text{both} & \text{if } d^0 = d^1 \end{array}$$

We obtain $2^m + \epsilon_t$ segments $a_0a_1a_2\ldots a_{m+t-1}$ (parametrized by the m last bits).
$a_0a_1\ldots a_{t-1}$ is *determined* by $a_t\ldots a_{m+t-1}$.

End of a Round of the Interior Algorithm

After $N-m$ steps, we hope to obtain (only) 2^m information words $a_0a_1\ldots a_{N-1}$:

$$a_0^{(1)}a_1^{(1)}\ldots a_{N-1}^{(1)},$$
$$a_0^{(2)}a_1^{(2)}\ldots a_{N-1}^{(2)},$$
$$\vdots$$
$$a_0^{(2^m)}a_1^{(2^m)}\ldots a_{N-1}^{(2^m)}$$

the prefixes of which are the candidates for the beginning of the original information bitstream. We note that a generous choice of N and a *decent error distribution* will probably make the cases of non-exclusion disappear in the last step.

Example

$$G(x) = \begin{pmatrix} 1+x \\ 1+x^2 \\ 1+x+x^2 \end{pmatrix} = \begin{pmatrix} 1 \\ 1 \\ 1 \end{pmatrix} + \begin{pmatrix} 1 \\ 0 \\ 1 \end{pmatrix} x + \begin{pmatrix} 0 \\ 1 \\ 1 \end{pmatrix} x^2.$$

The formula for the computation of the tth code frame is then:

$$\mathbf{c}_t = a_t \begin{pmatrix} 1 \\ 1 \\ 1 \end{pmatrix} + a_{t-1} \begin{pmatrix} 1 \\ 0 \\ 1 \end{pmatrix} + a_{t-2} \begin{pmatrix} 0 \\ 1 \\ 1 \end{pmatrix}.$$

Now the first eight frames of the received bitstream:

$$\begin{pmatrix} \mathbf{v}_0\ \mathbf{v}_1\ \mathbf{v}_2\ \mathbf{v}_3\ \mathbf{v}_4\ \mathbf{v}_5\ \mathbf{v}_6\ \mathbf{v}_7 \\ \\ 1\ 1\ 1\ 1\ 1\ 1\ 1\ 1 \\ 1\ 1\ 0\ 0\ 0\ 0\ 0\ 0 \\ 1\ 1\ 1\ 1\ 0\ 1\ 0\ 1 \end{pmatrix}$$

We shall fix $N = 3n = 9$.

Convention We note that the Hamming distance[5] is additive in the following sense:

$$d(\mathbf{v}_{(k+1)}, \mathbf{c}_{(k+1)}) = d(\mathbf{v}_{(k)}\mathbf{v}_{k+1}, \mathbf{c}_{(k)}\mathbf{c}_{k+1}) = d(\mathbf{v}_{(k)}, \mathbf{c}_{(k)}) + d(\mathbf{v}_{k+1}, \mathbf{c}_{k+1}).$$

This will permit us to write, for example:

$$\mathbf{c} = (4)\begin{matrix} 0 \\ 1 \\ 0 \end{matrix}$$

We have replaced all the frames of $\mathbf{c}$, except the last, by their distance from the corresponding segment of $\mathbf{v}$. The distance of $\mathbf{c}$ from the entire segment of $\mathbf{v}$ will then be equal to 4 + the number of positions where $\begin{matrix}0\\1\\0\end{matrix}$ differs from the terminal frame of the current segment of $\mathbf{v}$.

Step no 1:

$$\mathbf{a}_0\ a_1\ a_2 \qquad \mathbf{c}^0 = \begin{matrix}0\,0\,0\\0\,0\,0\\0\,0\,0\end{matrix} \quad d^0 = 8$$

$$\begin{matrix}0\ \ 0\ \ 0\\1\ \ 0\ \ 0\end{matrix} \qquad \mathbf{c}^1 = \begin{matrix}1\,1\,0\\1\,0\,1\\1\,1\,1\end{matrix} \quad d^1 = 3 \qquad |100\text{ survives}|$$

---

[5] Indeed, our innocent distance between binary words (vectors) has a name!

$$\mathbf{c}^0 = \begin{matrix} 0\ 0\ 1 \\ 0\ 0\ 1 \\ 0\ 0\ 1 \end{matrix} \quad d^0 = 7$$

$\mathbf{a}_0\ a_1\ a_2$

|101 survives|

0 0 1
1 0 1

$$\mathbf{c}^1 = \begin{matrix} 1\ 1\ 1 \\ 1\ 0\ 0 \\ 1\ 1\ 0 \end{matrix} \quad d^1 = 2$$

$$\mathbf{c}^0 = \begin{matrix} 0\ 1\ 1 \\ 0\ 1\ 0 \\ 0\ 1\ 1 \end{matrix} \quad d^0 = 3$$

$\mathbf{a}_0\ a_1\ a_2$

|010 survives|

0 1 0
1 1 0

$$\mathbf{c}^1 = \begin{matrix} 1\ 0\ 1 \\ 1\ 1\ 1 \\ 1\ 0\ 0 \end{matrix} \quad d^1 = 4$$

$$\mathbf{c}^0 = \begin{matrix} 0\ 1\ 0 \\ 0\ 1\ 1 \\ 0\ 1\ 0 \end{matrix} \quad d^0 = 6$$

$\mathbf{a}_0\ a_1\ a_2$

|111 survives|

0 1 1
1 1 1

$$\mathbf{c}^1 = \begin{matrix} 1\ 0\ 0 \\ 1\ 1\ 0 \\ 1\ 0\ 1 \end{matrix} \quad d^1 = 3$$

Step no 2:

$$\mathbf{c}^0 = (3) \begin{matrix} 0 \\ 0 \\ 0 \end{matrix} \quad d^0 = 5$$

$a_0\ \mathbf{a}_1\ a_2\ a_3$

|We save both|

1 **0** 0 0
0 **1** 0 0

$$\mathbf{c}^1 = (3) \begin{matrix} 0 \\ 1 \\ 1 \end{matrix} \quad d^1 = 5$$

$$\mathbf{c}^0 = (3) \begin{matrix} 1 \\ 1 \\ 1 \end{matrix} \quad d^0 = 4$$

$a_0\ \mathbf{a}_1\ a_2\ a_3$

|We save both|

1 **0** 0 1
0 **1** 0 1

$$\mathbf{c}^1 = (3) \begin{matrix} 1 \\ 0 \\ 0 \end{matrix} \quad d^1 = 4$$

$$\mathbf{c}^0 = (2) \begin{matrix} 1 \\ 0 \\ 1 \end{matrix} \quad d^0 = 2$$

$a_0\ \mathbf{a}_1\ a_2\ a_3$

|1010 survives|

1 **0** 1 0
1 **1** 1 0

$$\mathbf{c}^1 = (3) \begin{matrix} 1 \\ 1 \\ 0 \end{matrix} \quad d^1 = 5$$

$$a_0 \; \mathbf{a_1} \; a_2 \; a_3 \qquad \mathbf{c}^0 = (2) \begin{matrix} 0 \\ 1 \\ 0 \end{matrix} \quad d^0 = 5$$

$$
\begin{matrix}
1 & \mathbf{0} & 1 & 1 \\
1 & \mathbf{1} & 1 & 1
\end{matrix}
\qquad
\mathbf{c}^1 = (3) \begin{matrix} 0 \\ 0 \\ 1 \end{matrix} \quad d^1 = 4
$$

|1111 survives|

Step no 3:

$$a_0 \; a_1 \; \mathbf{a_2} \; a_3 \; a_4 \qquad \mathbf{c}^0 = (5) \begin{matrix} 0 \\ 0 \\ 0 \end{matrix} \quad d^0 = 6$$

$$
\begin{matrix}
1 & 0 & \mathbf{0} & 0 & 0 \\
0 & 1 & \mathbf{0} & 0 & 0 \\
1 & 0 & \mathbf{1} & 0 & 0
\end{matrix}
\qquad
\mathbf{c}^1 = (2) \begin{matrix} 0 \\ 1 \\ 1 \end{matrix} \quad d^1 = 5
$$

|10100 survives|

$$a_0 \; a_1 \; \mathbf{a_2} \; a_3 \; a_4 \qquad \mathbf{c}^0 = (5) \begin{matrix} 1 \\ 1 \\ 1 \end{matrix} \quad d^0 = 7$$

$$
\begin{matrix}
1 & 0 & \mathbf{0} & 0 & 1 \\
0 & 1 & \mathbf{0} & 0 & 1 \\
1 & 0 & \mathbf{1} & 0 & 1
\end{matrix}
\qquad
\mathbf{c}^1 = (2) \begin{matrix} 1 \\ 0 \\ 0 \end{matrix} \quad d^1 = 2
$$

|10101 survives|

$$a_0 \; a_1 \; \mathbf{a_2} \; a_3 \; a_4 \qquad \mathbf{c}^0 = (4) \begin{matrix} 1 \\ 0 \\ 1 \end{matrix} \quad d^0 = 5$$

$$
\begin{matrix}
1 & 0 & \mathbf{0} & 1 & 0 \\
0 & 1 & \mathbf{0} & 1 & 0 \\
1 & 1 & \mathbf{1} & 1 & 0
\end{matrix}
\qquad
\mathbf{c}^1 = (4) \begin{matrix} 1 \\ 1 \\ 0 \end{matrix} \quad d^1 = 5
$$

| We save the three|

$$a_0 \; a_1 \; \mathbf{a_2} \; a_3 \; a_4 \qquad \mathbf{c}^0 = (4) \begin{matrix} 0 \\ 1 \\ 0 \end{matrix} \quad d^0 = 6$$

$$
\begin{matrix}
1 & 0 & \mathbf{0} & 1 & 1 \\
0 & 1 & \mathbf{0} & 1 & 1 \\
1 & 1 & \mathbf{1} & 1 & 1
\end{matrix}
\qquad
\mathbf{c}^1 = (4) \begin{matrix} 0 \\ 0 \\ 1 \end{matrix} \quad d^1 = 6
$$

|We save the three|

Step no 4:

$$a_0 \; a_1 \; a_2 \; \mathbf{a_3} \; a_4 \; a_5 \qquad \mathbf{c}^0 = (5) \begin{matrix} 0 \\ 0 \\ 0 \end{matrix} \quad d^0 = 7$$

$$
\begin{matrix}
1 & 0 & 1 & \mathbf{0} & 0 & 0 \\
1 & 0 & 0 & \mathbf{1} & 0 & 0 \\
0 & 1 & 0 & \mathbf{1} & 0 & 0 \\
1 & 1 & 1 & \mathbf{1} & 0 & 0
\end{matrix}
\qquad
\mathbf{c}^1 = (5) \begin{matrix} 0 \\ 1 \\ 1 \end{matrix} \quad d^1 = 7
$$

|We save the four|

$$\begin{array}{l} \\ a_0\ a_1\ a_2\ \mathbf{a_3}\ a_4\ a_5 \end{array} \quad \mathbf{c}^0 = (5)\ \begin{array}{c}1\\1\\1\end{array} \quad d^0 = 6$$

```
1 0 1 0 0 1                              |We save the four|
1 0 0 1 0 1
0 1 0 1 0 1        c¹ = (5) 0   d¹ = 6
1 1 1 1 0 1
```

$$\mathbf{c}^1 = (5)\ \begin{array}{c}1\\0\\0\end{array} \quad d^1 = 6$$

$$a_0\ a_1\ a_2\ \mathbf{a_3}\ a_4\ a_5 \quad \mathbf{c}^0 = (2)\ \begin{array}{c}1\\0\\1\end{array} \quad d^0 = 2$$

```
1 0 1 0 1 0                              |101010 survives|
1 0 0 1 1 0
0 1 0 1 1 0
1 1 1 1 1 0
```

$$\mathbf{c}^1 = (6)\ \begin{array}{c}1\\0\end{array} \quad d^1 = 8$$

$$a_0\ a_1\ a_2\ \mathbf{a_3}\ a_4\ a_5 \quad \mathbf{c}^0 = (2)\ \begin{array}{c}0\\1\\0\end{array} \quad d^0 = 5$$

```
1 0 1 0 1 1                              |101011 survives|
1 0 0 1 1 1
0 1 0 1 1 1
1 1 1 1 1 1
```

$$\mathbf{c}^1 = (6)\ \begin{array}{c}0\\1\end{array} \quad d^1 = 7$$

Step no 5:

$$a_0\ a_1\ a_2\ a_3\ \mathbf{a_4}\ a_5\ a_6 \quad \mathbf{c}^0 = (7)\ \begin{array}{c}0\\0\\0\end{array} \quad d^0 = 8$$

```
1 0 1 0 0 0 0                         1 |1010100 survives|
1 0 0 1 0 0 0
0 1 0 1 0 0 0
1 1 1 1 0 0 0
1 0 1 0 1 0 0
```

$$\mathbf{c}^1 = (2)\ \begin{array}{c}0\\1\\1\end{array} \quad d^1 = 5$$

$$a_0\ a_1\ a_2\ a_3\ \mathbf{a_4}\ a_5\ a_6 \quad \mathbf{c}^0 = (7)\ \begin{array}{c}1\\1\end{array} \quad d^0 = 9$$

```
1 0 1 0 0 0 1                           |1010101 survives|
1 0 0 1 0 0 1
0 1 0 1 0 0 1
1 1 1 1 0 0 1
1 0 1 0 1 0 1
```

$$\mathbf{c}^1 = (2)\ \begin{array}{c}1\\0\\0\end{array} \quad d^1 = 2$$

$$a_0\ a_1\ a_2\ a_3\ \mathbf{a_4}\ a_5\ a_6 \quad \mathbf{c}^0 = (6)\ \begin{array}{c}1\\0\end{array} \quad d^0 = 7$$

```
1 0 1 0 0 1 0                           |1010110 survives|
1 0 0 1 0 1 0
0 1 0 1 0 1 0
1 1 1 1 0 1 0
1 0 1 0 1 1 0
```

$$\mathbf{c}^1 = (5)\ \begin{array}{c}1\\1\\0\end{array} \quad d^1 = 6$$

$$a_0\ a_1\ a_2\ a_3\ \mathbf{a_4}\ a_5\ a_6 \quad \mathbf{c}^0 = (6)\ \begin{array}{c}0\\1\\0\end{array} \quad d^0 = 8$$

```
1 0 1 0 0 1 1                           |1010111 survives|
1 0 0 1 0 1 1
0 1 0 1 0 1 1
1 1 1 1 0 1 1
1 0 1 0 1 1 1
```

$$\mathbf{c}^1 = (5)\ \begin{array}{c}0\\1\end{array} \quad d^1 = 7$$

We observe: After five steps, there remain correctly 4 ($=2^m$, for $m = 2$) (preliminary) candidates $a_0a_1a_2a_3a_4a_5a_6$ for the actual prefix of the information bitstream. Moreover, they have a *common prefix* (of maximal length), i.e. $a_0a_1a_2a_3a_4 = 10101$.

Once more: The four terminal couples a_5a_6 are of no significance for the decoding; they only enumerate our candidates.

Exercises

(1) Carry out the sixth step of our example above. The list of candidates that you will obtain is the following: 10101000   10101010   10101001   10101011

(2) Show that a common prefix of *all* the (preliminary) candidates produced by a step of the interior algorithm will remain a *common* prefix for all the following steps.

We shall take up again three decoding examples that have been treated at the end of the preceding section – but now in the context of the Viterbi algorithm.

(3) Begin with $G(x) = \begin{pmatrix} 1 + x^2 \\ 1 + x + x^2 \end{pmatrix} = \begin{pmatrix} 1 \\ 1 \end{pmatrix} + \begin{pmatrix} 0 \\ 1 \end{pmatrix} x + \begin{pmatrix} 1 \\ 1 \end{pmatrix} x^2$ and

v_0 v_1 v_2 v_3 v_4 v_5 v_6 v_7 v_8 v_9 v_{10} v_{11} v_{12} v_{13} v_{14} v_{15}

```
0 1 1 1 1 1 1 0 1 1 1  1  1  0  1  1
1 0 0 1 0 0 0 1 1 0 1  1  1  0  0  1
```

Carry out the first ten steps of the Viterbi algorithm and recover the common prefix of the candidates in the final list.

[*Answer*: The common prefix: 1100]

(4) Then

$$G(x) = \begin{pmatrix} 1 + x^2 + x^3 \\ 1 + x + x^3 \\ 1 + x + x^3 \\ 1 + x + x^2 + x^3 \end{pmatrix} = \begin{pmatrix} 1 \\ 1 \\ 1 \\ 1 \end{pmatrix} + \begin{pmatrix} 0 \\ 1 \\ 1 \\ 1 \end{pmatrix} x + \begin{pmatrix} 1 \\ 0 \\ 0 \\ 1 \end{pmatrix} x^2 + \begin{pmatrix} 1 \\ 1 \\ 1 \\ 1 \end{pmatrix} x^3$$

as well as the first 20 frames of the received bitstream:

v_0 v_1 v_2 v_3 v_4 v_5 v_6 v_7 v_8 v_9 v_{10} v_{11} v_{12} v_{13} v_{14} v_{15} v_{16} v_{17} v_{18} v_{19}

```
1 0 0 0 1 0 0 1 1 0 0  1  1  1  0  1  0  0  1  0
0 1 1 1 1 1 0 0 0 1 1  1  1  0  0  1  0  1  1  0
0 1 1 1 1 1 0 0 0 0 0  1  1  1  0  1  0  1  1  0
0 1 0 1 1 1 1 0 1 1 0  1  1  1  0  0  1  0  1  0
```

Go through the Viterbi algorithm, step-by-step, until the list of candidates reveals a common prefix.

[*Answer*: You need 16 steps. The common prefix: 101100111000]

(5) Consider, once more, the convolutional code of exercise (4). A constant information bitstream $a_0a_1a_2a_3 \cdots = 1111\ldots$ gives rise to an essentially constant code bitstream :

c_0 c_1 c_2 c_3 c_4 c_5

```
1 1 0 1 1 1
1 0 0 1 1 1  etc.
1 0 0 1 1 1
1 0 1 0 0 0
```

Inflict a periodic error of three bits onto every fourth frame:

$$\mathbf{e}_0 = \mathbf{e}_4 = \mathbf{e}_8 = \mathbf{e}_{12} = \cdots = \begin{pmatrix} 1 \\ 1 \\ 1 \\ 0 \end{pmatrix}, \qquad \mathbf{e}_k = \begin{pmatrix} 0 \\ 0 \\ 0 \\ 0 \end{pmatrix} \quad \text{else.}$$

This gives the following bitstream of received frames:

$\mathbf{v}_0$	$\mathbf{v}_1$	$\mathbf{v}_2$	$\mathbf{v}_3$	$\mathbf{v}_4$	$\mathbf{v}_5$	$\mathbf{v}_6$	$\mathbf{v}_7$	$\mathbf{v}_8$	$\mathbf{v}_9$	$\mathbf{v}_{10}$	$\mathbf{v}_{11}$	$\mathbf{v}_{12}$	$\mathbf{v}_{13}$	$\mathbf{v}_{14}$	$\mathbf{v}_{15}$	$\mathbf{v}_{16}$
0	1	0	1	0	1	1	1	0	1	1	1	0	1	1	1	0
0	0	0	1	0	1	1	1	0	1	1	1	0	1	1	1	0
0	0	0	1	0	1	1	1	0	1	1	1	0	1	1	1	0
1	0	1	0	0	0	0	0	0	0	0	0	0	0	0	0	0

We note: $\mathbf{d}^* = 7$, hence the number of binary errors over four successive frames is correctly bounded by $\frac{1}{2}(\mathbf{d}^* - 1)$.

(a) Carry out the Viterbi algorithm until the 22th step.
(b) Show that the interior algorithm will *never* terminate with a list of candidates which have a *common prefix* (there will always be a candidate beginning with the bit 0).

$\Big[$ *Help.* You should show: Let at step n^o $4k - 2$, $k \geq 3$, the list of candidates (with their distances from $\mathbf{v}_{(4k)}$) be the following:

a000 d−2   d100 d
x001 d   e101 d
b010 d   f110 d−2
c011 d−1   g111 d

then, at step n^o $4k + 2$, the list of candidates (with their distances from $\mathbf{v}_{(4k+4)}$) will be:

c0110000 d+1 a0001100 d+3

$\boxed{\begin{matrix} \text{x0111001} \\ \text{e1011001} \end{matrix}}$ d+3 a0001101 d+3

a0001010 d+3 g1111110 d+1
f1100011 d+2 g1111111 d+3

Here, a, b, c, d, e, f, g are binary words, x will be a *list* of binary words. If then x is "infected" by a prefix beginning with the bit 0, then the algorithm will never get rid of it. $\Big]$

The Interior Algorithm and the Decoding

Now, we want to seriously attack the following natural question:

Let $a_0^* a_1^* \cdots a_t^* \cdots$ be the original information bitstream.

Does the initial segment $a_0^* a_1^* \cdots a_{N-1}^*$ of the correct information bitstream appear as *one of the candidates* $a_0 a_1 \ldots a_{N-1}$ in the final list of the interior algorithm?

The answer should be the following – taking in account the well-functioning of the down-to-earth decoding under this hypothesis:

Proposition *Let $m = \deg G(x)$.*

If the number of binary errors over $m + 1$ successive frames never exceeds the threshold $\frac{1}{2}(\mathbf{d}^* - 1)$ (where $\mathbf{d}^* \equiv$ the minimum distance of the convolutional code), then the beginning of the original information bitstream will appear in the final list of the interior algorithm.

Proof Formal preliminaries: Put, for an arbitrary binary word $\mathbf{a}$:

$\|\mathbf{a}\| \equiv$ the weight of $\mathbf{a}$ $\equiv$ the number of non-zero binary positions of $\mathbf{a}$.

We have trivially: $\|\mathbf{a}_1 \mathbf{a}_2\| = \|\mathbf{a}_1\| + \|\mathbf{a}_2\|$ (the weight is additive with respect to concatenation).

The associated Hamming distance:

$d(\mathbf{a}, \mathbf{b}) = \|\mathbf{a} + \mathbf{b}\| \equiv$ the number of binary positions where $\mathbf{a}$ and $\mathbf{b}$ differ ($\mathbf{a}$ and $\mathbf{b}$ are supposed to have the same format: two vectors in the same space over $\mathbb{F}_2$).

Our situation: An $(n, 1)$-convolutional code $\mathcal{C}$, given by its generator polynomial $G(x) = \mathbf{g}_0 + \mathbf{g}_1 x + \mathbf{g}_2 x^2 + \cdots + \mathbf{g}_m x^m$

We note: $\mathbf{G} = \mathbf{g}_0 \mathbf{g}_1 \ldots \mathbf{g}_m \mathbf{0}\mathbf{0} \cdots \in \mathcal{C}$

Recall: $\mathbf{d}^* = \min_{\mathbf{c} \in \mathcal{C}} \{\|\mathbf{c}_0 \mathbf{c}_1 \ldots \mathbf{c}_m\| : \mathbf{c}_0 \neq \mathbf{0}\}$ Clearly: $\mathbf{d}^* \leq \|\mathbf{G}\| = \|\mathbf{g}_0 \mathbf{g}_1 \ldots \mathbf{g}_m\|$.

Our data:

The original information bitstream: $a_0^* a_1^* a_2^* \ldots a_t^* \ldots$

The associated code bitstream: $\mathbf{c}_0^* \mathbf{c}_1^* \mathbf{c}_2^* \ldots \mathbf{c}_t^* \cdots$

The bitstream of received frames: $\mathbf{v}_0 \mathbf{v}_1 \mathbf{v}_2 \ldots \mathbf{v}_t \ldots$

The error bitstream: $\mathbf{e}_0 \mathbf{e}_1 \mathbf{e}_2 \ldots \mathbf{e}_t \ldots$

We shall make use of the formal power series notation: $a^*(x), \mathbf{c}^*(x), \mathbf{v}(x), \mathbf{e}(x)$.

Thus we will have: $\begin{aligned}\mathbf{c}^*(x) &= a^*(x)G(x) \\ \mathbf{v}(x) &= \mathbf{c}^*(x) + \mathbf{e}(x)\end{aligned}$

Our hypothesis: Let $\varepsilon_t = \|\mathbf{e}_t \mathbf{e}_{t+1} \cdots \mathbf{e}_{t+m}\|$ for $t \geq 0$.

Then $\boxed{2\varepsilon_t + 1 \leq \mathbf{d}^* \quad \text{for } t \geq 0}$

[*Example*: For $G(x) = \begin{pmatrix} 1 + x \\ 1 + x^2 \\ 1 + x + x^2 \end{pmatrix}$ we got $\mathbf{d}^* = 5$.

In this case, our hypothesis becomes: There are not more than two binary errors over three successive frames – which have nine binary positions.]

Let us show: Under our hypothesis (the error bitstream is "locally thin") $a_0^* a_1^* a_2^* \cdots a_{N-1}^*$ will be in the final list of the interior algorithm.

We will show: $\boxed{a_0^* a_1^* \ldots a_{m+t-1}^* \quad \text{survives at the } t\text{th step of the algorithm.}}$

Recursion over t: $\boxed{t = 1:}$

Let $a_0^* a_1^* \cdots a_m^*$ be the correct beginning of the information bitstream, and let $\overline{a}_0^* a_1^* \ldots a_m^*$ be its alternative: $\overline{a}_0^* \equiv a_0^* + 1 \bmod 2$. $\mathbf{c}^* = \mathbf{c}_0^* \mathbf{c}_1^* \ldots \mathbf{c}_m^*$ the initial segment of the correct code bitstream $\overline{\mathbf{c}}^* = \overline{\mathbf{c}}_0^* \overline{\mathbf{c}}_1^* \ldots \overline{\mathbf{c}}_m^*$ the initial segment of the alternative code bitstream.

We have: $\overline{\mathbf{c}}^* = \mathbf{c}^* + \mathbf{G}$ hence: $\overline{\mathbf{c}}^*$ differs from $\mathbf{c}^*$ in $\|\mathbf{G}\|$ positions (and $\|\mathbf{G}\| \geq \mathbf{d}^*$).

Now: $d(\mathbf{c}^*, \mathbf{v}_{(m)}) = \varepsilon_0$ (since $\mathbf{v}_0 \mathbf{v}_1 \cdots \mathbf{v}_m = \mathbf{c}_0^* \mathbf{c}_1^* \cdots \mathbf{c}_m^* + \mathbf{e}_0 \mathbf{e}_1 \cdots \mathbf{e}_m$).

Then: $d(\overline{\mathbf{c}}^*, \mathbf{v}_{(m)}) > \varepsilon_0$ (and $\overline{a}_0^* a_1^* \ldots a_m^*$ will be eliminated at the first step).

Otherwise we would have: $\mathbf{d}^* \leq \|\mathbf{G}\| = d(\mathbf{c}^*, \overline{\mathbf{c}}^*) \leq 2\varepsilon_0$, a contradiction.

Finally: $a_0^* a_1^* \ldots a_m^*$ survives at the first step.

$\boxed{t \mapsto t + 1:}$

Suppose that $a_0^* a_1^* \ldots a_{m+t-1}^*$ has survived at the tth step of the algorithm. Let us show that then $a_0^* a_1^* \ldots a_{m+t-1}^* a_{m+t}^*$ will survive at the $(t+1)$th step.

At any rate, since $a_0^* a_1^* \cdots a_{m+t-1}^*$ is an output of the tth step, $a_0^* a_1^* \cdots a_{m+t-1}^* a_{m+t}^*$ will be in the competition of the $(t+1)$th step. The alternative $b_0 b_1 \ldots b_{t-1} \overline{a}_t^* a_{t+1}^* \ldots a_{m+t}^*$ is obtained, in polynomial notation, by addition of the polynomial $\delta_0 + \delta_1 x + \delta_2 x^2 + \cdots \delta_{t-1} x^{t-1} + x^t$ (note that we don't really control the difference word $\delta_0 \delta_1 \ldots \delta_{t-1} 1$ of the two alternatives).

Consequently:

$$\overline{\mathbf{c}}^*(x) = \mathbf{c}^*(x) + (\delta_0 + \delta_1 x + \delta_2 x^2 + \ldots \delta_{t-1} x^{t-1} + x^t) G(x).$$

Consider $\mathbf{c}_{(m+t)}^* = \mathbf{c}_0^* \cdots \mathbf{c}_{t-1}^* \mathbf{c}_t^* \ldots \mathbf{c}_{m+t}^*$ and $\overline{\mathbf{c}}_{(m+t)}^* = \overline{\mathbf{c}}_0^* \ldots \overline{\mathbf{c}}_{t-1}^* \overline{\mathbf{c}}_t^* \ldots \overline{\mathbf{c}}_{m+t}^*$ and the difference word $\mathbf{d}_{(m+t)} = \mathbf{d}_0 \mathbf{d}_1 \ldots \mathbf{d}_{m+t-1} \mathbf{d}_{m+t}$ (we have to do with the $m+t+1$ frames which are the coefficients of the polynomial $(\delta_0 + \delta_1 x + \delta_2 x^2 + \ldots \delta_{t-1} x^{t-1} + x^t) G(x)$).

This yields: $\overline{\mathbf{c}}_{(m+t)}^* = \mathbf{c}_{(m+t)}^* + \mathbf{d}_{(m+t)}$.

And then: $d(\mathbf{c}_{(m+t)}^*, \mathbf{v}_{(m+t)}) = \| \mathbf{e}_0 \mathbf{e}_1 \ldots \mathbf{e}_{m+t} \|$ and, for the alternative,

$$d(\overline{\mathbf{c}}_{(m+t)}^*, \mathbf{v}_{(m+t)}) = \| \mathbf{e}_0 \mathbf{e}_1 \ldots \mathbf{e}_{m+t} + \mathbf{d}_0 \mathbf{d}_1 \ldots \mathbf{d}_{m+t} \|.$$

We must show:

Under the hypothesis "$\| \mathbf{e}_t \mathbf{e}_{t+1} \ldots \mathbf{e}_{t+m} \| \leq \frac{1}{2}(\mathbf{d}^* - 1)$ for $t \geq 0$ " we get:

$\| \mathbf{e}_0 \mathbf{e}_1 \ldots \mathbf{e}_{m+t} + \mathbf{d}_0 \mathbf{d}_1 \ldots \mathbf{d}_{m+t} \| \geq \| \mathbf{e}_0 \mathbf{e}_1 \ldots \mathbf{e}_{m+t} \|$

We shall encapsulate into our recursion a proof (by recursion on t) of the statement above.

We note that the level t of our (auxiliary) recursive structure consists of *all* binary polynomials $\delta_0 + \delta_1 x + \delta_2 x^2 + \ldots \delta_{t-1} x^{t-1} + x^t$ of degree t.

The beginning of the recursion (the degree $t = 0$) is nothing but the beginning of our "exterior" recursion (corresponding to the initial step $t = 1$).

For $t - 1 \mapsto t$, write $\mathbf{d}_0 \mathbf{d}_1 \ldots \mathbf{d}_{m+t} = \mathbf{d}_0' \mathbf{d}_1' \ldots \mathbf{d}_{m+t-1}' + \mathbf{0}_{(t-1)} \mathbf{G}$ according to $(\delta_0 + \delta_1 x + \delta_2 x^2 + \ldots \delta_{t-1} x^{t-1} + x^t) G(x) = (\delta_0 + \delta_1 x + \delta_2 x^2 + \ldots \delta_{t-1} x^{t-1}) G(x) + x^t G(x)$.

We have clearly:

$d(\mathbf{d}_0 \mathbf{d}_1 \ldots \mathbf{d}_{m+t}, \mathbf{d}_0' \mathbf{d}_1' \ldots \mathbf{d}_{m+t-1}') = d(\mathbf{d}_t \mathbf{d}_{t+1} \ldots \mathbf{d}_{m+t}, \mathbf{d}_t' \mathbf{d}_{t+1}' \ldots \mathbf{d}_{m+t-1}') = \| \mathbf{G} \|$

On the other hand:

$d(\mathbf{e}_0 \mathbf{e}_1 \ldots \mathbf{e}_{m+t}, \mathbf{d}_0' \mathbf{d}_1' \ldots \mathbf{d}_{m+t-1}') = d(\mathbf{e}_0 \mathbf{e}_1 \ldots \mathbf{e}_{m+t-1}, \mathbf{d}_0' \mathbf{d}_1' \ldots \mathbf{d}_{m+t-1}') + \| \mathbf{e}_{m+t} \| \geq \| \mathbf{e}_0 \mathbf{e}_1 \ldots \mathbf{e}_{m+t} \|$ by the recursion hypothesis.

If we replace the $\mathbf{d}_k'$ by the $\mathbf{d}_k$, we change, beginning with the index t, $\| \mathbf{G} \|$ binary positions of the last $m+1$ frames. But, $\mathbf{e}_t \mathbf{e}_{t+1} \ldots \mathbf{e}_{m+t}$ admits $\epsilon_t \leq \frac{1}{2}(\mathbf{d}^* - 1)$ non-zero binary positions. Its additive partner changes by addition of $\| \mathbf{G} \| > 2\epsilon_t$ 1-positions. Hence:

$\| \mathbf{e}_t \mathbf{e}_{t+1} \ldots \mathbf{e}_{m+t} + \mathbf{d}_t \mathbf{d}_{t+1} \ldots \mathbf{d}_{m+t} \| > \| \mathbf{e}_t \mathbf{e}_{t+1} \ldots \mathbf{e}_{m+t} + \mathbf{d}_t' \mathbf{d}_{t+1}' \ldots \mathbf{d}_{m+t-1}' \|$,

which yields our claim.

(The attentive reader will note that the recursion hypothesis of the "technical" part is independent of the recursion hypothesis of the "conceptual" part: $d'_0 d'_1 \ldots d'_{m+t-1}$ has nothing to do with the promotion of $a^*_0 a^*_1 \ldots a^*_{m+t-1}$ at the preceding step). $\square$

The Exterior Viterbi Algorithm

The exterior Viterbi algorithm receives at the end of each round of the interior algorithm its list of candidates (for the beginning of the information bitstream), chooses the *common prefix* of the candidates and declares it as the prefix of the correct information bitstream. Then, it reinitializes the situation, according to the procedure discussed for the reinitialization of the down-to-earth decoding.

We have seen, in an earlier exercise, that our standard hypothesis for a correct decoding (the error bitstream has to be "locally thin" in the sense made precise by the estimation $\epsilon_t \leq \frac{1}{2}(d^* - 1)$ for $t \geq 0$) *is not sufficient* in order to guarantee the existence of a common prefix of the candidates in the final list of the interior algorithm.

We probably need a supplementary hypothesis which clarifies the notion "locally thin" in a more restrictive way: We will have to add a condition of "discontinuity" for the error bitstream: the inspection window moving over the error bitstream should meet regularly "clean" sections – where there are no transmission errors.

But in practice the situation is much simpler. Since *the first step of the Viterbi algorithm contains the complete information for the down-to-earth decoding*, we can use it for a correct *default decoding* whenever the interior algorithm does not arrive at a list of candidates with a common prefix.

Exercises

(1) Consider the convolutional code with generator polynomial

$$G(x) = \begin{pmatrix} 1 + x^2 \\ 1 + x + x^2 \end{pmatrix} = \begin{pmatrix} 1 \\ 1 \end{pmatrix} + \begin{pmatrix} 0 \\ 1 \end{pmatrix} x + \begin{pmatrix} 1 \\ 1 \end{pmatrix} x^2$$

and the received bitstream

$$v_0 \; v_1 \; v_2 \; v_3 \; v_4 \; v_5 \; v_6 \; v_7 \; v_8 \; v_9 \; v_{10} \; v_{11} \; v_{12} \; v_{13} \; v_{14} \; v_{15}$$

$$
\begin{array}{cccccccccccccccc}
1 & 0 & 0 & 0 & 0 & 0 & 0 & 0 & 0 & 1 & 1 & 0 & 0 & 1 & 0 & 0 \\
0 & 1 & 0 & 1 & 0 & 0 & 1 & 1 & 1 & 0 & 0 & 1 & 0 & 0 & 0 & 1
\end{array}
$$

Our interior algorithm will have six steps.

(a) Find the first 10 bits $a_0 a_1 \cdots a_9$ of the information bitstream (two rounds of the interior algorithm!).

(b) Determine the corresponding error bitstream.

(2) Consider the convolutional code defined by

$$G(x) = \begin{pmatrix} 1 + x^2 + x^3 \\ 1 + x + x^3 \\ 1 + x + x^3 \\ 1 + x + x^2 + x^3 \end{pmatrix} = \begin{pmatrix} 1 \\ 1 \\ 1 \\ 1 \end{pmatrix} + \begin{pmatrix} 0 \\ 1 \\ 1 \\ 1 \end{pmatrix} x + \begin{pmatrix} 1 \\ 0 \\ 0 \\ 1 \end{pmatrix} x^2 + \begin{pmatrix} 1 \\ 1 \\ 1 \\ 1 \end{pmatrix} x^3$$

as well as the first 24 frames of the received bitstream:

v_0 v_1 v_2 v_3 v_4 v_5 v_6 v_7 v_8 v_9 v_{10} v_{11} v_{12} v_{13} v_{14} v_{15} v_{16} v_{17} v_{18} v_{19} v_{20} v_{21} v_{22} v_{23}

```
0 0 0 0 1 0 1 1 0 1 1 0 0 1 1 1 0 1 1 0 0 1 1 0
1 0 1 1 0 0 1 1 0 0 1 1 0 0 1 1 0 0 1 1 0 0 1 1
1 0 1 1 0 0 1 1 0 0 1 1 0 0 1 1 0 0 1 1 0 0 1 1
1 0 0 0 0 0 0 0 0 0 0 0 0 0 0 1 0 0 0 0 0 0 0 0
```

The interior algorithm will have 12 steps.

(a) Find the first 18 bits $a_0 a_1 \ldots a_{17}$ of the information bitstream (you need two rounds of the interior algorithm – with a reinitialization).

(b) Keep track of the transmission errors.

5

Data Reduction: Lossy Compression

In this chapter, we shall treat algorithmic decorrelation methods for digital signals, based almost exclusively on the tools of Linear Algebra.

So, we will be confronted with certain (invertible) linear transformations which permit a representation of the numerical data in a form appropriate for decisions of the kind *"let's save all important features and let's delete everything which can be neglected"*, while maintaining at the same time a strict quantitative control over the truncation effect.

In a more traditional mathematical language: We search for optimal approximations of our numerical data units, i.e. for approximations the quality of which is defined with the help of linear transformations that "unmask" the underlying digital information. In order to enter the subject by a familiar gate, we shall begin with a complement on the Discrete Fourier Transform, more precisely on its action in trigonometric interpolation. This will permit us to introduce the basic notions of digital filtering in an ideal outset "where everything works neatly".

But it is not the DFT which is the typical decorrelation transformation in digital signal theory.

The center of our interest during this chapter will be *digital image processing*. This makes the Discrete Cosine Transform (JPEG) and the Discrete Wavelet Transform (JPEG 2000) appear.

The Discrete Cosine Transform is a classical real orthogonal transformation of signal theory, a kind of real part of the DFT, which has to be appreciated as the simplest member of the great family named the Karhunen-Loève Transform.[1] Its 2D version, acting on matrices, is best understood ad hoc: It preserves the Euclidean matrix norm (the "discrete energy") and diagonalizes constant matrices. As a consequence, it clusters the numerically significant information of matrices with small numerical variation in the left upper corner of the transformed matrix, thus creating the desired situation for efficient quantization. The Discrete Wavelet Transform is, exactly like the KLT, a whole family of transformations. In their operational versions, 2D Discrete Wavelet Transforms act via matrix conjugation on matrix schemes – completely similar to the action of the 2D Discrete Cosine Transform. Now, we have lost orthogonality, but we have won pictorial significance. More precisely: The

---

[1] There is a surprising uneasiness in literature to associate the Discrete Cosine Transform with the KLT – why?

matrix versions of the considered Discrete Wavelet Transform describe the action of a couple of digital filters – the low-pass filter computing local means, the high-pass filter annihilating highly regular progressions. Thus, in 2D version, the transformed matrix scheme splits into four sections LL, LH, HL, HH, according to the possible horizontally/vertically combined actions of the filters. The section LL consists of 2D local means, i.e. it is nothing but a lower resolution (of smaller size) of the initial matrix scheme; the other three sections have no real pictorial meaning – they contain the numerical details for faithful reconstruction.

All this is filter bank theory; it can (and will) be explained without reference to wavelets. What about the wavelets supporting the whole discrete machinery? Roughly speaking, the fact that a two channel filter bank is indeed a Discrete Wavelet Transform has an effect on the quality of the reconstruction: after courageous quantization, we will then be able to keep distortion reasonable. At any rate, compared with usual (more or less acoustic) signal theory, we work here in a rather delicate environment: The theory of the digital image is an *autonomous discrete theory* (which is not derived from some continuous theory). In principle, the spatial variable should not be stressed by the standard "time/frequency" duality (the reader will observe that we avoid spectral arguments as long as possible, and use them only in purely technical considerations). So, the *No signal theory without Fourier transform* – paradigm should be seriously relativized. On the other hand, the design criteria for our best transformations are precisely imported from these (conceptually) invalid domains – and finally yield extremely satisfactory results.

5.1 DFT, Passband Filtering and Digital Filtering

Recall   The Sampling Theorem for elementary periodic signals can also be stated as a trigonometric interpolation theorem in the following way:

Let $f_0, f_1, \ldots, f_{n-1}$ be $n = 2m$ complex values.

Then there exists a *unique* balanced trigonometric polynomial of degree $\leq m$
$f(t) = \frac{a_0}{2} + a_1 \cdot \cos 2\pi t + b_1 \cdot \sin 2\pi t + a_2 \cdot \cos 4\pi t + b_2 \cdot \sin 4\pi t + \cdots + \frac{a_m}{2} \cdot \cos 2\pi mt$
with $f(\frac{k}{n}) = f_k$ $0 \leq k \leq n - 1$.

Note that in the spirit of this chapter, the Discrete Fourier Transform of order n – which associates with the vector of n equidistant sample values the coefficient vector for the complex notation of $f(t)$ – indeed carries out a kind of decorrelation:

The action of F_n *uncovers* the frequential information contained in the list of given (time domain) values. The natural (lossy) compression thus proposed would be based on a quantization consisting of the annihilation of certain frequencies. So, we enter the domain of (digital) filtering.

Our first experiences there will be made in the context of the extremely well-behaving periodic theory.

In this section, we shall essentially treat passband (low-pass, high-pass) filtering on elementary periodic signals, which will be done as digital filtering on the sequences of their sample values.

Mathematically, we only have to efficiently combine the Convolution Theorem with the Sampling Theorem.

Filtering by (Circular) Convolution

Let us first fix the language and the formal setting of our presentation. As often in Mathematics, a rich dictionary does not necessarily mean a rich underlying structure. Things are still rather simple.

We shall begin with a

Definition *A discrete n-periodic signal is an (infinite) sequence $z = (z_k)_{k \in \mathbb{Z}}$ of complex numbers such that $z_{k+n} = z_k$ for all $k \in \mathbb{Z}$.*

Such a sequence is completely determined by the segment $(z_0, z_1, \ldots, z_{n-1})$.

Notation $z = \| : z_0, z_1, \ldots, z_{n-1} : \|$.

Example Let $z(t)$ be a (continuous) periodic signal, of period P.

Sample to the order n (i.e. at n equidistant nodes in any interval of length P). We obtain a discrete n-periodic signal $z = (z_k)_{k \in \mathbb{Z}}$, where $z_k = z(\frac{k}{n}P)$, $k \in \mathbb{Z}$.

Note that a priori there is *no relation* between the sampling frequency n and the period P of $x(t)$.

Observation   The discrete n-periodic signals add, and are multiplied by complex scalars, in an obvious way (they are *functions* defined on $\mathbb{Z}$).

Let $\mathbf{\Pi_n}$ be the (complex) vector space of all discrete n-periodic signals. The standard basis of $\mathbf{\Pi_n}$ is given by

$e_0 = \| : 1, 0, 0, \ldots, 0 : \|,$

$e_1 = \| : 0, 1, 0, \ldots, 0 : \|,$

.

.

.

$e_{n-1} = \| : 0, 0, 0, \ldots, 1 : \|.$

Hence: $z = \| : z_0, z_1, \ldots, z_{n-1} : \| \in \mathbf{\Pi_n}$ can be uniquely written as a linear combination of the e_i, $0 \leq i \leq n - 1$:

$z = z_0 e_0 + z_1 e_1 + \cdots + z_{n-1} e_{n-1}.$

Observation   The (circular) convolution product $*: \mathbb{C}^n \times \mathbb{C}^n \to \mathbb{C}^n$ extends naturally to a convolution product $*: \mathbf{\Pi_n} \times \mathbf{\Pi_n} \to \mathbf{\Pi_n}$:

$x = \| : x_0, x_1, \ldots, x_{n-1} : \|, y = \| : y_0, y_1, \ldots, y_{n-1} : \|$

$z = x * y = \| : z_0, z_1, \ldots, z_{n-1} : \|$ is given by

$$z_k = \sum_{j=0}^{n-1} x_j y_{k-j}, \qquad 0 \leq k \leq n - 1.$$

Moreover, the cyclic permutation $\sigma : \mathbb{C}^n \to \mathbb{C}^n$ defined by

$$\sigma \begin{pmatrix} z_0 \\ z_1 \\ . \\ z_{n-1} \end{pmatrix} = \begin{pmatrix} z_{n-1} \\ z_0 \\ . \\ z_{n-2} \end{pmatrix}$$

extends to a right-shift transformation on $\mathbf{\Pi_n}$:

$\sigma \| : z_0, z_1, \ldots, z_{n-1} : \| = \| : z_{-1}, z_0, \ldots, z_{n-2} : \|$

$\Big[$If $z = \| : z_0, z_1, \ldots, z_{n-1} : \|$ represents an (equidistant) sampling of the P-periodic signal $z(t)$, then $\sigma z = \| : z_{-1}, z_0, \ldots, z_{n-2} : \|$ represents the corresponding sampling of the signal $z(t) = z(t - \tau)$, with $\tau = \frac{1}{n} \cdot P\Big]$

Now we are ready to state and to prove the principal result of this section.

Definition *An invariant linear filter (acting) on the discrete n-periodic signals is a $\mathbb{C}$-linear transformation*

$$T : \mathbf{\Pi_n} \to \mathbf{\Pi_n}$$

which is insensitive to (discrete) time shifts, in the following sense:

$$T(\sigma z) = \sigma T(z) \text{ for all } z \in \mathbf{\Pi_n}.$$

Remark Let $h = \| : h_0, h_1, \ldots, h_{n-1} : \|$ be a fixed n-periodic sequence.

Then the mapping $T_h : \mathbf{\Pi_n} \to \mathbf{\Pi_n}$

given by

$$T_h(z) = z * h$$

is an invariant linear filter.

Note that $h = T_h(e_0) = e_0 * h$, i.e. h is the image of the elementary impulse $e_0 = \| : 1, 0, \ldots, 0 : \|$ by the filter T_h: h is its *impulse response*.

The following proposition shows that there are no other invariant linear filters.

Proposition *Let $T : \mathbf{\Pi_n} \to \mathbf{\Pi_n}$ be an invariant linear filter.*

Consider $h = T(e_0)$,

i.e. $\| : h_0, h_1, \ldots, h_{n-1} : \| = T(\| : 1, 0, \ldots, 0 : \|)$, the impulse response of the filter.

Then we have

$$T(x) = x * h \quad \text{for all} \quad x \in \mathbf{\Pi_n}$$

(the filtering is done by convolution with $h = T(e_0)$).

Proof Consider $h = T(e_0) = \| : h_0, h_1, \ldots, h_{n-1} : \|$.

But $e_1 = \sigma e_0$, $e_2 = \sigma^2 e_0$, ..., $e_{n-1} = \sigma^{n-1} e_0$ and $T(\sigma x) = \sigma T(x)$; hence

$T(e_1) = \| : h_{n-1}, h_0, \ldots, h_{n-2} : \|$.

$T(e_2) = \| : h_{n-2}, h_{n-1}, \ldots, h_{n-3} : \|$.

.

.

$T(e_{n-1}) = \| : h_1, h_2, \ldots, h_0 : \|$.

For $x = \| : x_0, x_1, \ldots, x_{n-1} : \| = x_0 e_0 + x_1 e_1 + \ldots + x_{n-1} e_{n-1}$ we obtain:

$T(x) = x_0 T(e_0) + x_1 T(e_1) + \cdots + x_{n-1} T(e_{n-1}) =$

$\qquad x_0 \| : h_0, h_1, \ldots, h_{n-1} : \|$

$\qquad + x_1 \| : h_{n-1}, h_0, \ldots, h_{n-2} : \|$

$\qquad .$

$\qquad x_{n-1} \| : h_1, h_2, \ldots, h_0 : \|$

$\qquad = \| : \sum x_m h_{-m}, \sum x_m h_{1-m}, \ldots, \sum x_m h_{n-1-m} : \| = x * h$ as claimed.

$\qquad\qquad\qquad\qquad\qquad\qquad\qquad\qquad\qquad\qquad\qquad\qquad\qquad\qquad\qquad\qquad$ $\square$

Exercises

(1) Let $T : \mathbf{\Pi_8} \to \mathbf{\Pi_8}$ be the smoothing filter defined by
$$T(x) = y \quad \Longleftrightarrow \quad y_k = \tfrac{1}{4}x_{k-1} + \tfrac{1}{2}x_k + \tfrac{1}{4}x_{k+1} \qquad k \in \mathbb{Z}$$
(a) Show that T is linear and invariant.
(b) Find the impulse response $\quad h = \| : h_0, h_1, \ldots, h_7 : \|$.

(2) Let $\quad \Delta : \mathbf{\Pi_n} \to \mathbf{\Pi_n} \quad$ be the filter of "discrete differentiation" defined by
$$\Delta(x) = y \quad \Longleftrightarrow \quad y_k = x_k - x_{k-1} \qquad k \in \mathbb{Z}$$
(a) Show that Δ is linear and invariant.
(b) Find the impulse response $\quad d = \| : d_0, d_1, \ldots, d_{n-1} : \|$.

Digital Filtering of Sampled Periodic Signals

Situation   Consider elementary periodic signals of the form
$$x(t) = \tfrac{a_0}{2} + a_1 \cdot cos2\pi t + b_1 \cdot sin2\pi t + a_2 \cdot cos4\pi t + b_2 \cdot sin4\pi t + \ldots \tfrac{a_m}{2} \cdot cos2\pi mt$$
(i.e. balanced trigonometric polynomials of degree $\leq m$).
We are interested in linear filterings
$$x(t) \longmapsto \tilde{x}(t) = \tfrac{\tilde{a}_0}{2} + \tilde{a}_1 \cdot \cos 2\pi t + \tilde{b}_1 \cdot sin2\pi t + \tilde{a}_2 \cdot \cos 4\pi t + \tilde{b}_2 \cdot sin4\pi t + \ldots \tfrac{\tilde{a}_m}{2} \cdot \cos 2\pi mt$$
which factor through digital filterings in the following sense:

$$x(t) \qquad\qquad \longrightarrow \qquad\qquad \tilde{x}(t)$$

	sampling	reconstruction	$\uparrow$
	to the order	by trigonometric	
$\downarrow$	$n = 2m$	interpolation	

$$\begin{pmatrix} x_0 \\ x_1 \\ \cdot \\ \cdot \\ x_{n-1} \end{pmatrix} \qquad \begin{array}{c} \text{digital filtering} \\ \longrightarrow \\ \text{of the sample values} \end{array} \qquad \begin{pmatrix} \tilde{x}_0 \\ \tilde{x}_1 \\ \cdot \\ \cdot \\ \tilde{x}_{n-1} \end{pmatrix}$$

Let us look more closely at the situation:

So, let $x(t) = \tfrac{a_0}{2} + a_1 \cdot cos2\pi t + b_1 \cdot sin2\pi t + a_2 \cdot cos4\pi t + b_2 \cdot sin4\pi t + \ldots \tfrac{a_m}{2} \cdot cos2\pi mt$ be a (balanced) trigonometric polynomial of degree $\leq m$. Sample to the order $n = 2m$:

$$\begin{pmatrix} x_0 \\ x_1 \\ \cdot \\ x_{n-1} \end{pmatrix} = \begin{pmatrix} x(0) \\ x(\tfrac{1}{n}) \\ \cdot \\ x(\tfrac{n-1}{n}) \end{pmatrix}$$

An invariant linear filter on the sample values will act via (circular) convolution:

Let $h = \| : h_0, h_1, l \ldots, h_{n-1} : \|$ be the impulse response which defines our invariant linear filter; then

$$\begin{pmatrix} \tilde{x}_0 \\ \tilde{x}_1 \\ \cdot \\ \cdot \\ \tilde{x}_{n-1} \end{pmatrix} = \begin{pmatrix} h_0 \\ h_1 \\ \cdot \\ \cdot \\ h_{n-1} \end{pmatrix} * \begin{pmatrix} x(0) \\ x(\frac{1}{n}) \\ \cdot \\ \cdot \\ x(\frac{n-1}{n}) \end{pmatrix}.$$

$\tilde{x}(t)$ will be the (balanced) trigonometric interpolator of degree $\leq m$ for the vector

$$\begin{pmatrix} \tilde{x}_0 \\ \tilde{x}_1 \\ \cdot \\ \cdot \\ \tilde{x}_{n-1} \end{pmatrix}$$

We know, by the Sampling Theorem:

If $x(t) = \frac{c_{-m}}{2}e^{-2\pi i m t} + \sum_{-m+1 \leq \nu \leq m-1} c_\nu e^{2\pi i \nu t} + \frac{c_m}{2}e^{2\pi i L m t}$

and $\tilde{x}(t) = \frac{\tilde{c}_{-m}}{2}e^{-2\pi i m t} + \sum_{-m+1 \leq \nu \leq m-1} \tilde{c}_\nu e^{2\pi i \nu t} + \frac{\tilde{c}_m}{2}e^{2\pi i m t}$

then

$$\frac{1}{n}F_n \begin{pmatrix} x_0 \\ x_1 \\ \cdot \\ \cdot \\ \cdot \\ x_{n-1} \end{pmatrix} = \begin{pmatrix} c_0 \\ c_1 \\ \cdot \\ c_{m-1} \\ c_m = c_{-m} \\ c_{-m+1} \\ \cdot \\ c_{-1} \end{pmatrix} \quad \text{and} \quad \frac{1}{n}F_n \begin{pmatrix} \tilde{x}_0 \\ \tilde{x}_1 \\ \cdot \\ \cdot \\ \cdot \\ \tilde{x}_{n-1} \end{pmatrix} = \begin{pmatrix} \tilde{c}_0 \\ \tilde{c}_1 \\ \cdot \\ \tilde{c}_{m-1} \\ \tilde{c}_m = \tilde{c}_{-m} \\ \tilde{c}_{-m+1} \\ \cdot \\ \tilde{c}_{-1} \end{pmatrix}.$$

Hence:

$$\begin{pmatrix} \tilde{c}_0 \\ \tilde{c}_1 \\ \cdot \\ \tilde{c}_{m-1} \\ \tilde{c}_m = \tilde{c}_{-m} \\ \tilde{c}_{-m+1} \\ \cdot \\ \tilde{c}_{-1} \end{pmatrix} = \begin{pmatrix} H_0 \cdot c_0 \\ H_1 \cdot c_1 \\ \cdot \\ H_{m-1} \cdot c_{m-1} \\ H_m \cdot c_m \\ H_{m+1} \cdot c_{-m+1} \\ \cdot \\ H_{n-1} \cdot c_{-1} \end{pmatrix}$$

where $\begin{pmatrix} H_0 \\ H_1 \\ \cdot \\ H_{n-1} \end{pmatrix} = F_n \begin{pmatrix} h_0 \\ h_1 \\ \cdot \\ h_{n-1} \end{pmatrix}$

Once more:

$$\tilde{c}_0 = H_0 \cdot c_0$$
$$\tilde{c}_1 = H_1 \cdot c_1$$
$$\vdots$$
$$\tilde{c}_{-1} = H_{n-1} \cdot c_{-1}.$$

Let us sum up: Let $h = \| : h_0, h_1, \ldots, h_{n-1} : \|$ be the impulse response of the digital filter acting on the sample values of our elementary periodic signals

($n = 2m$, where m is the maximum degree of the considered trigonometric polynomials).

Then the analog filter defined by

sampling $\longrightarrow$ *digital filtering* $\longrightarrow$ *reconstruction (of the analog nature) by trigono-metric interpolation*

is a "multiplicator on the complex amplitudes".

More precisely: Let $\begin{pmatrix} H_0 \\ H_1 \\ \cdot \\ H_{n-1} \end{pmatrix} = F_n \begin{pmatrix} h_0 \\ h_1 \\ \cdot \\ h_{n-1} \end{pmatrix}$.

Then

$$\tilde{x}(t) = \frac{\tilde{c}_{-m}}{2} e^{-2\pi imt} + \sum_{-m+1 \leq \nu \leq m-1} \tilde{c}_\nu e^{2\pi i\nu t} + \frac{\tilde{c}_m}{2} e^{2\pi imt}$$

is obtained from

$$x(t) = \frac{c_{-m}}{2} e^{-2\pi imt} + \sum_{-m+1 \leq \nu \leq m-1} c_\nu e^{2\pi i\nu t} + \frac{c_m}{2} e^{2\pi imt}$$

by means of simple multiplications on the coefficients:

$$\tilde{c}_0 = H_0 \cdot c_0$$
$$\tilde{c}_1 = H_1 \cdot c_1$$
$$\vdots$$
$$\tilde{c}_{-1} = H_{n-1} \cdot c_{-1}$$

In particular: Every passband (low-pass, high-pass) filter is obtained this way:

Put $H_\nu = H_{n-\nu} = 0$ *for the frequencies ν you would like to annihilate, and* $H_\nu = H_{n-\nu} = 1$ *for the frequencies ν you would like to preserve.*

Exercises

(1) Consider the smoothing filter $T : \mathbf{\Pi_8} \to \mathbf{\Pi_8}$ defined by $T(\| : x_0, x_1, \ldots x_7 : \|) = \| : y_0, y_1, \ldots y_7 : \|)$ with $y_k = \frac{1}{4}x_{k-1} + \frac{1}{2}x_k + \frac{1}{4}x_{k+1}$ $k \in \mathbb{Z}$.

Following the procedure described above, we shall construct a linear filter on the elementary periodic signals (more precisely: on the balanced trigonometric polynomials of degree ≤ 4) in the following way:

Sampling to the order 8, digital filtering (by T) of the sample values, reconstruction (of the analog situation) via trigonometric interpolation.

A signal

$x(t) = \frac{a_0}{2} + a_1 \cdot cos2\pi t + b_1 \cdot sin2\pi t + a_2 \cdot cos4\pi t + b_2 \cdot sin4\pi t + a_3 \cdot cos6\pi t + b_3 \cdot sin6\pi t + \frac{a_4}{2} \cdot cos8\pi t$ will be transformed into

$\tilde{x}(t) = \frac{\tilde{a}_0}{2} + \tilde{a}_1 \cdot cos2\pi t + \tilde{b}_1 \cdot sin2\pi t + \tilde{a}_2 \cdot cos4\pi t + \tilde{b}_2 \cdot sin4\pi t + \tilde{a}_3 \cdot cos6\pi t + \tilde{b}_3 \cdot sin6\pi t + \frac{\tilde{a}_4}{2} \cdot cos8\pi t$.

(a) Write the coefficients of $\tilde{x}(t)$ in function of the coefficients of $x(t)$.

(b) What happens if you iterate the operation of T?

(2) We consider the elementary signals of the form

$x(t) = \frac{a_0}{2} + a_1 \cdot cos2\pi t + b_1 \cdot sin2\pi t + a_2 \cdot cos4\pi t + b_2 \cdot sin4\pi t + a_3 \cdot cos6\pi t + b_3 \cdot sin6\pi t + \frac{a_4}{2} \cdot cos8\pi t$.

Construct a low-pass filter by means of a digital filter: Sampling to the order 8, digital filtering of the discrete 8-periodic signal, trigonometric interpolation of the output values.

The result of the filtering of
$$x(t) = \frac{a_0}{2} + a_1 \cdot cos2\pi t + b_1 \cdot sin2\pi t + a_2 \cdot cos4\pi t + b_2 \cdot sin4\pi t + a_3 \cdot cos6\pi t + b_3 \cdot sin6\pi t + \frac{a_4}{2} \cdot cos8\pi t$$
should be
$$\tilde{x}(t) = \frac{\tilde{a}_0}{2} + \tilde{a}_1 \cdot cos2\pi t + \tilde{b}_1 \cdot sin2\pi t + \tilde{a}_2 \cdot cos4\pi t + \tilde{b}_2 \cdot sin4\pi t + \tilde{a}_3 \cdot cos6\pi t + \tilde{b}_3 \cdot sin6\pi t + \frac{\tilde{a}_4}{2} \cdot cos8\pi t = \frac{a_0}{2} + a_1 \cdot cos2\pi t + b_1 \cdot sin2\pi t + a_2 \cdot cos4\pi t + b_2 \cdot sin4\pi t.$$

(a) Find the impulse response $h = \| : h_0, h_1, \ldots, h_7 : \|$ of the digital filter to be constructed.

(b) A little test: Sample $x(t) = cos6\pi t$ to the order 8, filter the sample values by T_h and verify that the result is indeed the zero vector.

(3) Consider the elementary signals of the form
$$x(t) = \frac{a_0}{2} + a_1 \cdot cos2\pi t + b_1 \cdot sin2\pi t + a_2 \cdot cos4\pi t + b_2 \cdot sin4\pi t + a_3 \cdot cos6\pi t + b_3 \cdot sin6\pi t + \frac{a_4}{2} \cdot cos8\pi t.$$
A channel has the undesirable effect to weaken the high frequencies. We shall model it as a linear filter which associates with $x(t)$ the signal
$$y(t) = \frac{a_0}{2} + a_1 \cdot cos2\pi t + b_1 \cdot sin2\pi t + \frac{a_2}{2} \cdot cos4\pi t + \frac{b_2}{2} \cdot sin4\pi t + \frac{a_3}{4} \cdot cos6\pi t + \frac{b_3}{4} \cdot sin6\pi t + \frac{a_4}{16} \cdot cos8\pi t.$$
We would like to reconstruct the initial signal by means of a digital filter: Sampling of $y(t)$ to the order 8, digital filtering of the discrete 8-periodic signal, reconstruction of the signal $x(t)$ via trigonometric interpolation.
Compute the impulse response $h = \| : h_0, h_1, \ldots, h_7 : \|$ of the digital filter to be constructed.

(4) We consider the elementary periodic signals of the form
$$x(t) = \frac{a_0}{2} + a_1 cos2\pi t + b_1 sin2\pi t + a_2 cos4\pi t + b_2 sin4\pi t + a_3 cos6\pi t + b_3 sin6\pi t + \frac{a_4}{2} cos8\pi t.$$
Construct a (high-pass with phase shift) filter
$$x(t) \longmapsto \tilde{x}(t) = a_2 cos4\pi(t - \tfrac{1}{8}) + b_2 sin4\pi(t - \tfrac{1}{8}) + a_3 cos6\pi(t - \tfrac{1}{8}) + b_3 sin6\pi(t - \tfrac{1}{8}) + \frac{a_4}{2} cos8\pi(t - \tfrac{1}{8})$$
by means of a digital filter: Sampling to the order 8, filtering of the sample values, trigonometric interpolation.
Find h, the impulse response of the digital filter. Which is the vector H of the multiplicators on the coefficients (of the complex notation) of the considered signals?

5.2 The Discrete Cosine Transform

In this section we shall treat the transformation which is perhaps the most popular in the big family of the *orthogonal* transformations which serve in signal theory. On the one hand, the Discrete Cosine Transform (DCT) is a kind of correctly defined real part of the Discrete Fourier Transform, on the other hand, it is an (almost trivial) member of the family of the Karhunen–Loève transformations, covering the case of highest correlation. Under this aspect, it will often be taken as the *predefinable* substitute for the other members of this family, which are algorithmically almost inaccessible, at least in real time problems.

As it has already been pointed out, the 2D version of the DCT – acting in digital image compression – is perhaps best appreciated without reference to its 1D constituent: It simply preserves the Euclidean matrix norm and diagonalizes constant matrices – two basic properties which, conjointly, are sufficient to explain its great interest for decorrelation problems in image compression.

5.2.1 Functional Description of the DCT

In a first approach, we shall be concerned with the properties of the DCT which stem from its connection with the Discrete Fourier Transform. The more conceptual properties which explain the frequent appearance of the DCT in engineering will be treated in the later sections.

Definition and Fundamental Properties of the DCT

Definition *The Discrete Cosine Transform of order n,*

$$
C_n : \quad \begin{pmatrix} \mathbb{R}^n \\ \begin{pmatrix} x_0 \\ \vdots \\ x_{n-1} \end{pmatrix} \end{pmatrix} \longrightarrow \begin{pmatrix} \mathbb{R}^n, \\ \begin{pmatrix} X_0 \\ \vdots \\ X_{n-1} \end{pmatrix} \end{pmatrix}.
$$

is defined by

$$
X_0 = \frac{1}{\sqrt{n}} \sum_{0 \le k \le n-1} x_k
$$
$$
X_j = \sqrt{\frac{2}{n}} \sum_{0 \le k \le n-1} x_k \cdot \cos \frac{j(2k+1)}{2n} \pi \qquad 1 \le j \le n-1.
$$

In matrix notation:

$$
C_n = \sqrt{\frac{2}{n}} (r_j \cdot \cos \frac{j(2k+1)}{2n} \pi)_{0 \le j, k \le n-1}
$$

with $r_j = \begin{cases} \frac{1}{\sqrt{2}} & \text{for } j = 0 \\ 1 & \text{else} \end{cases}$

$j \equiv$ the row index
$k \equiv$ the column index

Example

$$
C_4 = \frac{1}{\sqrt{2}} \begin{pmatrix} \frac{1}{\sqrt{2}} & \frac{1}{\sqrt{2}} & \frac{1}{\sqrt{2}} & \frac{1}{\sqrt{2}} \\ \cos\frac{\pi}{8} & \cos\frac{3\pi}{8} & -\cos\frac{3\pi}{8} & -\cos\frac{\pi}{8} \\ \cos\frac{\pi}{4} & -\cos\frac{\pi}{4} & -\cos\frac{\pi}{4} & \cos\frac{\pi}{4} \\ \cos\frac{3\pi}{8} & -\cos\frac{\pi}{8} & \cos\frac{\pi}{8} & -\cos\frac{3\pi}{8} \end{pmatrix}
$$

Proposition $C_n : \quad \mathbb{R}^n \quad \longrightarrow \quad \mathbb{R}^n$ *is an orthogonal transformation, i.e. the matrix C_n is invertible, and $C_n^{-1} = C_n^t$ (the inverse matrix is equal to the transposed matrix). More explicitly, the inverse DCT is given by the formulas*

$$
x_j = \frac{1}{\sqrt{n}} X_0 + \sqrt{\frac{2}{n}} \sum_{1 \le k \le n-1} X_k \cdot \cos\frac{(2j+1)k}{2n}\pi \qquad 0 \le j \le n-1.
$$

Proof This is the theme of the following section "DCT and DFT". □

Remarks (1) The most natural matrix for the DCT of order n would be the
matrix

$$C_n^* = \left(\cos \frac{j(2k+1)}{2n} \pi \right)_{0 \le j,k \le n-1}.$$

The normalizing coefficients $\sqrt{\frac{2}{n}} \cdot r_j$ are necessary in order to correctly
obtain $C_n \cdot C_n^t = I_n \equiv$ the unit matrix of order n.

Actually, we have $\quad C_n^* \cdot C_n^{*t} = \frac{n}{2} \begin{pmatrix} 2 & 0 & 0 & \ldots & 0 \\ 0 & 1 & 0 & \ldots & 0 \\ 0 & 0 & 1 & \ldots & 0 \\ \vdots & \vdots & \vdots & \ldots & \vdots \\ 0 & 0 & 0 & \ldots & 1 \end{pmatrix}.$

(2) The orthogonality of the matrix C_n can also be expressed in the following way:
The columns of C_n are a orthonormal basis of the $\mathbb{R}^n$.
The argument is simple: We have, for every matrix $A \in \mathbb{R}^{n \times n}$:

$$A^t \cdot A = \begin{pmatrix} <c_1,c_1> & <c_1,c_2> & \ldots & <c_1,c_n> \\ <c_2,c_1> & <c_2,c_2> & \ldots & <c_2,c_n> \\ \vdots & \vdots & & \vdots \\ <c_n,c_1> & <c_n,c_2> & \ldots & <c_n,c_n> \end{pmatrix},$$

i.e. the product of the transposed matrix with the considered matrix is equal to
the matrix of the inner products of the columns c_j, $1 \le j \le n$, of the matrix A.
We shall return to this aspect in a while.

Exercises

(1) Verify that $\quad C_4 \cdot C_4^t = I_4 = \begin{pmatrix} 1 & 0 & 0 & 0 \\ 0 & 1 & 0 & 0 \\ 0 & 0 & 1 & 0 \\ 0 & 0 & 0 & 1 \end{pmatrix}.$

(2) Write down explicitly the matrix $C_8 \in \mathbb{R}^{8 \times 8}$, taking in account the following
numerical values:
$c_1 = \cos \frac{\pi}{16} = 0.9808$ $c_2 = \cos \frac{\pi}{8} = 0.9239$ $c_3 = \cos \frac{3\pi}{16} = 0.8315$ $c_4 = \cos \frac{\pi}{4} = 0.7071$
$c_5 = \cos \frac{5\pi}{16} = 0.5556$ $c_6 = \cos \frac{3\pi}{8} = 0.3827$ $c_7 = \cos \frac{7\pi}{16} = 0.1951$

Answer

$$C_8 = \frac{1}{2} \cdot \begin{pmatrix} \frac{1}{\sqrt{2}} & \frac{1}{\sqrt{2}} & \frac{1}{\sqrt{2}} & \frac{1}{\sqrt{2}} & \frac{1}{\sqrt{2}} & \frac{1}{\sqrt{2}} & \frac{1}{\sqrt{2}} & \frac{1}{\sqrt{2}} \\ \cos\frac{1}{16}\pi & \cos\frac{3}{16}\pi & \cos\frac{5}{16}\pi & \cos\frac{7}{16}\pi & \cos\frac{9}{16}\pi & \cos\frac{11}{16}\pi & \cos\frac{13}{16}\pi & \cos\frac{15}{16}\pi \\ \cos\frac{2}{16}\pi & \cos\frac{6}{16}\pi & \cos\frac{10}{16}\pi & \cos\frac{14}{16}\pi & \cos\frac{18}{16}\pi & \cos\frac{22}{16}\pi & \cos\frac{26}{16}\pi & \cos\frac{30}{16}\pi \\ \cos\frac{3}{16}\pi & \cos\frac{9}{16}\pi & \cos\frac{15}{16}\pi & \cos\frac{21}{16}\pi & \cos\frac{27}{16}\pi & \cos\frac{33}{16}\pi & \cos\frac{39}{16}\pi & \cos\frac{45}{16}\pi \\ \cos\frac{4}{16}\pi & \cos\frac{12}{16}\pi & \cos\frac{20}{16}\pi & \cos\frac{28}{16}\pi & \cos\frac{36}{16}\pi & \cos\frac{44}{16}\pi & \cos\frac{52}{16}\pi & \cos\frac{60}{16}\pi \\ \cos\frac{5}{16}\pi & \cos\frac{15}{16}\pi & \cos\frac{25}{16}\pi & \cos\frac{35}{16}\pi & \cos\frac{45}{16}\pi & \cos\frac{55}{16}\pi & \cos\frac{65}{16}\pi & \cos\frac{75}{16}\pi \\ \cos\frac{6}{16}\pi & \cos\frac{18}{16}\pi & \cos\frac{30}{16}\pi & \cos\frac{42}{16}\pi & \cos\frac{54}{16}\pi & \cos\frac{66}{16}\pi & \cos\frac{78}{16}\pi & \cos\frac{90}{16}\pi \\ \cos\frac{7}{16}\pi & \cos\frac{21}{16}\pi & \cos\frac{35}{16}\pi & \cos\frac{49}{16}\pi & \cos\frac{63}{16}\pi & \cos\frac{77}{16}\pi & \cos\frac{91}{16}\pi & \cos\frac{105}{16}\pi \end{pmatrix}$$

$$= \frac{1}{2} \cdot \begin{pmatrix} c_4 & c_4 & c_4 & c_4 & c_4 & c_4 & c_4 & c_4 \\ c_1 & c_3 & c_5 & c_7 & -c_7 & -c_5 & -c_3 & -c_1 \\ c_2 & c_6 & -c_6 & -c_2 & -c_2 & -c_6 & c_6 & c_2 \\ c_3 & -c_7 & -c_1 & -c_5 & c_5 & c_1 & c_7 & -c_3 \\ c_4 & -c_4 & -c_4 & c_4 & c_4 & -c_4 & -c_4 & c_4 \\ c_5 & -c_1 & c_7 & c_3 & -c_3 & -c_7 & c_1 & -c_5 \\ c_6 & -c_2 & c_2 & -c_6 & -c_6 & c_2 & -c_2 & c_6 \\ c_7 & -c_5 & c_3 & -c_1 & c_1 & -c_3 & c_5 & -c_7 \end{pmatrix}.$$

DCT and DFT

First observation:

Consider $F_n = \left(\omega_n^{jk}\right)_{0 \le j,k \le n-1}$, the matrix of the DFT of order n. Its real part $\operatorname{Re} F_n = \left(\cos \frac{2jk\pi}{n}\right)_{0 \le j,k \le n-1}$ is, unfortunately, a very bad candidate for an orthogonal transformation: For $n \ge 3$, $\operatorname{Re} F_n$ is *not* invertible.

Exercise

(a) Show that $\operatorname{Re} F_4$ is not invertible.
(b) Generalize: Show that $\operatorname{Re} F_n$ is not invertible, for $n \ge 3$.

We shall see that, up to a certain (modest) algebraic deformation, the DCT is the correct "real part" of the DFT.

Choose $n = 2m$ even.

First two *auxiliary definitions*:

(i) Put, for $\quad x = \begin{pmatrix} x_0 \\ x_1 \\ \vdots \\ x_{n-1} \end{pmatrix} \in \mathbb{R}^n$

$$\check{x} = \begin{pmatrix} \check{x}_0 \\ \check{x}_1 \\ \vdots \\ \check{x}_{n-1} \end{pmatrix} = \begin{pmatrix} x_0 \\ x_2 \\ \cdot \\ x_{n-2} \\ x_{n-1} \\ \cdot \\ x_3 \\ x_1 \end{pmatrix} = P_n \begin{pmatrix} x_0 \\ x_1 \\ \vdots \\ x_{n-1} \end{pmatrix}$$

(with the obvious permutation matrix $P_n \in \mathbb{R}^{n \times n}$).

Explicitly:

$$\begin{aligned} \check{x}_j &= x_{2j} \\ \check{x}_{n-j-1} &= x_{2j+1} \end{aligned} \qquad 0 \le j \le m-1$$

(ii) Let $\quad \Omega_{4n} \in \mathbb{C}^{n \times n} \quad$ be the following diagonal matrix:

$$\Omega_{4n} = \begin{pmatrix} 1 & 0 & 0 & \cdots & 0 \\ 0 & \omega_{4n} & 0 & \cdots & 0 \\ 0 & 0 & \omega_{4n}^2 & \cdots & 0 \\ & & & \ddots & \\ 0 & 0 & & \cdots & \omega_{4n}^{(n-1)} \end{pmatrix}$$

(with $\omega_{4n} = \cos \frac{\pi}{2n} - \mathrm{i} \sin \frac{\pi}{2n}$).

Definition $T_n = \Omega_{4n} F_n P_n \in \mathbb{C}^{n \times n}$.

Proposition $\operatorname{Re} T_n = \left(\cos \frac{j(2k+1)}{2n} \pi \right)_{0 \leq j, k \leq n-1}$
($j \equiv$ *row index*, $k \equiv$ *column index*).

Proof Write
$$
\begin{pmatrix} z_0 \\ z_1 \\ \vdots \\ z_{n-1} \end{pmatrix} = T_n \begin{pmatrix} x_0 \\ x_1 \\ \vdots \\ x_{n-1} \end{pmatrix}
$$

i.e.
$$
z_j = \omega_{4n}^j \sum_{0 \leq k \leq n-1} \check{x}_k \omega_n^{jk} \qquad 0 \leq j \leq n-1
$$

Hence:

$$
z_j = \omega_{4n}^j \cdot \left(\sum_{0 \leq k \leq m-1} x_{2k} \omega_n^{jk} + \sum_{0 \leq k \leq m-1} x_{2k+1} \omega_n^{j(n-1-k)} \right)
$$

$$
= \omega_{4n}^j \cdot \sum_{0 \leq k \leq m-1} x_{2k} \omega_n^{jk} + \omega_{4n}^j \cdot \sum_{0 \leq k \leq m-1} x_{2k+1} \omega_n^{-j} \omega_n^{-jk}
$$

$$
= \sum_{0 \leq k \leq m-1} x_{2k} \omega_{4n}^{j(4k+1)} + \sum_{0 \leq k \leq m-1} x_{2k+1} \omega_{4n}^{-j(4k+3)}
$$

We obtain:
$$
\operatorname{Re} z_j = \sum_{0 \leq k \leq m-1} x_{2k} \cdot \cos \frac{j(4k+1)}{2n} \pi + \sum_{0 \leq k \leq m-1} x_{2k+1} \cdot \cos \frac{j(4k+3)}{2n} \pi =
$$
$$
\sum_{0 \leq k \leq n-1} x_k \cdot \cos \frac{j(2k+1)}{2n} \pi \qquad 0 \leq j \leq n-1
$$
which gives our claim. $\square$

Complementary observation: With the notations of the proposition, we have:
$$
\operatorname{Re} z_{n-j} = -\operatorname{Im} z_j \qquad 1 \leq j \leq n-1.
$$

Proof $x = \begin{pmatrix} x_0 \\ x_1 \\ \vdots \\ x_{n-1} \end{pmatrix} \in \mathbb{R}^n, \quad y = F_n(x) \quad \Rightarrow \quad y_{n-j} = \bar{y}_j \qquad 1 \leq j \leq n-1.$

On the other hand, $\omega_{4n}^{n-j} = -i \omega_{4n}^{-j}$ (since $\omega_{4n}^n = -i$).
Hence: $z_{n-j} = -i \bar{z}_j \qquad 1 \leq j \leq n-1$.
Then: $\operatorname{Re} z_{n-j} = \frac{1}{2}(z_{n-j} + \bar{z}_{n-j}) = \frac{1}{2}(-i \bar{z}_j + i z_j) = \frac{i}{2}(z_j - \bar{z}_j) = -\operatorname{Im} z_j$
$1 \leq j \leq n-1$. $\square$

Algorithmic consequence:

Fast computation of $X^* = C_n^*(x)$ for $x = \begin{pmatrix} x_0 \\ x_1 \\ \vdots \\ x_{n-1} \end{pmatrix} \in \mathbb{R}^n, \quad n = 2^r$

$$
C_n^* = \left(\cos \frac{j(2k+1)}{2n} \pi \right)_{0 \leq j, k \leq n-1} \qquad j \equiv \text{row index} \quad k \equiv \text{column index}
$$

(1) Passage of x to $\check{x} = P_n(x)$ (permutation of the positions).

(2) Computation of $y = F_n(\check{x})$ by FFT.
(3) Multiplication of the components y_j of y by the w_{4n}^j, $0 \le j \le n - 1$. We obtain z.

$$
(4) \text{ Output } X^* = \begin{pmatrix} z_0 \\ Rez_1 \\ . \\ Rez_m \\ -Imz_{m-1} \\ . \\ -Imz_1 \end{pmatrix} . \quad m = \tfrac{n}{2} = 2^{r-1}.
$$

We insist: The vector x and the vector X^* are *real*.

Exercise

Compute $X^* = C_8^*(x)$ for

$$
x = \frac{1}{2} \begin{pmatrix} 1 \\ 0 \\ 1 \\ 0 \\ 2 \\ 0 \\ 1 \\ 1 \end{pmatrix} \quad \text{and for} \quad x = \begin{pmatrix} 1 \\ 2 \\ 3 \\ 4 \\ 5 \\ 6 \\ 7 \\ 8 \end{pmatrix} .
$$

At the end of this section, we shall compute the matrix C_n^{-1}, as promised.

Lemma $T_n = \Omega_{4n} F_n P_n$ is invertible, and $T_n^{-1} = \frac{1}{n} \overline{T}_n^t$
(the matrix $\overline{T}_n^t$ is the transposed conjugate matrix of the matrix T_n).

Proof We have: $\Omega_{4n}^{-1} = \overline{\Omega}_{4n}$ (inversion of the roots of unity on the diagonal)
and $F_n^{-1} = \frac{1}{n} \overline{F}_n$ and $P_n^{-1} = P_n^t$ (for *every* permutation matrix $P \in \mathbb{R}^{n \times n}$ we have $P^{-1} = P^t$).
Hence: $T_n^{-1} = P_n^{-1} F_n^{-1} \Omega_{4n}^{-1} = \frac{1}{n} P_n^t \overline{F}_n \overline{\Omega}_{4n} = \frac{1}{n} P_n^t \overline{F}_n^t \overline{\Omega}_{4n}^t = \frac{1}{n} \overline{(\Omega_{4n} F_n P_n)}^t = \frac{1}{n} \overline{T}_n^t$. $\square$

Notation $I_{11} = \begin{pmatrix} 1\,0\,...\,0 \\ 0\,0\,...\,0 \\ \vdots\,\vdots\quad\vdots \\ 0\,0\,...\,0 \end{pmatrix} \in \mathbb{R}^{n \times n}$ $\qquad J_n = \begin{pmatrix} 1\,0\,...\,0 \\ 0\,0\,...\,1 \\ 0\,...\,1\,0 \\ \vdots\quad.\quad\vdots \\ 0\,1\,...\,0 \end{pmatrix} \in \mathbb{R}^{n \times n}$

Recall $J_n = \frac{1}{n} F_n^2$ (a previous exercise).

Lemma $T_n T_n^t + \overline{T_n T_n^t} = 2n I_{11}$

$$\text{Proof } T_n T_n^t = \Omega_{4n} F_n P_n P_n^t F_n \Omega_{4n} = \Omega_{4n} F_n^2 \Omega_{4n} = n \Omega_{4n} J_n \Omega_{4n} = n \begin{pmatrix} 1 & 0 & \dots & 0 \\ 0 & 0 & \dots & -i \\ \vdots & \vdots & & 0 \\ 0 & -i & \dots & 0 \end{pmatrix}$$

$$\text{Hence: } T_n T_n^t + \overline{T_n T_n^t} = 2n I_{11} = \begin{pmatrix} 2n & 0 & \dots & 0 \\ 0 & \dots & & 0 \\ \vdots & & & \vdots \\ 0 & \dots & \dots & 0 \end{pmatrix} \qquad \square .$$

Proposition Let $C_n^* = \text{Re } T_n$.

$$\text{Then } C_n^* C_n^{*t} = \tfrac{n}{2}(I + I_{11}) = \tfrac{n}{2} \begin{pmatrix} 2 & 0 & 0 & \dots & 0 \\ 0 & 1 & 0 & \dots & 0 \\ 0 & 0 & 1 & \dots & 0 \\ \vdots & & & & \vdots \\ 0 & \dots & & \dots & 1 \end{pmatrix} .$$

$\text{Proof } C_n^* = \tfrac{1}{2}(T_n + \overline{T}_n) \ \ C_n^{*t} = \tfrac{1}{2}(T_n^t + \overline{T}_n^t) \ \ C_n^* C_n^{*t} = \tfrac{1}{4}(T_n T_n^t + \overline{T_n T_n^t} + T_n \overline{T}_n^t + \overline{T}_n T_n^t) = \tfrac{1}{4}(T_n T_n^t + \overline{T_n T_n^t} + T_n \overline{T}_n^t + \overline{T_n \overline{T}_n^t}) = \tfrac{1}{4}(2n I_{11} + 2n I) = \tfrac{n}{2}(I + I_{11}) \quad \square .$

Corollary Let

$$l_0 = (1, 1, \dots, 1)$$
$$l_1 = \left(\cos \tfrac{\pi}{2n}, \dots, \cos \tfrac{(2n-1)\pi}{2n}\right)$$
$$l_2 = \left(\cos \tfrac{2\pi}{2n}, \dots, \cos \tfrac{2(2n-1)\pi}{2n}\right)$$
$$\vdots$$
$$l_{n-1} = \left(\cos \tfrac{(n-1)\pi}{2n}, \dots, \cos \tfrac{(n-1)(2n-1)\pi}{2n}\right)$$

be the n rows of C_n^*.

$$\text{Then } \begin{cases} l_0 l_0^t = n \\ l_j l_j^t = \tfrac{n}{2} \text{ for } 1 \le j \le n-1 \\ l_j l_k^t = 0 \text{ for } j \ne k. \end{cases}$$

This yields:

$$C_n = \begin{pmatrix} \tfrac{1}{\sqrt{n}} l_0 \\ \sqrt{\tfrac{2}{n}} l_1 \\ \vdots \\ \sqrt{\tfrac{2}{n}} l_{n-1} \end{pmatrix} \qquad \text{is an orthogonal matrix: } C_n^{-1} = C_n^t,$$

and proves the inversion formulas of the earlier section.

Exercises

(1) Associate with $\quad x = \begin{pmatrix} x_0 \\ x_1 \\ x_2 \\ x_3 \end{pmatrix} \in \mathbb{R}^4 \quad$ its "mirror twin" $\quad \hat{x} = \begin{pmatrix} x_0 \\ \vdots \\ x_3 \\ x_3 \\ \vdots \\ x_0 \end{pmatrix} \in \mathbb{R}^8.$

Put $\quad \hat{F}_4(x) = F_8(\hat{x})$.
(a) Write down the matrix $\quad \hat{F}_4 \in \mathbb{C}^{8 \times 4}$.
(b) Show that

$$C_4 = \frac{1}{2\sqrt{2}} \begin{pmatrix} \frac{1}{\sqrt{2}} & 0 & 0 & 0 \\ 0 & 1 & 0 & 0 \\ 0 & 0 & 1 & 0 \\ 0 & 0 & 0 & 1 \end{pmatrix} (\Omega_{16} \quad \mathbf{0})\hat{F}_4$$

(c) Generalize (replace 4 by arbitrary n).
(2) Find the identity which compares, for $x, y \in \mathbb{R}^4$, $C_4(x * y)$ and $C_4(x) \cdot C_4(y)$ (where the product in the first term is the circular convolution product, and the product of the second term is meant component by component).
Generalize (replace 4 by arbitrary n).
In the sequel, we shall denote by X_k the kth component of the vector $C_n(x)$, where $n = 2, 4, 8, \ldots$.

Notation $\quad c_k^j = \cos \frac{j\pi}{k} \quad s_k^j = \sin \frac{j\pi}{k}$

(3) Put $\quad \alpha = \frac{1}{\sqrt{2}}, \quad \beta = c_8^1, \quad \delta = s_8^1 \quad$ and $\quad \sigma = c_{16}^1, \quad \epsilon = c_{16}^3, \quad \mu = s_{16}^3, \quad \tau = s_{16}^1$.

Let $\quad \hat{T}_4 = \begin{pmatrix} 1 & 1 & 1 & 1 \\ \alpha & -\alpha & \alpha & -\alpha \\ \beta & -\delta & -\beta & \delta \\ \delta & \beta & -\delta & -\beta \end{pmatrix} \quad$ and $\quad \hat{D}_4 = \begin{pmatrix} \sigma & \mu & -\tau & \epsilon \\ \mu & \tau & -\epsilon & \sigma \\ \epsilon & -\sigma & \mu & \tau \\ \tau & \epsilon & \sigma & \mu \end{pmatrix}$

(a) Show that $\quad \begin{pmatrix} X_0 \\ X_2 \\ X_1 \\ X_3 \end{pmatrix} = \hat{T}_4 \begin{pmatrix} x_0 \\ x_2 \\ x_3 \\ x_1 \end{pmatrix}$

(b) Verify that $\quad \begin{pmatrix} X_0 \\ X_4 \\ X_2 \\ X_6 \\ X_1 \\ X_5 \\ X_3 \\ X_7 \end{pmatrix} = \begin{pmatrix} \hat{T}_4 & \hat{T}_4 \\ \hat{D}_4 & -\hat{D}_4 \end{pmatrix} \begin{pmatrix} x_0 \\ x_2 \\ x_4 \\ x_6 \\ x_7 \\ x_5 \\ x_3 \\ x_1 \end{pmatrix}$

Let $\hat{T}_{2^n}$ be the matrix of the Discrete Cosine Transform to the order 2^n in *permuted version*: The input vector will be in "increasing even positions – decreasing odd positions" ordering, and the output vector will be in bit-inverted enumeration; we deduce that

$$\widehat{T_8} = \begin{pmatrix} \widehat{T_4} & \widehat{T_4} \\ \widehat{D_4} & -\widehat{D_4} \end{pmatrix}.$$

(4) Put $\varphi_k = \frac{(4k+1)\pi}{8}$ for $k = 0, 1, 2, 3$.

(a) Write $\widehat{T_4}$ in function of the $\cos j\varphi_k$, j even, $0 \le k \le 3$.
(b) Write $\widehat{D_4}$ in function of the $\cos j\varphi_k$, j odd, $0 \le k \le 3$.
(c) Using the identities $\cos(2m+1)\varphi = 2\cos(2m)\varphi \cdot \cos\varphi - \cos(2m-1)\varphi$, you should obtain

$$\widehat{D_4} = \begin{pmatrix} \cos\varphi_0 & \cdots \\ \cos\varphi_0 \cdot (1 + 2\cos 4\varphi_0 - 2\cos 2\varphi_0) & \cdots \\ \cos\varphi_0 \cdot (-1 + 2\cos 2\varphi_0) & \cdots \\ \cos\varphi_0 \cdot (-1 - 2\cos 4\varphi_0 + 2\cos 2\varphi_0 + 2\cos 6\varphi_0) & \cdots \end{pmatrix}.$$

(d) Verify the following equality: $\widehat{D_4} = P_4' L_4 P_4' \widehat{T_4} Q_4$
with

$$P_4' = \begin{pmatrix} 1 & 0 & 0 & 0 \\ 0 & 0 & 1 & 0 \\ 0 & 1 & 0 & 0 \\ 0 & 0 & 0 & 1 \end{pmatrix},$$

$$L_4 = \begin{pmatrix} 1 & 0 & 0 & 0 \\ -1 & 2 & 0 & 0 \\ 1 & -2 & 2 & 0 \\ -1 & 2 & -2 & 2 \end{pmatrix},$$

$$Q_4 = \begin{pmatrix} \cos\varphi_0 & 0 & 0 & 0 \\ 0 & \cos\varphi_1 & 0 & 0 \\ 0 & 0 & \cos\varphi_2 & 0 \\ 0 & 0 & 0 & \cos\varphi_3 \end{pmatrix}.$$

(e) Generalize $(\widehat{T_{2^n}} \longrightarrow \widehat{T_{2^{n+1}}})$.

Orthogonal Matrices and Hermitian Matrices

Consider first the $\mathbb{R}^n$ (i.e. the vector space of real n-tuples, in column notation), with its Euclidean structure defined by the inner product

$$\langle x, y \rangle = x^t \cdot y = x_0 y_0 + x_1 y_1 + \cdots + x_{n-1} y_{n-1} \quad \text{for} \quad x = \begin{pmatrix} x_0 \\ \vdots \\ x_{n-1} \end{pmatrix}, \quad y = \begin{pmatrix} y_0 \\ \vdots \\ y_{n-1} \end{pmatrix}$$

The Euclidean norm of the $\mathbb{R}^n$: $\| x \| = \sqrt{\langle x, x \rangle} = \sqrt{x_0^2 + x_1^2 + \cdots + x_{n-1}^2}$
     If the vector x has a geometrical meaning, we shall speak of *length*; if the vector x is the list of the sample values of a signal, we shall speak of *energy*.

Recall   We have, for every matrix $A \in \mathbb{R}^{n \times n}$ and every couple $x, y \in \mathbb{R}^n$:
$\langle Ax, y \rangle = \langle x, A^t y \rangle$
     This elementary identity is the basic tool for every transfer between the arithmetical properties and the geometrical properties of the matrices.

First, we shall deal with orthogonal matrices:

For a matrix $Q \in \mathbb{R}^{n \times n}$ the following properties are equivalent:

(1) Q is an orthogonal matrix, i.e. Q is invertible, and $Q^{-1} = Q^t$ (the inverse matrix is the transposed matrix).
(2) $\| Qx \| = \| x \|$ for all $x \in \mathbb{R}^n$:
the linear transformation "multiplication by Q" preserves the norm (the length, the energy).
(3) The columns of $Q = (\mathbf{c}_0, \mathbf{c}_1, \dots, \mathbf{c}_{n-1})$ are an orthonormal basis of the $\mathbb{R}^n$:
$$\langle \mathbf{c}_i, \mathbf{c}_j \rangle = \begin{cases} 1 \text{ for } i = j. \\ 0 \quad \text{else.} \end{cases}$$

Exercise Prove the equivalence of these three statements.

The generalization of the orthogonality property from the real domain to the complex domain needs some care.

First, for the $\mathbb{C}^n$, the vector space of columns with n complex components, a $\mathbb{C}$-bilinear inner product (obtained by simple matrix multiplication of a row with a column) is not natural, since the definition of a *real* norm – in formal analogy with the case of the $\mathbb{R}^n$ – does not work.

Hence we are obliged to accept the following *definition*:

$$\langle z, w \rangle = z_0 \overline{w}_0 + z_1 \overline{w}_1 + \cdots + z_{n-1} \overline{w}_{n-1} \quad \text{for} z = \begin{pmatrix} z_0 \\ z_1 \\ \vdots \\ z_{n-1} \end{pmatrix}, \quad w = \begin{pmatrix} w_0 \\ w_1 \\ \vdots \\ w_{n-1} \end{pmatrix} \in \mathbb{C}^n.$$

(note that we have lost bilinearity and symmetry:
$\langle z, \lambda w \rangle = \overline{\lambda} \langle z, w \rangle$, $\langle w, z \rangle = \overline{\langle z, w \rangle}$!). But we thus obtain the right norm:

$$\| z \| = \sqrt{\langle z, z \rangle} = \sqrt{x_0^2 + y_0^2 + x_1^2 + y_1^2 + \dots x_{n-1}^2 + y_{n-1}^2}$$

$$\text{for} \quad z = \begin{pmatrix} z_0 \\ z_1 \\ \vdots \\ z_{n-1} \end{pmatrix} = \begin{pmatrix} x_0 + iy_0 \\ x_1 + iy_1 \\ \vdots \\ x_{n-1} + iy_{n-1} \end{pmatrix},$$

identifying the $\mathbb{C}^n$ with the $\mathbb{R}^{2n}$, in a natural way.

Now we have, for $A \in \mathbb{C}^{n \times n}$: $\langle Az, w \rangle = \langle z, \overline{A}^t w \rangle$ for all $z, w \in \mathbb{C}^n$.

The natural generalization of the orthogonality in the domain of real matrices is the notion of a *Hermitian matrix* in the complex domain:

For a matrix $Q \in \mathbb{C}^{n \times n}$ the following properties are equivalent:

(1) Q is a Hermitian matrix, i.e. Q is invertible, and $Q^{-1} = \overline{Q}^t$ (the inverse matrix is the transposed conjugate matrix).
(2) $\| Qz \| = \| z \|$ for all $z \in \mathbb{C}^n$:
The linear transformation "multiplication by Q" preserves the norm (the length, the energy).
(3) The columns of $Q = (\mathbf{c}_0, \mathbf{c}_1, \dots, \mathbf{c}_{n-1})$ are an orthonormal basis of the $\mathbb{C}^n$.

Example Until now, we have met only two types of *Hermitian matrices*:

(a) $\mathbf{F}_n = \frac{1}{\sqrt{n}}\left(\omega_n^{jk}\right)$

(note that we need the factor $\frac{1}{\sqrt{n}}$ in order to get a symmetric situation: We thus obtain $\mathbf{F}_n^{-1} = \frac{1}{\sqrt{n}}\left(\omega_n^{-jk}\right) = \overline{\mathbf{F}}_n$; on the other hand, this factor is highly unpleasant on account of the recursion formulas for the algorithm FFT – this explains its appearance at this late stage of the discussion).

(b) $\mathbf{T}_n = \Omega_{4n}\mathbf{F}_n P_n$ with the matrix $\mathbf{F}_n$ above.

Let us point out, once more, that the DCT matrices are *orthogonal*:

(c) $C_n = \sqrt{\frac{2}{n}}\left(r_j \cdot \cos\frac{j(2k+1)}{2n}\pi\right)$ with $r_j = \begin{cases} \frac{1}{\sqrt{2}} & \text{for } j = 0 \\ 1 & \text{else} \end{cases}$

Then $C_n^{-1} = C_n^t$.

Exercises

(1) Show the orthogonality of the following two matrices:

$$Q = \frac{1}{3}\begin{pmatrix} -1 & 2 & 2 \\ 2 & 2 & -1 \\ -2 & 1 & -2 \end{pmatrix} \in \mathbb{R}^{3\times 3},$$

$$Q = \frac{1}{2}\begin{pmatrix} 1 & 1 & -1 & -1 \\ 1 & -1 & -1 & 1 \\ 1 & -1 & 1 & -1 \\ 1 & 1 & 1 & 1 \end{pmatrix} \in \mathbb{R}^{4\times 4}.$$

(2) Let n be even, and consider the matrix $Q = (q_{jk})_{0\le j,k\le n-1} \in \mathbb{R}^{n\times n}$, defined by

$$q_{jk} = \begin{cases} \sqrt{\frac{2}{n}} \cdot \cos\frac{(j+2)k\pi}{n} & \text{for } j = 0, 2, \ldots n-4, \\ \sqrt{\frac{2}{n}} \cdot \sin\frac{(j+1)k\pi}{n} & \text{for } j = 1, 3, \ldots n-3, \\ \frac{1}{\sqrt{n}} \cdot \cos k\pi & \text{for } j = n-2, \\ \frac{1}{\sqrt{n}} & \text{for } j = n-1, \end{cases}$$

Show that Q is an orthogonal matrix.

(3) The *Walsh–Hadamard* transform.

$N = 2^n$

Associate with the binary notation of j: $0 \le j \le N-1$ the vector $\overrightarrow{j} \in \mathbb{F}_2^n$:

$j = \beta_{n-1}2^{n-1} + \ldots \beta_1 2 + \beta_0$ will be identified with $\overrightarrow{j} = \begin{pmatrix} \beta_{n-1} \\ \vdots \\ \beta_0 \end{pmatrix} \in \mathbb{F}_2^n$.

Let us define: $h_n(j, k) = (-1)^{\langle \overrightarrow{j}, \overrightarrow{k}\rangle}$

We obtain the matrix $H(n) = \left(h_n(j, k)\right)_{0\le j,k\le N-1} \in \mathbb{R}^{N\times N}$

(a) Write explicitly the matrices $H(1)$, $H(2)$ and $H(3)$.

(b) We observe: $H(2) = \begin{pmatrix} H(1) & H(1) \\ H(1) & -H(1) \end{pmatrix}$ and $H(3) = \begin{pmatrix} H(2) & H(2) \\ H(2) & -H(2) \end{pmatrix}$.

Show that $H(n+1) = \begin{pmatrix} H(n) & H(n) \\ H(n) & -H(n) \end{pmatrix}$ for $n \ge 1$.

(c) Show that $\quad H(n)H(n) = N \cdot I_N$.
The matrix $\quad \frac{1}{\sqrt{N}}H(n) = 2^{-\frac{n}{2}}H(n) \quad$ is then a (symmetric) orthogonal matrix, for all $n \geq 1$.
(d) Let $M \in \mathbb{F}_2^{8 \times 3}$ be the matrix the rows of which are the 8 binary notations of $0 = 000, 1 = 001, \ldots, 7 = 111$.
Compute $MM^t \in \mathbb{F}_2^{8 \times 8}$.
What is the relation of MM^t to the matrix $H(3) \in \mathbb{R}^{8 \times 8}$?

The Fundamental Property of the DCT

Recall: Diagonalization of Real Symmetric Matrices

Let $A \in \mathbb{R}^{n \times n}$ be a symmetric matrix $(A = A^t)$.
Then there exists an orthonormal basis $\{\mathbf{q}_0, \mathbf{q}_1, \ldots \mathbf{q}_{n-1}\}$ of the $\mathbb{R}^n$ and $\lambda_0, \lambda_1, \ldots \lambda_{n-1} \in \mathbb{R}$ (not necessarily distinct) with
$A\mathbf{q}_0 = \lambda_0\mathbf{q}_0$
$A\mathbf{q}_1 = \lambda_1\mathbf{q}_1$
$\vdots$
$A\mathbf{q}_{n-1} = \lambda_{n-1}\mathbf{q}_{n-1}.$

Language $\{\mathbf{q}_0, \mathbf{q}_1, \ldots \mathbf{q}_{n-1}\}$ is an orthonormal basis of eigenvectors of A, corresponding to the eigenvalues $\lambda_0, \lambda_1, \ldots \lambda_{n-1}$.

Convention In the sequel, we shall always suppose $\lambda_0 \geq \lambda_1 \geq \cdots \geq \lambda_{n-1}$.
(λ_0 will thus be the maximum eigenvalue...).

Matrix version of the statement above:
Let $A \in \mathbb{R}^{n \times n}$ be a symmetric matrix; then there exists an orthogonal matrix $Q \in \mathbb{R}^{n \times n}$

and a diagonal matrix $D = \begin{pmatrix} \lambda_0 & \cdots & 0 \\ 0 & \lambda_1 & \cdots & 0 \\ \vdots & & \ddots & 0 \\ 0 & \cdots & 0 & \lambda_{n-1} \end{pmatrix}$

with $A = QDQ^t$ i.e. $D = Q^tAQ$.
The argument: Let $Q = (\mathbf{q}_0, \mathbf{q}_1, \ldots \mathbf{q}_{n-1}) \in \mathbb{R}^{n \times n}$ (column notation)
Then: $A = QDQ^t \Longleftrightarrow AQ = QD \Longleftrightarrow A\mathbf{q}_0 = \lambda_0\mathbf{q}_0, A\mathbf{q}_1 = \lambda_1\mathbf{q}_1, \ldots$
$A\mathbf{q}_{n-1} = \lambda_{n-1}\mathbf{q}_{n-1}.$

Important Remark The computation of the eigenvalues $\lambda_0, \lambda_1, \ldots \lambda_{n-1}$ of a real symmetric matrix A demands generally a certain amount of hard work in numerical (matrix) analysis (iterative methods). Here we have an *essentially non-algebraic* problem.
If we *know* the eigenvalues $\lambda_0, \lambda_1, \ldots \lambda_{n-1}$ of A, then the computation of $Q = (\mathbf{q}_0, \mathbf{q}_1, \ldots \mathbf{q}_{n-1}) \in \mathbb{R}^{n \times n}$ is a purely algebraic problem:
Resolution of the linear systems $Ax = \lambda_i x$, $0 \leq i \leq n - 1$, and orthonormalization of the "fundamental solutions".

Fundamental Property of the DCT

$$A = \begin{pmatrix} 1\ 1\ \dots\ 1 \\ 1\ 1\ \dots\ 1 \\ \vdots\ \vdots \quad \vdots \\ 1\ 1\ \dots\ 1 \end{pmatrix} \in \mathbb{R}^{n \times n}$$

$C_n \equiv$ the matrix of the DCT of order n

Then:

$$AC_n^t = \begin{pmatrix} \sqrt{n}\ 0\ \dots\ 0 \\ \sqrt{n}\ 0\ \dots\ 0 \\ \vdots \qquad 0 \\ \sqrt{n}\ 0\ \dots\ 0 \end{pmatrix} = C_n^t \begin{pmatrix} n\ 0\ \dots\ 0 \\ 0\ 0 \\ \vdots \quad 0 \quad \vdots \\ 0 \qquad \dots\ 0 \end{pmatrix}$$

Hence:

$$A = C_n^t \begin{pmatrix} n\ 0\ \dots\ 0 \\ 0\ 0 \\ \vdots \quad 0 \ \vdots \\ 0 \qquad \dots\ 0 \end{pmatrix} C_n$$

with (clearly) $\lambda_0 = n$, $\lambda_1 = \lambda_2 = \cdots = \lambda_{n-1} = 0$.

The columns of C_n^t are an orthonormal basis of the $\mathbb{R}^n$, composed of eigenvectors for A.

Exercises

(1) Let $C_8 \in \mathbb{R}^{8 \times 8}$ the matrix of the DCT to the order 8.

Compute the (symmetric) matrix $B = C_8 \begin{pmatrix} 8\ 0\ \cdots\ 0 \\ 0\ 0\ \cdots\ 0 \\ \vdots \\ 0 \qquad \cdots\ 0 \end{pmatrix} C_8^t$.

(2) Consider the matrix

$$Q_8 = \begin{pmatrix}
\frac{1}{2\sqrt{2}} & \frac{1}{2\sqrt{2}} & \frac{1}{2\sqrt{2}} & \frac{1}{2\sqrt{2}} & \frac{1}{2\sqrt{2}} & \frac{1}{2\sqrt{2}} & \frac{1}{2\sqrt{2}} & \frac{1}{2\sqrt{2}} \\
-\frac{7}{2\sqrt{14}} & \frac{1}{2\sqrt{14}} & \frac{1}{2\sqrt{14}} & \frac{1}{2\sqrt{14}} & \frac{1}{2\sqrt{14}} & \frac{1}{2\sqrt{14}} & \frac{1}{2\sqrt{14}} & \frac{1}{2\sqrt{14}} \\
0 & -\frac{6}{\sqrt{42}} & \frac{1}{\sqrt{42}} & \frac{1}{\sqrt{42}} & \frac{1}{\sqrt{42}} & \frac{1}{\sqrt{42}} & \frac{1}{\sqrt{42}} & \frac{1}{\sqrt{42}} \\
0 & 0 & -\frac{5}{\sqrt{30}} & \frac{1}{\sqrt{30}} & \frac{1}{\sqrt{30}} & \frac{1}{\sqrt{30}} & \frac{1}{\sqrt{30}} & \frac{1}{\sqrt{30}} \\
0 & 0 & 0 & -\frac{4}{2\sqrt{5}} & \frac{1}{2\sqrt{5}} & \frac{1}{2\sqrt{5}} & \frac{1}{2\sqrt{5}} & \frac{1}{2\sqrt{5}} \\
0 & 0 & 0 & 0 & -\frac{3}{2\sqrt{3}} & \frac{1}{2\sqrt{3}} & \frac{1}{2\sqrt{3}} & \frac{1}{2\sqrt{3}} \\
0 & 0 & 0 & 0 & 0 & -\frac{2}{\sqrt{6}} & \frac{1}{\sqrt{6}} & \frac{1}{\sqrt{6}} \\
0 & 0 & 0 & 0 & 0 & 0 & -\frac{1}{\sqrt{2}} & \frac{1}{\sqrt{2}}
\end{pmatrix}.$$

Show that it is orthogonal and that it diagonalizes the *same* matrix $A = \begin{pmatrix} 1\ \cdots\ 1 \\ \vdots \quad \vdots \\ 1\ \cdots\ 1 \end{pmatrix}$ as the DCT C_8.

(3) Let $A = (\mathbf{c}_0, \ldots \mathbf{c}_{n-1})$, $B = \begin{pmatrix} \mathbf{l}_0 \\ \vdots \\ \mathbf{l}_{n-1} \end{pmatrix} \in \mathbb{R}^{n \times n}$ (notation in n-tuples of

columns and of rows).

Show that $A \cdot B = \mathbf{c}_0 \mathbf{l}_0 + \mathbf{c}_1 \mathbf{l}_1 + \cdots + \mathbf{c}_{n-1} \mathbf{l}_{n-1}$.

(Every $\mathbf{c}_i \mathbf{l}_i \in \mathbb{R}^{n \times n}$ $0 \le i \le n-1$!).

(4) Let $Q = (\mathbf{q}_0, \mathbf{q}_1, \cdots \mathbf{q}_{n-1}) \in \mathbb{R}^{n \times n}$ be an orthogonal matrix, and let

$$D = \begin{pmatrix} \lambda_0 & & \cdots & & 0 \\ 0 & \lambda_1 & \cdots & & 0 \\ \vdots & & \ddots & & 0 \\ 0 & \cdots & & 0 & \lambda_{n-1} \end{pmatrix} \in \mathbb{R}^{n \times n} \text{ be a diagonal matrix.}$$

Show that $QDQ^{\mathrm{t}} = \lambda_0 \mathbf{q}_0 \mathbf{q}_0^{\mathrm{t}} + \lambda_1 \mathbf{q}_1 \mathbf{q}_1^{\mathrm{t}} + \cdots + \lambda_{n-1} \mathbf{q}_{n-1} \mathbf{q}_{n-1}^{\mathrm{t}}$.

Recall   The matrix $\mathbf{q}_0 \mathbf{q}_0^{\mathrm{t}} \in \mathbb{R}^{n \times n}$ is the matrix of the orthogonal projection on the straight line $\mathcal{D} = \mathbb{R}\mathbf{q}_0$.

(5) Consider the matrix $S = \begin{pmatrix} 0 & \frac{1}{\sqrt{2}} & 0 & 0 \\ \frac{1}{\sqrt{2}} & 0 & \frac{1}{2} & 0 \\ 0 & \frac{1}{2} & 0 & \frac{1}{\sqrt{2}} \\ 0 & 0 & \frac{1}{\sqrt{2}} & 0 \end{pmatrix} \in \mathbb{R}^{4 \times 4}$.

Let $T_n(x) = cos(n\theta)$, $x = cos\theta$, be the nth Tchebyshev polynomial, $n \ge 0$.

Put $x_k = cos\frac{k\pi}{3}$, $k = 0, 1, 2, 3$.

Let us define

$$\mathbf{v}_k = \begin{pmatrix} \frac{T_0(x_k)}{\sqrt{2}} \\ T_1(x_k) \\ T_2(x_k) \\ T_3(x_k) \end{pmatrix}, \qquad k = 0, 1, 2, 3.$$

Show that $S\mathbf{v}_k = x_k \mathbf{v}_k$ $0 \le k \le 3$.

Is $\{\mathbf{v}_0, \mathbf{v}_1, \mathbf{v}_2, \mathbf{v}_3\}$ an orthonormal basis of the $\mathbb{R}^4$?

Complements on the Eigenvalues of the Real Symmetric Matrices

Approximation of the Eigenvalues

In order to insist on the extra-algebraic invasion in the diagonalization procedure of a real symmetric matrix, we shall give an example of an iterative method which determines the eigenvalues of a matrix $A = A^{\mathrm{t}}$: The *Jacobi algorithm* (1846) for the diagonalization of a real symmetric matrix $A \in \mathbb{R}^{n \times n}$:

Basic idea: Construct a sequence of similar matrices, by conjugation with certain simple rotation matrices which annihilate, at every step, the predominant off-diagonal couple in the current matrix (fly swatter principle).

The iteration: $A_0 = A$, $A_k = R_k A_{k-1} R_k^{\mathrm{t}}$, $k \ge 1$,

where the rotation matrix R_k is determined as follows:

One aims at annihilating the off-diagonal coefficient of maximum absolute value in the (symmetric) matrix A_{k-1}.

[Note that during the iterations the position may again become non-zero, but will be of lesser importance...].

Let then $a_{pq}^{(k-1)}$ be the critical element of the matrix A_{k-1} (sometimes, you will have to make a choice).

R_k will correspond to a rotation in the plane (p, q), and the angle θ is chosen such as to annihilate $a_{pq}^{(k-1)}$.

More precisely (suppose $p < q$): $R_k = \begin{pmatrix} r_{00} & r_{01} & \cdots & r_{0,n-1} \\ r_{10} & r_{11} & \cdots & r_{1,n-1} \\ \vdots & & & \vdots \\ r_{n-1,0} & r_{n-1,1} & \cdots & r_{n-1,n-1} \end{pmatrix}$

with
$$r_{pp} = r_{qq} = \cos\theta$$
$$r_{pq} = -r_{qp} = \sin\theta$$
$$r_{ii} = 1 \quad i \neq p, q$$
$$r_{ij} = 0 \quad \text{else}$$

Note that the matrix A_k differs from the matrix A_{k-1} only at positions concerning the rows and columns of indices p and q.

The modified values will then be:

$$a_{ip}^{(k)} = a_{pi}^{(k)} = a_{ip}^{(k-1)}\cos\theta + a_{iq}^{(k-1)}\sin\theta$$
$$a_{iq}^{(k)} = a_{qi}^{(k)} = -a_{ip}^{(k-1)}\sin\theta + a_{iq}^{(k-1)}\cos\theta$$
$$i \neq p, q$$

$$a_{pp}^{(k)} = a_{pp}^{(k-1)}\cos^2\theta + 2a_{pq}^{(k-1)}\cos\theta\sin\theta + a_{qq}^{(k-1)}\sin^2\theta$$

$$a_{qq}^{(k)} = a_{pp}^{(k-1)}\sin^2\theta - 2a_{pq}^{(k-1)}\cos\theta\sin\theta + a_{qq}^{(k-1)}\cos^2\theta$$

$$a_{pq}^{(k)} = a_{qp}^{(k)} = (a_{qq}^{(k-1)} - a_{pp}^{(k-1)})\cos\theta\sin\theta + a_{pq}^{(k-1)}(\cos^2\theta - \sin^2\theta)$$

In order to obtain $a_{pq}^{(k)} = a_{qp}^{(k)} = 0$, we have to choose θ this way:

$$\tan 2\theta = \frac{2a_{pq}^{(k-1)}}{a_{pp}^{(k-1)} - a_{qq}^{(k-1)}} \quad with \; |\,\theta\,| \leq \frac{\pi}{4}$$

(If $a_{pp}^{(k-1)} = a_{qq}^{(k-1)}$, one chooses $\theta = \pm\frac{\pi}{4}$,   where the sign is the sign of $a_{pq}^{(k-1)}$).

The convergence:

$$A_k \quad \longmapsto \quad D = \begin{pmatrix} \lambda_0 & 0 & \cdots \\ 0 & \lambda_1 & \cdots \\ \vdots & \cdots & \lambda_{n-1} \end{pmatrix} \quad for \quad k \to \infty.$$

Note that $\lambda_0, \lambda_1, \ldots \lambda_{n-1}$ are not only the eigenvalues of the matrix A, but also those of every "intermediate" symmetric matrix A_k.

Write now $A_k = \begin{pmatrix} a_{00}^{(k)} & 0 & \cdots \\ 0 & a_{11}^{(k)} & \cdots \\ \vdots & \cdots & \ddots \end{pmatrix} + E_k$,

where E_k coincides with the matrix A_k, except on the diagonal where it is identically zero.

Let us show that $\|E_k\| \longrightarrow 0$ for $k \to \infty$.

$\|E_k\| \equiv$ the Euclidean norm of E_k as a vector of length n^2.

We have:
$$\sum_{i \neq p,q}((a_{ip}^{(k)})^2 + (a_{iq}^{(k)})^2) = \sum_{i \neq p,q}((a_{ip}^{(k-1)})^2 + (a_{iq}^{(k-1)})^2)$$

Hence:
$$\|E_k\|^2 = \|E_{k-1}\|^2 - 2(a_{pq}^{(k-1)})^2 \qquad (\ a_{pq}^{(k)} = a_{qp}^{(k)} = 0 \ !).$$
But $| \ a_{pq}^{(k-1)} \ |$ is maximum for E_{k-1}:
$$\|E_k\|^2 \le (1 - \tfrac{2}{n^2-n})\|E_{k-1}\|^2 \le (1 - \tfrac{2}{n^2-n})^k\|E_0\|^2.$$
This gives the convergence, as claimed.

Exercises

(1) Let
$$A = \begin{pmatrix} \tfrac{1}{2} & 0 & -\tfrac{1}{2}\sqrt{3} \\ 0 & 1 & 0 \\ -\tfrac{1}{2}\sqrt{3} & 0 & \tfrac{3}{2} \end{pmatrix}$$

(a) Find, following the Jacobi algorithm (which reduces here to its first step), the eigenvalues $\lambda_0 \ge \lambda_1 \ge \lambda_2$ of A.

(b) Give an orthonormal basis $\{\mathbf{q}_0, \mathbf{q}_1, \mathbf{q}_2\}$ of the $\mathbb{R}^3$, composed of eigenvectors for A, corresponding to the eigenvalues $\lambda_0 \ge \lambda_1 \ge \lambda_2$ of A.

(2) Let

$$A = \begin{pmatrix} 3 & 1 & -1 & 1 \\ 1 & 3 & -1 & 1 \\ -1 & -1 & 3 & -1 \\ 1 & 1 & -1 & 3 \end{pmatrix} \in \mathbb{R}^{4 \times 4}.$$

(a) Beginning with $A = A_0$, compute A_2, the second iterate of the Jacobi algorithm towards the diagonalization of A.

(b) Find an orthonormal basis $\{\mathbf{q}_0, \mathbf{q}_1, \mathbf{q}_2, \mathbf{q}_3\}$ of the $\mathbb{R}^4$, composed of eigenvectors for A. (We observe: $\lambda = 2$ is clearly an eigenvalue of A).

Extremal Properties of Eigenvalues

Let $A \in \mathbb{R}^{n \times n}$ be a symmetric matrix.

Consider the quadric $\quad q_A(x) = x^t A x = \sum_{i,j=0}^{n-1} a_{ij} x_i x_j$

for $\quad x = \begin{pmatrix} x_0 \\ \vdots \\ x_{n-1} \end{pmatrix} \in \mathbb{R}^n, \quad$ with $\quad A = (a_{ij})_{0 \le i,j \le n-1}.$

Let $\lambda_0 \ge \lambda_1 \ge \cdots \ge \lambda_{n-1}$ be the eigenvalues of A, and let $\{\mathbf{q}_0, \mathbf{q}_1, \ldots \mathbf{q}_{n-1}\}$ be an orthonormal basis of eigenvectors for (the multiplication by) A.

We are interested in extremal values of $q_A(x)$ on the unit sphere $S^{n-1} = \{x \in \mathbb{R}^n : \|x\| = 1\}$.

The first important result of this section is the following:

Proposition *The situation as above.*

Then: $\quad \max\{q_A(x) : \|x\| = 1\} = \lambda_0$, *and this maximum is attained for $x = \mathbf{q}_0$.*

Impose now orthogonality constraints:

$\max\{q_A(x) : \|x\| = 1$ *and* $\langle x, \mathbf{q}_0 \rangle = 0\} = \lambda_1$, *and this maximum is attained for* $x = \mathbf{q}_1$.

In general, we have, for $0 \leq s \leq n-2$:

$\max\{q_A(x) : \|x\| = 1 \text{ and } \langle x, q_0 \rangle = \langle x, q_1 \rangle = \ldots = \langle x, q_s \rangle = 0\} = \lambda_{s+1}$, *and this maximum is attained for* $x = q_{s+1}$.

Proof Let $Q = (q_0, q_1, \ldots q_{n-1}) \in \mathbb{R}^{n \times n}$ be the orthogonal matrix the columns of which constitute an orthonormal basis of eigenvectors for A.

$$D = \begin{pmatrix} \lambda_0 & & \cdots & & 0 \\ 0 & \lambda_1 & \cdots & & 0 \\ \vdots & & \ddots & & 0 \\ 0 & & \cdots & 0 & \lambda_{n-1} \end{pmatrix} = Q^t A Q.$$

Put $y = Q^t x$ (hence $x = Qy$).
We know: $\|x\| = 1 \iff \|y\| = 1$ (orthogonality of Q).

We have, with $x = Qy$ and $y = \begin{pmatrix} y_0 \\ \vdots \\ y_{n-1} \end{pmatrix}$:

$q_A(x) = x^t A x = (Qy)^t A(Qy) = y^t Q^t A Q y = y^t D y = \lambda_0 y_0^2 + \lambda_1 y_1^2 + \cdots + \lambda_{n-1} y_{n-1}^2$

Hence: $\max\{q_A(x) : \|x\| = 1\} = \max\{\sum_{i=0}^{n-1} \lambda_i y_i^2 : \sum_{i=0}^{n-1} y_i^2 = 1\} = \lambda_0$
since $\sum_{i=0}^{n-1} \lambda_i y_i^2 \leq \lambda_0 \sum_{i=0}^{n-1} y_i^2 \leq \lambda_0$

and the value λ_0 is attained for $y = e_0 = \begin{pmatrix} 1 \\ 0 \\ \vdots \\ 0 \end{pmatrix}$.   But: $Q e_0 = q_0$,   i.e. our

maximum is attained for $x = q_0$.
Now consider the supplementary constraint $\langle x, q_0 \rangle = 0$ (orthogonality to q_0).
It means simply that $\langle y, e_0 \rangle = 0$,   i.e. that $y_0 = 0$.
This implies:
$\max\{q_A(x) : \|x\| = 1 \text{ and } \langle x, q_0 \rangle = 0\} = \max\{\sum_{i=0}^{n-1} \lambda_i y_i^2 : \sum_{i=0}^{n-1} y_i^2 = 1 \text{ and } y_0 = 0\} = \max\{\sum_{i=1}^{n-1} \lambda_i y_i^2 : \sum_{i=1}^{n-1} y_i^2 = 1\} = \lambda_1$
(the same argument as before).

This maximum is attained for $y = e_1 = \begin{pmatrix} 0 \\ 1 \\ 0 \\ \vdots \\ 0 \end{pmatrix}$; now, $Q e_1 = q_1$, hence the

maximum is attained for $x = q_1$.
The general case is proved in the same manner.   □

In order to obtain stronger results on extremal properties of the eigenvalues, we shall be obliged to consider a whole variety of real symmetric matrices associated with our matrix $A \in \mathbb{R}^{n \times n}$.

Definition $1 \leq m \leq n$. *Let* $U_m \in \mathbb{R}^{n \times m}$ *be a matrix of m columns* $u_0, u_1, \ldots u_{m-1}$, *which are an orthonormal basis of* $L(U_m) = \sum_{i=0}^{m-1} \mathbb{R} u_i$.

The matrix $U_m^t A U_m \in \mathbb{R}^{m \times m}$ will be called a section of A.

Example $A = \begin{pmatrix} a_{00} & a_{01} & a_{02} & a_{03} & a_{04} \\ a_{10} & a_{11} & a_{12} & a_{13} & a_{14} \\ a_{20} & a_{21} & a_{22} & a_{23} & a_{24} \\ a_{30} & a_{31} & a_{32} & a_{33} & a_{34} \\ a_{40} & a_{41} & a_{42} & a_{43} & a_{44} \end{pmatrix}, \in \mathbb{R}^{5 \times 5}$ $U_3 = \begin{pmatrix} 1 & 0 & 0 \\ 0 & 0 & 0 \\ 0 & 0 & 1 \\ 0 & 0 & 0 \\ 0 & 1 & 0 \end{pmatrix}$,

$U_3^t A U_3 = \begin{pmatrix} a_{00} & a_{04} & a_{02} \\ a_{40} & a_{44} & a_{42} \\ a_{20} & a_{24} & a_{22} \end{pmatrix} \in \mathbb{R}^{3 \times 3}$.

Lemma *Let $U_m^t A U_m$ be a section of A.*
Then the eigenvalues of $U_m^t A U_m$ are contained in the interval $I = [\lambda_{n-1}, \lambda_0]$.

Proof Put $B = U_m^t A U_m$.

One verifies easily: $\frac{q_B(z)}{\|z\|^2} = \frac{q_A(U_m z)}{\|U_m z\|^2}$ for all $z \in \mathbb{R}^m - \{0\}$.

(note that (the multiplication by) U_m preserves the lengths, since $U_m^t U_m = I_m$).

But: The set $\{\frac{q_A(x)}{\|x\|^2} : x \in \mathbb{R}^n - \{0\}\}$ is equal to the interval $I = [\lambda_{n-1}, \lambda_0]$ (*Exercise*).

In applying this auxiliary result as well to the matrix B as to the matrix A, we obtain the claim of our lemma. □

The principal result of this section is the following proposition:

Proposition *Let $U_m = (\mathbf{u}_0, \mathbf{u}_1, \ldots, \mathbf{u}_{m-1}) \in \mathbb{R}^{n \times m}$ be a matrix of m orthonormal columns. Then: $\min\{q_A(x) : x \in L(U_m) \cap S^{n-1}\} \le \lambda_{m-1}$*
Make now vary U_m (but keep $m \ge 1$ fixed); then
$\max_{U_m} \min\{q_A(x) : x \in L(U_m) \cap S^{n-1}\} = \lambda_{m-1}$

Proof Write $A = QDQ^t$, with $Q = (Q_m, Q'_m)$ and $Q_m = (\mathbf{q}_0, \mathbf{q}_1, \ldots, \mathbf{q}_{m-1})$.

Then the second claim of our proposition is – modulo the first one – easy to accept, since $\min\{q_A(x) : x \in L(Q_m) \cap S^{n-1}\} = \lambda_{m-1}$

As to the first claim, we have to show: There exists $x_0 \in L(U_m) - \{0\}$ such that $\frac{q_A(x_0)}{\|x_0\|^2} \le \lambda_{m-1}$.

Now consider the linear system $U_m u = \theta \mathbf{q}_{m-1} + Q'_m v$. Note that we deal with n equations in $n+1$ unknowns: $u \in \mathbb{R}^m, v \in \mathbb{R}^{n-m}, \theta \in \mathbb{R}$. So, let (u_0, θ_0, v_0) be a non-trivial solution.

$x_0 = U_m u_0$ has to be non-zero, since $\mathbf{q}_{m-1} \notin L(Q'_m)$.

Write $x_0 = \alpha_{m-1} \mathbf{q}_{m-1} + \alpha_m \mathbf{q}_m + \ldots \alpha_{n-1} \mathbf{q}_{n-1}$.

This gives:

$$\frac{q_A(x_0)}{\|x_0\|^2} = \frac{\alpha_{m-1}^2 \lambda_{m-1} + \alpha_m^2 \lambda_m + \cdots + \alpha_{n-1}^2 \lambda_{n-1}}{\alpha_{m-1}^2 + \alpha_m^2 + \cdots + \alpha_{n-1}^2} \le \lambda_{m-1}$$

and proves our proposition. □

Exercise

Show similarly:

$\max\{q_A(x) : x \in L(U_m) \cap S^{n-1}\} \geq \lambda_{n-m}$
and $\min_{U_m} \max\{q_A(x) : x \in L(U_m) \cap S^{n-1}\} = \lambda_{n-m}$.

(**Help:** Write $Q = (Q_{n-m}, Q'_{n-m})$, with $Q_{n-m} = (\mathbf{q}_0, \mathbf{q}_1, \ldots, \mathbf{q}_{n-m-1})$ and consider the linear system $U_m u = Q_{n-m} v + \theta \mathbf{q}_{n-m}$.
Moreover, observe that $\max\{q_A(x) : x \in L(Q'_{n-m}) \cap S^{n-1}\} = \lambda_{n-m}$).

Now let us look at some consequences which will be needed in the sequel.
First, recall:
Let $A \in \mathbb{R}^{n \times n}$ be a symmetric matrix, $B = U_m^t A U_m \in \mathbb{R}^{m \times m}$ a section of A,
and let

$\lambda_0 \geq \lambda_1 \geq \ldots \geq \lambda_{n-1}$ be the eigenvalues of A,
$\lambda'_0 \geq \lambda'_1 \geq \ldots \geq \lambda'_{m-1}$ the eigenvalues of B,
then: $[\lambda'_{m-1}, \lambda'_0] \subset [\lambda_{n-1}, \lambda_0]$. It seems that one can do better:

Observation   In the situation above, we have, more precisely:

$\lambda_0 \geq \lambda'_0 \geq \lambda_{n-m}$,
$\lambda_1 \geq \lambda'_1 \geq \lambda_{n-m+1}$,

$\vdots$

$\lambda_{m-1} \geq \lambda'_{m-1} \geq \lambda_{n-1}$.

Proof Our minimax results immediately give what we want: For $1 \leq s \leq m$ we have actually:

$\max_{V_s} \min\{q_A(x) : x \in L(V_s) \cap S^{n-1}\} = \lambda_{s-1}$.
$\max_{V'_s} \min\{q_B(x) : x \in L(V'_s) \cap S^{n-1}\} = \lambda'_{s-1}$.
Hence: $\lambda'_{s-1} \leq \lambda_{s-1}$.
On the other hand
$\min_{V_s} \max\{q_A(x) : x \in L(V_s) \cap S^{n-1}\} = \lambda_{n-s}$.
$\min_{V'_s} \max\{q_B(x) : x \in L(V'_s) \cap S^{n-1}\} = \lambda'_{m-s}$.
Hence: $\lambda'_{m-s} \geq \lambda_{n-s}$.
This provides our claim. $\square$

We are particularly interested in the following estimation:

Consequence $\mathrm{tr}(U_m^t A U_m) \leq \lambda_0 + \lambda_1 + \ldots + \lambda_{m-1}$
and this bound is attained for $U_m = Q_m = (\mathbf{q}_0, \mathbf{q}_1, \ldots \mathbf{q}_{m-1})$.
When choosing $U_m = (\mathbf{e}_0, \mathbf{e}_1, \ldots \mathbf{e}_{m-1}) \in \mathbb{R}^{n \times m}$, we obtain:
$a_{00} + a_{11} + \ldots + a_{m-1,m-1} \leq \lambda_0 + \lambda_1 + \ldots + \lambda_{m-1}$ $1 \leq m \leq n$.

Exercises

(1) A symmetric matrix $A \in \mathbb{R}^{n \times n}$ is *positive definite* $\Longleftrightarrow$ the quadric

$$q_A(x) = x^t A x = \sum_{i,j=0}^{n-1} a_{ij} x_i x_j$$

only takes, for non-zero x, positive values (the eigenvalues of A are all positive).

Accept the following result:

$$A = A^t \text{ is positive definite} \iff \text{there exists } L = \begin{pmatrix} c_{00} & 0 & \cdots & 0 \\ c_{10} & c_{11} & & 0 \\ \vdots & & \ddots & \\ c_{n-1,0} & c_{n-1,1} & & c_{n-1,n-1} \end{pmatrix}$$

subtriangular, with $A = L \cdot L^t$ (extraction of a "matrix root").

(a) The matrix L (if it exists) can be computed column by column from the identity $A = L \cdot L^t$

(Cholesky algorithm). We obtain at the same time a criterion for the decision whether A is positive definite.

Find the formulas for the coefficients of L in case when $A \in \mathbb{R}^{3 \times 3}$.

(b) Let $q(x, y, z) = x^2 + y^2 + z^2 + xy + xz + yz$.

Find the symmetric matrix $A \in \mathbb{R}^{3 \times 3}$ such that $q(x, y, z) = (\,x \ y \ z\,)A \begin{pmatrix} x \\ y \\ z \end{pmatrix}$.

Show, by the method of Cholesky, that A (hence q) is positive definite.

Finally, diagonalize A: Find the orthogonal matrix $Q \in \mathbb{R}^{3 \times 3}$ with

$$Q^t A Q = D = \begin{pmatrix} \lambda_0 & 0 & 0 \\ 0 & \lambda_1 & 0 \\ 0 & 0 & \lambda_2 \end{pmatrix}.$$

(2) Let A be a positive definite symmetric matrix, $A = L \cdot L^t$ its Cholesky decomposition.

True or false: The eigenvalues of L are the roots of the eigenvalues of A.

(3) True or false: A symmetric matrix with (at least) one negative element on the diagonal admits (at least) one negative eigenvalue.

(4) Let A be a positive definite symmetric matrix. Show that there exists a positive definite symmetric matrix C such that $A = C^2$.

5.2.2 The 2D DCT

The real transformations in traditional digital signal theory are almost exclusively given by orthogonal matrices acting on (sample) vectors. The orthogonality is commonly considered as essential in order to guarantee the "preservation of energy" – in more mathematical terms: one concentrates on isometries (preserving the Euclidean norms). It is above all in digital image processing that we are confronted with the problem of finding isometries acting on 2D schemes, i.e. on the matrices of the sample values of a digital image component. The mathematical construction which resolves easily this problem is the theme of this section.

Tensor Products of Linear Transformations

Tensor products of vectors.

Definition $x = \begin{pmatrix} x_0 \\ \vdots \\ x_{n-1} \end{pmatrix}$, $y = \begin{pmatrix} y_0 \\ \vdots \\ y_{n-1} \end{pmatrix} \in \mathbb{R}^n,$

$$x \otimes y = xy^t = \begin{pmatrix} x_0 y_0 & x_0 y_1 & \cdots & x_0 y_{n-1} \\ x_1 y_0 & x_1 y_1 & \cdots & x_1 y_{n-1} \\ \vdots & \vdots & & \vdots \\ x_{n-1} y_0 & x_{n-1} y_1 & \cdots & x_{n-1} y_{n-1} \end{pmatrix} \in \mathbb{R}^{n \times n}.$$

Remarks (1) Let e_i be the ith standard unit vector, $0 \le i \le n-1$.
Then $e_i \otimes e_j = E_{ij} \equiv$ the matrix which is uniformly zero except at the position (i, j), where the entry is equal to 1.
(2) Every matrix $X = (x_{ij}) \in \mathbb{R}^{n \times n}$ can be written in a unique way as follows

$$X = \sum_{0 \le i, j \le n-1} x_{ij} e_i \otimes e_j.$$

(3) (Exercise) Let $X \in \mathbb{R}^{n \times n}$ be a non-zero matrix; then there exists $x, y \in \mathbb{R}^n$ such that $X = x \otimes y \iff$ rg $X = 1$ (all the rows (columns) of X are proportional).

Definition of a Euclidean structure on $\mathbb{R}^{n \times n}$:

The inner product:
$$\langle X, Y \rangle = \sum_{0 \le i, j \le n-1} x_{ij} y_{ij}.$$

The Euclidean norm:

$$\|X\| = \sqrt{\langle X, X \rangle} = \sqrt{\sum_{0 \le i, j \le n-1} x_{ij}^2}.$$

Observation $\langle X, Y \rangle = \text{tr}(XY^t)$
(Recall: The trace of a matrix is the sum of the elements on the main diagonal).

Consequence $\langle x \otimes y, x' \otimes y' \rangle = \langle x, x' \rangle \langle y, y' \rangle$ for $x, x', y, y' \in \mathbb{R}^n$.
In particular: $\|x \otimes y\| = \|x\| \cdot \|y\|$.

Proof $\langle x \otimes y, x' \otimes y' \rangle = \text{tr}(xy^t(x'y'^t)^t) = \text{tr}(xy^t y' x'^t) = \text{tr}(xx'^t)\langle y, y' \rangle = \langle x, x' \rangle \langle y, y' \rangle.$

Exercises

(1) Let $\{q_0, q_1\}$ be the orthonormal basis of the $\mathbb{R}^2$ given by

$$q_0 = \frac{1}{\sqrt{2}} \begin{pmatrix} 1 \\ 1 \end{pmatrix} \qquad q_1 = \frac{1}{\sqrt{2}} \begin{pmatrix} 1 \\ -1 \end{pmatrix}.$$

   (a) Compute the four matrices $q_0 \otimes q_0, q_0 \otimes q_1, q_1 \otimes q_0, q_1 \otimes q_1$.
   (b) Show that these four matrices are an orthonormal basis of the $\mathbb{R}^{2 \times 2}$.
(2) Let $\{q_0, q_1, \ldots q_{n-1}\}$ be an orthonormal basis of the $\mathbb{R}^n$.
Consider the n^2 matrices $q_0 \otimes q_0, q_0 \otimes q_1, \ldots, q_{n-1} \otimes q_{n-1}$.
Show that these n^2 matrices are an orthonormal basis of the $\mathbb{R}^{n \times n}$.
(3) Let $Q \in \mathbb{R}^{n \times n}$ be an *orthogonal* matrix. Show that $\langle QXQ^t, Y \rangle = \langle X, Q^t Y Q \rangle$ for all $X, Y \in \mathbb{R}^{n \times n}$.
Deduce that the transformation $X \longmapsto QXQ^t$ is an isometry of the $\mathbb{R}^{n \times n}$.

Tensor Products of Linear Transformations

Consider two matrices $A, B \in \mathbb{R}^{n \times n}$.

Write $A = (\mathbf{c}_0, \mathbf{c}_1, \dots, \mathbf{c}_{n-1})$ and $B = (\mathbf{c}_0', \mathbf{c}_1', \dots, \mathbf{c}_{n-1}')$
(notation in column vectors).
Let us define $T_{A \otimes B} : \quad \mathbb{R}^{n \times n} \longrightarrow \mathbb{R}^{n \times n}$ by

$$T_{A \otimes B}(X) = \sum_{0 \le i,j \le n-1} x_{ij}(A e_i \otimes B e_j) = \sum_{0 \le i,j \le n-1} x_{ij} \mathbf{c}_i \otimes \mathbf{c}_j'.$$

Note that this is well-defined, since a linear transformation is uniquely determined by (linear extension on) the images of a basis; here:
$T_{A \otimes B}(E_{ij}) = \mathbf{c}_i \otimes \mathbf{c}_j' \quad 0 \le i, j \le n - 1.$
But:

$$\mathbf{c}_i \otimes \mathbf{c}_j' = \begin{pmatrix} c_{0i}c_{0j}' & c_{0i}c_{1j}' & \cdots & c_{0i}c_{n-1,j}' \\ c_{1i}c_{0j}' & c_{1i}c_{1j}' & \cdots & c_{1i}c_{n-1,j}' \\ \vdots & \vdots & & \vdots \\ c_{n-1,i}c_{0j}' & c_{n-1,i}c_{1j}' & \cdots & c_{n-1,i}c_{n-1,j}' \end{pmatrix}.$$

This gives finally, for $Y = T_{A \otimes B}(X)$,
$y_{kl} = \sum_{0 \le i,j \le n-1} c_{ki}x_{ij}c_{lj}', \qquad 0 \le k, l \le n - 1.$

Observation $\boxed{T_{Q \otimes Q}(X) = QXQ^{\mathrm{t}}}$ (Q need not be orthogonal).

Proof Write $T_{Q \otimes Q}(X) = (y_{kl})_{0 \le k,l \le n-1}$ i.e. $y_{kl} = \sum_{0 \le i,j \le n-1} c_{ki}x_{ij}c_{lj}$,
$0 \le k, l \le n - 1.$
But $\sum_{0 \le i \le n-1} c_{ki}x_{ij} = (QX)_{kj}$,   hence $\sum_{0 \le i,j \le n-1} c_{ki}x_{ij}c_{lj} = (QXQ^t)_{kl}$ as claimed.

Remark Clearly, the observation above eliminates the tensor products that we have introduced so carefully. But certain (conceptual) properties of the 2D versions, inherited from the characteristic properties of the 1D transformations from which they are derived, are best understood in tensor formalism. This will be a recurrent implicit argument in the sequel.

Exercises

(1) Let $A, B, C, D \in \mathbb{R}^{n \times n}$ and let I_n be the unit matrix of order n.
  (a) Show that $T_{AC \otimes BD} = T_{A \otimes B} \circ T_{C \otimes D}$ and that $T_{I_n \otimes I_n} = Id_{\mathbb{R}^{n \times n}}$.
  (b) Deduce that, for invertible A, B, the transformation $T_{A \otimes B}$ is invertible, and we have:
  $$T_{A \otimes B}^{-1} = T_{A^{-1} \otimes B^{-1}}.$$

(2) $T_{Q \otimes Q} \begin{pmatrix} 2 & -3 \\ -1 & 1 \end{pmatrix} = \begin{pmatrix} -\frac{1}{2} & \frac{3}{2} \\ -\frac{1}{2} & \frac{7}{2} \end{pmatrix}.$
  Find the orthogonal matrices $Q \in \mathbb{R}^{2 \times 2}$, which are thus determined.

(3) For $A = (a_{ij})$, $B = (b_{kl}) \in \mathbb{R}^{n \times n}$ let us define their tensor product (their *Kronecker product*, in old-fashioned language) $A \otimes B \in \mathbb{R}^{n^2 \times n^2}$ by

$$A \otimes B = \begin{pmatrix} a_{00}B & a_{01}B & \cdots & a_{0,n-1}B \\ a_{10}B & a_{11}B & \cdots & a_{1,n-1}B \\ \vdots & \vdots & & \vdots \\ a_{n-1,0}B & a_{n-1,1}B & \cdots & a_{n-1,n-1}B \end{pmatrix}$$

(we have to deal with a matrix of n^2 exemplaries of the second factor B, weighted by the coefficients of the first factor A).

(a) $A = \begin{pmatrix} 1 & -2 \\ -1 & 1 \end{pmatrix}$, $B = \begin{pmatrix} 1 & 2 \\ 3 & -1 \end{pmatrix}$.

Compute $A \otimes B$ and $B \otimes A \in \mathbb{R}^{4 \times 4}$.

(b) Write, for $A = \begin{pmatrix} a_{00} & a_{01} \\ a_{10} & a_{11} \end{pmatrix}$ and $B = \begin{pmatrix} b_{00} & b_{01} \\ b_{10} & b_{11} \end{pmatrix} \in \mathbb{R}^{2 \times 2}$ the matrix

$A \otimes B \in \mathbb{R}^{4 \times 4}$, with its 16 coefficients.

(c) What is the relation between the transformation $T_{A \otimes B} : \mathbb{R}^{n \times n} \longrightarrow \mathbb{R}^{n \times n}$ and the matrix $A \otimes B \in \mathbb{R}^{n^2 \times n^2}$?

(Help: The question is finally the following: How to write a matrix $X \in \mathbb{R}^{n \times n}$ as a column vector $X^c \in \mathbb{R}^{n^2}$ in order to dispose of the following equivalence: $Y = T_{A \otimes B}(X) \iff Y^c = (A \otimes B)X^c?$).

(4) Let $H(n) \in \mathbb{R}^{2^n \times 2^n}$ be the matrix of the (non-normalized) Walsh–Hadamard transform to the order 2^n. Show that $H(n) = H(1) \otimes H(1) \otimes \cdots \otimes H(1)$ (n times).

The 2D DCT

The DCT acting on 8 × 8 matrices

Recall The DCT of order 8 $\quad C_8 : \mathbb{R}^8 \longrightarrow \mathbb{R}^8$

$$\begin{pmatrix} x_0 \\ \vdots \\ x_7 \end{pmatrix} \longmapsto \begin{pmatrix} y_0 \\ \vdots \\ y_7 \end{pmatrix}$$

is given by

$$y_j = \frac{r_j}{2} \cdot \sum_{k=0}^{7} x_k \cdot \cos \frac{j(2k+1)\pi}{16} \qquad 0 \le j \le 7$$

$$\left[\text{with} \quad r_j = \begin{cases} \frac{1}{\sqrt{2}} & \text{for } j = 0 \\ 1 & \text{else} \end{cases} \right]$$

The inverse DCT C_8^{-1} is obtained by transposition:

$$x_j = \sum_{k=0}^{7} \frac{r_k}{2} \cdot y_k \cdot \cos \frac{(2j+1)k\pi}{16} \qquad 0 \le j \le 7$$

Consequence of the previous section:

The 2D DCT $\quad T_{C\otimes C} : \mathbb{R}^{8\times 8} \quad \longrightarrow \quad \mathbb{R}^{8\times 8}$

$$X = \begin{pmatrix} x_{00} & \cdots & x_{07} \\ \vdots & & \vdots \\ x_{70} & \cdots & x_{77} \end{pmatrix} \longmapsto Y = \begin{pmatrix} y_{00} & \cdots & y_{07} \\ \vdots & & \vdots \\ y_{70} & \cdots & y_{77} \end{pmatrix}$$

is given by the formulas

$$y_{mn} = \frac{r_m}{2}\frac{r_n}{2} \cdot \sum_{j=0}^{7}\sum_{k=0}^{7} x_{jk} \cdot \cos\frac{(2j+1)m\pi}{16} \cdot \cos\frac{(2k+1)n\pi}{16}, \qquad 0 \le m,n \le 7.$$

The inverse transformation $T_{C\otimes C}^{-1}$ is then obtained this way:

$$x_{mn} = \sum_{j=0}^{7}\frac{r_j}{2} \cdot \sum_{k=0}^{7}\frac{r_k}{2} \cdot y_{jk} \cdot \cos\frac{j(2m+1)\pi}{16} \cdot \cos\frac{k(2n+1)\pi}{16}, \qquad 0 \le m,n \le 7.$$

Exercises

(1) Compute $T_{C\otimes C}$
$$\begin{pmatrix} 1&1&1&1&1&1&1&1 \\ 1&1&1&1&1&1&1&1 \\ 1&1&1&1&1&1&1&1 \\ 1&1&1&1&1&1&1&1 \\ 1&1&1&1&1&1&1&1 \\ 1&1&1&1&1&1&1&1 \\ 1&1&1&1&1&1&1&1 \\ 1&1&1&1&1&1&1&1 \end{pmatrix}.$$

(2) Compute $T_{C\otimes C}$
$$\begin{pmatrix} 0&0&0&0&0&0&0&0 \\ 0&1&2&3&4&5&6&7 \\ 0&2&4&6&8&10&12&14 \\ 0&3&6&9&12&15&18&21 \\ 0&4&8&12&16&20&24&28 \\ 0&5&10&15&20&25&30&35 \\ 0&6&12&18&24&30&36&42 \\ 0&7&14&21&28&35&42&49 \end{pmatrix}.$$

(3) Determine the $X \in \mathbb{R}^{8\times 8}$ such that $T_{C\otimes C}(X) = \begin{pmatrix} \|X\| & 0 & \cdots & 0 \\ 0 & 0 & \cdots & 0 \\ \vdots & \vdots & & \vdots \\ 0 & 0 & \cdots & 0 \end{pmatrix}$

("the energy of the transformed matrix is concentrated in the leading position").

The DCT in JPEG

The most popular standard in image compression, JPEG ("Joint Photographic Experts Group"), is designed to work in several modes.

The mode in which we are interested here is the mode of *lossy compression*. And it is exactly the 2D DCT which paves the way for the suppression of (secondary) information.

More precisely: The operational scheme of the standard JPEG – in lossy mode – consists of four steps:

(1) *Sampling*: Digital representation of the image (mostly photographic); the *data units* will be 8×8 matrices "covering" in three digital layers[2] image "tiles" of 8×8 pixels.
(2) (Local) intervention of the 2D DCT to the order 8×8: decorrelation of the numerical data, data unit by data unit.
(3) Quantization: Annihilation of the "light" coefficients in the transformed matrices.
(4) Compaction of the quantified data via Huffman coding (or via arithmetic coding).

Why is the Discrete Cosine Transform particularly appropriate to the decorrelation of the considered numerical data?

The following section will give a precise answer to this question.

But the pragmatic practitioner can also reason like this:

We know that
$$
T_{C \otimes C} \begin{pmatrix} 1\,1\,1\,1\,1\,1\,1\,1 \\ 1\,1\,1\,1\,1\,1\,1\,1 \\ 1\,1\,1\,1\,1\,1\,1\,1 \\ 1\,1\,1\,1\,1\,1\,1\,1 \\ 1\,1\,1\,1\,1\,1\,1\,1 \\ 1\,1\,1\,1\,1\,1\,1\,1 \\ 1\,1\,1\,1\,1\,1\,1\,1 \\ 1\,1\,1\,1\,1\,1\,1\,1 \end{pmatrix} = \begin{pmatrix} 8\,0\,0\,0\,0\,0\,0\,0 \\ 0\,0\,0\,0\,0\,0\,0\,0 \\ 0\,0\,0\,0\,0\,0\,0\,0 \\ 0\,0\,0\,0\,0\,0\,0\,0 \\ 0\,0\,0\,0\,0\,0\,0\,0 \\ 0\,0\,0\,0\,0\,0\,0\,0 \\ 0\,0\,0\,0\,0\,0\,0\,0 \\ 0\,0\,0\,0\,0\,0\,0\,0 \end{pmatrix}.
$$

Hence, a constant data unit is transformed into one leading coefficient which is its "energy" (the DC coefficient) and into 63 zeros (the AC coefficients).

More generally, consider an arbitrary data unit: It consists of an 8×8 matrix of 64 sample values, where every single value encodes the intensity of a colour component (or of a derived coordinate) associated with a pixel.

The scheme of the sample values:

$x_{00}\ x_{01}\ \dots\ x_{07}$
$x_{10}\ x_{11}\ \dots\ x_{17}$

$\vdots\quad\vdots\qquad\vdots$

$x_{70}\ x_{71}\ \dots\ x_{77}$

After 2D DCT of order 8×8, we obtain the transformed scheme

$y_{00}\ y_{01}\ \dots\ y_{07}$
$y_{10}\ y_{11}\ \dots\ y_{17}$

$\vdots\quad\vdots\qquad\vdots$

$y_{70}\ y_{71}\ \dots\ y_{77}$

We know: $\sum x_{ij}^2 = \sum y_{kl}^2$ (the DCT preserves the Euclidean norm).

---

[2] According to the chosen colour system (we suppose a transformation RGB $\longrightarrow$ YCbCr).

Now write

$$
\begin{pmatrix}
x_{00} & x_{01} & \cdots & x_{07} \\
x_{10} & x_{11} & \cdots & x_{17} \\
\vdots & \vdots & & \vdots \\
x_{70} & x_{71} & \cdots & x_{77}
\end{pmatrix}
=
\begin{pmatrix}
m & m & \cdots & m \\
m & m & \cdots & m \\
\vdots & \vdots & & \vdots \\
m & m & \cdots & m
\end{pmatrix}
+
\begin{pmatrix}
\delta_{00} & \delta_{01} & \cdots & \delta_{07} \\
\delta_{10} & \delta_{11} & \cdots & \delta_{17} \\
\vdots & \vdots & & \vdots \\
\delta_{70} & \delta_{71} & \cdots & \delta_{77}
\end{pmatrix},
$$

where $m = \frac{1}{8} y_{00}$ is the arithmetical mean of the x_{ij}, and the δ_{ij} are precisely the deviations from this average. This yields, after transformation, the decomposition

$$
\begin{pmatrix}
y_{00} & y_{01} & \cdots & y_{07} \\
y_{10} & y_{11} & \cdots & y_{17} \\
\vdots & \vdots & & \vdots \\
y_{70} & y_{71} & \cdots & y_{77}
\end{pmatrix}
=
\begin{pmatrix}
8m & 0 & \cdots & 0 \\
0 & 0 & \cdots & 0 \\
\vdots & \vdots & & \vdots \\
0 & 0 & \cdots & 0
\end{pmatrix}
+
\begin{pmatrix}
0 & y_{01} & \cdots & y_{07} \\
y_{10} & y_{11} & \cdots & y_{17} \\
\vdots & \vdots & & \vdots \\
y_{70} & y_{71} & \cdots & y_{77}
\end{pmatrix}.
$$

As a consequence, the DC coefficient of the transformed scheme only depends on the average of the sample values, whereas the AC coefficients only depend on the deviations of the sample values from this average.

This roughly explains the names *direct current/alternative current* for the coefficients.

Note that (since the DCT preserves the Euclidean norm) a small mean square value of the AC (transform) coefficients equals a small mean square value of the sample variations.

Example

43 44 46 51 57 65 72 76	512.3 −15.2 6.6 2.9 −1.8 −0.5 −0.3 0
44 45 46 50 56 63 69 72	−79.3 −51.8 7.5 0.6 0.6 0 −1 −0.6
47 47 46 49 55 60 64 65	45 −10.5 0.8 −2.5 0.7 0.4 0.6 0.3
53 53 52 52 54 57 59 59	−6.6 5.3 −1.1 0.2 0.1 0.1 0.2 −0.6
62 62 59 58 58 58 57 55	0.8 −0.7 0.3 −1.1 −0.3 0.3 0.1 0
75 73 71 68 67 66 62 60	0 0.4 0.1 −0.7 0.1 0.2 −0.3 0
85 85 83 80 77 74 72 70	0.3 −0.8 0.1 1.5 0.3 −0.1 0.8 0.2
92 91 91 87 85 83 81 80	−1.2 0.1 −0.3 0.3 0.1 −0.2 −1.1 −0.1

Considering the example above as generic, we can state: If the values x_{ij} are "numerically connected", then we can expect that the values y_{kl} are numerically important in the leading positions of the left upper corner of the matrix transform scheme, i.e. that we have to do with a value distribution which has approximately the form

```
A B B C C D D D
B B C C D D D D
B C C D D D D D
C C D D D D D D
C D D D D D D D'
D D D D D D D D
D D D D D D D D
D D D D D D D D
```

where A, B, C, D indicate, in decreasing order, the numerical weight of the position.

Note that this situation seems to be a pictorial commonplace: a data unit describes a local region of 8×8 pixels in the given digital image. Now, in photographic images, local uniformity of colour (or of the same Grey intensity) should be, in general, frequent: there will be sky, walls, doors, red cheeks, lawn ...

All this will be transformed into data units which are "numerically homogeneous".

Exercise

Let A be the 8×8 matrix all coefficients of which are equal to 128, and let X be a 8×8 matrix such that $\|A - X\| = 16$ (we admit an average deviation of size 2 per coefficient). Let Y be the 2D DCT image of X. Show: We have for the DC coefficient: $1008 \leq y_{00} \leq 1040$ and for the AC coefficients: $\sum y_{ij}^2 \leq 256$ – i.e. in average $\mid y_{ij} \mid \leq 2$.

The compaction algorithms (Huffman coding or arithmetic coding), which work in the final step of the JPEG compression, need, in order to be efficient, a great *statistical imbalance* of the data (of the letters to be treated). In other words: We need an important invasion of zeros in the DCT data units.

This leads to a first (somewhat naïve) idea of clever information suppression:

If we annihilate the 48 coefficients of the transformed scheme which are outside the upper left 4×4 square, we will lower – for an image zone which is not too "agitated" – the Euclidean norm (the energy) of the entire scheme perhaps around 5%. After retransformation of the "truncated scheme" we shall obtain a slightly changed initial square: The new numerical values of the 64 coefficients will differ from the initial values in average 5% each. If the colour palette is tolerant, this will not change drastically the quality of the image.

The annihilation of the "light" coefficients will actually be done more carefully: One uses *quantization tables*:

$$q_{00} \ q_{01} \ \cdots \ q_{07}$$
$$q_{10} \ q_{11} \ \cdots \ q_{17}$$

$$\vdots \quad \vdots \qquad \vdots$$

$$q_{70} \ q_{71} \ \cdots \ q_{77}$$

The "planing down" is then done when passing to the *quantized coefficients*:

$$\tilde{y}_{jk} = \left[\frac{y_{jk}}{q_{jk}} \right],$$

where $[\,]$ means the round to the nearest integer. Two examples of quantization tables:

A. *Luminance*:

```
16 11 10 16  24  40  51  61
12 12 14 19  26  58  60  55
14 13 16 24  40  57  69  56
14 17 22 29  51  87  80  62
18 22 37 56  68 109 103  77
24 35 55 64  81 104 113  92
49 64 78 87 103 121 120 101
72 92 95 98 112 100 103  99
```

(Note that the luminance Y varies a priori[3] between 0 and 255, then will be renormalized between 16 – the Black threshold – and 235).

B. *Chrominance*:

17 18 24 47 99 99 99 99
18 21 26 66 99 99 99 99
24 26 56 99 99 99 99 99
47 66 99 99 99 99 99 99
99 99 99 99 99 99 99 99
99 99 99 99 99 99 99 99
99 99 99 99 99 99 99 99
99 99 99 99 99 99 99 99

(the chrominance – Cb and Cr – varies a priori between -127.5 and 127.5, but will be renormalized between 16 and 240, with 128 as "zero value").

These tables have been established empirically.

Example In the following example, we consider the treatment of a data unit of luminance sample values. Note that the luminance component will be the *only* component in the treatment of a Black and White digital photo (the chrominance components concern the colours). We shall display the compression as well as the decompression (the encoding and the decoding in the compaction step will remain in a black box – we refer the reader to the first chapter of this book).

The Grey sample values

139 144 149 153 155 155 155 155
144 151 153 156 159 156 156 156
150 155 160 163 158 156 156 156
159 161 162 160 160 159 159 159
159 160 161 162 162 155 155 155
161 161 161 161 160 157 157 157
162 162 161 163 162 157 157 157
162 162 161 161 163 158 158 158

The DCT coefficients

1259.6 -1.0 -12.1 -5.2 2.1 -1.7 -2.7 1.3
-22.6 -17.5 -6.2 -3.2 -2.9 -0.1 0.4 -1.2
-10.9 -9.3 -1.6 1.5 0.2 -0.9 -0.6 -0.1
-7.1 -1.9 0.2 1.5 0.9 -0.1 0.0 0.3
-0.6 -0.8 1.5 1.6 -0.1 -0.7 0.6 1.3
1.8 -0.2 1.6 -0.3 -0.8 1.5 1.0 -1.0
-1.3 -0.4 -0.3 -1.5 -0.5 1.7 1.1 -0.8
-2.6 1.6 -3.8 -1.8 1.9 1.2 -0.6 -0.4

The quantization table

16 11 10 16 24 40 51 61
12 12 14 19 26 58 60 55
14 13 16 24 40 57 69 56
14 17 22 29 51 87 80 62
18 22 37 56 68 109 103 77
24 35 55 64 81 104 113 92
49 64 78 87 103 121 120 101
72 92 95 98 112 100 103 99

The quantized coefficients

79 0 -1 0 0 0 0 0
-2 -1 0 0 0 0 0 0
-1 -1 0 0 0 0 0 0
-1 0 0 0 0 0 0 0
0 0 0 0 0 0 0 0
0 0 0 0 0 0 0 0
0 0 0 0 0 0 0 0
0 0 0 0 0 0 0 0

---

[3] A priori: computed, together with blue and red chrominance, via linear transformation, from the colour coordinates Red–Green–Blue, which vary between 0 and 255.

The dequantized coefficients	The reconstructed Grey values
1264 0 −10 0 0 0 0 0	142 144 147 150 152 153 154 154
−24 −12 0 0 0 0 0 0	149 150 153 155 156 157 156 156
−14 −13 0 0 0 0 0 0	157 158 159 161 161 160 159 158
−14 0 0 0 0 0 0 0	162 162 163 163 162 160 158 157
0 0 0 0 0 0 0 0	162 162 162 162 161 158 156 155
0 0 0 0 0 0 0 0	160 161 161 161 160 158 156 154
0 0 0 0 0 0 0 0	160 160 161 162 161 160 158 157
0 0 0 0 0 0 0 0	160 161 163 164 164 163 161 160

Let us explain the operations that we have carried out:

(1) The table of the Grey sample values X describes a relatively uniform region: The sample values don't vary sensibly.
(2) The DCT coefficients: We have computed $Y = C_8 X C_8^t$.
(3) The quantized coefficients: We have divided every coefficient of the matrix Y by the corresponding coefficient of the quantization table, then we have rounded to the next integer.
    −The table of the quantized AC coefficients is read sequentially, in zigzag:
    $0, -2, -1, -1, -1, 0, 0, -1, -1, 0, 0, \ldots$ and will be encoded by a compaction algorithm (Huffman or arithmetic coding). The DC coefficients are encoded separately, along the data units. This ends the *compression*.
    The *decompression* begins with the decoding of the compacted data: One reconstructs the table of the quantized coefficients.
(4) The table of the reconstructed DCT coefficients: We have multiplied every quantized coefficient by the corresponding quantizer $(1, 264 = 79 \cdot 16, -24 = -2 \cdot 12$, etc.).
(5) The matrix X' of the reconstructed Grey values: Let Y' be the matrix of the reconstructed DCT coefficients; then $X' = C_8^t Y' C_8$.

Let us finish this section on the intervention of the 2D DCT in JPEG with a

Remark The DCT can justly be considered as a decorrelation transformation which does not change the character of the data that it treats: What is temporal, remains temporal, what is spatial, remains spatial. The next section, treating the DCT as a KLT, will reinforce this statement. Conceptually, there are then no frequencies indexing the positions in the DCT coefficient table. On the other hand, as we have already pointed out, the *direct current/alternative current* terminology seems to be perfectly adapted. Moreover, the DCT is intimately linked to the Discrete Fourier Transform which decorrelates time domain data into frequency domain data. Will there be altogether "high frequencies" in DCT tables?
    Let us briefly look at this question.
    As an exercise, we shall replace, in an implementation of JPEG, the matrix C_8 of the DCT by the following matrix Q_8:

$$Q_8 = \begin{pmatrix}
\frac{1}{2\sqrt{2}} & \frac{1}{2\sqrt{2}} & \frac{1}{2\sqrt{2}} & \frac{1}{2\sqrt{2}} & \frac{1}{2\sqrt{2}} & \frac{1}{2\sqrt{2}} & \frac{1}{2\sqrt{2}} & \frac{1}{2\sqrt{2}} \\
-\frac{7}{2\sqrt{14}} & \frac{1}{2\sqrt{14}} & \frac{1}{2\sqrt{14}} & \frac{1}{2\sqrt{14}} & \frac{1}{2\sqrt{14}} & \frac{1}{2\sqrt{14}} & \frac{1}{2\sqrt{14}} & \frac{1}{2\sqrt{14}} \\
0 & -\frac{6}{\sqrt{42}} & \frac{1}{\sqrt{42}} & \frac{1}{\sqrt{42}} & \frac{1}{\sqrt{42}} & \frac{1}{\sqrt{42}} & \frac{1}{\sqrt{42}} & \frac{1}{\sqrt{42}} \\
0 & 0 & -\frac{5}{\sqrt{30}} & \frac{1}{\sqrt{30}} & \frac{1}{\sqrt{30}} & \frac{1}{\sqrt{30}} & \frac{1}{\sqrt{30}} & \frac{1}{\sqrt{30}} \\
0 & 0 & 0 & -\frac{4}{2\sqrt{5}} & \frac{1}{2\sqrt{5}} & \frac{1}{2\sqrt{5}} & \frac{1}{2\sqrt{5}} & \frac{1}{2\sqrt{5}} \\
0 & 0 & 0 & 0 & -\frac{3}{2\sqrt{3}} & \frac{1}{2\sqrt{3}} & \frac{1}{2\sqrt{3}} & \frac{1}{2\sqrt{3}} \\
0 & 0 & 0 & 0 & 0 & -\frac{2}{\sqrt{6}} & \frac{1}{\sqrt{6}} & \frac{1}{\sqrt{6}} \\
0 & 0 & 0 & 0 & 0 & 0 & -\frac{1}{\sqrt{2}} & \frac{1}{\sqrt{2}}
\end{pmatrix}.$$

This matrix is orthogonal and it diagonalizes the *same* matrix $A = \begin{pmatrix} 1 \dots 1 \\ \vdots \quad \vdots \\ 1 \dots 1 \end{pmatrix}$ as

the DCT C_8.

We get an equivalent decorrelation transform, but every "frequential flair" has now disappeared.

What about the effect of either transformation on samples of high variation?

Example Consider the sample matrix

$$\begin{matrix}
0 & 250 & 0 & 250 & 0 & 250 & 0 & 250 \\
250 & 0 & 250 & 0 & 250 & 0 & 250 & 0 \\
0 & 250 & 0 & 250 & 0 & 250 & 0 & 250 \\
250 & 0 & 250 & 0 & 250 & 0 & 250 & 0 \\
0 & 250 & 0 & 250 & 0 & 250 & 0 & 250 \\
250 & 0 & 250 & 0 & 250 & 0 & 250 & 0 \\
0 & 250 & 0 & 250 & 0 & 250 & 0 & 250 \\
250 & 0 & 250 & 0 & 250 & 0 & 250 & 0
\end{matrix}$$

We get, for the C_8 – values and for the Q_8 – values, the two matrices:

1000	0	0	0	0	0	0	0
0	−32.5	0	−38.3	0	−57.4	0	−163.3
0	0	0	0	0	0	0	0
0	−38.3	0	−45.2	0	−67.6	0	−192.6
0	0	0	0	0	0	0	0
0	−57.4	0	−67.6	0	−101.2	0	−288.3
0	0	0	0	0	0	0	0
0	−163.3	0	−192.6	0	−288.3	0	−821.1

1000	0	0	0	0	0	0
0	−142.9	123.7	−146.4	119.5	−154.3	109.1	−189
0	123.7	−107.1	126.8	−103.5	133.6	−94.5	163.7
0	−146.4	126.8	−150	122.5	−158.1	111.8	−193.6
0	119.5	−103.5	122.5	−100	129.1	−91.3	158.1
0	−154.3	133.6	−158.1	129.1	−166.7	117.9	−204.1
0	109.1	−94.5	111.8	−91.3	117.9	−83.3	144.3
0	−189	163.7	−193.6	158.1	−204.1	144.3	−250

We note: Whereas the Q_8 – transform matrix reflects faithfully the value oscillation of the sample matrix, the C_8 – transform matrix displays this information differently: the "high frequency" AC coefficients in the last row (and column) are plainly dominant.

So, we have to accept: Despite the geometrical background of our decorrelation transform – find principal axes for a highly degenerate quadratic form – the DCT version allows, by its Fourier descent, a kind of traditional frequency domain description.

By the way: Looking at the performance of JPEG(Q_8), compared with the performance of JPEG(C_8), we shall realize a small advantage of JPEG(C_8), which stems from the fact that the quantization tables have been established for the DCT.

Exercises

(1) Compute the transformed matrix of

$$
\begin{bmatrix}
139 & 144 & 149 & 153 & 155 & 155 & 155 & 155 \\
144 & 151 & 153 & 156 & 159 & 156 & 156 & 156 \\
150 & 155 & 160 & 163 & 158 & 156 & 156 & 156 \\
159 & 161 & 162 & 160 & 160 & 159 & 159 & 159 \\
159 & 160 & 161 & 162 & 162 & 155 & 155 & 155 \\
161 & 161 & 161 & 161 & 160 & 157 & 157 & 157 \\
162 & 162 & 161 & 163 & 162 & 157 & 157 & 157 \\
162 & 162 & 161 & 161 & 163 & 158 & 158 & 158
\end{bmatrix}
$$

in JPEG(Q_8) version.

Quantize according to the luminance quantization table of JPEG(C_8).

The transformed scheme

$$
\begin{bmatrix}
1259.6 & 8.9 & 2.7 & -1.5 & -6.2 & -8 & 0 & 0 \\
20.6 & -10 & -8.6 & -6 & -4.2 & -3 & 0 & 0 \\
13.9 & -9.4 & -5.2 & -5 & -3.7 & -0.7 & 0 & 0 \\
7.6 & -7.5 & -4.3 & -0.1 & 2.7 & -1.7 & 0 & 0 \\
-0.6 & -2.2 & -0.8 & 0.3 & -2.4 & -3.1 & 0 & 0 \\
3.3 & -1.2 & -0.5 & 1 & 1.5 & 2 & 0 & 0 \\
2 & -0.1 & -0.1 & 0.7 & 0 & -1.4 & 0 & 0 \\
0.5 & 0.2 & 0.2 & 0.3 & 1.9 & 0 & 0 & 0
\end{bmatrix}
$$

The quantized coefficients

$$
\begin{bmatrix}
79 & 1 & 0 & 0 & 0 & 0 & 0 & 0 \\
2 & -1 & -1 & 0 & 0 & 0 & 0 & 0 \\
1 & -1 & 0 & 0 & 0 & 0 & 0 & 0 \\
1 & 0 & 0 & 0 & 0 & 0 & 0 & 0 \\
0 & 0 & 0 & 0 & 0 & 0 & 0 & 0 \\
0 & 0 & 0 & 0 & 0 & 0 & 0 & 0 \\
0 & 0 & 0 & 0 & 0 & 0 & 0 & 0 \\
0 & 0 & 0 & 0 & 0 & 0 & 0 & 0
\end{bmatrix}
$$

(2) Compare the transforms of

$$
\begin{bmatrix}
30 & 30 & 30 & 30 & 30 & 30 & 30 & 30 \\
60 & 60 & 60 & 60 & 60 & 60 & 60 & 60 \\
90 & 90 & 90 & 90 & 90 & 90 & 90 & 90 \\
120 & 120 & 120 & 120 & 120 & 120 & 120 & 120 \\
150 & 150 & 150 & 150 & 150 & 150 & 150 & 150 \\
180 & 180 & 180 & 180 & 180 & 180 & 180 & 180 \\
210 & 210 & 210 & 210 & 210 & 210 & 210 & 210 \\
240 & 240 & 240 & 240 & 240 & 240 & 240 & 240
\end{bmatrix}
$$

in JPEG(C_8) and in JPEG(Q_8).

The transformed scheme in JPEG(C_8)

$$
\begin{bmatrix}
1080 & 0 & 0 & 0 & 0 & 0 & 0 & 0 \\
-546.6 & 0 & 0 & 0 & 0 & 0 & 0 & 0 \\
0 & 0 & 0 & 0 & 0 & 0 & 0 & 0 \\
-57.1 & 0 & 0 & 0 & 0 & 0 & 0 & 0 \\
0 & 0 & 0 & 0 & 0 & 0 & 0 & 0 \\
-17 & 0 & 0 & 0 & 0 & 0 & 0 & 0 \\
0 & 0 & 0 & 0 & 0 & 0 & 0 & 0 \\
-4.3 & 0 & 0 & 0 & 0 & 0 & 0 & 0
\end{bmatrix}
$$

The transformed scheme in JPEG(Q_8)

$$
\begin{bmatrix}
1080 & 0 & 0 & 0 & 0 & 0 & 0 & 0 \\
317.5 & 0 & 0 & 0 & 0 & 0 & 0 & 0 \\
275 & 0 & 0 & 0 & 0 & 0 & 0 & 0 \\
232.4 & 0 & 0 & 0 & 0 & 0 & 0 & 0 \\
189.7 & 0 & 0 & 0 & 0 & 0 & 0 & 0 \\
147 & 0 & 0 & 0 & 0 & 0 & 0 & 0 \\
103.9 & 0 & 0 & 0 & 0 & 0 & 0 & 0 \\
60 & 0 & 0 & 0 & 0 & 0 & 0 & 0
\end{bmatrix}
$$

5.2.3 The Karhunen–Loève Transform and the DCT

This section furnishes a conceptual complement for the assessment of the DCT. The 1D Discrete Cosine Transform will be presented as the most ordinary member[4] of the family of orthogonal transformations reassembled under the generic name *KLT*.

The Karhunen–Loève Transform (KLT) is the solution of the following problem: Find a universal method of Linear Algebra which, in the outset of probabilistic signal theory, works efficiently for the suppression of secondary information (redundant or not), and which is optimal in a natural mathematical sense, inspired by a Euclidian (i.e. mean square error) formalism.

Our language will thus be that of random variables. It should be pointed out that in this grey zone of descriptive Applied Mathematics all concrete algorithmic glory depends on an efficient statistical evaluation of the data.

The Karhunen–Loève Transform of a Random Vector

Now we will have to refer to our knowledge of the diagonalization of real symmetric matrices, in a language of probabilistic modelling.

Diagonalization of the Covariance Matrix

The situation:

Let $\quad \mathbf{X} = \begin{pmatrix} X_0 \\ X_1 \\ \vdots \\ X_{n-1} \end{pmatrix} \quad$ be a vector of n random variables.

Its mean is defined by

$$\mathbf{m}_X = \begin{pmatrix} m_0 \\ m_1 \\ \vdots \\ m_{n-1} \end{pmatrix} = \begin{pmatrix} E(X_0) \\ E(X_1) \\ \vdots \\ E(X_{n-1}) \end{pmatrix},$$

where $E(X_j)$ is the expectation of the random variable X_j $\quad 0 \le j \le n-1$. Its *covariance matrix*:

$$\Gamma_X = \begin{pmatrix} \sigma_{00} & \sigma_{01} & \cdots & \sigma_{0,n-1} \\ \sigma_{10} & \sigma_{11} & \cdots & \sigma_{1,n-1} \\ \vdots & & & \vdots \\ \sigma_{n-1,0} & \sigma_{n-1,1} & \cdots & \sigma_{n-1,n-1} \end{pmatrix},$$

where $\sigma_{ij} = E((X_i - m_i)(X_j - m_j))$, $0 \le i, j \le n-1$.

In order to avoid formal pedantries, we shall always tacitly assume that the mean m_X and the covariance matrix Γ_X are well-defined.

Remark Concerning the expectation: At this rather general stage, any reasonable definition of a mean is acceptable, provided it satisfies the following two conditions:

(i) Linearity: $E(\lambda_1 X_1 + \lambda_2 X_2) = \lambda_1 E(X_1) + \lambda_2 E(X_2)$.

---

[4] This is not a common opinion in signal theory.

(ii) Monotony: $X \leq Y \Longrightarrow E(X) \leq E(Y)$.

Lemma *The matrix* Γ_X *is positive semidefinite:*

$$u^t \Gamma_X u \geq 0 \qquad for\ all \quad u = \begin{pmatrix} u_0 \\ \vdots \\ u_{n-1} \end{pmatrix} \in \mathbb{R}^n.$$

Proof $\left(u_0, u_1, \ldots, u_{n-1}\right) \left(E((X_i - m_i)(X_j - m_j))\right) \begin{pmatrix} u_0 \\ u_1 \\ \vdots \\ u_{n-1} \end{pmatrix}$

$= E((\sum_{i=0}^{n-1} u_i (X_i - m_i))^2) \geq 0.$

Consequence The eigenvalues of the matrix Γ_X (which is a real symmetric matrix) are all non-negative.

Diagonalize Γ_X:

There exists a orthogonal matrix $Q \in \mathbb{R}^{n \times n}$ such that

$$\Gamma_X = Q \begin{pmatrix} \sigma_0^2 & & \\ 0 & \ddots & \\ 0 & & \sigma_{n-1}^2 \end{pmatrix} Q^t,$$

where

(i) $\sigma_0^2 \geq \sigma_1^2 \geq \cdots \geq \sigma_{n-1}^2 \geq 0$ are the eigenvalues of Γ_X
(ii) the columns $\mathbf{q}_0, \mathbf{q}_1, \ldots, \mathbf{q}_{n-1}$ of Q are an orthonormal basis of the $\mathbb{R}^n$, composed of eigenvectors for Γ_X.

Definition *With the notations introduced above, put*

$$\mathbf{Z} = \begin{pmatrix} Z_0 \\ Z_1 \\ \vdots \\ Z_{n-1} \end{pmatrix} = Q^t \cdot (\mathbf{X} - \mathbf{m}_X) = Q^t \begin{pmatrix} X_0 - m_0 \\ X_1 - m_1 \\ \vdots \\ X_{n-1} - m_{n-1} \end{pmatrix}.$$

Z *is called the KLT of the random vector* **X**.

Formal observation Let $\mathbf{X} = \begin{pmatrix} X_0 \\ X_1 \\ \vdots \\ X_{n-1} \end{pmatrix}$ be a vector of n random variables.

$A \in \mathbb{R}^{n \times n}, b \in \mathbb{R}^n$

Then:

(1) $\mathbf{m}_{AX+b} = A\mathbf{m}_X + b$
(2) $\Gamma_{AX+b} = A\Gamma_X A^t$

Consequence $\mathbf{m}_Z = 0$ and $\Gamma_Z = \begin{pmatrix} \sigma_0^2 & & & \\ & \sigma_1^2 & & 0 \\ & & \ddots & \\ 0 & & & \sigma_{n-1}^2 \end{pmatrix}$.

In particular, σ_j^2 is the variance of the random variable Z_j, $0 \le j \le n-1$.

Auxiliary result Let Y be a random variable such that $\sigma_Y^2 = 0$ (the variance of Y is zero).

Then $Y = m_Y$ almost surely (this is a consequence of Tchebyshev's inequality). So, we can affirm:

If $\sigma_{n-r}^2 = \cdots = \sigma_{n-1}^2 = 0$ (the rank of $\Gamma_\mathbf{X}$ is equal to $\mathbf{n-r}$)
then $\mathbf{Z_{n-r}} = \cdots = \mathbf{Z_{n-1}} = 0$ almost surely.

The number of the "degrees of freedom" of a random vector $\mathbf{X}$ is thus given by the *rank* of its covariance matrix $\Gamma_\mathbf{X}$.

The Optimality of the KLT

Let us first point out the distinctive particularity of the KLT:

Every random vector $\mathbf{X}$ has *its own* KLT, which depends on (the diagonalization of) its covariance matrix $\Gamma_\mathbf{X}$. We insist: In the transformation equation
$\mathbf{Z} = Q^t(\mathbf{X} - \mathbf{m}_X)$, the coefficients of the transformation matrix Q^t *depend on* $\mathbf{X}$).

In what sense is the KLT optimal?

First, recall:

Let $f(\theta)$ be a (continuous) periodic signal, and let $\sigma_N f(\theta)$ be its Fourier polynomial to the order N.

Then $\sigma_N f(\theta)$ *minimizes* a certain Euclidian distance:

$\sigma_N f(\theta)$ is the orthogonal projection of $f(\theta)$ (in an appropriate space of functions) onto the subspace of the trigonometric polynomials of degree $\le N$.

Here, we are in an *analogous situation.*

Consider the vector space Υ^n of n-component random vectors $\mathbf{X} = \begin{pmatrix} X_0 \\ \vdots \\ X_{n-1} \end{pmatrix}$,

with $\mathbf{m}_X = 0$ (modulo the subspace of those which are almost surely zero).

$\Big[$Our *general hypothesis*: The mean $\mathbf{m}_X$ of $\mathbf{X}$ and its covariance matrix Γ_X will always be well-defined. Note, moreover, that the KLT acts only on *zero mean* random vectors.$\Big]$

Introduce now the following inner product:

$$\langle \mathbf{X}, \mathbf{Y} \rangle = E\left(\sum_{i=0}^{n-1} X_i Y_i\right) \quad \text{for}\quad \mathbf{X} = \begin{pmatrix} X_0 \\ \vdots \\ X_{n-1} \end{pmatrix}, \mathbf{Y} = \begin{pmatrix} Y_0 \\ \vdots \\ Y_{n-1} \end{pmatrix} \in \Upsilon^n.$$

We obtain for the associated Euclidean norm:
$\|\mathbf{X}\|^2 = \langle \mathbf{X}, \mathbf{X} \rangle = E(\sum_{i=0}^{n-1} X_i^2) = \text{tr}\Gamma_X$,
where the trace of the covariance matrix Γ_X is equal to the sum of its eigenvalues.

Now fix $\mathbf{X} = \begin{pmatrix} X_0 \\ \vdots \\ X_{n-1} \end{pmatrix}$. Let $A = \Gamma_X$ be the covariance matrix of $\mathbf{X}$.

Choose $m \equiv$ *the order of truncation* $1 \le m \le n$.

We are interested in *orthonormal representations* to the order m

$$\mathbf{Z} = \begin{pmatrix} Z_0 \\ \vdots \\ Z_{m-1} \end{pmatrix} = U_m^t \mathbf{X} = U_m^t \begin{pmatrix} X_0 \\ \vdots \\ X_{n-1} \end{pmatrix} \text{ of } \mathbf{X},$$

where $U_m = (\mathbf{u}_0, \mathbf{u}_1, \ldots, \mathbf{u}_{m-1}) \in \mathbb{R}^{n \times m}$ is a matrix with m orthonormal columns (of length n).

Note that to every orthonormal representation $\mathbf{X} \longmapsto \mathbf{Z} = U_m^t \mathbf{X}$ corresponds a *section* $U_m^t A U_m \in \mathbb{R}^{m \times m}$ of the covariance matrix $A = \Gamma_X$ of $\mathbf{X}$.

In order to compare the random vector $\mathbf{X} = \begin{pmatrix} X_0 \\ \vdots \\ X_{n-1} \end{pmatrix}$

with its "orthonormal truncation to the order m" $\mathbf{Z} = U_m^t \mathbf{X}$, we shall use the natural *isometric section* $\Upsilon^m \longrightarrow \Upsilon^n$ given by (the multiplication with) U_m.

$\left[\text{The identity } U_m^t U_m = I_m \text{ implies that } \|U_m \mathbf{Z}\| = \|\mathbf{Z}\| \text{ for all } \mathbf{Z} \in \Upsilon^m \right]$

The result which states the optimality of the KLT is the following:

Proposition *The situation as described above; then we have:*
$\|\mathbf{X} - U_m \mathbf{Z}\|$ *is minimum for* $U_m = Q_m = (\mathbf{q}_0, \mathbf{q}_1, \ldots, \mathbf{q}_{m-1})$, *where* $\{\mathbf{q}_0, \mathbf{q}_1, \cdots, \mathbf{q}_{m-1}\}$ *is an orthonormal basis of eigenvectors, corresponding to the eigenvalues* $\sigma_0^2 \ge \sigma_1^2 \ge \cdots \ge \sigma_{n-1}^2 \ge 0$ *of* $A = \Gamma_X$.

Proof $\|\mathbf{X} - U_m \mathbf{Z}\|^2 = \langle \mathbf{X}, \mathbf{X} \rangle - 2\langle \mathbf{X}, U_m \mathbf{Z} \rangle + \langle U_m \mathbf{Z}, U_m \mathbf{Z} \rangle$.

But $\langle \mathbf{X}, U_m \mathbf{Z} \rangle = \langle \mathbf{X}, U_m U_m^t \mathbf{X} \rangle = \langle U_m^t \mathbf{X}, U_m^t \mathbf{X} \rangle$ and $\langle U_m \mathbf{Z}, U_m \mathbf{Z} \rangle = \langle U_m^t \mathbf{X}, U_m^t \mathbf{X} \rangle$.

Hence: $\|\mathbf{X} - U_m \mathbf{Z}\|^2 = \|\mathbf{X}\|^2 - \|U_m^t \mathbf{X}\|^2$.

Finally: $\|\mathbf{X} - U_m \mathbf{Z}\|^2 = \operatorname{tr}\Gamma_X - \operatorname{tr}(U_m^t \Gamma_X U_m) = \sum_{i=0}^{n-1} \sigma_i^2 - \sum_{i=0}^{m-1} \sigma_i'^2$, where $\sigma_0^2 \ge \sigma_1^2 \ge \cdots \ge \sigma_{n-1}^2 \ge 0$ are the eigenvalues of $A = \Gamma_X$ and $\sigma_0'^2 \ge \sigma_1'^2 \ge \cdots \ge \sigma_{m-1}'^2 \ge 0$ are the eigenvalues of $B = U_m^t \Gamma_X U_m$.

Our previous results on the extremal properties of the eigenvalues permit us to conclude.

Let us sum up:

Let $\mathbf{X} = \begin{pmatrix} X_0 \\ \vdots \\ X_{n-1} \end{pmatrix}$ be a random vector with $\mathbf{m}_X = \begin{pmatrix} 0 \\ \vdots \\ 0 \end{pmatrix}$.

Let $\mathbf{Z} = Q^t \mathbf{X}$ be its KLT.
Then:

(1) The components of $\mathbf{Z}$ are completely *decorrelated*:
$\sigma_{ij} = \operatorname{cov}(Z_i, Z_j) = 0 \quad$ for $\quad i \ne j$

(2) Consider, for each *fixed* m, $1 \leq m \leq n$, the transformations by *orthonormal truncation* to the order m $\mathbf{X} \longmapsto U_m^t \mathbf{X}$ (as explained above). Then the difference of the "energy norms" $\|\mathbf{X}\|^2 - \|U_m^t \mathbf{X}\|^2$ is *minimized* by the *Karhunen–Loève truncation* $\mathbf{X} \longmapsto Q_m^t \mathbf{X}$.

The DCT – a Karhunen–Loève Transform

The Discrete Cosine Transform is the simplest KLT. In order to understand its strong position in concrete applications, recall the meaning of the notion "random variable": Algorithmic information on the concrete values of the signals is replaced by (hopefully sufficient) information on the *behaviour* of the signals, thus permitting satisfactory predictions. You have to accept an environment where precise information comes from descriptive estimation, together with statistical evaluation.

Now, the KLT of a random vector $\mathbf{X}$ is *parametrized* by its *current argument*, i.e. it is of the form $\mathbf{Z} = \mathrm{KLT}_X(\mathbf{X})$.

In other words, in order to be able to transform an actual candidate $\mathbf{X}$, we need a certain amount of statistical information which allows the computation of its covariance matrix (the diagonalization of which demands a pretty supplementary numerical effort).

So, the KLT is, alas, *not predefinable* (and thus *without interest* for massive practical real time applications).

But there is a small family of random vectors which has a *common* predetermined KLT:

Recall Let $\mathbf{X} = \begin{pmatrix} X_0 \\ \vdots \\ X_{n-1} \end{pmatrix}$ be a random vector such that $\Gamma_X = \begin{pmatrix} 1\,1\dots1 \\ 1\,1\dots1 \\ \vdots\;\vdots\quad\vdots \\ 1\,1\dots1 \end{pmatrix}$.

Then $\Gamma_X = C_n^t \begin{pmatrix} n\,0\dots0 \\ 0\,0\quad0 \\ \vdots\quad\ddots \\ 0\quad\dots0 \end{pmatrix} C_n$

We are manifestly in the situation of a KLT with $Q^t = C_n$.

Let then $\mathbf{Z} = C_n(\mathbf{X} - \mathbf{m}_X) = C_n \begin{pmatrix} X_0 - m_0 \\ \vdots \\ X_{n-1} - m_{n-1} \end{pmatrix}$.

We have: $\Gamma_Z = \begin{pmatrix} n\,0\dots0 \\ 0\,0\quad0 \\ \vdots\quad\ddots\;\vdots \\ 0\quad\dots0 \end{pmatrix}$.

Hence we obtain: $\mathbf{Z} = \begin{pmatrix} Z_0 \\ 0 \\ \vdots \\ 0 \end{pmatrix}$ almost surely.

Remark It is immediate – by retransformation of the result above – that a random vector $\mathbf{X}$ the covariance matrix Γ_X of which is constant, is necessarily (simply) repetitive:

$$\mathbf{X} = \begin{pmatrix} X_0 \\ X_0 \\ \vdots \\ X_0 \end{pmatrix}.$$

Let us sum up: The DCT is the KLT of an extremal case (the case of "highest correlation"). It is predefinable.

In practical applications, one can always use the DCT as a *brute force approximation* of the KLTs which actually ought to intervene, hoping that it will honestly replace them, at least in constellations of high correlation. More precisely: One hopes that the value $\|\mathbf{X}\|^2 - \|C_m \mathbf{X}\|^2$ will not be too far away from the minimum $\|\mathbf{X}\|^2 - \|Q_m^t \mathbf{X}\|^2$ given by the KLT of $\mathbf{X}$.

Note that there exist rigorous approximation results.

Exercises

Our theme will be the *Discrete Sine Transform*.

(1) Some auxiliary results.

   Recall   Four trigonometrical identities:

   $$\sum\nolimits_{j=1}^{n} \sin(jx) = \frac{\sin \frac{nx}{2} \cdot \sin \frac{(n+1)x}{2}}{\sin \frac{x}{2}}$$

   $$\sum\nolimits_{j=1}^{n} \cos(jx) = \frac{\cos \frac{nx}{2} \cdot \sin \frac{(n+1)x}{2}}{\sin \frac{x}{2}}$$

   $$\sum\nolimits_{j=1}^{n} \sin^2(jx) = \frac{n}{2} - \frac{\cos(n+1)x \cdot \sin(nx)}{2 \sin x}$$

   $$\sum\nolimits_{j=1}^{n} \cos^2(jx) = \frac{n+2}{2} + \frac{\cos(n+1)x \cdot \sin(nx)}{2 \sin x}$$

   Making use of the formulas above, show that

   $$\sum\nolimits_{j=1}^{n} \sin \frac{jk\pi}{n+1} \cdot \sin \frac{jl\pi}{n+1} = \begin{cases} 0 & \text{for} \quad k \neq l, \\ \frac{n+1}{2} & \text{for} \quad k = l, \end{cases} \quad 1 \leq k, l \leq n.$$

   (Do not forget: $\sin x \cdot \sin y = \frac{1}{2}(\cos(x-y) - \cos(x+y))$.)

(2) Consider the matrix $\mathbf{S}_2 = \sqrt{\frac{2}{3}} \begin{pmatrix} \sin \frac{\pi}{3} & \sin \frac{2\pi}{3} \\ \sin \frac{2\pi}{3} & \sin \frac{4\pi}{3} \end{pmatrix} = \frac{1}{\sqrt{2}} \begin{pmatrix} 1 & 1 \\ 1 & -1 \end{pmatrix}.$

   (a) Show that $\mathbf{S}_2$ is an orthogonal (symmetric) matrix.

   (b) Let $T_2(a) = \begin{pmatrix} 1 & -a \\ -a & 1 \end{pmatrix} \in \mathbb{R}^{2 \times 2}.$

Compute the matrix $\quad D_2 = \mathbf{S}_2 T_2(a)\mathbf{S}_2$.

(c) Show that $\quad \lambda_1 = 1 - 2a\cos\frac{\pi}{3} = 1 - a \quad$ and
$$\lambda_2 = 1 - 2a\cos\frac{2\pi}{3} = 1 + a$$
are the two eigenvalues of $T_2(a)$, and find an orthonormal basis $\{q_1, q_2\}$ of eigenvectors for $T_2(a)$.

(3) Consider now the matrix $\quad \mathbf{S}_4 = \sqrt{\frac{2}{5}} \begin{pmatrix} \sin\frac{\pi}{5} & \sin\frac{2\pi}{5} & \sin\frac{3\pi}{5} & \sin\frac{4\pi}{5} \\ \sin\frac{2\pi}{5} & \sin\frac{4\pi}{5} & \sin\frac{6\pi}{5} & \sin\frac{8\pi}{5} \\ \sin\frac{3\pi}{5} & \sin\frac{6\pi}{5} & \sin\frac{9\pi}{5} & \sin\frac{12\pi}{5} \\ \sin\frac{4\pi}{5} & \sin\frac{8\pi}{5} & \sin\frac{12\pi}{5} & \sin\frac{16\pi}{5} \end{pmatrix}$.

Put $\alpha = \sin\frac{\pi}{5}$, $\beta = \sin\frac{2\pi}{5}$.

(a) Write $\mathbf{S}_4$ in function of α and of β.

(b) Show that $\mathbf{S}_4$ is an orthogonal (symmetric) matrix.

(c) Let $T_4(a) = \begin{pmatrix} 1 & -a & 0 & 0 \\ -a & 1 & -a & 0 \\ 0 & -a & 1 & -a \\ 0 & 0 & -a & 1 \end{pmatrix} \in \mathbb{R}^{4\times 4}$.

Compute $\quad D_4 = \mathbf{S}_4 T_4(a)\mathbf{S}_4 \in \mathbb{R}^{4\times 4}$ and verify that
$\lambda_1 = 1 - 2a\cos\frac{\pi}{5}$,
$\lambda_2 = 1 - 2a\cos\frac{2\pi}{5}$,
$\lambda_3 = 1 - 2a\cos\frac{3\pi}{5}$,
$\lambda_4 = 1 - 2a\cos\frac{4\pi}{5}$
are the eigenvalues of $T_4(a)$.

Conclusion *The four columns of $\mathbf{S}_4$ are an orthonormal basis of eigenvectors for every matrix $T_4(a)$, $a \in \mathbb{R}$.*

(4) Generalize: Let $n \geq 2$ and let $\quad \mathbf{S}_n = \sqrt{\frac{2}{n+1}}\left(\sin\frac{jk}{n+1}\pi\right)_{1\leq j,k\leq n}$

(a) Show that $\mathbf{S}_n$ is an orthogonal (symmetric) matrix.

(b) Let $T_n(a) = \begin{pmatrix} 1 & -a & 0 & 0 & \cdots & 0 \\ -a & 1 & -a & 0 & \cdots & 0 \\ 0 & -a & 1 & -a & & 0 \\ \vdots & \vdots & & \ddots & & \vdots \\ & & & & 1 & -a \\ 0 & 0 & \cdots & 0 & -a & 1 \end{pmatrix}$ be the $n \times n$ 3-band matrix, as indicated.

Show that $\quad \mathbf{S}_n T_n(a)\mathbf{S}_n = \begin{pmatrix} \lambda_1 & 0 & \cdots & 0 \\ 0 & \lambda_2 & \cdots & 0 \\ \vdots & & \ddots & \vdots \\ 0 & \cdots & 0 & \lambda_n \end{pmatrix}$,

where $\lambda_k = 1 - 2a\cos\frac{k\pi}{n+1} \qquad 1 \leq k \leq n$.

Consequence Let $\mathbf{X} = \begin{pmatrix} X_0 \\ \vdots \\ X_{n-1} \end{pmatrix}$ be a random vector such that $\Gamma_X = T_n(a)$,

for a certain $a \in \mathbb{R}$.

Then the Discrete Sine Transform is the associated KLT.

Attention   In signal processing, the situation which calls for the Discrete Sine Transform, is initially not a genuine vector setting. In other words, it does not stem from parallel thinking, but rather from modeling of sequential structures:

$$\mathbf{X} = \begin{pmatrix} X_0 \\ \vdots \\ X_{n-1} \end{pmatrix} \qquad \text{will be the initial segment of a } \textit{process} \text{ (which has been trun-}$$

cated after n steps).

In case of a stationary first-order Markov process, the matrix $T_n(a)$ comes up in a non-causal set-up of the situation (and is *not* a covariance matrix). On the other hand, the matrix $T_n(a)^{-1}$ will be, up to a scalar factor, the covariance matrix of the residual process $\mathbf{Y}$ obtained by a decomposition of $\mathbf{X}$ in a deterministic part (the prediction) and a non-deterministic part (the prediction error) which is precisely $\mathbf{Y}$.

The Discrete Sine Transform is then the KLT of $\mathbf{Y}$ (if $\mathbf{S}_n$ diagonalizes $T_n(a)$, then $\mathbf{S}_n$ diagonalizes also $T_n(a)^{-1}$).

Now let us look at some clarifying exercises:
(5) Algebraic prelude.

$$\text{Let} \quad A = \begin{pmatrix} 1 & \rho & \rho^2 & \rho^3 \\ \rho & 1 & \rho & \rho^2 \\ \rho^2 & \rho & 1 & \rho \\ \rho^3 & \rho^2 & \rho & 1 \end{pmatrix} \in \mathbb{R}^{4\times 4} \qquad \text{with} \quad \rho \neq \pm 1.$$

Show that A is then invertible, and that $\quad A^{-1} = \frac{1}{1-\rho^2} \begin{pmatrix} 1 & -\rho & 0 & 0 \\ -\rho & 1+\rho^2 & -\rho & 0 \\ 0 & -\rho & 1+\rho^2 & -\rho \\ 0 & 0 & -\rho & 1 \end{pmatrix}$

(it is a symmetric three-band matrix).
(6) First conceptual invasion.

Recall   A *white noise* is given by a sequence of random variables (or of random vectors) $(\epsilon_n)_{n\in\mathbb{Z}}$, with zero mean: $E(\epsilon_n) = 0, n \in \mathbb{Z}$, of same variance: σ^2, and mutually non-correlated: $\text{cov}(\epsilon_m, \epsilon_n) = 0 \quad \text{for } m \neq n$.
One often tacitly assumes them *Gaussian*: $\epsilon_n \sim N(0,1)$ for all $n \in \mathbb{Z}$ ($N(0,1)$ $\equiv$ the Gaussian law of mean 0 and variance 1).
Now consider a first-order (stationary) autoregressive process:
$X_n = \rho \cdot X_{n-1} + \epsilon_n, \qquad n \geq 1 \quad \text{with} \quad |\rho| < 1.$
Our hypotheses:
 (i) $(\epsilon_n)_{n\geq 1}$ is a white noise.
 (ii) $E(X_n) = 0$ for all $n \geq 0$.
 (iii) $\text{cov}(X_n, X_{n+h})$ only depends on $h \in \mathbb{Z}$.
 (iv) $\text{cov}(X_m, \epsilon_n) = 0$ for $m < n$.
The term $\rho \cdot X_{n-1}$ corresponds to a (determinist) prediction, and the term ϵ_n describes the (non-determinist) prediction error.
It seems that this is the simplest model for speech signals and 1D scan lines of images.

$$\text{Let now} \quad \mathbf{X} = \begin{pmatrix} X_0 \\ \vdots \\ X_{n-1} \end{pmatrix} \qquad \text{be a truncation to the order } n \text{ of such a process.}$$

Show that

$$\Gamma_X = \frac{\sigma^2}{1-\rho^2} \begin{pmatrix} 1 & \rho & \rho^2 & \cdots & \rho^{n-1} \\ \rho & 1 & \rho & \cdots & \rho^{n-2} \\ \rho^2 & \rho & 1 & \cdots & \rho^{n-3} \\ \vdots & \vdots & \vdots & & \vdots \\ \rho^{n-1} & \rho^{n-2} & \rho^{n-3} & \cdots & 1 \end{pmatrix}$$

(where σ^2 is the variance of the considered white noise).

(7) The non-causal formalism.
(a) Verify that the process $(X_n)_{n\geq 0}$ of the preceding exercise can also be written in the following form:

$$X_n = a \cdot (X_{n-1} + X_{n+1}) + e_n, \qquad n \geq 1,$$

where $a = \frac{\rho}{1+\rho^2}$ and $e_n = \frac{1}{1+\rho^2}(\epsilon_n - \rho\epsilon_{n+1})$.
(b) Show that $E(e_n) = 0$ and that $\beta^2 = E(e_n^2) = \frac{\sigma^2}{1+\rho^2}$ for $n \geq 1$.

(c) Show that we have, for $\mathbf{e} = \begin{pmatrix} e_1 \\ \vdots \\ e_n \end{pmatrix}$:

$$\Gamma_{\mathbf{e}} = \beta^2 \begin{pmatrix} 1 & -\frac{\rho}{1+\rho^2} & 0 & \cdots & 0 \\ -\frac{\rho}{1+\rho^2} & 1 & -\frac{\rho}{1+\rho^2} & \cdots & 0 \\ \vdots & & \ddots & & \vdots \\ 0 & 0 & \cdots & -\frac{\rho}{1+\rho^2} & 1 \end{pmatrix}$$

(a symmetric three-band matrix).

(8) Consider the matrix $T_4(a) = \begin{pmatrix} 1 & -a & 0 & 0 \\ -a & 1 & -a & 0 \\ 0 & -a & 1 & -a \\ 0 & 0 & -a & 1 \end{pmatrix} \in \mathbb{R}^{4\times 4}$

(where $a = \frac{\rho}{1+\rho^2}$ and $|\rho| < 1$).
For which values of ρ is the matrix $T_4(a)$ invertible? Compute $T_4^{-1}(a)$.

(9) Still the situation of Exercise 7.
(a) Verify the identity

$$T_4(a) \begin{pmatrix} X_1 \\ X_2 \\ X_3 \\ X_4 \end{pmatrix} = \begin{pmatrix} e_1 \\ e_2 \\ e_3 \\ e_4 \end{pmatrix} + \begin{pmatrix} aX_0 \\ 0 \\ 0 \\ aX_5 \end{pmatrix}.$$

(b) Put $\mathbf{X}_b = T_4^{-1}(a)\mathbf{b}$ with $\mathbf{b} = \begin{pmatrix} aX_0 \\ 0 \\ 0 \\ aX_5 \end{pmatrix}$.

$\mathbf{X}_b$, the *boundary response*, is the determinist prediction which resolves the equations
$X_n = a(X_{n-1} + X_{n+1})$ according to the boundary conditions given by

$$\mathbf{b} = \begin{pmatrix} aX_0 \\ 0 \\ 0 \\ aX_5 \end{pmatrix}.$$

Find the four components of $\mathbf{X}_b$ in function of X_0, X_5, and a.

(c) Now consider $\mathbf{Y} = \mathbf{X} - \mathbf{X}_b = T_4^{-1}(a) \begin{pmatrix} e_1 \\ e_2 \\ e_3 \\ e_4 \end{pmatrix}$.

Show that $\mathbf{X}_b$ and $\mathbf{Y}$ are *orthogonal* and that $\Gamma_Y = \beta^2 T_4^{-1}(a)$.

5.3 Filter Banks and Discrete Wavelet Transform

The practical interest of the transformations which will be presented in this last section comes from their appearance in JPEG 2000 – which is the successor of JPEG (just like AES is the successor of DES). So, we are still in search of *decorrelation methods* for our numerical data: the role of the DCT will now be occupied by the Discrete Wavelet Transform, which will efficiently create a kind of data hierarchy, according to a *high resolution/low resolution* display of the digital image. The objective will be, as before, to establish the appropriate numerical structure for quantization, i.e. for an intelligent suppression of "what seems to be unimportant".

Our presentation will be similar to that of the preceding section: First, we shall give a functional description ("how it works"), then we shall deal with design arguments ("why it works well"), i.e. with the underlying Mathematics.

5.3.1 Two Channel Filter Banks

Let us first point out that the *Discrete Wavelet Transform* (DWT) is actually *a whole family of transforms* – exactly as in the case of the KLT.

In order to understand the algorithmic functioning of a Discrete Wavelet Transform in image compression, it is almost preferable, as a first approach, not to know what a wavelet is (more precisely: what a multi-resolution analysis is). For the novice, the (two channel) filter bank formalism seems to be the right access: One faces simple and robust (exclusively algebraic) mathematical structures, the algorithmic aspects of which are easily assimilated. As a first intuitive orientation: the filter banks are to Discrete Wavelet Transforms what formal power series are to convergent power series.

Perfect Reconstruction Filter Banks

Deconvolution

Recall. In the first section of this chapter, we have encountered the mathematical formalism for the treatment of digital n-periodic signals (which, for $n > 2m$, are precisely the *faithful* samplings to the order n of trigonometric polynomials of degree $\leq m$).

Our first result was the following: consider, for an invariant linear filter T on the discrete n-periodic signals, $h = T(e_0)$, the *impulse response* of the filter, i.e. $\| : h_0, h_1, \ldots, h_{n-1} : \| = T(\| : 1, 0, \ldots, 0 : \|)$. Then $T(x) = x * h$ for all input signals x (the filtering is done via convolution with $h = T(e_0)$).

The associativity of the convolution product means that the impulse response of a composition of two invariant linear filters is precisely the convolution product of their individual impulse responses; clearly, e_0 is the impulse response of the identity filter.

In the domain of the n-periodic digital signal, the *perfect reconstruction* (or *deconvolution*, i.e. the invertibility of an invariant linear filter) has become a familiar topic for us:

Consider $h = \| : h_0, h_1, \ldots, h_{n-1} : \|$, then there exists $g = \| : g_0, g_1, \ldots, g_{n-1} : \|$ with $g * h = h * g = e_0$ $\iff$ every component of

$$\hat{h} = F_n(h) = \begin{pmatrix} \hat{h}_0 \\ \hat{h}_1 \\ \vdots \\ \hat{h}_{n-1} \end{pmatrix}$$

is non-zero.

In this case,

$$\begin{pmatrix} g_0 \\ g_1 \\ \vdots \\ g_{n-1} \end{pmatrix} = F_n^{-1} \begin{pmatrix} \frac{1}{\hat{h}_0} \\ \frac{1}{\hat{h}_1} \\ \vdots \\ \frac{1}{\hat{h}_{n-1}} \end{pmatrix}.$$

Hence, in the theory of n-periodic digital signals, there are a lot of perfect reconstruction filters. Unfortunately, the periodic formalism is of minor practical interest.

The *non-periodic* digital signal theory gets its natural Hilbert space formalization as follows:

Consider then $l^2(\mathbb{Z})$, the vector space of (complex-valued) sequences $\mathbf{f} = (f[k])_{k \in \mathbb{Z}}$ such that

$$\sum_{k \in \mathbb{Z}} | f[k] |^2 < \infty$$

We insist: $f[k] \equiv$ the value of $\mathbf{f}$ "at the moment k" (the discrete time "is" $\mathbb{Z}$).

That is our Hilbert space[5] of "finite energy" discrete signals.

For the practitioner, this formal setting is clearly too general. So, consider the subspace of the digital signals with finite support:

$$l_c^2(\mathbb{Z}) = \{\mathbf{f} = (f[k])_{k \in \mathbb{Z}} : f[k] = 0 \quad \text{for almost all } k\}$$

A digital signal is an element of $l_c^2(\mathbb{Z})$ whenever it takes only a finite number of non-zero values. For formal convenience, let us introduce a variable t, which will be the "discret-time carrier": an element $\mathbf{f} \in l_c^2(\mathbb{Z})$ thus becomes a Laurent polynomial $\mathbf{f} = \sum f[k]t^k$

Note that we also shall have negative powers of t!)

---

[5] Recall the definitions of the Chap. 3. At this stage, we would not substantially make use of the inner product and of the norm.

Example Consider **f** with

$$f[k] = 0 \qquad \text{for } k \le -3,$$
$$f[-2] = -1, \quad f[-1] = 2,$$
$$f[0] = 1, \qquad f[1] = 0,$$
$$f[2] = -3, \qquad f[3] = 1, \text{ and}$$
$$f[k] = 0 \qquad \text{for } k \ge 4.$$

Then $\mathbf{f} = -t^{-2} + 2t^{-1} + 1 - 3t^2 + t^3$.

Attention The variable t *does not take values*. Its unique mission is to position the values of the digital signals in time (or in space). [6]

The *convolution product* $\mathbf{f} * \mathbf{g}$ of two digital signals with finite support **f** and **g** is defined by

$$\mathbf{f} * \mathbf{g}[n] = \sum_{k+j=n} f[k]g[j] = \sum_k f[k]g[n-k] = \sum_k f[n-k]g[k].$$

In other words: the polynomial notation of the convolution product is the polynomial product (of the notations) of the factors. As in the case of the periodic digital signals, we consider the right-shift operator σ:

$$(\sigma \mathbf{f})[k] = f[k-1].$$

Clearly, this operator corresponds to the multiplication by t on the polynomial notations.

Exercise

For $n \in \mathbb{Z}$, let $\mathbf{e}_n$ be the characteristic function of $\{n\} \subset \mathbb{Z}$.

In other words,

$$\mathbf{e}_n[k] = \begin{cases} 1, & k = n, \\ 0, & \text{else.} \end{cases}$$

(a) Show that $\{\mathbf{e}_n, n \in \mathbb{Z}\}$ is a basis of the vector space $l_c^2(\mathbb{Z})$.
(b) Show that $\mathbf{e}_m * \mathbf{e}_n = \mathbf{e}_{m+n}$.
(c) Show that $\sigma \mathbf{f} = \mathbf{e}_1 * \mathbf{f}$.

Remark The multiplication by t shifts the values of the signals one position to the right, while translating the discrete time one unit to the left ($k \mapsto k-1$). If you want to repair this formal discordance, replace t by z^{-1}. And you have got the z-transform of the considered digital signal!

But let us return to our theme: the deconvolution – now in the non-periodic case. Let then T be an invariant linear filter on the finite support digital signals (T commutes with the operator σ), and let $\mathbf{h} = T(\mathbf{e}_0)$ be the impulse response of T.

Then $T(\mathbf{f}) = \mathbf{f} * \mathbf{h} = \mathbf{h} * \mathbf{f}$ (the filtering is a simple convolution with **h**). The proof is the same as in the n-periodic case (*Exercise*). We insist: in polynomial notation, a linear and time-invariant filtering is nothing but the multiplication by a fixed

---

[6] We have already encountered this situation when formalizing convolutional codes. But, in binary arithmetic, there was no temptation to evaluate the variable.

polynomial **h**. The problem of deconvolution in the (finite support) non-periodic case now reads like this:

Given a (Laurent) polynomial **h**, find a (Laurent) polynomial **g** with **hg** = **gh** = 1.

Exercise

The only Laurent polynomials which are invertible for the multiplication (of Laurent polynomials) are the monomials of the form αt^n, $n \in \mathbb{Z}$, $\alpha \neq 0$.

Consequence The only invariant linear filters on finite support digital signals which admit a perfect reconstruction (i.e. which are invertible as operators on finite support digital signals) are the shift operators (combined with scalar multiplication).[7] How can we repair this shortage of invertible filters?

First Appearance of a DWT

Arithmetical observation In $\mathbb{Z}$, only the two integers 1 and -1 are invertible for the multiplication. In order to get a lot of invertibles for the multiplication in integer arithmetic, you must either pass to arithmetic modulo n or pass to integer matrix arithmetic. For example, the multiplicative group $GL_2(\mathbb{Z})$ of the 2×2 integer matrices of determinant ± 1 is sufficiently rich; look:

$$\begin{pmatrix} 3 & 4 \\ 2 & 3 \end{pmatrix} \begin{pmatrix} 3 & -4 \\ -2 & 3 \end{pmatrix} = \begin{pmatrix} 3 & -4 \\ -2 & 3 \end{pmatrix} \begin{pmatrix} 3 & 4 \\ 2 & 3 \end{pmatrix} = \begin{pmatrix} 1 & 0 \\ 0 & 1 \end{pmatrix} \quad \text{etc.}$$

Looking for analogies in discrete signal theory, we observe that the n-periodic case (corresponding to arithmetic modulo n) is indeed richer in perfect reconstruction filters than the non-periodic case. But what about similarities when passing to matrices with polynomial coefficients – and how can we interpret a matrix impulse response? Let us begin with an example which will be important for the continuation.

Exercise

Consider the two matrices

$$A = \begin{pmatrix} \frac{3}{4} - \frac{1}{8}(t + t^{-1}) & \frac{1}{4}(1+t) \\ -\frac{1}{4}(1 + t^{-1}) & \frac{1}{2} \end{pmatrix} \quad \text{and} \quad S = \begin{pmatrix} 1 & -\frac{1}{2}(1+t) \\ \frac{1}{2}(1 + t^{-1}) & \frac{3}{2} - \frac{1}{4}(t + t^{-1}) \end{pmatrix}$$

Verify that $AS = SA = \begin{pmatrix} 1 & 0 \\ 0 & 1 \end{pmatrix}$. We have got two "matrix impulse responses", which are mutual inverses. So, we are in a situation of perfect reconstruction. But there remains the question: What is the action of a matrix impulse response on a digital signal (of finite support)?

---

[7] Traditionally, deconvolution will produce an *infinite length* inverse, which has to undergo Procrustean truncation – this yields a kind of imperfect reconstruction.

A 2×2 matrix transforms *couples* into *couples*. Hence we have to create a filtering situation, where the input signal enters via *two subbands* and where the output signal leaves in *two channels*. Actually, for the *analysis matrix* A, the two input channels will be virtual, due to a certain *interleaved lecture* of the input signal, whereas the two output channels will be real – there will be two separate subbands. For the *synthesis matrix* S, the two input channels will be real, whereas the two output channels will be virtual: one reconstructs the initial signal in its interleaved lecture.

More precisely:

Let $\mathbf{x} = \sum x[n]t^n$ be the input signal for A (in polynomial notation). Introduce an (even/odd) interleaved lecture which creates two virtual subbands:

$$\mathbf{x}_0 = \sum x[2n]t^n$$

$$\mathbf{x}_1 = \sum x[2n+1]t^n$$

(Pay attention to discrete-time (or discrete-space) positioning: we carry out a *down-sampling* of factor 2).

One obtains

$$\begin{pmatrix} \mathbf{y}_0 \\ \mathbf{y}_1 \end{pmatrix} = A \begin{pmatrix} \mathbf{x}_0 \\ \mathbf{x}_1 \end{pmatrix},$$

where

$$\mathbf{y}_0 = \sum y_0[n]t^n = \sum y[2n]t^n$$

$$\mathbf{y}_1 = \sum y_1[n]t^n = \sum y[2n+1]t^n$$

are the low-pass and high-pass components (take this as a definition) of the output signal

$$\mathbf{y} = \sum y[n]t^n \quad \text{of } A$$

(here, it is the unifying interleaved notation which is a little bit artificial...).

Look at our example:

$$\begin{pmatrix} \mathbf{y}_0 \\ \mathbf{y}_1 \end{pmatrix} = \begin{pmatrix} \sum y[2n]t^n \\ \sum y[2n+1]t^n \end{pmatrix} = \begin{pmatrix} \frac{3}{4} - \frac{1}{8}(t+t^{-1}) & \frac{1}{4}(1+t) \\ -\frac{1}{4}(1+t^{-1}) & \frac{1}{2} \end{pmatrix} \begin{pmatrix} \sum x[2n]t^n \\ \sum x[2n+1]t^n \end{pmatrix},$$

$$\mathbf{y}_0 = \sum y[2n]t^n = \sum \frac{3}{4}x[2n]t^n - \sum \frac{1}{8}x[2n]t^{n+1} - \sum \frac{1}{8}x[2n]t^{n-1}$$
$$+ \sum \frac{1}{4}x[2n+1]t^n + \sum \frac{1}{4}x[2n+1]t^{n+1}.$$

$$\mathbf{y}_1 = \sum y[2n+1]t^n = -\sum \frac{1}{4}x[2n]t^n - \sum \frac{1}{4}x[2n]t^{n-1} + \sum \frac{1}{2}x[2n+1]t^n.$$

Comparing the coefficients of t^n in both cases, we obtain:

$$y[2n] = -\tfrac{1}{8}x[2n-2] + \tfrac{1}{4}x[2n-1] + \tfrac{3}{4}x[2n] + \tfrac{1}{4}x[2n+1] - \tfrac{1}{8}x[2n+2]$$
$$y[2n+1] = -\tfrac{1}{4}x[2n] + \tfrac{1}{2}x[2n+1] - \tfrac{1}{4}x[2n+2]$$

In non-interleaved version we have, for the low-pass component $\mathbf{y}_0$ and for the high-pass component $\mathbf{y}_1$ of the output signal:

$$\mathbf{y}_0[n] = \{ -\tfrac{1}{8}x[k-2] + \tfrac{1}{4}x[k-1] + \tfrac{3}{4}x[k] + \tfrac{1}{4}x[k+1] - \tfrac{1}{8}x[k+2] \}_{k=2n},$$

$$\mathbf{y}_1[n] = \{ -\tfrac{1}{4}x[k-1] + \tfrac{1}{2}x[k] - \tfrac{1}{4}x[k+1] \}_{k=2n+1}.$$

We note: Finally, we have got rid of the matrix formalism. The matrix A is replaced by two ordinary filters with impulse responses[8]

$$\mathbf{h}_0^t = (h_0^t[-2], h_0^t[-1], h_0^t[0], h_0^t[1], h_0^t[2]) = (-\frac{1}{8}, \frac{1}{4}, \frac{3}{4}, \frac{1}{4}, -\frac{1}{8})$$

and

$$\mathbf{h}_1^t = (h_1^t[-1], h_1^t[0], h_1^t[1]) = (-\frac{1}{4}, \frac{1}{2}, -\frac{1}{4})$$

The input signal $\mathbf{x} = (x[n])$ splits into a low-pass component $\mathbf{x} * \mathbf{h}_0^t$ and a high-pass component $\mathbf{x} * \mathbf{h}_1^t$, which give, by down-sampling of a factor 2 – in the sense mentioned above – the definite values of the two subbands $\mathbf{y}_0$ and $\mathbf{y}_1$.

The two components $\mathbf{y}_0$ and $\mathbf{y}_1$ admit a unified lecture, in *interleaved version*, as $\mathbf{y} = (y[n])$ with $y[2n] = y_0[n]$ and $y[2n+1] = y_1[n]$.

Exercise

In the situation above, we know that there is perfect reconstruction:

$$\begin{pmatrix} \mathbf{x}_0 \\ \mathbf{x}_1 \end{pmatrix} = \begin{pmatrix} \sum x[2n]t^n \\ \sum x[2n+1]t^n \end{pmatrix} = \begin{pmatrix} 1 & -\tfrac{1}{2}(1+t) \\ \tfrac{1}{2}(1+t^{-1}) & \tfrac{3}{2} - \tfrac{1}{4}(t+t^{-1}) \end{pmatrix} \begin{pmatrix} \sum y[2n]t^n \\ \sum y[2n+1]t^n \end{pmatrix}.$$

Find the two impulse responses $\mathbf{s}_0$ and $\mathbf{s}_1$ such that

$$\mathbf{x}_0 = \mathbf{y} * \mathbf{s}_0 \quad \text{and} \quad \mathbf{x}_1 = \mathbf{y} * \mathbf{s}_1.$$

$$[\mathbf{s}_0 = (s_0[-1], s_0[0], s_0[1]) = (-\frac{1}{2}, 1, -\frac{1}{2}),$$

$$\mathbf{s}_1 = (s_1[-2], s_1[-1], s_1[0], s_1[1], s_1[2]) = (-\frac{1}{4}, \frac{1}{2}, \frac{3}{2}, \frac{1}{2}, -\frac{1}{4})]$$

This is our first example of a two channel *filter bank* (or: a subband transform).

Note that we have committed a "formal clumsiness" concerning our two reconstruction filters: as well $\mathbf{s}_0$ as $\mathbf{s}_1$ take their input values in zigzag lecture from the *two* subbands of our filter bank; on the other hand, they reconstruct the initial signal $\mathbf{x}$ in two (virtual) separate subbands. Conceptually, the reconstruction should be done by two (impulse responses of) synthesis filters $\mathbf{g}_0^t$ and $\mathbf{g}_1^t$, defined *separately* on the low-pass channel and on the high-pass channel, and such that the values of the initial signal are obtained by simple addition of the results of these two filterings.

More concretely, we aim at a reconstruction of the form

$$x[k] = \sum_i y[2i]g_0^t[k-2i] + \sum_i y[2i+1]g_1^t[k-2i-1]$$

We note: in interleaved reading, $\mathbf{g}_0^t$ and $\mathbf{g}_1^t$ will act alternately on the values of $\mathbf{y}$.

---

[8] $()^t$ for "translated" will distinguish the interleaved notation from the traditional notation.

Exercise

Show that in the case of our example, the two synthesis filters $\mathbf{g}_0^t$ *and* $\mathbf{g}_1^t$ *are*

$$\mathbf{g}_0^t = (g_0^t[-1], g_0^t[0], g_0^t[1]) = (\frac{1}{2}, 1, \frac{1}{2})$$

and

$$\mathbf{g}_1^t = (g_1^t[-2], g_1^t[-1], g_1^t[0], g_1^t[1], g_1^t[2]) = (-\frac{1}{4}, -\frac{1}{2}, \frac{3}{2}, -\frac{1}{2}, -\frac{1}{4}).$$

Now let us display the usual scheme of a *two channel filter bank*:

$$x[k] \longrightarrow \boxed{*h_0} \rightarrow \boxed{\downarrow 2} \rightarrow y_0[n] \rightarrow \boxed{\uparrow 2} \rightarrow \boxed{*g_0} \rightarrow$$
$$\downarrow$$
$$\rightarrow \boxed{*h_1} \rightarrow \boxed{\downarrow 2} \rightarrow y_1[n] \rightarrow \boxed{\uparrow 2} \rightarrow \boxed{*g_1} \rightarrow \oplus \longrightarrow x[k]$$

Analysis filter bank   Subbands   Synthesis filter bank
$\downarrow 2$: down-sampling of a factor 2
$\uparrow 2$: up-sampling of a factor 2

(In the first case, we only sample for even arguments ($f_k = f[2k]$),
in the second case, we put the sample values on even positions while filling with
zeros on odd positions ($f_{2k} = f[k]$, $f_{2k+1} = 0$).)

Attention The formalism according to the scheme above *differs* slightly from the
formalism that we have adopted. The interleaved notation and the down-sampling
(up-sampling) of factor 2 harmonize for h_0 and g_0. In the case of h_1 and g_1, there
will be, for counting arguments, a shift of one position.

$\Big[$ Let us explain, following our example; we get

$$\mathbf{y}_1[n] = \{ -\frac{1}{4}x[k-1] + \frac{1}{2}x[k] - \frac{1}{4}x[k+1] \}_{k=2n+1}, \quad \text{i.e.}$$

$$\mathbf{y}_1[n] = \{ -\frac{1}{4}x[k] + \frac{1}{2}x[k+1] - \frac{1}{4}x[k+2] \}_{k=2n}$$

This gives

$\mathbf{h}_1 = (h_1[-2], h_1[-1], h_1[0]) = (-\frac{1}{4}, \frac{1}{2}, -\frac{1}{4})$ in traditional notation.$\Big]$

Exercise

Find, in the case of our example, the traditional notation for $\mathbf{g}_1$.

Answer $\mathbf{g}_1 = (-\frac{1}{4}, -\frac{1}{2}, \frac{3}{2}, -\frac{1}{2}, -\frac{1}{4}) = (g_1[-1], g_1[0], g_1[1], g_1[2], g_1[3])$.

Let us sum up: the traditional notation for h_1 begins one position lower, that
for g_1 begins one position higher than the notation of the interleaved formalism.

For algorithmic arguments, the interleaved formalism is much more natural –
and we shall adopt it in the sequel. The traditional formalism is more natural when
using standard transformations (the z-transform, the Fourier transform), since it
decomposes our subband transforms nicely into its primitive constituents.

Exercises

We shall try to find the natural generalizations for our generic example. All impulse responses are to be read in interleaved formalism.

(1) Let $(h_0^t[k])$ and $(h_1^t[k])$ be the analysis filters of a two channel filter bank:

$$y[2n] = \sum_k h_0^t[k]x[2n-k], \quad y[2n+1] = \sum_k h_1^t[k]x[2n+1-k]$$

Let

$$A = \begin{pmatrix} \sum a_k t^k & \sum b_k t^k \\ \sum c_k t^k & \sum d_k t^k \end{pmatrix}$$

be the corresponding analysis matrix.
Show that then

$$a_k = h_0^t[2k] \qquad b_k = h_0^t[2k-1],$$

$$c_k = h_1^t[2k+1] \quad d_k = h_1^t[2k].$$

Convention Throughout the rest of this section, we shall always tacitly assume to be in *real* signal theory.

(2) The matrix A is invertible (we get perfect reconstruction) $\Longleftrightarrow$ $\det A$ is invertible (as a Laurent polynomial) $\Longleftrightarrow$ $\det A = \alpha t^n$ for $\alpha \in \mathbb{R}^*, n \in \mathbb{Z}$. Show that a given analysis filter bank (h_0^t, h_1^t) admits a synthesis filter bank (g_0^t, g_1^t) for perfect reconstruction $\Longleftrightarrow$ $\sum_k h_0^t[k] \cdot (-1)^k h_1^t[2n-k]$ is non-zero for *precisely one* value of n.
Note that we can always suppose that $\det A = \alpha$ for $\alpha \in \mathbb{R}^*$.
We will have then: $\alpha = \sum_k h_0^t[k] \cdot (-1)^k h_1^t[-k]$.
(The argument is simple: Multiply each of the two rows of the matrix A by an appropriate power of t, and the determinant will become "constant").
In the sequel, we always shall suppose to be in this situation.

(3) Assuming that $\det A = \alpha \in \mathbb{R}^*$, the synthesis matrix has to be

$$S = \frac{1}{\alpha} \begin{pmatrix} \sum d_k t^k & -\sum b_k t^k \\ -\sum c_k t^k & \sum a_k t^k \end{pmatrix}.$$

Conclude, for s_0, s_1 with $x_0 = y * s_0$ and $x_1 = y * s_1$ that

$$s_0[2k] = \frac{1}{\alpha}d_k, \qquad s_0[2k-1] = -\frac{1}{\alpha}b_k,$$
$$s_1[2k+1] = -\frac{1}{\alpha}c_k, \qquad s_1[2k] = \frac{1}{\alpha}a_k.$$

(4) In the situation of the preceding exercise, let us go a little bit further:
Show that the synthesis impulse responses g_0^t and g_1^t of a perfect reconstruction filter bank are obtained from the analysis impulse responses h_0^t and h_1^t as follows:

$$g_0^t[n] = \frac{1}{\alpha}(-1)^n h_1^t[n],$$

$$g_1^t[n] = \frac{1}{\alpha}(-1)^n h_0^t[n],$$

where $\alpha = \sum_k h_0^t[k] \cdot (-1)^k h_1^t[-k]$.

(5) Consider an analysis filter bank given by $(\mathbf{h}_0^t, \mathbf{h}_1^t)$, where

$$\mathbf{h}_0^t = (h_0^t[-4], h_0^t[-3], h_0^t[-2], h_0^t[-1], h_0^t[0], h_0^t[1], h_0^t[2], h_0^t[3], h_0^t[4])$$

$$= \left(\frac{1}{32}, -\frac{1}{64}, -\frac{5}{64}, \frac{17}{64}, \frac{19}{32}, \frac{17}{64}, -\frac{5}{64}, -\frac{1}{64}, \frac{1}{32} \right),$$

$$\mathbf{h}_1^t = (h_1^t[-3], h_1^t[-2], h_1^t[-1], h_1^t[0], h_1^t[1], h_1^t[2], h_1^t[3])$$

$$= \left(\frac{3}{64}, -\frac{1}{32}, -\frac{19}{64}, \frac{9}{16}, -\frac{19}{64}, -\frac{1}{32}, \frac{3}{64} \right).$$

Verify if the couple $(\mathbf{h}_0^t, \mathbf{h}_1^t)$ satisfies the conditions for the analysis impulse responses of a perfect reconstruction filter bank.

(6) Let

$$\mathbf{h}_0^t = (h_0^t[-3], h_0^t[-2], h_0^t[-1], h_0^t[0], h_0^t[1], h_0^t[2], h_0^t[3])$$

$$= \left(-\frac{3}{280}, -\frac{3}{56}, \frac{73}{280}, \frac{17}{28}, \frac{73}{280}, -\frac{3}{56}, -\frac{3}{280} \right),$$

$$\mathbf{h}_1^t = (h_1^t[-2], h_1^t[-1], h_1^t[0], h_1^t[1], h_1^t[2]) = \left(-\frac{1}{20}, -\frac{1}{4}, \frac{3}{5}, -\frac{1}{4}, -\frac{1}{20} \right).$$

Verify if the couple $(\mathbf{h}_0^t, \mathbf{h}_1^t)$ satisfies the conditions for the analysis impulse responses of a perfect reconstruction filter bank.

(7) Let R_1 be the real root, and let R_2 and R_3 be the two other (conjugate) roots of the equation $R(x) = 8 - \frac{29}{2}x + 10x^2 - \frac{5}{2}x^3 = 0$.

Let us define:

$$h_0^t[0] = \frac{5}{128}(R_1 - 1)(14 + 24R_2R_3 - 16(R_2 + R_3)),$$

$$h_0^t[1] = h_0^t[-1] = \frac{5}{128}(R_1 - 1)(12 + 16R_2R_3 - 14(R_2 + R_3)),$$

$$h_0^t[2] = h_0^t[-2] = \frac{5}{128}(R_1 - 1)(8 + 4R_2R_3 - 8(R_2 + R_3)),$$

$$h_0^t[3] = h_0^t[-3] = \frac{5}{128}(R_1 - 1)(4 - 2(R_2 + R_3)),$$

$$h_0^t[4] = h_0^t[-4] = \frac{10}{128}(R_1 - 1);$$

then:

$$h_1^t[0] = \frac{3R_1 - 2}{8(R_1 - 1)},$$

$$h_1^t[1] = h_1^t[-1] = -\frac{8R_1 - 7}{32(R_1 - 1)},$$

$$h_1^t[2] = h_1^t[-2] = -\frac{2 - R_1}{16(R_1 - 1)},$$

$$h_1^t[3] = h_1^t[-3] = \frac{1}{32(R_1 - 1)}.$$

(a) Verify that

$$\sum h_0^t[n] = 1, \qquad\qquad \sum h_1^t[n] = 0,$$
$$\sum (-1)^n h_0^t[n] = 0, \qquad \sum (-1)^n h_1^t[n] = 1.$$

(b) Verify that $(\mathbf{h}_0^t, \mathbf{h}_1^t)$ satisfy the conditions for the analysis impulse responses of a perfect reconstruction filter bank.

(8) We return to the situation of the exercise (4), with $\alpha = 1$ (this is a modest normalization).

Let us define two systems of vectors $(\mathbf{e}_n)_{n \in \mathbb{Z}}$ and $(\tilde{\mathbf{e}}_n)_{n \in \mathbb{Z}}$ as follows:

$$e_n[k] = \begin{cases} g_0^t[k - n] & n \equiv 0 \bmod 2, \\ g_1^t[k - n] & n \equiv 1 \bmod 2, \end{cases}$$

$$\tilde{e}_n[k] = \begin{cases} h_0^t[n - k] & n \equiv 0 \bmod 2, \\ h_1^t[n - k] & n \equiv 1 \bmod 2. \end{cases}$$

(Pay attention to the time-inverted progressions !)
Show that

(a) $\langle \mathbf{e}_m, \tilde{\mathbf{e}}_n \rangle = \begin{cases} 1, & m = n \\ 0 & \text{else} \end{cases}$ (inner product of $l^2(\mathbb{Z})$).

(b) Every $\mathbf{x} \in l_c^2(\mathbb{Z})$ admits a unique representation $\mathbf{x} = \sum_n \langle \mathbf{x}, \tilde{\mathbf{e}}_n \rangle \mathbf{e}_n$.

$\Big[$ Help:(a) For the inner products $\langle \mathbf{e}_m, \tilde{\mathbf{e}}_n \rangle$, with m and n of the same parity, you should make use of the reconstruction identity

$$\sum_k h_0^t[k] \cdot (-1)^k h_1^t[2n - k] = \begin{cases} 1, & \text{for } n = 0, \\ 0, & \text{else.} \end{cases}$$

For the inner products $\langle \mathbf{e}_m, \tilde{\mathbf{e}}_n \rangle$ with m and n of unequal parity, you should use the following auxiliary result:

$\sum_k (-1)^k \cdot a[k] a[n - k] = 0$ for all sequences $(a[k])_{k \in \mathbb{Z}}$ and every *odd* n .

(b) Here, you only deal with a different notation of the identity

$$x[k] = \sum_l y_0[l] g_0^t[k - 2l] + \sum_l y_1[l] g_1^t[k - (2l + 1)] \Big]$$

The Discrete Wavelet Transform – version JPEG 2000

In this section, we shall give a purely functional presentation of the Discrete Wavelet Transform as a kind of *decorrelation kit* for the digital data in JPEG 2000. Actually, the DWT will appear as (a certain type of) two channel *filter bank*.

We shall give a rather rigid definition of the type of filter bank which really interests the practitioner in digital image processing – and shall call it, *by misuse of language*, a Discrete Wavelet Transform. The criteria which allow recovering such a filter bank as a true Discrete Wavelet Transform – i.e. as a collection of operations making sense in a continuous world – will be discussed in the last section of this chapter.

Note that the two filter banks imposed by the first version of JPEG 2000 – the DWT 5/3 spline and the DWT 9/7 CDF – are *true* Discrete Wavelet Transforms.

The 1D Formalism

We shall restart almost from scratch with the presentation of a two channel filter bank, while insisting upon the aspect "treat the data by a window of fixed length", which is – a priori – hostile to convolutional thinking.

So, consider the following situation:

$x[n] \equiv x[0]x[1]x[2]\ldots x[N-1]$ a sequence of N digital values. We shall suppose N to be even, often $N = 1, 2, 4, 8, \ldots$, i.e. a power of 2. We shall associate with the sequence $x[n]$ two sequences $y_0[n]$ and $y_1[n]$, the length of which will be half of that of $x[n]$:

$$y_0[n] \equiv \text{the "low-pass" subband sequence}$$

$$y_1[n] \equiv \text{the "high-pass" subband sequence}$$

In order to obtain a correspondence of sequences of the *same* length, pass now to the *interleaved sequence* $y[n]$, with

$$y[2n] = y_0[n],$$

$$y[2n+1] = y_1[n].$$

Example $x[0]\ x[1]\ x[2]\ x[3]\ x[4]\ x[5]\ x[6]\ x[7]$
becomes
$y[0]\ y[1]\ y[2]\ y[3]\ y[4]\ y[5]\ y[6]\ y[7],$
$y_0[0]\ y_1[0]\ y_0[1]\ y_1[1]\ y_0[2]\ y_1[2]\ y_0[3]\ y_1[3].$

In the sequel, each of the particular values $y[n]$ will depend, via some fixed linear combination, on a "numerical neighbourhood" of the corresponding value $x[n]$, for example on $x[n-2]x[n-1]x[n]x[n+1]x[n+2]$.

We shall get formal evaluation problems at the boundaries.

The remedy will be a representation by *symmetric extension* [9]:

$\check{x}[n] = x[n], \qquad 0 \le n \le N-1,$
$\check{x}[-n] = \check{x}[n],$
$\check{x}[N-1-n] = \check{x}[N-1+n].$

Example $x[0]\ x[1]\ x[2]\ x[3]$
becomes
$\ldots\ \check{x}[-2]\ \check{x}[-1]\ \check{x}[0]\ \check{x}[1]\ \check{x}[2]\ \check{x}[3]\ \check{x}[4]\ \check{x}[5]\ \ldots$
$\ldots\ \ x[2]\quad\ x[1]\quad\ \ x[0]\ x[1]\ x[2]\ x[3]\ x[2]\ x[1]\ \ldots$

---

[9] It is more frequent that a digital image will be cut up along "numerical plateaux" than along "numerical fractures": this yields the idea of a symmetric prolongation.

How shall we pass concretely from the sequence $x[n]$ to the sequence $y[n]$, and (reconstruction!) from the sequence $y[n]$ to the sequence $x[n]$? The relations which define the *analysis* transformation and the *synthesis* transformation will be formulated in symmetric extensions (i.e. on infinite periodic sequences):

$$\breve{y}[n] = \sum_{i \in \mathbb{Z}} h^t_{n \bmod 2}[i] \; \breve{x}[n - i] \quad \text{(analysis)}$$

$$\breve{x}[n] = \sum_{i \in \mathbb{Z}} \breve{y}[i] g^t_{i \bmod 2}[n - i] \quad \text{(synthesis)}$$

Attention The sequences $x[n]$ vary – hence the sequences $y[n]$ vary with them (the symmetries of $\breve{y}[n]$ will be the same as those of $\breve{x}[n]$). On the other hand, the four finite sequences $h^t_0[n]$, $h^t_1[n]$, $g^t_0[n]$, $g^t_1[n]$ are *constant* (they define "sliding windows" of weighting coordinates) and characteristic for the considered *filter bank*.

We insist:
The analysis filters:

$h^t_0[n] \equiv$ the low-pass analysis impulse response,
$h^t_1[n] \equiv$ the high-pass analysis impulse response.

The synthesis filters:

$g^t_0[n] \equiv$ the low-pass synthesis impulse response,
$g^t_1[n] \equiv$ the high-pass synthesis impulse response.

Concerning the $()^t \equiv$ *translated* notation:
In a conventional-filter-bank formalism (we mentioned it in the preceding section, without further specification) we have four impulse responses h_0, h_1, g_0, g_1, which are related to the translated variants in the following way:

$$h^t_0[k] = h_0[k], \qquad g^t_0[k] = g_0[k],$$
$$h^t_1[k] = h_1[k - 1], \qquad g^t_1[k] = g_1[k + 1].$$

The translated formalism has been chosen in order to obtain a natural situation relative to the interleaved notation. In the design of (two channel) filter banks for image compression we shall insist upon
Linear phase:

$$h^t_0[n] = h^t_0[-n],$$
$$h^t_1[n] = h^t_1[-n],$$
$$g^t_0[n] = g^t_0[-n],$$
$$g^t_1[n] = g^t_1[-n].$$

Hence our (translated) impulse responses always will have *odd length* and will be *symmetric* with respect to $n = 0$.
The *fundamental relations* giving the invertibility of the analysis transform:

$$\sum_k h^t_0[k] \cdot (-1)^k h^t_1[2n - k] = \begin{cases} \alpha \in \mathbb{R}^*, & \text{for } n = 0, \\ 0, & \text{else.} \end{cases} \qquad (\mathcal{R})$$

We shall then *necessarily* have (cf. the exercises at the end of the last section):

$$g_0^t[n] = \tfrac{1}{\alpha}(-1)^n h_1^t[n],$$

$$g_1^t[n] = \tfrac{1}{\alpha}(-1)^n h_0^t[n],$$

where α is given by the condition of perfect reconstruction ($\mathcal{R}$) above, and is nothing but the determinant of the analysis matrix A considered in the preceding section. By the way, we have, in a more frequential language (*Exercise*, for the incredulous):

$$\alpha = \tfrac{1}{2}(h_0^{dc} h_1^{nyq} + h_1^{dc} h_0^{nyq}) \qquad \text{with} \qquad \begin{aligned} h^{dc} &= \sum h^t[n] \\ h^{nyq} &= \sum (-1)^n h^t[n] \end{aligned}$$

For the filter bank design (in the restricted meaning of this section), you thus only have to specify the (impulse responses of the) analysis filters $h_0^t[n]$ and $h_1^t[n]$, satisfying the relations of perfect reconstruction ($\mathcal{R}$). All the remainder will then be determined.

Note We shall try to obtain, in general:

$\quad h_0^{dc} = 1 \qquad h_1^{nyq} = 1 \qquad$ (this is a normalization)
$\quad h_1^{dc} = 0 \qquad h_0^{nyq} = 0 \qquad$ (this is a conceptual necessity)

We insist: the cancellation condition is necessary in order to guarantee that our purely arithmetical constructions have a chance to stem from the wavelet universe (that they can be derived from operations in a continuous world).

Finally, we shall have, in most cases: $\alpha = \tfrac{1}{2}$.

Commentary

The condition $h_0^{dc} = 1$ means that the low-pass analysis filter computes local averages – it will be a kind of integrator.

The condition $h_1^{dc} = 0$ means that the high-pass analysis filter computes generalized slopes – il will be a kind of differentiator.

The information in the low-pass subband more or less concerns numerical regularities.

The information in the high-pass subband rather concerns numerical irregularities.

Example The DWT 5/3 spline.

$\quad h_0^t = (-\tfrac{1}{8}, \tfrac{1}{4}, \tfrac{3}{4}, \tfrac{1}{4}, -\tfrac{1}{8}), \qquad h_1^t = (-\tfrac{1}{4}, \tfrac{1}{2}, -\tfrac{1}{4}).$

That's our "generic" example of the preceding section.

We will search for the matrix representation of the analysis filter on sequences of length 8:

How can we obtain $\begin{pmatrix} y[0] \\ \vdots \\ y[7] \end{pmatrix}$ in function of $\begin{pmatrix} x[0] \\ \vdots \\ x[7] \end{pmatrix}$?

The general formula splits into low-pass filtering and into high-pass filtering in the following way:

$$y[2n] = \sum_{i=2n-2}^{2n+2} \breve{x}[i] h_0^t[2n - i],$$

$$y[2n + 1] = \sum_{i=2n}^{2n+2} \breve{x}[i] h_1^t[2n + 1 - i]$$

More explicitly:

$$y[0] = \breve{x}[-2]h_0^t[2] + \breve{x}[-1]h_0^t[1] + \breve{x}[0]h_0^t[0] + \breve{x}[1]h_0^t[-1] + \breve{x}[2]h_0^t[-2],$$
$$y[1] = \breve{x}[0]h_1^t[1] + \breve{x}[1]h_1^t[0] + \breve{x}[2]h_1^t[-1],$$
$$y[2] = \breve{x}[0]h_0^t[2] + \breve{x}[1]h_0^t[1] + \breve{x}[2]h_0^t[0] + \breve{x}[3]h_0^t[-1] + \breve{x}[4]h_0^t[-2],$$
$$y[3] = \breve{x}[2]h_1^t[1] + \breve{x}[3]h_1^t[0] + \breve{x}[4]h_1^t[-1],$$
$$y[4] = \breve{x}[2]h_0^t[2] + \breve{x}[3]h_0^t[1] + \breve{x}[4]h_0^t[0] + \breve{x}[5]h_0^t[-1] + \breve{x}[6]h_0^t[-2],$$
$$y[5] = \breve{x}[4]h_1^t[1] + \breve{x}[5]h_1^t[0] + \breve{x}[6]h_1^t[-1],$$
$$y[6] = \breve{x}[4]h_0^t[2] + \breve{x}[5]h_0^t[1] + \breve{x}[6]h_0^t[0] + \breve{x}[7]h_0^t[-1] + \breve{x}[8]h_0^t[-2],$$
$$y[7] = \breve{x}[6]h_1^t[1] + \breve{x}[7]h_1^t[0] + \breve{x}[8]h_1^t[-1],$$

But: $\breve{x}[-2] = x[2], \quad \breve{x}[-1] = x[1], \quad$ and $\quad \breve{x}[8] = x[6].$

This gives (the impulse responses are symmetric with respect to $n = 0$):

$$\begin{pmatrix} y[0] \\ y[1] \\ y[2] \\ y[3] \\ y[4] \\ y[5] \\ y[6] \\ y[7] \end{pmatrix} = \begin{pmatrix} h_0^t[0] & 2h_0^t[1] & 2h_0^t[2] & 0 & 0 & 0 & 0 & 0 \\ h_1^t[1] & h_1^t[0] & h_1^t[1] & 0 & 0 & 0 & 0 & 0 \\ h_0^t[2] & h_0^t[1] & h_0^t[0] & h_0^t[1] & h_0^t[2] & 0 & 0 & 0 \\ 0 & 0 & h_1^t[1] & h_1^t[0] & h_1^t[1] & 0 & 0 & 0 \\ 0 & 0 & h_0^t[2] & h_0^t[1] & h_0^t[0] & h_0^t[1] & h_0^t[2] & 0 \\ 0 & 0 & 0 & 0 & h_1^t[1] & h_1^t[0] & h_1^t[1] & 0 \\ 0 & 0 & 0 & 0 & h_0^t[2] & h_0^t[1] & h_0^t[0] + h_0^t[2] & h_0^t[1] \\ 0 & 0 & 0 & 0 & 0 & 0 & 2h_1^t[1] & h_1^t[0] \end{pmatrix} \begin{pmatrix} x[0] \\ x[1] \\ x[2] \\ x[3] \\ x[4] \\ x[5] \\ x[6] \\ x[7] \end{pmatrix}.$$

Now let us substitute the concrete values; we obtain:

$$\begin{pmatrix} y[0] \\ y[1] \\ y[2] \\ y[3] \\ y[4] \\ y[5] \\ y[6] \\ y[7] \end{pmatrix} = \begin{pmatrix} \frac{3}{4} & \frac{1}{2} & -\frac{1}{4} & 0 & 0 & 0 & 0 & 0 \\ -\frac{1}{4} & \frac{1}{2} & -\frac{1}{4} & 0 & 0 & 0 & 0 & 0 \\ -\frac{1}{8} & \frac{1}{4} & \frac{3}{4} & \frac{1}{4} & -\frac{1}{8} & 0 & 0 & 0 \\ 0 & 0 & -\frac{1}{4} & \frac{1}{2} & -\frac{1}{4} & 0 & 0 & 0 \\ 0 & 0 & -\frac{1}{8} & \frac{1}{4} & \frac{3}{4} & \frac{1}{4} & -\frac{1}{8} & 0 \\ 0 & 0 & 0 & 0 & -\frac{1}{4} & \frac{1}{2} & -\frac{1}{4} & 0 \\ 0 & 0 & 0 & 0 & -\frac{1}{8} & \frac{1}{4} & \frac{5}{8} & \frac{1}{4} \\ 0 & 0 & 0 & 0 & 0 & 0 & -\frac{1}{2} & \frac{1}{2} \end{pmatrix} \begin{pmatrix} x[0] \\ x[1] \\ x[2] \\ x[3] \\ x[4] \\ x[5] \\ x[6] \\ x[7] \end{pmatrix}.$$

Consider, as a first example, a *constant* input sequence: $x[n] \equiv 1$.   We obtain:

$$\begin{pmatrix} y[0] \\ y[1] \\ y[2] \\ y[3] \\ y[4] \\ y[5] \\ y[6] \\ y[7] \end{pmatrix} = \begin{pmatrix} 1 \\ 0 \\ 1 \\ 0 \\ 1 \\ 0 \\ 1 \\ 0 \end{pmatrix} \quad \text{i.e.} \quad \begin{pmatrix} y_0[0] \\ y_0[1] \\ y_0[2] \\ y_0[3] \end{pmatrix} = \begin{pmatrix} 1 \\ 1 \\ 1 \\ 1 \end{pmatrix} \quad \text{and} \quad \begin{pmatrix} y_1[0] \\ y_1[1] \\ y_1[2] \\ y_1[3] \end{pmatrix} = \begin{pmatrix} 0 \\ 0 \\ 0 \\ 0 \end{pmatrix}.$$

Refine a little bit: consider an input sequence in linear progression:
$x[n] = n + 1, 0 \le n \le 7$.
This gives

$$\begin{pmatrix} y[0] \\ y[1] \\ y[2] \\ y[3] \\ y[4] \\ y[5] \\ y[6] \\ y[7] \end{pmatrix} = \begin{pmatrix} 1 \\ 0 \\ 3 \\ 0 \\ 5 \\ 0 \\ 7.25 \\ 0.5 \end{pmatrix} \quad \text{i.e.} \quad \begin{pmatrix} y_0[0] \\ y_0[1] \\ y_0[2] \\ y_0[3] \end{pmatrix} = \begin{pmatrix} 1 \\ 3 \\ 5 \\ 7.25 \end{pmatrix} \quad \text{and} \quad \begin{pmatrix} y_1[0] \\ y_1[1] \\ y_1[2] \\ y_1[3] \end{pmatrix} = \begin{pmatrix} 0 \\ 0 \\ 0 \\ 0.5 \end{pmatrix}.$$

We note: the low-pass channel displays the regularity of the sequence (a linear progression). The high-pass channel displays the absence of (numerical) gaps in the considered sequence. There are "numerical deviations" due to the treatment of the boundaries by symmetric extension.

Now let us pass to the *synthesis* transform (inversion of the analysis transform).

We have to make explicit the reconstruction formula

$\check{x}[n] = \sum_{i \in \mathbb{Z}} \check{y}[i] g^t_{i \bmod 2}[n - i].$

Recall $\begin{aligned} g^t_0[n] &= 2 \cdot (-1)^n h^t_1[n] \\ g^t_1[n] &= 2 \cdot (-1)^n h^t_0[n] \end{aligned}$

In our case:

$$g^t_0 = (\tfrac{1}{2}, 1, \tfrac{1}{2}),$$

$$g^t_1 = (-\tfrac{1}{4}, -\tfrac{1}{2}, \tfrac{3}{2}, -\tfrac{1}{2}, -\tfrac{1}{4}).$$

When evaluating the reconstruction formula position by position, we obtain:

$$\begin{pmatrix} x[0] \\ x[1] \\ x[2] \\ x[3] \\ x[4] \\ x[5] \\ x[6] \\ x[7] \end{pmatrix} = 2 \cdot \begin{pmatrix} h^t_1[0] & -2h^t_1[1] & 0 & 0 & 0 & 0 & 0 & 0 \\ -h^t_1[1] & h^t_0[0]+h^t_0[2] & -h^t_1[1] & h^t_0[2] & 0 & 0 & 0 & 0 \\ 0 & -h^t_0[1] & h^t_1[0] & -h^t_0[1] & 0 & 0 & 0 & 0 \\ 0 & h^t_0[2] & -h^t_1[1] & h^t_0[0] & -h^t_1[1] & h^t_0[2] & 0 & 0 \\ 0 & 0 & 0 & -h^t_0[1] & h^t_1[0] & -h^t_0[1] & 0 & 0 \\ 0 & 0 & 0 & h^t_0[2] & -h^t_1[1] & h^t_0[0] & -h^t_1[1] & h^t_0[2] \\ 0 & 0 & 0 & 0 & 0 & -h^t_0[1] & h^t_1[0] & -h^t_0[1] \\ 0 & 0 & 0 & 0 & 0 & 2h^t_0[2] & -2h^t_1[1] & h^t_0[0] \end{pmatrix} \begin{pmatrix} y[0] \\ y[1] \\ y[2] \\ y[3] \\ y[4] \\ y[5] \\ y[6] \\ y[7] \end{pmatrix}.$$

5.3 Filter Banks and Discrete Wavelet Transform   329

This gives concretely:

$$
\begin{pmatrix} x[0] \\ x[1] \\ x[2] \\ x[3] \\ x[4] \\ x[5] \\ x[6] \\ x[7] \end{pmatrix}
=
\begin{pmatrix}
1 & -1 & 0 & 0 & 0 & 0 & 0 & 0 \\
\frac{1}{2} & \frac{5}{4} & \frac{1}{2} & -\frac{1}{4} & 0 & 0 & 0 & 0 \\
0 & -\frac{1}{2} & 1 & -\frac{1}{2} & 0 & 0 & 0 & 0 \\
0 & -\frac{1}{4} & \frac{1}{2} & \frac{3}{2} & \frac{1}{2} & -\frac{1}{4} & 0 & 0 \\
0 & 0 & 0 & -\frac{1}{2} & 1 & -\frac{1}{2} & 0 & 0 \\
0 & 0 & 0 & -\frac{1}{4} & \frac{1}{2} & \frac{3}{2} & \frac{1}{2} & -\frac{1}{4} \\
0 & 0 & 0 & 0 & 0 & -\frac{1}{2} & 1 & -\frac{1}{2} \\
0 & 0 & 0 & 0 & 0 & -\frac{1}{2} & 1 & \frac{3}{2}
\end{pmatrix}
\begin{pmatrix} y[0] \\ y[1] \\ y[2] \\ y[3] \\ y[4] \\ y[5] \\ y[6] \\ y[7] \end{pmatrix}.
$$

For the incredulous: verify that the product of the synthesis matrix with the analysis matrix is indeed the unit matrix.

More modestly: we clearly have:

$$
\begin{pmatrix} 1 \\ 2 \\ 3 \\ 4 \\ 5 \\ 6 \\ 7 \\ 8 \end{pmatrix}
=
\begin{pmatrix}
1 & -1 & 0 & 0 & 0 & 0 & 0 & 0 \\
\frac{1}{2} & \frac{5}{4} & \frac{1}{2} & -\frac{1}{4} & 0 & 0 & 0 & 0 \\
0 & -\frac{1}{2} & 1 & -\frac{1}{2} & 0 & 0 & 0 & 0 \\
0 & -\frac{1}{4} & \frac{1}{2} & \frac{3}{2} & \frac{1}{2} & -\frac{1}{4} & 0 & 0 \\
0 & 0 & 0 & -\frac{1}{2} & 1 & -\frac{1}{2} & 0 & 0 \\
0 & 0 & 0 & -\frac{1}{4} & \frac{1}{2} & \frac{3}{2} & \frac{1}{2} & -\frac{1}{4} \\
0 & 0 & 0 & 0 & 0 & -\frac{1}{2} & 1 & -\frac{1}{2} \\
0 & 0 & 0 & 0 & 0 & -\frac{1}{2} & 1 & \frac{3}{2}
\end{pmatrix}
\begin{pmatrix} 1 \\ 0 \\ 3 \\ 0 \\ 5 \\ 0 \\ 7.25 \\ 0.5 \end{pmatrix}.
$$

Exercises

(1) Let $(a[k])_{-N \leq k \leq N}$ be an arbitrary finite sequence.
    Show then that $\sum_k (-1)^k a[k] a[n-k] = 0$ for all *odd* n.

(2) $\mathbf{h}_0^t = (-\frac{1}{8}, \frac{1}{4}, \frac{3}{4}, \frac{1}{4}, -\frac{1}{8})$ (symmetric, of length 5).
    Write the equations $(\mathcal{R})$ – with $\alpha = \frac{1}{2}$ – as a linear system for the computation of $\mathbf{h}_1^t = (h_1^t[-1], h_1^t[0], h_1^t[1])$ (symmetric, of length 3) such that $(\mathbf{h}_0^t, \mathbf{h}_1^t)$ will be the analysis filters of a perfect reconstruction filter bank, and recover by this method the DWT 5/3 spline.

(3) Let $\mathbf{h}_0^t = (h_0^t[-4], \ldots, h_0^t[4])$ be a (low-pass) analysis impulse response, symmetric, of length 9 (preferably with $h_0^{dc} = 1$ and $h_0^{nyq} = 0$). Consider the following linear system:

$$h_0^t[0]X - 2h_0^t[1]Y + 2h_0^t[2]Z - 2h_0^t[3]U = \alpha (= \tfrac{1}{2}),$$
$$h_0^t[2]X - (h_0^t[1] + h_0^t[3])Y + (h_0^t[0] + h_0^t[4])Z - h_0^t[1]U = 0,$$
$$h_0^t[4]X - h_0^t[3]Y + h_0^t[2]Z - h_0^t[1]U = 0,$$
$$h_0^t[4]Z - h_0^t[3]U = 0.$$

Show that there exists $\mathbf{h}_1^t = (h_1^t[-3], \ldots, h_1^t[3])$, a (high-pass) analysis impulse response, symmetric, of length 7 such that the ensuing analysis filter bank admits perfect reconstruction $\Longleftrightarrow$ the linear system above admits a solution $(X, Y, Z, U) = (h_1^t[0], h_1^t[1], h_1^t[2], h_1^t[3])$.

(4) Consider the one-parameter family of symmetric (high-pass) impulse responses of the form
    $$\mathbf{h}_1^t = (h_1^t[-2], h_1^t[-1], h_1^t[0], h_1^t[1], h_1^t[2])$$
    with
    $$h_1^t[0] = a$$
    $$h_1^t[\pm 1] = -\tfrac{1}{4}$$
    $$h_1^t[\pm 2] = \tfrac{1}{4} - \tfrac{a}{2}$$

We search for symmetric (low-pass) impulse responses[10] of the form

$$\mathbf{h}_0^t = (h_0^t[-3], h_0^t[-2], h_0^t[-1], h_0^t[0], h_0^t[1], h_0^t[2], h_0^t[3]),$$

such that $(\mathbf{h}_0^t, \mathbf{h}_1^t)$ are the analysis filters of a perfect reconstruction filter bank.

(a) Show that the low-pass partner is obtained when solving the following linear system:

$$\begin{pmatrix} 2a & 1 & 1-2a & 0 \\ 1-2a & 1 & 4a & 1 \\ 0 & 0 & 1-2a & 1 \\ 1 & -2 & 2 & -2 \end{pmatrix} \begin{pmatrix} h_0[0] \\ h_0[1] \\ h_0[2] \\ h_0[3] \end{pmatrix} = \begin{pmatrix} 1 \\ 0 \\ 0 \\ 0 \end{pmatrix}.$$

(b) The system above admits a unique solution for every parameter value $a \neq \frac{1}{4}$. Give the solution in parametrized form.

$$\left[\textbf{Answer:}^{11} \quad \begin{array}{ll} h_0[0] = \frac{1+4a}{4(4a-1)} & h_0[1] = \frac{-8a^2+18a-5}{8(4a-1)} \\ h_0[2] = \frac{4a-3}{8(4a-1)} & h_0[3] = \frac{(1-2a)(3-4a)}{8(4a-1)} \end{array} \right]$$

(c) For $a = \frac{3}{5}$ write explicitly $(\mathbf{h}_0^t, \mathbf{h}_1^t)$ and $(\mathbf{g}_0^t, \mathbf{g}_1^t)$.

The 2D Formalism

Initial situation We dispose of a DWT (a two channel filter bank).

Objective Define the associated (D level) 2D DWT.
    Let $x[\mathbf{n}] = x[n_1, n_2]$, $0 \leq n_1, n_2 \leq N-1$
be a matrix[12] of N^2 sample values (i.e. a component of a digital image).
    Our 2D DWT will transform it into an *interleaved matrix* of the following four-subband structure:

$$y[2n_1 + b_1, 2n_2 + b_2] = y_{b_1 b_2}[n_1, n_2] \qquad \text{with} \qquad \begin{array}{l} b_1, b_2 \in \{0,1\} \\ 0 \leq n_1, n_2 \leq \frac{1}{2}N - 1 \end{array}$$

Recall Our N should preferably be a power of 2:
                2, 4, 8, 16, 32,...

How shall we get the matrix $y[\mathbf{n}]$ from the matrix $x[\mathbf{n}]$?
    First, we apply the given DWT N times separately on the N *columns* of $x[\mathbf{n}]$. Call the resulting matrix $y'[\mathbf{n}]$. Then, in order to obtain $y[\mathbf{n}]$, apply our DWT N times separately on the N *rows* of $y'[\mathbf{n}]$.

---

[10] Do not forget the condition $\sum(-1)^k h_0[k] = 0$.
[11] Recall that $\mathbf{h}_0^t = \mathbf{h}_0$.
[12] When dealing with matrices of sample values, we shall often skip the matrix brackets.

Observation   Support from matrix algebra.

Consider the matrix representation of our DWT: $\mathbf{y} = T \cdot \mathbf{x}$

Then we shall simply have, in 2D: $\boxed{\mathbf{Y} = T \cdot \mathbf{X} \cdot T^{\mathrm{t}}}$.

We insist: the matrix $\mathbf{Y} = y[\mathbf{n}]$ will be a triple matrix product, with the matrix $\mathbf{X} = x[\mathbf{n}]$ as its central factor.

(Generic) *scheme*:

$$
\begin{array}{llllllll}
x[0,0] & x[0,1] & x[0,2] & x[0,3] & x[0,4] & x[0,5] & x[0,6] & x[0,7] \\
x[1,0] & x[1,1] & x[1,2] & x[1,3] & x[1,4] & x[1,5] & x[1,6] & x[1,7] \\
x[2,0] & x[2,1] & x[2,2] & x[2,3] & x[2,4] & x[2,5] & x[2,6] & x[2,7] \\
x[3,0] & x[3,1] & x[3,2] & x[3,3] & x[3,4] & x[3,5] & x[3,6] & x[3,7] \\
x[4,0] & x[4,1] & x[4,2] & x[4,3] & x[4,4] & x[4,5] & x[4,6] & x[4,7] \\
x[5,0] & x[5,1] & x[5,2] & x[5,3] & x[5,4] & x[5,5] & x[5,6] & x[5,7] \\
x[6,0] & x[6,1] & x[6,2] & x[6,3] & x[6,4] & x[6,5] & x[6,6] & x[6,7] \\
x[7,0] & x[7,1] & x[7,2] & x[7,3] & x[7,4] & x[7,5] & x[7,6] & x[7,7]
\end{array}
$$

becomes, after transformation of the columns,

$$
\begin{array}{lllllllll}
L & y'[0,0] & y'[0,1] & y'[0,2] & y'[0,3] & y'[0,4] & y'[0,5] & y'[0,6] & y'[0,7] \\
H & y'[1,0] & y'[1,1] & y'[1,2] & y'[1,3] & y'[1,4] & y'[1,5] & y'[1,6] & y'[1,7] \\
L & y'[2,0] & y'[2,1] & y'[2,2] & y'[2,3] & y'[2,4] & y'[2,5] & y'[2,6] & y'[2,7] \\
H & y'[3,0] & y'[3,1] & y'[3,2] & y'[3,3] & y'[3,4] & y'[3,5] & y'[3,6] & y'[3,7] \\
L & y'[4,0] & y'[4,1] & y'[4,2] & y'[4,3] & y'[4,4] & y'[4,5] & y'[4,6] & y'[4,7] \\
H & y'[5,0] & y'[5,1] & y'[5,2] & y'[5,3] & y'[5,4] & y'[5,5] & y'[5,6] & y'[5,7] \\
L & y'[6,0] & y'[6,1] & y'[6,2] & y'[6,3] & y'[6,4] & y'[6,5] & y'[6,6] & y'[6,7] \\
H & y'[7,0] & y'[7,1] & y'[7,2] & y'[7,3] & y'[7,4] & y'[7,5] & y'[7,6] & y'[7,7]
\end{array}
$$

and finally

$$
\begin{array}{ccccccccc}
 & L & H & L & H & L & H & L & H \\
\end{array}
$$

$$
\begin{array}{lllllllll}
L & y[0,0] & y[0,1] & y[0,2] & y[0,3] & y[0,4] & y[0,5] & y[0,6] & y[0,7] \\
H & y[1,0] & y[1,1] & y[1,2] & y[1,3] & y[1,4] & y[1,5] & y[1,6] & y[1,7] \\
L & y[2,0] & y[2,1] & y[2,2] & y[2,3] & y[2,4] & y[2,5] & y[2,6] & y[2,7] \\
H & y[3,0] & y[3,1] & y[3,2] & y[3,3] & y[3,4] & y[3,5] & y[3,6] & y[3,7] \\
L & y[4,0] & y[4,1] & y[4,2] & y[4,3] & y[4,4] & y[4,5] & y[4,6] & y[4,7] \\
H & y[5,0] & y[5,1] & y[5,2] & y[5,3] & y[5,4] & y[5,5] & y[5,6] & y[5,7] \\
L & y[6,0] & y[6,1] & y[6,2] & y[6,3] & y[6,4] & y[6,5] & y[6,6] & y[6,7] \\
H & y[7,0] & y[7,1] & y[7,2] & y[7,3] & y[7,4] & y[7,5] & y[7,6] & y[7,7]
\end{array}
$$

$L \equiv$ low-pass

$H \equiv$ high-pass

Now separate the four interleaved matrices:

$\mathrm{LL} \equiv$ low-pass sub-band

$$
\begin{array}{llll}
y_{00}[0,0] & y_{00}[0,1] & y_{00}[0,2] & y_{00}[0,3] \\
y_{00}[1,0] & y_{00}[1,1] & y_{00}[1,2] & y_{00}[1,3] \\
y_{00}[2,0] & y_{00}[2,1] & y_{00}[2,2] & y_{00}[2,3] \\
y_{00}[3,0] & y_{00}[3,1] & y_{00}[3,2] & y_{00}[3,3]
\end{array}
=
\begin{array}{llll}
y[0,0] & y[0,2] & y[0,4] & y[0,6] \\
y[2,0] & y[2,2] & y[2,4] & y[2,6] \\
y[4,0] & y[4,2] & y[4,4] & y[4,6] \\
y[6,0] & y[6,2] & y[6,4] & y[6,6].
\end{array}
$$

HL ≡ horizontally high-pass sub-band

$$
\begin{matrix}
y_{01}[0,0] & y_{01}[0,1] & y_{01}[0,2] & y_{01}[0,3] \\
y_{01}[1,0] & y_{01}[1,1] & y_{01}[1,2] & y_{01}[1,3] \\
y_{01}[2,0] & y_{01}[2,1] & y_{01}[2,2] & y_{01}[2,3] \\
y_{01}[3,0] & y_{01}[3,1] & y_{01}[3,2] & y_{01}[3,3]
\end{matrix}
=
\begin{matrix}
y[0,1] & y[0,3] & y[0,5] & y[0,7] \\
y[2,1] & y[2,3] & y[2,5] & y[2,7] \\
y[4,1] & y[4,3] & y[4,5] & y[4,7] \\
y[6,1] & y[6,3] & y[6,5] & y[6,7]
\end{matrix}.
$$

LH ≡ vertically high-pass sub-band

$$
\begin{matrix}
y_{10}[0,0] & y_{10}[0,1] & y_{10}[0,2] & y_{10}[0,3] \\
y_{10}[1,0] & y_{10}[1,1] & y_{10}[1,2] & y_{10}[1,3] \\
y_{10}[2,0] & y_{10}[2,1] & y_{10}[2,2] & y_{10}[2,3] \\
y_{10}[3,0] & y_{10}[3,1] & y_{10}[3,2] & y_{10}[3,3]
\end{matrix}
=
\begin{matrix}
y[1,0] & y[1,2] & y[1,4] & y[1,6] \\
y[3,0] & y[3,2] & y[3,4] & y[3,6] \\
y[5,0] & y[5,2] & y[5,4] & y[5,6] \\
y[7,0] & y[7,2] & y[7,4] & y[7,6]
\end{matrix}.
$$

HH ≡ high-pass sub-band

$$
\begin{matrix}
y_{11}[0,0] & y_{11}[0,1] & y_{11}[0,2] & y_{11}[0,3] \\
y_{11}[1,0] & y_{11}[1,1] & y_{11}[1,2] & y_{11}[1,3] \\
y_{11}[2,0] & y_{11}[2,1] & y_{11}[2,2] & y_{11}[2,3] \\
y_{11}[3,0] & y_{11}[3,1] & y_{11}[3,2] & y_{11}[3,3]
\end{matrix}
=
\begin{matrix}
y[1,1] & y[1,3] & y[1,5] & y[1,7] \\
y[3,1] & y[3,3] & y[3,5] & y[3,7] \\
y[5,1] & y[5,3] & y[5,5] & y[5,7] \\
y[7,1] & y[7,3] & y[7,5] & y[7,7]
\end{matrix}.
$$

Attention The notation for the horizontally high-pass and vertically high-pass sub-bands is (perhaps) surprising; you might have expected a "transposed" notation.

Example Back to our DWT 5/3 spline
(the name refers to the length of the impulse responses and to a design criterion).
First:

$$
\begin{matrix}
1\,1\,1\,1\,1\,1\,1\,1 \\
1\,1\,1\,1\,1\,1\,1\,1 \\
1\,1\,1\,1\,1\,1\,1\,1 \\
1\,1\,1\,1\,1\,1\,1\,1 \\
1\,1\,1\,1\,1\,1\,1\,1 \\
1\,1\,1\,1\,1\,1\,1\,1 \\
1\,1\,1\,1\,1\,1\,1\,1 \\
1\,1\,1\,1\,1\,1\,1\,1
\end{matrix}
\longmapsto
\begin{matrix}
1\,0\,1\,0\,1\,0\,1\,0 \\
0\,0\,0\,0\,0\,0\,0\,0 \\
1\,0\,1\,0\,1\,0\,1\,0 \\
0\,0\,0\,0\,0\,0\,0\,0 \\
1\,0\,1\,0\,1\,0\,1\,0 \\
0\,0\,0\,0\,0\,0\,0\,0 \\
1\,0\,1\,0\,1\,0\,1\,0 \\
0\,0\,0\,0\,0\,0\,0\,0
\end{matrix}
$$

A constant matrix is "condensed" into LL.
HL, LH and HH are zero.
Then:

$$
\begin{matrix}
1\,1\,1\,1\,1\,1\,1\,1 \\
2\,2\,2\,2\,2\,2\,2\,2 \\
3\,3\,3\,3\,3\,3\,3\,3 \\
4\,4\,4\,4\,4\,4\,4\,4 \\
5\,5\,5\,5\,5\,5\,5\,5 \\
6\,6\,6\,6\,6\,6\,6\,6 \\
7\,7\,7\,7\,7\,7\,7\,7 \\
8\,8\,8\,8\,8\,8\,8\,8
\end{matrix}
\longmapsto
\begin{matrix}
1 & 0 & 1 & 0 & 1 & 0 & 1 & 0 \\
0 & 0 & 0 & 0 & 0 & 0 & 0 & 0 \\
3 & 0 & 3 & 0 & 3 & 0 & 3 & 0 \\
0 & 0 & 0 & 0 & 0 & 0 & 0 & 0 \\
5 & 0 & 5 & 0 & 5 & 0 & 5 & 0 \\
0 & 0 & 0 & 0 & 0 & 0 & 0 & 0 \\
7.25 & 0 & 7.25 & 0 & 7.25 & 0 & 7.25 & 0 \\
0.5 & 0 & 0.5 & 0 & 0.5 & 0 & 0.5 & 0
\end{matrix}
$$

It is instructive to have a look on the four interleaved matrices.

$$
LL \equiv
\begin{matrix}
1 & 1 & 1 & 1 \\
3 & 3 & 3 & 3 \\
5 & 5 & 5 & 5 \\
7.25 & 7.25 & 7.25 & 7.25
\end{matrix}
\qquad
HL \equiv
\begin{matrix}
0 & 0 & 0 & 0 \\
0 & 0 & 0 & 0 \\
0 & 0 & 0 & 0 \\
0 & 0 & 0 & 0
\end{matrix}
$$

$$
LH \equiv
\begin{matrix}
0 & 0 & 0 & 0 \\
0 & 0 & 0 & 0 \\
0 & 0 & 0 & 0 \\
0.5 & 0.5 & 0.5 & 0.5
\end{matrix}
\qquad
HH \equiv
\begin{matrix}
0 & 0 & 0 & 0 \\
0 & 0 & 0 & 0 \\
0 & 0 & 0 & 0 \\
0 & 0 & 0 & 0
\end{matrix}
$$

LL is an approximate lower resolution of the initial gradation.

The vertically high-pass sub-band LH is (timidly) non-zero; this stems from the fact that our gradation is vertically non-constant. It is amusing that this absolutely satisfactory information comes from a mathematical ugliness: the oblique treatment of a linear progression because of its symmetric extension![13]

Exercises

(1) *Recall*: the 2D DWT 5/3 spline (to the order 8) transforms

$$
X_1 =
\begin{matrix}
1 & 1 & 1 & 1 & 1 & 1 & 1 & 1 \\
1 & 1 & 1 & 1 & 1 & 1 & 1 & 1 \\
1 & 1 & 1 & 1 & 1 & 1 & 1 & 1 \\
1 & 1 & 1 & 1 & 1 & 1 & 1 & 1 \\
1 & 1 & 1 & 1 & 1 & 1 & 1 & 1 \\
1 & 1 & 1 & 1 & 1 & 1 & 1 & 1 \\
1 & 1 & 1 & 1 & 1 & 1 & 1 & 1 \\
1 & 1 & 1 & 1 & 1 & 1 & 1 & 1
\end{matrix}
\quad \text{into} \quad
Y_1 =
\begin{matrix}
1 & 0 & 1 & 0 & 1 & 0 & 1 & 0 \\
0 & 0 & 0 & 0 & 0 & 0 & 0 & 0 \\
1 & 0 & 1 & 0 & 1 & 0 & 1 & 0 \\
0 & 0 & 0 & 0 & 0 & 0 & 0 & 0 \\
1 & 0 & 1 & 0 & 1 & 0 & 1 & 0 \\
0 & 0 & 0 & 0 & 0 & 0 & 0 & 0 \\
1 & 0 & 1 & 0 & 1 & 0 & 1 & 0 \\
0 & 0 & 0 & 0 & 0 & 0 & 0 & 0
\end{matrix}
$$

and

$$
X_2 =
\begin{matrix}
1 & 1 & 1 & 1 & 1 & 1 & 1 & 1 \\
2 & 2 & 2 & 2 & 2 & 2 & 2 & 2 \\
3 & 3 & 3 & 3 & 3 & 3 & 3 & 3 \\
4 & 4 & 4 & 4 & 4 & 4 & 4 & 4 \\
5 & 5 & 5 & 5 & 5 & 5 & 5 & 5 \\
6 & 6 & 6 & 6 & 6 & 6 & 6 & 6 \\
7 & 7 & 7 & 7 & 7 & 7 & 7 & 7 \\
8 & 8 & 8 & 8 & 8 & 8 & 8 & 8
\end{matrix}
\quad \text{into} \quad
Y_2 =
\begin{matrix}
1 & 0 & 1 & 0 & 1 & 0 & 1 & 0 \\
0 & 0 & 0 & 0 & 0 & 0 & 0 & 0 \\
3 & 0 & 3 & 0 & 3 & 0 & 3 & 0 \\
0 & 0 & 0 & 0 & 0 & 0 & 0 & 0 \\
5 & 0 & 5 & 0 & 5 & 0 & 5 & 0 \\
0 & 0 & 0 & 0 & 0 & 0 & 0 & 0 \\
7.25 & 0 & 7.25 & 0 & 7.25 & 0 & 7.25 & 0 \\
0.5 & 0 & 0.5 & 0 & 0.5 & 0 & 0.5 & 0
\end{matrix}
$$

---

[13] Pay attention to pictorial interpretations of the "compressed image": it is the vertically high-pass sub-band LH which will show a small horizontal bar on uniform Grey – your conclusion?

Deduce the transform of

$$X_3 = \begin{array}{cccccccc}
-0.375 & -0.25 & -0.125 & 0 & 0.125 & 0.25 & 0.375 & 0.5 \\
-0.375 & -0.25 & -0.125 & 0 & 0.125 & 0.25 & 0.375 & 0.5 \\
-0.375 & -0.25 & -0.125 & 0 & 0.125 & 0.25 & 0.375 & 0.5 \\
-0.375 & -0.25 & -0.125 & 0 & 0.125 & 0.25 & 0.375 & 0.5 \\
-0.375 & -0.25 & -0.125 & 0 & 0.125 & 0.25 & 0.375 & 0.5 \\
-0.375 & -0.25 & -0.125 & 0 & 0.125 & 0.25 & 0.375 & 0.5 \\
-0.375 & -0.25 & -0.125 & 0 & 0.125 & 0.25 & 0.375 & 0.5 \\
-0.375 & -0.25 & -0.125 & 0 & 0.125 & 0.25 & 0.375 & 0.5.
\end{array}$$

(2) Consider an analysis filter bank $(\mathbf{h}_0^t, \mathbf{h}_1^t)$ such that $\sum h_1^t[k] = 0$.
Then show that the transform of a vertical gradation (all rows are constant) necessarily satisfies $\mathrm{HL} \equiv 0$ and $\mathrm{HH} \equiv 0$ (the horizontally high-pass sub-band HL and the high-pass sub-band HH are uniformly zero).
Give, in case of the DWT 5/3 spline to the order 8, an example of an 8×8 scheme which is not a gradation (i.e. which is a matrix of rank ≥ 2) and such that its transform satisfies $\mathrm{HL} \equiv 0$, $\mathrm{LH} \equiv 0$ and $\mathrm{HH} \equiv 0$.
(3) Compute, with the 2D DWT 5/3 spline to the order 8, the second and the third transform iterate of

$$X = \begin{array}{cccccccc}
1 & 1 & 1 & 1 & 1 & 1 & 1 & 1 \\
2 & 2 & 2 & 2 & 2 & 2 & 2 & 2 \\
3 & 3 & 3 & 3 & 3 & 3 & 3 & 3 \\
4 & 4 & 4 & 4 & 4 & 4 & 4 & 4 \\
5 & 5 & 5 & 5 & 5 & 5 & 5 & 5 \\
6 & 6 & 6 & 6 & 6 & 6 & 6 & 6 \\
7 & 7 & 7 & 7 & 7 & 7 & 7 & 7 \\
8 & 8 & 8 & 8 & 8 & 8 & 8 & 8
\end{array}$$

$\left[\text{Answer:}\right.$

$$\begin{array}{cccccccc}
1 & 0 & 1 & 0 & 1 & 0 & 1 & 0 \\
0 & 0 & 0 & 0 & 0 & 0 & 0 & 0 \\
3 & 0 & 3 & 0 & 3 & 0 & 3 & 0 \\
0 & 0 & 0 & 0 & 0 & 0 & 0 & 0 \\
5 & 0 & 5 & 0 & 5 & 0 & 5 & 0 \\
0 & 0 & 0 & 0 & 0 & 0 & 0 & 0 \\
7.25 & 0 & 7.25 & 0 & 7.25 & 0 & 7.25 & 0 \\
0.5 & 0 & 0.5 & 0 & 0.5 & 0 & 0.5 & 0
\end{array} \longmapsto$$

$$\begin{array}{cccccccc}
0 & 0 & 0 & 0 & 0 & 0 & 0 & 0 \\
-0.5 & 0.5 & -0.5 & 0.5 & -0.5 & 0.5 & -0.5 & 0.5 \\
0.75 & -0.75 & 0.75 & -0.75 & 0.75 & -0.75 & 0.75 & -0.75 \\
-1 & 1 & -1 & 1 & -1 & 1 & -1 & 1 \\
1.23 & -1.23 & 1.23 & -1.23 & 1.23 & -1.23 & 1.23 & -1.23 \\
-1.53 & 1.53 & -1.53 & 1.53 & -1.53 & 1.53 & -1.53 & 1.53 \\
2.02 & -2.02 & 2.02 & -2.02 & 2.02 & -2.02 & 2.02 & -2.02 \\
-1.69 & 1.69 & -1.69 & 1.69 & -1.69 & 1.69 & -1.69 & 1.69
\end{array}\right]$$

$$\begin{array}{cccccccc}
0 & 0.44 & 0 & 0.44 & 0 & 0.44 & 0 & 0.44 \\
0 & 0.44 & 0 & 0.44 & 0 & 0.44 & 0 & 0.44 \\
0 & -0.03 & 0 & -0.03 & 0 & -0.03 & 0 & -0.03 \\
0 & 1 & 0 & 1 & 0 & 1 & 0 & 1 \\
0 & 0.05 & 0 & 0.05 & 0 & 0.05 & 0 & 0.05 \\
0 & 1.58 & 0 & 1.58 & 0 & 1.58 & 0 & 1.58 \\
0 & -0.3 & 0 & -0.3 & 0 & -0.3 & 0 & -0.3 \\
0 & 1.85 & 0 & 1.85 & 0 & 1.85 & 0 & 1.85
\end{array} \longmapsto$$

(3) The general philosophy is that the low-pass sub-band LL contains a lower (but rather faithful) resolution of the digital image (we have carefully computed local means).

Give an example (of a *non-constant* 8×8 matrix) such that the low-pass sub-band LL is constant (by the 2D DWT 5/3 spline to the order 8): all the digital information will be "in the details".

$$
\begin{bmatrix}
200 & 0 & 200 & 0 & 200 & 0 & 200 & 0 \\
0 & 200 & 0 & 200 & 0 & 200 & 0 & 200 \\
200 & 0 & 200 & 0 & 200 & 0 & 200 & 0 \\
0 & 200 & 0 & 200 & 0 & 200 & 0 & 200 \\
200 & 0 & 200 & 0 & 200 & 0 & 200 & 0 \\
0 & 200 & 0 & 200 & 0 & 200 & 0 & 200 \\
200 & 0 & 200 & 0 & 200 & 0 & 200 & 0 \\
0 & 200 & 0 & 200 & 0 & 200 & 0 & 200
\end{bmatrix}
\longmapsto
\begin{bmatrix}
100 & 0 & 100 & 0 & 100 & 0 & 100 & 0 \\
0 & 100 & 0 & 100 & 0 & 100 & 0 & 100 \\
100 & 0 & 100 & 0 & 100 & 0 & 100 & 0 \\
0 & 100 & 0 & 100 & 0 & 100 & 0 & 100 \\
100 & 0 & 100 & 0 & 100 & 0 & 100 & 0 \\
0 & 100 & 0 & 100 & 0 & 100 & 0 & 100 \\
100 & 0 & 100 & 0 & 100 & 0 & 100 & 0 \\
0 & 100 & 0 & 100 & 0 & 100 & 0 & 100
\end{bmatrix}
$$

(4) Consider the DWT 7/5 Burt[14] given by

$$
\mathbf{h}_0^t = (h_0^t[-3], h_0^t[-2], h_0^t[-1], h_0^t[0], h_0^t[1], h_0^t[2], h_0^t[3])
$$

$$
= \left(-\frac{3}{280}, -\frac{3}{56}, \frac{73}{280}, \frac{17}{28}, \frac{73}{280}, -\frac{3}{56}, -\frac{3}{280} \right),
$$

$$
\mathbf{h}_1^t = (h_1^t[-2], h_1^t[-1], h_1^t[0], h_1^t[1], h_1^t[2]) = \left(-\frac{1}{20}, -\frac{1}{4}, \frac{3}{5}, -\frac{1}{4}, -\frac{1}{20} \right),
$$

$$
\mathbf{g}_0^t = (g_0^t[-2], g_0^t[-1], g_0^t[0], g_0^t[1], g_0^t[2]) = \left(-\frac{1}{10}, \frac{1}{2}, \frac{6}{5}, \frac{1}{2}, -\frac{1}{10} \right),
$$

$$
\mathbf{g}_1^t = (g_1^t[-3], g_1^t[-2], g_1^t[-1], g_1^t[0], g_1^t[1], g_1^t[2], g_1^t[3])
$$

$$
= \left(\frac{3}{140}, -\frac{3}{28}, -\frac{73}{140}, \frac{17}{14}, -\frac{73}{280}, -\frac{3}{28}, \frac{3}{140} \right).
$$

Write the analysis and synthesis matrices to the order 8 and compute the transformed scheme of the gradation

```
1 1 1 1 1 1 1 1
2 2 2 2 2 2 2 2
3 3 3 3 3 3 3 3
4 4 4 4 4 4 4 4
5 5 5 5 5 5 5 5
6 6 6 6 6 6 6 6
7 7 7 7 7 7 7 7
8 8 8 8 8 8 8 8
```

---

[14] We deal with the most distinguished member of a family of filter banks.

The two matrices:

$$\begin{pmatrix}
\frac{17}{28} & \frac{73}{140} & -\frac{3}{28} & -\frac{3}{140} & 0 & 0 & 0 & 0 \\[4pt]
-\frac{1}{4} & \frac{11}{20} & -\frac{1}{4} & -\frac{1}{20} & 0 & 0 & 0 & 0 \\[4pt]
-\frac{3}{56} & \frac{1}{4} & \frac{17}{28} & \frac{73}{280} & -\frac{3}{56} & -\frac{3}{280} & 0 & 0 \\[4pt]
0 & -\frac{1}{20} & -\frac{1}{4} & \frac{3}{5} & -\frac{1}{4} & -\frac{1}{20} & 0 & 0 \\[4pt]
0 & -\frac{3}{280} & -\frac{3}{56} & \frac{73}{280} & \frac{17}{28} & \frac{73}{280} & -\frac{3}{56} & -\frac{3}{280} \\[4pt]
0 & 0 & 0 & -\frac{1}{20} & -\frac{1}{4} & \frac{3}{5} & -\frac{1}{4} & -\frac{1}{20} \\[4pt]
0 & 0 & 0 & -\frac{3}{280} & -\frac{3}{56} & \frac{1}{4} & \frac{31}{56} & \frac{73}{280} \\[4pt]
0 & 0 & 0 & 0 & 0 & -\frac{1}{10} & -\frac{1}{2} & \frac{3}{5}
\end{pmatrix},$$

$$\begin{pmatrix}
\frac{6}{5} & -\frac{73}{70} & -\frac{1}{5} & \frac{3}{70} & 0 & 0 & 0 & 0 \\[4pt]
\frac{1}{2} & \frac{31}{28} & \frac{1}{2} & -\frac{3}{28} & 0 & 0 & 0 & 0 \\[4pt]
-\frac{1}{10} & -\frac{1}{2} & \frac{6}{5} & -\frac{73}{140} & -\frac{1}{10} & \frac{3}{140} & 0 & 0 \\[4pt]
0 & -\frac{3}{28} & \frac{1}{2} & \frac{17}{14} & \frac{1}{2} & -\frac{3}{28} & 0 & 0 \\[4pt]
0 & \frac{3}{140} & -\frac{1}{10} & -\frac{73}{140} & \frac{6}{5} & -\frac{73}{140} & -\frac{1}{10} & \frac{3}{140} \\[4pt]
0 & 0 & 0 & -\frac{3}{28} & \frac{1}{2} & \frac{17}{14} & \frac{1}{2} & -\frac{3}{28} \\[4pt]
0 & 0 & 0 & \frac{3}{140} & -\frac{1}{10} & -\frac{1}{2} & \frac{11}{10} & -\frac{73}{140} \\[4pt]
0 & 0 & 0 & 0 & 0 & -\frac{3}{14} & 1 & \frac{17}{14}
\end{pmatrix}.$$

The transformed matrix:
$$\begin{bmatrix}
1.24 & 0 & 1.24 & 0 & 1.24 & 0 & 1.24 & 0 \\
-0.1 & 0 & -0.1 & 0 & -0.1 & 0 & -0.1 & 0 \\
2.98 & 0 & 2.98 & 0 & 2.98 & 0 & 2.98 & 0 \\
0 & 0 & 0 & 0 & 0 & 0 & 0 & 0 \\
5 & 5 & 5 & 5 & 5 & 5 & 5 & 5 \\
0 & 0 & 0 & 0 & 0 & 0 & 0 & 0 \\
7.15 & 0 & 7.15 & 0 & 7.15 & 0 & 7.15 & 0 \\
0.7 & 0 & 0.7 & 0 & 0.7 & 0 & 0.7 & 0
\end{bmatrix}$$

(5) We aim for a (rough) comparison between the performances of the DWT 5/3 spline and of the DWT 7/5 Burt to the order 8^{15} with respect to ensuing compression (after appropriate quantization).

(a) If the coefficients of the 8×8 scheme Y_0 vary between $-\frac{1}{2}$ and $\frac{1}{2}$, what will be the bounds of variation for the transform coefficients of Y_1, in case of the DWT 5/3 spline and of the DWT 7/5 Burt?

(b) Consider the following four test matrices:

---

[15] Concerning the order 8: we suppose tacitly a transformation of *square* sample schemes. This is by no means necessary – and often not even wanted: a rectangular scheme Y_0 of m rows and n columns will be multiplied (as a matrix) by T_m from the left and by T_n^t from the right – where T_m and T_n are the matrices to the order m and n of the considered analysis filter bank.

$$A_0 \equiv \begin{bmatrix}
-\tfrac{1}{4} & -\tfrac{5}{16} & -\tfrac{3}{8} & -\tfrac{7}{16} & -\tfrac{1}{2} & -\tfrac{7}{16} & -\tfrac{3}{8} & -\tfrac{5}{16} \\[4pt]
-\tfrac{3}{16} & -\tfrac{15}{64} & -\tfrac{9}{32} & -\tfrac{21}{64} & -\tfrac{3}{8} & -\tfrac{21}{64} & -\tfrac{9}{32} & -\tfrac{15}{64} \\[4pt]
-\tfrac{1}{8} & -\tfrac{5}{32} & -\tfrac{3}{16} & -\tfrac{7}{32} & -\tfrac{1}{4} & -\tfrac{7}{32} & -\tfrac{3}{16} & -\tfrac{5}{32} \\[4pt]
-\tfrac{1}{16} & -\tfrac{5}{64} & -\tfrac{3}{32} & -\tfrac{7}{64} & -\tfrac{1}{8} & -\tfrac{7}{64} & -\tfrac{3}{32} & -\tfrac{5}{64} \\[4pt]
0 & 0 & 0 & 0 & 0 & 0 & 0 & 0 \\[4pt]
\tfrac{1}{16} & \tfrac{5}{64} & \tfrac{3}{32} & \tfrac{7}{64} & \tfrac{1}{8} & \tfrac{7}{64} & \tfrac{3}{32} & \tfrac{5}{64} \\[4pt]
\tfrac{1}{8} & \tfrac{5}{32} & \tfrac{3}{16} & \tfrac{7}{32} & \tfrac{1}{4} & \tfrac{7}{32} & \tfrac{3}{16} & \tfrac{5}{32} \\[4pt]
\tfrac{3}{16} & \tfrac{15}{64} & \tfrac{9}{32} & \tfrac{21}{64} & \tfrac{3}{8} & \tfrac{21}{64} & \tfrac{9}{32} & \tfrac{15}{64}
\end{bmatrix}$$

$$B_0 \equiv \begin{bmatrix}
\tfrac{3}{8} & \tfrac{3}{8} & \tfrac{3}{8} & \tfrac{3}{8} & \tfrac{3}{8} & \tfrac{3}{8} & \tfrac{3}{8} & \tfrac{3}{8} \\[4pt]
\tfrac{3}{8} & \tfrac{1}{4} & \tfrac{1}{4} & \tfrac{1}{4} & \tfrac{1}{4} & \tfrac{1}{4} & \tfrac{1}{4} & \tfrac{3}{8} \\[4pt]
\tfrac{3}{8} & \tfrac{1}{4} & \tfrac{1}{8} & 0 & 0 & \tfrac{1}{8} & \tfrac{1}{4} & \tfrac{3}{8} \\[4pt]
\tfrac{3}{8} & \tfrac{1}{4} & \tfrac{1}{8} & 0 & 0 & \tfrac{1}{8} & \tfrac{1}{4} & \tfrac{3}{8} \\[4pt]
\tfrac{3}{8} & \tfrac{1}{4} & \tfrac{1}{4} & \tfrac{1}{4} & \tfrac{1}{4} & \tfrac{1}{4} & \tfrac{1}{4} & \tfrac{3}{8} \\[4pt]
\tfrac{3}{8} & \tfrac{3}{8} & \tfrac{3}{8} & \tfrac{3}{8} & \tfrac{3}{8} & \tfrac{3}{8} & \tfrac{3}{8} & \tfrac{3}{8} \\[4pt]
\tfrac{3}{8} & \tfrac{3}{8} & \tfrac{3}{8} & \tfrac{3}{8} & \tfrac{3}{8} & \tfrac{3}{8} & \tfrac{3}{8} & \tfrac{3}{8} \\[4pt]
\tfrac{1}{8} & \tfrac{1}{8} & \tfrac{1}{8} & \tfrac{1}{8} & \tfrac{1}{8} & \tfrac{1}{8} & \tfrac{1}{8} & \tfrac{1}{8}
\end{bmatrix}$$

$$C_0 \equiv \begin{bmatrix}
-\tfrac{1}{2} & \tfrac{3}{8} & \tfrac{1}{8} & \tfrac{1}{8} & 0 & -\tfrac{1}{8} & -\tfrac{1}{4} & -\tfrac{3}{8} \\[4pt]
-\tfrac{3}{8} & -\tfrac{1}{2} & \tfrac{3}{8} & \tfrac{1}{8} & \tfrac{1}{8} & 0 & -\tfrac{1}{8} & -\tfrac{1}{4} \\[4pt]
-\tfrac{1}{4} & -\tfrac{3}{8} & -\tfrac{1}{2} & \tfrac{3}{8} & \tfrac{1}{8} & \tfrac{1}{8} & 0 & -\tfrac{1}{8} \\[4pt]
-\tfrac{1}{8} & -\tfrac{1}{4} & -\tfrac{3}{8} & -\tfrac{1}{2} & \tfrac{3}{8} & \tfrac{1}{8} & \tfrac{1}{8} & 0 \\[4pt]
0 & -\tfrac{1}{8} & -\tfrac{1}{4} & -\tfrac{3}{8} & -\tfrac{1}{2} & \tfrac{3}{8} & \tfrac{1}{8} & \tfrac{1}{8} \\[4pt]
\tfrac{1}{8} & 0 & -\tfrac{1}{8} & -\tfrac{1}{4} & -\tfrac{3}{8} & -\tfrac{1}{2} & \tfrac{3}{8} & \tfrac{1}{8} \\[4pt]
\tfrac{1}{8} & \tfrac{1}{8} & 0 & -\tfrac{1}{8} & -\tfrac{1}{4} & -\tfrac{3}{8} & -\tfrac{1}{2} & \tfrac{3}{8} \\[4pt]
\tfrac{3}{8} & \tfrac{1}{4} & \tfrac{1}{8} & 0 & -\tfrac{1}{8} & -\tfrac{1}{4} & -\tfrac{3}{8} & -\tfrac{1}{2}
\end{bmatrix}$$

$$D_0 \equiv \begin{bmatrix}
0 & 0 & 0 & \tfrac{1}{2} & \tfrac{1}{2} & 0 & 0 & 0 \\[4pt]
0 & 0 & 0 & \tfrac{1}{2} & \tfrac{1}{2} & 0 & 0 & 0 \\[4pt]
0 & 0 & 0 & \tfrac{1}{2} & \tfrac{1}{2} & 0 & 0 & 0 \\[4pt]
\tfrac{1}{2} & \tfrac{1}{2} & \tfrac{1}{2} & \tfrac{1}{2} & \tfrac{1}{2} & \tfrac{1}{2} & \tfrac{1}{2} & \tfrac{1}{2} \\[4pt]
0 & 0 & 0 & \tfrac{1}{2} & \tfrac{1}{2} & 0 & 0 & 0 \\[4pt]
0 & 0 & 0 & \tfrac{1}{2} & \tfrac{1}{2} & 0 & 0 & 0 \\[4pt]
0 & 0 & 0 & \tfrac{1}{2} & \tfrac{1}{2} & 0 & 0 & 0
\end{bmatrix}$$

First, find their transforms A_1, B_1, C_1 and D_1 by the DWT 5/3 spline as well as by the DWT 7/5 Burt.

We obtain (the coefficients have been duly rounded):

$$A_1 \equiv \begin{bmatrix}
-0.25 & 0 & -0.38 & 0 & -0.5 & 0 & -0.36 & 0.03 \\
0 & 0 & 0 & 0 & 0 & 0 & 0 & 0 \\
-0.13 & 0 & -0.19 & 0 & -0.25 & 0 & -0.18 & 0.02 \\
0 & 0 & 0 & 0 & 0 & 0 & 0 & 0 \\
0 & 0 & 0 & 0 & 0 & 0 & 0 & 0 \\
0 & 0 & 0 & 0 & 0 & 0 & 0 & 0 \\
0.14 & 0 & 0.21 & 0 & 0.28 & 0 & 0.2 & -0.02 \\
0.03 & 0 & 0.05 & 0 & 0.06 & 0 & 0.04 & 0
\end{bmatrix}$$

$$A_1' \equiv \begin{bmatrix}
-0.25 & 0.01 & -0.35 & -0.01 & -0.46 & -0.01 & -0.34 & 0.04 \\
-0.01 & 0 & -0.01 & 0 & -0.01 & 0 & -0.01 & 0 \\
-0.13 & 0 & -0.19 & 0 & -0.25 & 0 & -0.19 & 0.02 \\
0 & 0 & 0 & 0 & 0 & 0 & 0 & 0 \\
0 & 0 & 0 & 0 & 0 & 0 & 0 & 0 \\
0 & 0 & 0 & 0 & 0 & 0 & 0 & 0 \\
0.14 & 0 & 0.2 & 0 & 0.26 & 0 & 0.2 & -0.02 \\
0.05 & 0 & 0.07 & 0 & 0.08 & 0 & 0.06 & -0.01
\end{bmatrix}$$

$$B_1 \equiv \begin{matrix}
0.36 & -0.02 & 0.37 & 0 & 0.38 & 0.01 & 0.36 & 0.02 \\
-0.02 & -0.02 & -0.01 & 0 & 0 & 0.01 & -0.02 & 0.02 \\
0.37 & -0.01 & 0.12 & 0 & 0.1 & -0.02 & 0.26 & 0.07 \\
0 & 0 & 0 & -0.01 & -0.03 & 0.01 & 0 & 0 \\
0.38 & 0 & 0.1 & -0.03 & -0.03 & 0 & 0.28 & 0.06 \\
0.01 & 0.01 & -0.02 & 0.01 & 0 & -0.02 & -0.01 & 0 \\
0.36 & -0.02 & 0.26 & 0 & 0.28 & -0.01 & 0.3 & 0.05 \\
0.02 & 0.02 & 0.07 & 0 & 0.06 & 0 & 0.05 & -0.03
\end{matrix}$$

$$B_1' \equiv \begin{matrix}
0.35 & -0.02 & 0.34 & 0 & 0.34 & 0 & 0.34 & 0.03 \\
-0.02 & -0.02 & 0 & 0 & 0.01 & 0.01 & -0.02 & 0.02 \\
0.34 & 0 & 0.15 & -0.01 & 0.11 & -0.03 & 0.26 & 0.09 \\
0 & 0 & -0.01 & -0.02 & -0.04 & 0.01 & 0 & 0 \\
0.34 & 0.01 & 0.11 & -0.04 & 0.01 & -0.01 & 0.26 & 0.09 \\
0 & 0.01 & -0.03 & 0.01 & -0.01 & -0.02 & -0.01 & 0.01 \\
0.34 & -0.02 & 0.26 & 0 & 0.26 & -0.01 & 0.3 & 0.06 \\
0.03 & 0.02 & 0.09 & 0 & 0.09 & 0.01 & 0.06 & -0.05
\end{matrix}$$

$$C_1 \equiv \begin{matrix}
-0.44 & 0.06 & 0.5 & -0.06 & -0.03 & 0 & -0.28 & -0.06 \\
-0.19 & -0.19 & 0.13 & -0.06 & -0.03 & 0 & 0 & 0 \\
-0.34 & -0.09 & -0.36 & 0.13 & 0.42 & -0.03 & -0.05 & -0.06 \\
0 & 0 & -0.09 & -0.19 & 0.13 & -0.06 & -0.03 & 0 \\
0 & 0 & -0.3 & -0.09 & -0.36 & 0.13 & 0.38 & -0.13 \\
0 & 0 & 0 & 0 & -0.09 & -0.19 & 0.09 & -0.13 \\
0.28 & 0 & 0.03 & 0 & -0.27 & -0.09 & -0.27 & 0.25 \\
0.06 & 0 & 0.06 & 0 & 0.06 & 0 & -0.06 & -0.25
\end{matrix}$$

$$C_1' \equiv \begin{matrix}
-0.33 & 0.01 & 0.32 & 0 & 0.03 & 0 & -0.24 & -0.09 \\
-0.18 & -0.22 & 0.12 & -0.03 & -0.01 & 0 & -0.01 & 0 \\
-0.35 & -0.11 & -0.25 & 0.12 & 0.29 & 0 & -0.02 & -0.09 \\
0.01 & 0.03 & -0.13 & -0.25 & 0.12 & -0.03 & 0 & -0.01 \\
-0.03 & 0.02 & -0.29 & -0.13 & -0.26 & 0.12 & 0.27 & -0.09 \\
0 & 0 & 0.01 & 0.03 & -0.13 & -0.25 & 0.11 & -0.09 \\
0.24 & 0.01 & 0.02 & 0.01 & -0.27 & -0.12 & -0.22 & 0.27 \\
0.09 & 0 & 0.09 & 0 & 0.1 & 0.06 & -0.12 & -0.31
\end{matrix}$$

$$D_1 \equiv \begin{matrix}
0 & 0 & 0.06 & 0.13 & 0.5 & -0.13 & -0.06 & 0 \\
0 & 0 & 0 & 0 & 0 & 0 & 0 & 0 \\
0.06 & 0 & 0.12 & 0.11 & 0.5 & -0.11 & 0.01 & 0 \\
0.13 & 0 & 0.11 & -0.03 & 0 & 0.03 & 0.14 & 0 \\
0.5 & 0 & 0.5 & 0 & 0.5 & 0 & 0.5 & 0 \\
-0.13 & 0 & -0.11 & 0.03 & 0 & -0.03 & -0.14 & 0 \\
-0.06 & 0 & 0.01 & 0.14 & 0.5 & -0.14 & -0.13 & 0 \\
0 & 0 & 0 & 0 & 0 & 0 & 0 & 0
\end{matrix}$$

$$D_1' \equiv \begin{array}{rrrrrrrr}
-0.02 & -0.03 & 0.1 & 0.18 & 0.43 & -0.15 & -0.04 & 0 \\
-0.03 & 0 & -0.02 & 0.01 & 0 & -0.01 & -0.03 & 0 \\
0.1 & -0.02 & 0.19 & 0.14 & 0.45 & -0.12 & 0.08 & 0 \\
0.18 & 0.01 & 0.14 & -0.06 & 0.02 & 0.05 & 0.10 & 0 \\
0.43 & 0 & 0.45 & 0.02 & 0.49 & -0.02 & 0.43 & 0 \\
-0.15 & -0.01 & -0.12 & 0.05 & -0.02 & -0.05 & -0.16 & 0 \\
-0.04 & -0.03 & 0.08 & 0.19 & 0.43 & -0.16 & -0.07 & 0 \\
0 & 0 & 0 & 0 & 0 & 0 & 0 & 0
\end{array}$$

Note the "numerical neighbourliness" of the matrices transformed by the two DWTs. $\Big]$

(c) In order to compare the compressed bit-rates in the two cases, we shall proceed as follows.

Our quantization will associate with every (rounded) coefficient an 8-bit byte, the first bit of which encodes the sign of the considered coefficient.

The compaction by arithmetic coding will be carried out on the bitstream of the $7 \times 64 = 448$ bits given by the (truncated) binary notations of the 64 coefficients in the matrix scheme to be treated.

(Example: $0.34 \leftrightarrow 0101011$).

First we need the value $p_0 = \frac{N_0}{448}$, where N_0 is the number of the zeros in the "source bitstream" of 448 bits.

Counting table ("statistically balanced" [16]):

Size of the coefficient a	Number of zeros in the seven significant bits
$0.5 \leq a$	$0 + 3$
$0.25 \leq a < 0.5$	$1 + 2.5$
$0.13 \leq a < 0.25$	$2 + 2$
$0.06 \leq a < 0.12$	$3 + 1.5$
$0.03 \leq a < 0.06$	$4 + 1$
$0.02 \leq a < 0.03$	$5 + 0.5$
$a = 0.01$	6
$a = 0$	7

We shall replace the compressed bit-rate by the inverse compression ratio, i.e. by (approximately) the entropy $H = -p_0 \text{Log}_2 p_0 - (1 - p_0) \text{Log}_2 (1 - p_0)$.

Compare – in our four test examples – the inverse compression ratio obtained when using alternatively the DWT 5/3 spline and the DWT 7/5 Burt.

(6) Effect of local quantization on the decompressed image.

We shall study the propagation – in synthesis – of a local error (at the position (i, j)). Choose four positions in the four sub-bands of the transformed image:

LL: $E_{4,4}$ HL: $E_{4,3}$

LH: $E_{3,4}$ HH: $E_{3,3}$

Compute, for the DWT 5/3 spline and for the DWT 7/5 Burt the four corresponding synthesized matrices.

We note: the unit matrices $E_{i,j}$ are the matrix products of a column unit vector e_i with a row unit vector e_j^t. As a consequence, their synthesis transforms will

---

[16] We suppose the "residual parts" – after the first 1 in the binary notation – to be of equal probability.

be the matrix product of the ith column of the synthesis matrix with the transposed jth column of the synthesis matrix.

$\Big[$ Let $S_{i,j}$ (and $S'_{i,j}$, respectively) be the *decompressed* matrices in question. Then we obtain (with rounded coefficients):

$$S_{4,4} \equiv \begin{bmatrix} 0 & 0 & 0 & 0 & 0 & 0 & 0 & 0 \\ 0 & 0 & 0 & 0 & 0 & 0 & 0 & 0 \\ 0 & 0 & 0 & 0 & 0 & 0 & 0 & 0 \\ 0 & 0 & 0 & 0.25 & 0.5 & 0.25 & 0 & 0 \\ 0 & 0 & 0 & 0.5 & 1 & 0.5 & 0 & 0 \\ 0 & 0 & 0 & 0.25 & 05 & 0.25 & 0 & 0 \\ 0 & 0 & 0 & 0 & 0 & 0 & 0 & 0 \\ 0 & 0 & 0 & 0 & 0 & 0 & 0 & 0 \end{bmatrix}$$

$$S'_{4,4} \equiv \begin{bmatrix} 0 & 0 & 0 & 0 & 0 & 0 & 0 & 0 \\ 0 & 0 & 0 & 0 & 0 & 0 & 0 & 0 \\ 0 & 0 & 0.01 & -0.05 & -0.12 & -0.05 & 0.01 & 0 \\ 0 & 0 & -0.05 & 0.25 & 0.6 & 0.25 & -0.05 & 0 \\ 0 & 0 & -0.12 & 0.6 & 1.44 & 0.6 & -0.12 & 0 \\ 0 & 0 & -0.05 & 0.25 & 0.6 & 0.25 & -0.05 & 0 \\ 0 & 0 & 0.01 & -0.05 & -0.12 & -0.05 & 0.01 & 0 \\ 0 & 0 & 0 & 0 & 0 & 0 & 0 & 0 \end{bmatrix}$$

$$S_{3,4} \equiv \begin{bmatrix} 0 & 0 & 0 & 0 & 0 & 0 & 0 & 0 \\ 0 & 0 & 0 & -0.13 & -0.25 & -0.13 & 0 & 0 \\ 0 & 0 & 0 & -0.25 & -0.5 & -0.25 & 0 & 0 \\ 0 & 0 & 0 & 0.75 & 1.5 & 0.75 & 0 & 0 \\ 0 & 0 & 0 & -0.25 & -0.5 & -0.25 & 0 & 0 \\ 0 & 0 & 0 & -0.13 & -0.25 & -0.13 & 0 & 0 \\ 0 & 0 & 0 & 0 & 0 & 0 & 0 & 0 \\ 0 & 0 & 0 & 0 & 0 & 0 & 0 & 0 \end{bmatrix}$$

$$S'_{3,4} \equiv \begin{bmatrix} 0 & 0 & 0 & 0.02 & 0.05 & 0.02 & 0 & 0 \\ 0 & 0 & 0.01 & -0.05 & -0.13 & -0.05 & 0.01 & 0 \\ 0 & 0 & 0.05 & -0.26 & -0.63 & -0.26 & 0.05 & 0 \\ 0 & 0 & -0.12 & 0.61 & 1.46 & 0.61 & -0.12 & 0 \\ 0 & 0 & 0.05 & -0.26 & -0.63 & -0.26 & 0.05 & 0 \\ 0 & 0 & 0.01 & -0.05 & -0.13 & -0.05 & 0.01 & 0 \\ 0 & 0 & 0 & 0.01 & 0.03 & 0.01 & 0 & 0 \\ 0 & 0 & 0 & 0 & 0 & 0 & 0 & 0 \end{bmatrix}$$

$$S_{4,3} \equiv \begin{bmatrix} 0 & 0 & 0 & 0 & 0 & 0 & 0 & 0 \\ 0 & 0 & 0 & 0 & 0 & 0 & 0 & 0 \\ 0 & 0 & 0 & 0 & 0 & 0 & 0 & 0 \\ 0 & -0.13 & -0.25 & 0.75 & -0.25 & -0.13 & 0 & 0 \\ 0 & -0.25 & -0.5 & 1.5 & -0.5 & -0.25 & 0 & 0 \\ 0 & -0.13 & -0.25 & 0.75 & -0.25 & -0.13 & 0 & 0 \\ 0 & 0 & 0 & 0 & 0 & 0 & 0 & 0 \\ 0 & 0 & 0 & 0 & 0 & 0 & 0 & 0 \end{bmatrix}$$

$$S'_{4,3} \equiv \begin{bmatrix} 0 & 0 & 0 & 0 & 0 & 0 & 0 & 0 \\ 0 & 0 & 0 & 0 & 0 & 0 & 0 & 0 \\ 0 & 0.01 & 0.05 & -0.12 & 0.05 & 0.01 & 0 & 0 \\ 0.02 & -0.05 & -0.26 & 0.61 & -0.26 & -0.05 & 0.01 & 0 \\ 0.05 & -0.13 & -0.63 & 1.46 & -0.63 & -0.13 & 0.03 & 0 \\ 0.02 & -0.05 & -0.26 & 0.61 & -0.26 & -0.05 & 0.01 & 0 \\ 0 & 0.01 & 0.05 & -0.12 & 0.05 & 0.01 & 0 & 0 \\ 0 & 0 & 0 & 0 & 0 & 0 & 0 & 0 \end{bmatrix}$$

$$S_{3,3} \equiv \begin{bmatrix} 0 & 0 & 0 & 0 & 0 & 0 & 0 & 0 \\ 0 & 0.06 & 0.13 & -0.38 & 0.13 & 0.06 & 0 & 0 \\ 0 & 0.13 & 0.25 & -0.75 & 0.25 & 0.13 & 0 & 0 \\ 0 & -0.38 & -0.75 & 2.25 & -0.75 & -0.38 & 0 & 0 \\ 0 & 0.13 & 0.25 & -0.75 & 0.25 & 0.13 & 0 & 0 \\ 0 & 0.06 & 0.13 & -0.38 & 0.13 & 0.06 & 0 & 0 \\ 0 & 0 & 0 & 0 & 0 & 0 & 0 & 0 \\ 0 & 0 & 0 & 0 & 0 & 0 & 0 & 0 \end{bmatrix}$$

$$S'_{3,3} \equiv \begin{bmatrix} 0 & 0 & -0.02 & 0.05 & -0.02 & 0 & 0 & 0 \\ 0 & 0.01 & 0.06 & -0.13 & 0.06 & 0.01 & 0 & 0 \\ -0.02 & 0.06 & 0.27 & -0.63 & 0.27 & 0.06 & -0.01 & 0 \\ 0.05 & -0.13 & -0.63 & 1.47 & -0.63 & -0.13 & 0.03 & 0 \\ -0.02 & 0.06 & 0.27 & -0.63 & 0.27 & 0.06 & -0.01 & 0 \\ 0 & 0.01 & 0.06 & -0.13 & 0.06 & 0.01 & 0 & 0 \\ 0 & 0 & -0.01 & 0.03 & -0.01 & 0 & 0 & 0 \\ 0 & 0 & 0 & 0 & 0 & 0 & 0 & 0 \end{bmatrix}$$

Commentary *Note that the transformed matrices only have an explicit pictorial meaning for the sub-band LL. The three other sub-bands LH, HL and HH bear information on details. Now, a local error (by local quantization) concerning merely descriptive data will be decompressed into a pictorial patch. This is not very serious, as far as everywhere quantization creates some modest uniform numerical deviation.* $\Big]$

Resolution levels – interleaved perspective

Let us continue with our *example* of the 2D DWT 5/3 spline, acting on the gradation

$$\begin{matrix}
1 & 1 & 1 & 1 & 1 & 1 & 1 & 1 \\
2 & 2 & 2 & 2 & 2 & 2 & 2 & 2 \\
3 & 3 & 3 & 3 & 3 & 3 & 3 & 3 \\
4 & 4 & 4 & 4 & 4 & 4 & 4 & 4 \\
5 & 5 & 5 & 5 & 5 & 5 & 5 & 5 \\
6 & 6 & 6 & 6 & 6 & 6 & 6 & 6 \\
7 & 7 & 7 & 7 & 7 & 7 & 7 & 7 \\
8 & 8 & 8 & 8 & 8 & 8 & 8 & 8
\end{matrix}$$

First some preliminary computations.

We are interested in matrix representations to the order 4, 2 (and 1) of the DWT 5/3 spline:

Applying the analysis formula to $h_0^t = (-\frac{1}{8}, \frac{1}{4}, \frac{3}{4}, \frac{1}{4}, -\frac{1}{8})$ and $h_1^t = (-\frac{1}{4}, \frac{1}{2}, -\frac{1}{4})$, we obtain, for the computation to the order 4,

$$\begin{pmatrix} y[0] \\ y[1] \\ y[2] \\ y[3] \end{pmatrix} = \begin{pmatrix} h_0^t[0] & 2h_0^t[1] & 2h_0^t[2] & 0 \\ h_1^t[1] & h_1^t[0] & h_1^t[1] & 0 \\ h_0^t[2] & h_0^t[1] & h_0^t[0] + h_0^t[2] & h_0^t[1] \\ 0 & 0 & 2h_1^t[1] & h_1^t[0] \end{pmatrix} \begin{pmatrix} x[0] \\ x[1] \\ x[2] \\ x[3] \end{pmatrix}$$

(you have only to copy the first four rows in the computation to the order 8, while observing that now $\check{x}[4] = x[2]$).

Concretely:

$$\begin{pmatrix} y[0] \\ y[1] \\ y[2] \\ y[3] \end{pmatrix} = \begin{pmatrix} \frac{3}{4} & \frac{1}{2} & -\frac{1}{4} & 0 \\ -\frac{1}{4} & \frac{1}{2} & -\frac{1}{4} & 0 \\ -\frac{1}{8} & \frac{1}{4} & \frac{5}{8} & \frac{1}{4} \\ 0 & 0 & -\frac{1}{2} & \frac{1}{2} \end{pmatrix} \begin{pmatrix} x[0] \\ x[1] \\ x[2] \\ x[3] \end{pmatrix}.$$

In the same spirit, we easily obtain the matrix representation to the order 2:

$$\begin{pmatrix} y[0] \\ y[1] \end{pmatrix} = \begin{pmatrix} h_0^t[0] + 2h_0^t[2] & 2h_0^t[1] \\ 2h_1^t[1] & h_1^t[0] \end{pmatrix} \begin{pmatrix} x[0] \\ x[1] \end{pmatrix}.$$

Concretely:

$$\begin{pmatrix} y[0] \\ y[1] \end{pmatrix} = \begin{pmatrix} \frac{1}{2} & \frac{1}{2} \\ -\frac{1}{2} & \frac{1}{2} \end{pmatrix} \begin{pmatrix} x[0] \\ x[1] \end{pmatrix}.$$

Finally, we will have, to the order 1:

$$y[0] = (2h_0^t[2] + 2h_0^t[1] + h_0^t[0]) \cdot x[0] = x[0].$$

For the corresponding synthesis transformations (to the order 4 and 2) we obtain:

$$\begin{pmatrix} x[0] \\ x[1] \\ x[2] \\ x[3] \end{pmatrix} = 2 \cdot \begin{pmatrix} h_1^t[0] & -2h_0^t[1] & 0 & 0 \\ -h_1^t[1] & h_0^t[0] + h_0^t[2] & -h_1^t[1] & h_0^t[2] \\ 0 & -h_0^t[1] & h_1^t[0] & -h_0^t[1] \\ 0 & 2h_0^t[2] & -2h_1^t[1] & h_0^t[0] \end{pmatrix} \begin{pmatrix} y[0] \\ y[1] \\ y[2] \\ y[3] \end{pmatrix}.$$

This concretely gives, in the case of order 4:

$$\begin{pmatrix} x[0] \\ x[1] \\ x[2] \\ x[3] \end{pmatrix} = \begin{pmatrix} 1 & -1 & 0 & 0 \\ \frac{1}{2} & \frac{5}{4} & \frac{1}{2} & -\frac{1}{4} \\ 0 & -\frac{1}{2} & 1 & -\frac{1}{2} \\ 0 & -\frac{1}{2} & 1 & \frac{3}{2} \end{pmatrix} \begin{pmatrix} y[0] \\ y[1] \\ y[2] \\ y[3] \end{pmatrix}.$$

The synthesis equations to the order 2, in matrix notation:

$$\begin{pmatrix} x[0] \\ x[1] \end{pmatrix} = 2 \cdot \begin{pmatrix} h_1^t[0] & -2h_0^t[1] \\ -2h_1^t[1] & h_0^t[0] + 2h_0^t[2] \end{pmatrix} \begin{pmatrix} y[0] \\ y[1] \end{pmatrix} = \begin{pmatrix} 1 & -1 \\ 1 & 1 \end{pmatrix} \begin{pmatrix} y[0] \\ y[1] \end{pmatrix}.$$

Now return to our initial gradation. Write

$$LL_0 = \begin{array}{cccccccc} 1 & 1 & 1 & 1 & 1 & 1 & 1 & 1 \\ 2 & 2 & 2 & 2 & 2 & 2 & 2 & 2 \\ 3 & 3 & 3 & 3 & 3 & 3 & 3 & 3 \\ 4 & 4 & 4 & 4 & 4 & 4 & 4 & 4 \\ 5 & 5 & 5 & 5 & 5 & 5 & 5 & 5 \\ 6 & 6 & 6 & 6 & 6 & 6 & 6 & 6 \\ 7 & 7 & 7 & 7 & 7 & 7 & 7 & 7 \\ 8 & 8 & 8 & 8 & 8 & 8 & 8 & 8 \end{array}$$

Then:

$$LL_1 \equiv \begin{array}{cccc} 1 & 1 & 1 & 1 \\ 3 & 3 & 3 & 3 \\ 5 & 5 & 5 & 5 \\ 7.25 & 7.25 & 7.25 & 7.25 \end{array} \qquad HL_1 \equiv \begin{array}{cccc} 0 & 0 & 0 & 0 \\ 0 & 0 & 0 & 0 \\ 0 & 0 & 0 & 0 \\ 0 & 0 & 0 & 0 \end{array}$$

$$LH_1 \equiv \begin{array}{cccc} 0 & 0 & 0 & 0 \\ 0 & 0 & 0 & 0 \\ 0 & 0 & 0 & 0 \\ 0.5 & 0.5 & 0.5 & 0.5 \end{array} \qquad HH_1 \equiv \begin{array}{cccc} 0 & 0 & 0 & 0 \\ 0 & 0 & 0 & 0 \\ 0 & 0 & 0 & 0 \\ 0 & 0 & 0 & 0 \end{array}$$

Apply now the 2D DWT to the order 4 to LL_1.

You have to evaluate the matrix product

$$\begin{pmatrix} \frac{3}{4} & \frac{1}{2} & -\frac{1}{4} & 0 \\ -\frac{1}{4} & \frac{1}{2} & -\frac{1}{4} & 0 \\ -\frac{1}{8} & \frac{1}{4} & \frac{5}{8} & \frac{1}{4} \\ 0 & 0 & -\frac{1}{2} & \frac{1}{2} \end{pmatrix} \begin{pmatrix} 1 & 1 & 1 & 1 \\ 3 & 3 & 3 & 3 \\ 5 & 5 & 5 & 5 \\ 7.25 & 7.25 & 7.25 & 7.25 \end{pmatrix} \begin{pmatrix} \frac{3}{4} & -\frac{1}{4} & -\frac{1}{8} & 0 \\ \frac{1}{2} & \frac{1}{2} & \frac{1}{4} & 0 \\ -\frac{1}{4} & -\frac{1}{4} & \frac{5}{8} & -\frac{1}{2} \\ 0 & 0 & \frac{1}{4} & \frac{1}{2} \end{pmatrix}.$$

The result:

$$\begin{pmatrix} 1 & 0 & 1 & 0 \\ 0 & 0 & 0 & 0 \\ \frac{89}{16} & 0 & \frac{89}{16} & 0 \\ \frac{9}{8} & 0 & \frac{9}{8} & 0 \end{pmatrix}.$$

This gives the four sub-bands

$$LL_2 \equiv \begin{array}{cc} 1 & 1 \\ \frac{89}{16} & \frac{89}{16} \end{array} \qquad HL_2 \equiv \begin{array}{cc} 0 & 0 \\ 0 & 0 \end{array}$$

$$LH_2 \equiv \begin{array}{cc} 0 & 0 \\ \frac{9}{8} & \frac{9}{8} \end{array} \qquad HH_2 \equiv \begin{array}{cc} 0 & 0 \\ 0 & 0 \end{array}$$

Finally, apply the 2D DWT to the order 2 to LL_2:

$$\begin{pmatrix} \frac{1}{2} & \frac{1}{2} \\ -\frac{1}{2} & \frac{1}{2} \end{pmatrix} \begin{pmatrix} 1 & 1 \\ \frac{89}{16} & \frac{89}{16} \end{pmatrix} \begin{pmatrix} \frac{1}{2} & -\frac{1}{2} \\ \frac{1}{2} & \frac{1}{2} \end{pmatrix} = \begin{pmatrix} \frac{105}{32} & 0 \\ \frac{73}{32} & 0 \end{pmatrix}.$$

We have obtained the following four sub-bands:

$$LL_3 \equiv \tfrac{105}{32}, \qquad HL_3 \equiv 0,$$

$$LH_3 \equiv \tfrac{73}{32}, \qquad HH_3 \equiv 0.$$

This ends the *three level resolution* of our given gradation.

Now let us pass to the *general situation*.

We dispose of a DWT (given by a filter bank). The ensuing D level, 2D DWT will function as follows:

$LL_0 = x[\mathbf{n}] \equiv$ (the component of) the digital image (considered as an $N \times N$ matrix, where N will be – for dichotomic convenience – a power of 2).

When applying the 2D DWT of order N to LL_0, we obtain

$$LL_1 = y_{00}^{(1)}[\mathbf{n}], \qquad HL_1 = y_{01}^{(1)}[\mathbf{n}],$$
$$LH_1 = y_{10}^{(1)}[\mathbf{n}], \qquad HH_1 = y_{11}^{(1)}[\mathbf{n}].$$

In interleaved version:

$$y^{(1)}[2n_1 + b_1, 2n_2 + b_2] = y_{b_1 b_2}^{(1)}[n_1, n_2], \quad 0 \le n_1, n_2 \le \frac{1}{2}N - 1.$$

Now, we apply the 2D DWT of order $\frac{1}{2}N$ to LL_1, and we get:

$$LL_2 = y_{00}^{(2)}[\mathbf{n}], \quad HL_2 = y_{01}^{(2)}[\mathbf{n}],$$
$$LH_2 = y_{10}^{(2)}[\mathbf{n}], \quad HH_2 = y_{11}^{(2)}[\mathbf{n}].$$

We note

We only have treated the positions of the interleaved matrix $y^{(1)}[\mathbf{n}]$ with coordinates (multiple of 2, multiple of 2).

We *replace* these entries, while *keeping* the others unchanged, which gives the interleaved matrix $y^{(2)}[\mathbf{n}]$ (it is still an $N \times N$ matrix), with

$$y^{(2)}[4n_1 + 2b_1, 4n_2 + 2b_2] = y_{b_1 b_2}^{(2)}[n_1, n_2], \quad 0 \le n_1, n_2 \le \frac{1}{4}N - 1.$$

Then, we apply the 2D DWT of order $\frac{1}{4}N$ to LL_2, and we obtain:

$$LL_3 = y_{00}^{(3)}[\mathbf{n}], \quad HL_3 = y_{01}^{(3)}[\mathbf{n}],$$
$$LH_3 = y_{10}^{(3)}[\mathbf{n}], \quad HH_3 = y_{11}^{(3)}[\mathbf{n}].$$

We note

We have only treated the positions of the interleaved matrix $y^{(2)}[\mathbf{n}]$ with coordinates (multiple of 4, multiple of 4).

We *replace* these entries, while *keeping* the others unchanged, which gives the interleaved matrix $y^{(3)}[\mathbf{n}]$ (this is still an $N \times N$ matrix) with

$$y^{(3)}[8n_1 + 4b_1, 8n_2 + 4b_2] = y_{b_1 b_2}^{(3)}[n_1, n_2], \quad 0 \le n_1, n_2 \le \frac{1}{8}N - 1$$

and we continue.

On the dth level, we have

$$LL_d = y_{00}^{(d)}[\mathbf{n}], \quad HL_d = y_{01}^{(d)}[\mathbf{n}],$$
$$LH_d = y_{10}^{(d)}[\mathbf{n}], \quad HH_d = y_{11}^{(d)}[\mathbf{n}],$$

with

$$y^{(d)}[2^d n_1 + 2^{d-1} b_1, 2^d n_2 + 2^{d-1} b_2] = y_{b_1 b_2}^{(d)}[n_1, n_2], \quad 0 \le n_1, n_2 \le \frac{1}{2^d}N - 1.$$

Return to our *example*, in interleaved version. The considered component of the digital image will hence be given by

$$x[\mathbf{n}] = \begin{matrix}
1 & 1 & 1 & 1 & 1 & 1 & 1 & 1 \\
2 & 2 & 2 & 2 & 2 & 2 & 2 & 2 \\
3 & 3 & 3 & 3 & 3 & 3 & 3 & 3 \\
4 & 4 & 4 & 4 & 4 & 4 & 4 & 4 \\
5 & 5 & 5 & 5 & 5 & 5 & 5 & 5 \\
6 & 6 & 6 & 6 & 6 & 6 & 6 & 6 \\
7 & 7 & 7 & 7 & 7 & 7 & 7 & 7 \\
8 & 8 & 8 & 8 & 8 & 8 & 8 & 8
\end{matrix}$$

Look now at the three levels of the treatment, in interleaved notation:

$$y^{(1)}[\mathbf{n}] = \begin{matrix}
1 & 0 & 1 & 0 & 1 & 0 & 1 & 0 \\
0 & 0 & 0 & 0 & 0 & 0 & 0 & 0 \\
3 & 0 & 3 & 0 & 3 & 0 & 3 & 0 \\
0 & 0 & 0 & 0 & 0 & 0 & 0 & 0 \\
5 & 0 & 5 & 0 & 5 & 0 & 5 & 0 \\
0 & 0 & 0 & 0 & 0 & 0 & 0 & 0 \\
\frac{29}{4} & 0 & \frac{29}{4} & 0 & \frac{29}{4} & 0 & \frac{29}{4} & 0 \\
\frac{1}{2} & 0 & \frac{1}{2} & 0 & \frac{1}{2} & 0 & \frac{1}{2} & 0
\end{matrix}$$

$$y^{(2)}[\mathbf{n}] = \begin{matrix}
1 & 0 & 0 & 0 & 1 & 0 & 0 & 0 \\
0 & 0 & 0 & 0 & 0 & 0 & 0 & 0 \\
0 & 0 & 0 & 0 & 0 & 0 & 0 & 0 \\
0 & 0 & 0 & 0 & 0 & 0 & 0 & 0 \\
\frac{89}{16} & 0 & 0 & 0 & \frac{89}{16} & 0 & 0 & 0 \\
0 & 0 & 0 & 0 & 0 & 0 & 0 & 0 \\
\frac{9}{8} & 0 & 0 & 0 & \frac{9}{8} & 0 & 0 & 0 \\
\frac{1}{2} & 0 & \frac{1}{2} & 0 & \frac{1}{2} & 0 & \frac{1}{2} & 0
\end{matrix}$$

$$y^{(3)}[\mathbf{n}] = \begin{matrix}
\frac{105}{32} & 0 & 0 & 0 & 0 & 0 & 0 & 0 \\
0 & 0 & 0 & 0 & 0 & 0 & 0 & 0 \\
0 & 0 & 0 & 0 & 0 & 0 & 0 & 0 \\
0 & 0 & 0 & 0 & 0 & 0 & 0 & 0 \\
\frac{73}{32} & 0 & 0 & 0 & 0 & 0 & 0 & 0 \\
0 & 0 & 0 & 0 & 0 & 0 & 0 & 0 \\
\frac{9}{8} & 0 & 0 & 0 & \frac{9}{8} & 0 & 0 & 0 \\
\frac{1}{2} & 0 & \frac{1}{2} & 0 & \frac{1}{2} & 0 & \frac{1}{2} & 0
\end{matrix}$$

Scheme of a D level, 2D DWT (in concrete applications, $D = 4, 5$).

LL_0	$\equiv$	A component of the digital image	
$\downarrow$			
LL_1	$\mathcal{R}_D$:	HL_1	Dth resolution
	LH_1	HH_1	level
$\downarrow$			
LL_2	$\mathcal{R}_{D-1}$:	HL_2	$(D-1)$th resolution
	LH_2	HH_2	level
$\downarrow$			
$\vdots$			
LL_{D-1}	$\mathcal{R}_2$:	HL_{D-1}	Second resolution
	LH_{D-1}	HH_{D-1}	level
$\downarrow$			
LL_D	$\mathcal{R}_1$:	HL_D	First resolution
	LH_D	HH_D	level
$\mathcal{R}_0$			

Language *The three (2D) sub-bands needed for the passage from the resolution* LL_r *to the resolution* LL_{r-1} *will be called the resolution levels* $\mathcal{R}_{D-r+1}$.

In order to get a first idea of the degree of influence which have the resolution levels for the faithful reconstruction of the image, start with

$$(\mathcal{R}_0, \mathcal{R}_1) \equiv \begin{matrix} \frac{105}{32} & 0 \\ \frac{73}{32} & 0 \end{matrix}$$

in our example of a three level resolution for a 8×8 gradation.

In $\mathcal{R}_2$ and $\mathcal{R}_3$, let us set LH_2 and LH_1 to zero: the vertically high-pass information is "skipped".

In other words, we have to apply the synthesis transform to the interleaved matrix

$$y'^{(3)} = \begin{matrix} \frac{105}{32} & 0 & 0 & 0 & 0 & 0 & 0 & 0 \\ 0 & 0 & 0 & 0 & 0 & 0 & 0 & 0 \\ 0 & 0 & 0 & 0 & 0 & 0 & 0 & 0 \\ 0 & 0 & 0 & 0 & 0 & 0 & 0 & 0 \\ \frac{73}{32} & 0 & 0 & 0 & 0 & 0 & 0 & 0 \\ 0 & 0 & 0 & 0 & 0 & 0 & 0 & 0 \\ 0 & 0 & 0 & 0 & 0 & 0 & 0 & 0 \\ 0 & 0 & 0 & 0 & 0 & 0 & 0 & 0 \end{matrix}$$

First, use the 2D synthesis transform to the order 2:

$$LL_2' = \begin{pmatrix} 1 & -1 \\ 1 & 1 \end{pmatrix} \begin{pmatrix} \frac{105}{32} & 0 \\ \frac{73}{32} & 0 \end{pmatrix} \begin{pmatrix} 1 & 1 \\ -1 & 1 \end{pmatrix} = \begin{pmatrix} 1 & 1 \\ \frac{89}{16} & \frac{89}{16} \end{pmatrix}$$

This gives – in interleaved version:

$$y'^{(2)} = \begin{pmatrix} 1 & 0 & 0 & 0 & 1 & 0 & 0 & 0 \\ 0 & 0 & 0 & 0 & 0 & 0 & 0 & 0 \\ 0 & 0 & 0 & 0 & 0 & 0 & 0 & 0 \\ 0 & 0 & 0 & 0 & 0 & 0 & 0 & 0 \\ \frac{89}{16} & 0 & 0 & 0 & \frac{89}{16} & 0 & 0 & 0 \\ 0 & 0 & 0 & 0 & 0 & 0 & 0 & 0 \\ 0 & 0 & 0 & 0 & 0 & 0 & 0 & 0 \\ 0 & 0 & 0 & 0 & 0 & 0 & 0 & 0 \end{pmatrix}$$

Now, we have to apply the 2D synthesis transform of order 4 (to the entries on the (even, even) positions of the interleaved matrix).

$$LL'_1 = \begin{pmatrix} 1 & -1 & 0 & 0 \\ \frac{1}{2} & \frac{5}{4} & \frac{1}{2} & -\frac{1}{4} \\ 0 & -\frac{1}{2} & 1 & -\frac{1}{2} \\ 0 & -\frac{1}{2} & 1 & \frac{3}{2} \end{pmatrix} \begin{pmatrix} 1 & 0 & 1 & 0 \\ 0 & 0 & 0 & 0 \\ \frac{89}{16} & 0 & \frac{89}{16} & 0 \\ 0 & 0 & 0 & 0 \end{pmatrix} \begin{pmatrix} 1 & \frac{1}{2} & 0 & 0 \\ -1 & \frac{5}{4} & -\frac{1}{2} & -\frac{1}{2} \\ 0 & \frac{1}{2} & 1 & 1 \\ 0 & -\frac{1}{4} & -\frac{1}{2} & \frac{3}{2} \end{pmatrix}$$

$$= \begin{pmatrix} 1 & 1 & 1 & 1 \\ \frac{105}{32} & \frac{105}{32} & \frac{105}{32} & \frac{105}{32} \\ \frac{89}{16} & \frac{89}{16} & \frac{89}{16} & \frac{89}{16} \\ \frac{89}{16} & \frac{89}{16} & \frac{89}{16} & \frac{89}{16} \end{pmatrix}.$$

This gives – in interleaved version:

$$y'^{(1)} = \begin{pmatrix} 1 & 0 & 1 & 0 & 1 & 0 & 1 & 0 \\ 0 & 0 & 0 & 0 & 0 & 0 & 0 & 0 \\ \frac{105}{32} & 0 & \frac{105}{32} & 0 & \frac{105}{32} & 0 & \frac{105}{32} & 0 \\ 0 & 0 & 0 & 0 & 0 & 0 & 0 & 0 \\ \frac{89}{16} & 0 & \frac{89}{16} & 0 & \frac{89}{16} & 0 & \frac{89}{16} & 0 \\ 0 & 0 & 0 & 0 & 0 & 0 & 0 & 0 \\ \frac{89}{16} & 0 & \frac{89}{16} & 0 & \frac{89}{16} & 0 & \frac{89}{16} & 0 \\ 0 & 0 & 0 & 0 & 0 & 0 & 0 & 0 \end{pmatrix}$$

Finally, we reconstruct the image: multiply

$$\begin{pmatrix} 1 & -1 & 0 & 0 & 0 & 0 & 0 & 0 \\ \frac{1}{2} & \frac{5}{4} & \frac{1}{2} & -\frac{1}{4} & 0 & 0 & 0 & 0 \\ 0 & -\frac{1}{2} & 1 & -\frac{1}{2} & 0 & 0 & 0 & 0 \\ 0 & -\frac{1}{4} & \frac{1}{2} & \frac{3}{2} & \frac{1}{2} & -\frac{1}{4} & 0 & 0 \\ 0 & 0 & 0 & -\frac{1}{2} & 1 & -\frac{1}{2} & 0 & 0 \\ 0 & 0 & 0 & -\frac{1}{4} & \frac{1}{2} & \frac{3}{2} & \frac{1}{2} & -\frac{1}{4} \\ 0 & 0 & 0 & 0 & 0 & -\frac{1}{2} & 1 & -\frac{1}{2} \\ 0 & 0 & 0 & 0 & 0 & -\frac{1}{2} & 1 & \frac{3}{2} \end{pmatrix} \begin{pmatrix} 1 & 0 & 1 & 0 & 1 & 0 & 1 & 0 \\ 0 & 0 & 0 & 0 & 0 & 0 & 0 & 0 \\ \frac{105}{32} & 0 & \frac{105}{32} & 0 & \frac{105}{32} & 0 & \frac{105}{32} & 0 \\ 0 & 0 & 0 & 0 & 0 & 0 & 0 & 0 \\ \frac{89}{16} & 0 & \frac{89}{16} & 0 & \frac{89}{16} & 0 & \frac{89}{16} & 0 \\ 0 & 0 & 0 & 0 & 0 & 0 & 0 & 0 \\ \frac{89}{16} & 0 & \frac{89}{16} & 0 & \frac{89}{16} & 0 & \frac{89}{16} & 0 \\ 0 & 0 & 0 & 0 & 0 & 0 & 0 & 0 \end{pmatrix}$$

$$= \begin{pmatrix} 1 & 0 & 1 & 0 & 1 & 0 & 1 & 0 \\ \frac{137}{64} & 0 & \frac{137}{64} & 0 & \frac{137}{64} & 0 & \frac{137}{64} & 0 \\ \frac{105}{32} & 0 & \frac{105}{32} & 0 & \frac{105}{32} & 0 & \frac{105}{32} & 0 \\ \frac{283}{64} & 0 & \frac{283}{64} & 0 & \frac{283}{64} & 0 & \frac{283}{64} & 0 \\ \frac{89}{16} & 0 & \frac{89}{16} & 0 & \frac{89}{16} & 0 & \frac{89}{16} & 0 \\ \frac{89}{16} & 0 & \frac{89}{16} & 0 & \frac{89}{16} & 0 & \frac{89}{16} & 0 \\ \frac{89}{16} & 0 & \frac{89}{16} & 0 & \frac{89}{16} & 0 & \frac{89}{16} & 0 \\ \frac{89}{16} & 0 & \frac{89}{16} & 0 & \frac{89}{16} & 0 & \frac{89}{16} & 0 \end{pmatrix}.$$

Then, multiply

$$
\begin{pmatrix}
1 & 0 & 1 & 0 & 1 & 0 & 1 & 0 \\
\frac{137}{64} & 0 & \frac{137}{64} & 0 & \frac{137}{64} & 0 & \frac{137}{64} & 0 \\
\frac{105}{32} & 0 & \frac{105}{32} & 0 & \frac{105}{32} & 0 & \frac{105}{32} & 0 \\
\frac{283}{64} & 0 & \frac{283}{64} & 0 & \frac{283}{64} & 0 & \frac{283}{64} & 0 \\
\frac{89}{16} & 0 & \frac{89}{16} & 0 & \frac{89}{16} & 0 & \frac{89}{16} & 0 \\
\frac{89}{16} & 0 & \frac{89}{16} & 0 & \frac{89}{16} & 0 & \frac{89}{16} & 0 \\
\frac{89}{16} & 0 & \frac{89}{16} & 0 & \frac{89}{16} & 0 & \frac{89}{16} & 0 \\
\frac{89}{16} & 0 & \frac{89}{16} & 0 & \frac{89}{16} & 0 & \frac{89}{16} & 0
\end{pmatrix}
\begin{pmatrix}
1 & \frac{1}{2} & 0 & 0 & 0 & 0 & 0 & 0 \\
-1 & \frac{5}{4} & -\frac{1}{2} & -\frac{1}{4} & 0 & 0 & 0 & 0 \\
0 & \frac{1}{2} & 1 & \frac{1}{2} & 0 & 0 & 0 & 0 \\
0 & -\frac{1}{4} & -\frac{1}{2} & \frac{3}{2} & -\frac{1}{2} & -\frac{1}{4} & 0 & 0 \\
0 & 0 & 0 & \frac{1}{2} & 1 & \frac{1}{2} & 0 & 0 \\
0 & 0 & 0 & -\frac{1}{4} & -\frac{1}{2} & \frac{3}{2} & -\frac{1}{2} & -\frac{1}{2} \\
0 & 0 & 0 & 0 & 0 & \frac{1}{2} & 1 & 1 \\
0 & 0 & 0 & 0 & 0 & -\frac{1}{4} & -\frac{1}{2} & \frac{3}{2}
\end{pmatrix}
=
$$

$$
\begin{pmatrix}
1 & 1 & 1 & 1 & 1 & 1 & 1 & 1 \\
\frac{137}{64} & \frac{137}{64} & \frac{137}{64} & \frac{137}{64} & \frac{137}{64} & \frac{137}{64} & \frac{137}{64} & \frac{137}{64} \\
\frac{105}{32} & \frac{105}{32} & \frac{105}{32} & \frac{105}{32} & \frac{105}{32} & \frac{105}{32} & \frac{105}{32} & \frac{105}{32} \\
\frac{283}{64} & \frac{283}{64} & \frac{283}{64} & \frac{283}{64} & \frac{283}{64} & \frac{283}{64} & \frac{283}{64} & \frac{283}{64} \\
\frac{89}{16} & \frac{89}{16} & \frac{89}{16} & \frac{89}{16} & \frac{89}{16} & \frac{89}{16} & \frac{89}{16} & \frac{89}{16} \\
\frac{89}{16} & \frac{89}{16} & \frac{89}{16} & \frac{89}{16} & \frac{89}{16} & \frac{89}{16} & \frac{89}{16} & \frac{89}{16} \\
\frac{89}{16} & \frac{89}{16} & \frac{89}{16} & \frac{89}{16} & \frac{89}{16} & \frac{89}{16} & \frac{89}{16} & \frac{89}{16} \\
\frac{89}{16} & \frac{89}{16} & \frac{89}{16} & \frac{89}{16} & \frac{89}{16} & \frac{89}{16} & \frac{89}{16} & \frac{89}{16}
\end{pmatrix}
=
\begin{pmatrix}
1 & 1 & 1 & 1 & 1 & 1 & 1 & 1 \\
2.14 & 2.14 & 2.14 & 2.14 & 2.14 & 2.14 & 2.14 & 2.14 \\
3.28 & 3.28 & 3.28 & 3.28 & 3.28 & 3.28 & 3.28 & 3.28 \\
4.42 & 4.42 & 4.42 & 4.42 & 4.42 & 4.42 & 4.42 & 4.42 \\
5.56 & 5.56 & 5.56 & 5.56 & 5.56 & 5.56 & 5.56 & 5.56 \\
5.56 & 5.56 & 5.56 & 5.56 & 5.56 & 5.56 & 5.56 & 5.56 \\
5.56 & 5.56 & 5.56 & 5.56 & 5.56 & 5.56 & 5.56 & 5.56 \\
5.56 & 5.56 & 5.56 & 5.56 & 5.56 & 5.56 & 5.56 & 5.56
\end{pmatrix}.
$$

We note: when annihilating the vertically high-pass information for two resolution levels, we have decompressed into a half-constant gradation.

Exercises

The DWT of Cohen–Daubechies–Fauveau

The DWT 9/7 CDF – which is, by the way, a constituent of the standard JPEG 2000/1 – is defined by the following two analysis impulse responses:

$$
h_0^t[-4] = h_0^t[4] = 0.026748757411,
$$
$$
h_0^t[-3] = h_0^t[3] = -0.016864118443,
$$
$$
h_0^t[-2] = h_0^t[2] = -0.078223266529,
$$
$$
h_0^t[-1] = h_0^t[1] = 0.266864118443,
$$
$$
h_0^t[0] = 0.602949018236,
$$
$$
h_1^t[-3] = h_1^t[3] = 0.045635881557,
$$
$$
h_1^t[-2] = h_1^t[2] = -0.028771763114,
$$
$$
h_1^t[-1] = h_1^t[1] = -0.295635881557,
$$
$$
h_1^t[0] = 0.557543526229.
$$

Note that the coefficients are approximations of irrational expressions (cf. Exercise (7) at the end of the filter bank section).

In the sequel, we shall consider the interleaved transforms $\mathbf{y}^{(4)}$ of matrix schemes $\mathbf{x} = \mathbf{y}^{(0)}$, in 16×16 format. The intervening filter banks will be the DWT 5/3 spline, the DWT 7/5 Burt and the DWT 9/7 CDF, respectively.

(1) For the three considered transforms, determine $\mathbf{y}^{(4)}$ in case where $\mathbf{x} = \mathbf{y}^{(0)}$ is constant.

(2) Find, for each of the three DWTs, the matrix scheme $\mathbf{x} = \mathbf{y}^{(0)}$ such that:

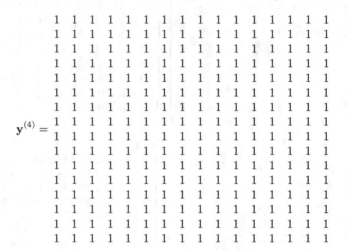

$$\mathbf{y}^{(4)} = \begin{matrix}
1 & 1 & 1 & 1 & 1 & 1 & 1 & 1 & 1 & 1 & 1 & 1 & 1 & 1 & 1 & 1 \\
1 & 1 & 1 & 1 & 1 & 1 & 1 & 1 & 1 & 1 & 1 & 1 & 1 & 1 & 1 & 1 \\
1 & 1 & 1 & 1 & 1 & 1 & 1 & 1 & 1 & 1 & 1 & 1 & 1 & 1 & 1 & 1 \\
1 & 1 & 1 & 1 & 1 & 1 & 1 & 1 & 1 & 1 & 1 & 1 & 1 & 1 & 1 & 1 \\
1 & 1 & 1 & 1 & 1 & 1 & 1 & 1 & 1 & 1 & 1 & 1 & 1 & 1 & 1 & 1 \\
1 & 1 & 1 & 1 & 1 & 1 & 1 & 1 & 1 & 1 & 1 & 1 & 1 & 1 & 1 & 1 \\
1 & 1 & 1 & 1 & 1 & 1 & 1 & 1 & 1 & 1 & 1 & 1 & 1 & 1 & 1 & 1 \\
1 & 1 & 1 & 1 & 1 & 1 & 1 & 1 & 1 & 1 & 1 & 1 & 1 & 1 & 1 & 1 \\
1 & 1 & 1 & 1 & 1 & 1 & 1 & 1 & 1 & 1 & 1 & 1 & 1 & 1 & 1 & 1 \\
1 & 1 & 1 & 1 & 1 & 1 & 1 & 1 & 1 & 1 & 1 & 1 & 1 & 1 & 1 & 1 \\
1 & 1 & 1 & 1 & 1 & 1 & 1 & 1 & 1 & 1 & 1 & 1 & 1 & 1 & 1 & 1 \\
1 & 1 & 1 & 1 & 1 & 1 & 1 & 1 & 1 & 1 & 1 & 1 & 1 & 1 & 1 & 1 \\
1 & 1 & 1 & 1 & 1 & 1 & 1 & 1 & 1 & 1 & 1 & 1 & 1 & 1 & 1 & 1 \\
1 & 1 & 1 & 1 & 1 & 1 & 1 & 1 & 1 & 1 & 1 & 1 & 1 & 1 & 1 & 1 \\
1 & 1 & 1 & 1 & 1 & 1 & 1 & 1 & 1 & 1 & 1 & 1 & 1 & 1 & 1 & 1 \\
1 & 1 & 1 & 1 & 1 & 1 & 1 & 1 & 1 & 1 & 1 & 1 & 1 & 1 & 1 & 1
\end{matrix}$$

(3) Knowing that the gradation:

$$\mathbf{x} = \mathbf{y}^{(0)} = \begin{matrix}
1 & 1 & 1 & 1 & 1 & 1 & 1 & 1 & 1 & 1 & 1 & 1 & 1 & 1 & 1 & 1 \\
2 & 2 & 2 & 2 & 2 & 2 & 2 & 2 & 2 & 2 & 2 & 2 & 2 & 2 & 2 & 2 \\
3 & 3 & 3 & 3 & 3 & 3 & 3 & 3 & 3 & 3 & 3 & 3 & 3 & 3 & 3 & 3 \\
4 & 4 & 4 & 4 & 4 & 4 & 4 & 4 & 4 & 4 & 4 & 4 & 4 & 4 & 4 & 4 \\
5 & 5 & 5 & 5 & 5 & 5 & 5 & 5 & 5 & 5 & 5 & 5 & 5 & 5 & 5 & 5 \\
6 & 6 & 6 & 6 & 6 & 6 & 6 & 6 & 6 & 6 & 6 & 6 & 6 & 6 & 6 & 6 \\
7 & 7 & 7 & 7 & 7 & 7 & 7 & 7 & 7 & 7 & 7 & 7 & 7 & 7 & 7 & 7 \\
8 & 8 & 8 & 8 & 8 & 8 & 8 & 8 & 8 & 8 & 8 & 8 & 8 & 8 & 8 & 8 \\
9 & 9 & 9 & 9 & 9 & 9 & 9 & 9 & 9 & 9 & 9 & 9 & 9 & 9 & 9 & 9 \\
10 & 10 & 10 & 10 & 10 & 10 & 10 & 10 & 10 & 10 & 10 & 10 & 10 & 10 & 10 & 10 \\
11 & 11 & 11 & 11 & 11 & 11 & 11 & 11 & 11 & 11 & 11 & 11 & 11 & 11 & 11 & 11 \\
12 & 12 & 12 & 12 & 12 & 12 & 12 & 12 & 12 & 12 & 12 & 12 & 12 & 12 & 12 & 12 \\
13 & 13 & 13 & 13 & 13 & 13 & 13 & 13 & 13 & 13 & 13 & 13 & 13 & 13 & 13 & 13 \\
14 & 14 & 14 & 14 & 14 & 14 & 14 & 14 & 14 & 14 & 14 & 14 & 14 & 14 & 14 & 14 \\
15 & 15 & 15 & 15 & 15 & 15 & 15 & 15 & 15 & 15 & 15 & 15 & 15 & 15 & 15 & 15 \\
16 & 16 & 16 & 16 & 16 & 16 & 16 & 16 & 16 & 16 & 16 & 16 & 16 & 16 & 16 & 16
\end{matrix}$$

is transformed into

(a) $\mathbf{y}^{(4)} =$
$$\begin{bmatrix}
5.57 & 0 & 0 & 0 & 0 & 0 & 0 & 0 & 0 & 0 & 0 & 0 & 0 & 0 & 0 & 0 \\
0 & 0 & 0 & 0 & 0 & 0 & 0 & 0 & 0 & 0 & 0 & 0 & 0 & 0 & 0 & 0 \\
0 & 0 & 0 & 0 & 0 & 0 & 0 & 0 & 0 & 0 & 0 & 0 & 0 & 0 & 0 & 0 \\
0 & 0 & 0 & 0 & 0 & 0 & 0 & 0 & 0 & 0 & 0 & 0 & 0 & 0 & 0 & 0 \\
0 & 0 & 0 & 0 & 0 & 0 & 0 & 0 & 0 & 0 & 0 & 0 & 0 & 0 & 0 & 0 \\
0 & 0 & 0 & 0 & 0 & 0 & 0 & 0 & 0 & 0 & 0 & 0 & 0 & 0 & 0 & 0 \\
0 & 0 & 0 & 0 & 0 & 0 & 0 & 0 & 0 & 0 & 0 & 0 & 0 & 0 & 0 & 0 \\
0 & 0 & 0 & 0 & 0 & 0 & 0 & 0 & 0 & 0 & 0 & 0 & 0 & 0 & 0 & 0 \\
4.57 & 0 & 0 & 0 & 0 & 0 & 0 & 0 & 0 & 0 & 0 & 0 & 0 & 0 & 0 & 0 \\
0 & 0 & 0 & 0 & 0 & 0 & 0 & 0 & 0 & 0 & 0 & 0 & 0 & 0 & 0 & 0 \\
0 & 0 & 0 & 0 & 0 & 0 & 0 & 0 & 0 & 0 & 0 & 0 & 0 & 0 & 0 & 0 \\
0 & 0 & 0 & 0 & 0 & 0 & 0 & 0 & 0 & 0 & 0 & 0 & 0 & 0 & 0 & 0 \\
2.28 & 0 & 0 & 0 & 0 & 0 & 0 & 2.28 & 0 & 0 & 0 & 0 & 0 & 0 & 0 & 0 \\
0 & 0 & 0 & 0 & 0 & 0 & 0 & 0 & 0 & 0 & 0 & 0 & 0 & 0 & 0 & 0 \\
1.13 & 0 & 0 & 0 & 1.13 & 0 & 0 & 0 & 1.13 & 0 & 0 & 0 & 1.13 & 0 & 0 & 0 \\
0.5 & 0 & 0.5 & 0 & 0.5 & 0 & 0.5 & 0 & 0.5 & 0 & 0.5 & 0 & 0.5 & 0 & 0.5 & 0
\end{bmatrix}$$
[DWT 5/3 spline],

(b) $\mathbf{y}^{(4)} =$
$$\begin{bmatrix}
5.93 & 0 & 0 & 0 & 0 & 0 & 0 & 0 & 0 & 0 & 0 & 0 & 0 & 0 & 0 & 0 \\
-0.1 & 0 & -0.1 & 0 & -0.1 & 0 & -0.1 & 0 & -0.1 & 0 & -0.1 & 0 & -0.1 & 0 & -0.1 & 0 \\
-0.27 & 0 & 0 & 0 & -0.27 & 0 & 0 & 0 & -0.27 & 0 & 0 & 0 & -0.27 & 0 & 0 & 0 \\
0 & 0 & 0 & 0 & 0 & 0 & 0 & 0 & 0 & 0 & 0 & 0 & 0 & 0 & 0 & 0 \\
-0.61 & 0 & 0 & 0 & 0 & 0 & 0 & 0 & -0.61 & 0 & 0 & 0 & 0 & 0 & 0 & 0 \\
0 & 0 & 0 & 0 & 0 & 0 & 0 & 0 & 0 & 0 & 0 & 0 & 0 & 0 & 0 & 0 \\
0 & 0 & 0 & 0 & 0 & 0 & 0 & 0 & 0 & 0 & 0 & 0 & 0 & 0 & 0 & 0 \\
0 & 0 & 0 & 0 & 0 & 0 & 0 & 0 & 0 & 0 & 0 & 0 & 0 & 0 & 0 & 0 \\
3.62 & 0 & 0 & 0 & 0 & 0 & 0 & 0 & 0 & 0 & 0 & 0 & 0 & 0 & 0 & 0 \\
0 & 0 & 0 & 0 & 0 & 0 & 0 & 0 & 0 & 0 & 0 & 0 & 0 & 0 & 0 & 0 \\
-0.01 & 0 & 0 & 0 & -0.01 & 0 & 0 & 0 & -0.01 & 0 & 0 & 0 & -0.01 & 0 & 0 & 0 \\
0 & 0 & 0 & 0 & 0 & 0 & 0 & 0 & 0 & 0 & 0 & 0 & 0 & 0 & 0 & 0 \\
3.01 & 0 & 0 & 0 & 0 & 0 & 0 & 0 & 3.01 & 0 & 0 & 0 & 0 & 0 & 0 & 0 \\
0 & 0 & 0 & 0 & 0 & 0 & 0 & 0 & 0 & 0 & 0 & 0 & 0 & 0 & 0 & 0 \\
1.49 & 0 & 0 & 0 & 1.49 & 0 & 0 & 0 & 1.49 & 0 & 0 & 0 & 1.49 & 0 & 0 & 0 \\
0.7 & 0 & 0.7 & 0 & 0.7 & 0 & 0.7 & 0 & 0.7 & 0 & 0.7 & 0 & 0.7 & 0 & 0.7 & 0
\end{bmatrix}$$
[DWT 7/5 Burt],

(c) $\mathbf{y}^{(4)} =$
$$\begin{bmatrix}
6.02 & 0 & 0 & 0 & 0 & 0 & 0 & 0 & 0 & 0 & 0 & 0 & 0 & 0 & 0 & 0 \\
0.12 & 0 & 0.12 & 0 & 0.12 & 0 & 0.12 & 0 & 0.12 & 0 & 0.12 & 0 & 0.12 & 0 & 0.12 & 0 \\
0.19 & 0 & 0 & 0 & 0.19 & 0 & 0 & 0 & 0.19 & 0 & 0 & 0 & 0.19 & 0 & 0 & 0 \\
0 & 0 & 0 & 0 & 0 & 0 & 0 & 0 & 0 & 0 & 0 & 0 & 0 & 0 & 0 & 0 \\
-0.04 & 0 & 0 & 0 & 0 & 0 & 0 & 0 & -0.04 & 0 & 0 & 0 & 0 & 0 & 0 & 0 \\
0 & 0 & 0 & 0 & 0 & 0 & 0 & 0 & 0 & 0 & 0 & 0 & 0 & 0 & 0 & 0 \\
0.01 & 0 & 0 & 0 & 0.01 & 0 & 0 & 0 & 0.01 & 0 & 0 & 0 & 0.01 & 0 & 0 & 0 \\
0 & 0 & 0 & 0 & 0 & 0 & 0 & 0 & 0 & 0 & 0 & 0 & 0 & 0 & 0 & 0 \\
3.49 & 0 & 0 & 0 & 0 & 0 & 0 & 0 & 0 & 0 & 0 & 0 & 0 & 0 & 0 & 0 \\
0 & 0 & 0 & 0 & 0 & 0 & 0 & 0 & 0 & 0 & 0 & 0 & 0 & 0 & 0 & 0 \\
-0.17 & 0 & 0 & 0 & -0.17 & 0 & 0 & 0 & -0.17 & 0 & 0 & 0 & -0.17 & 0 & 0 & 0 \\
0 & 0 & 0 & 0 & 0 & 0 & 0 & 0 & 0 & 0 & 0 & 0 & 0 & 0 & 0 & 0 \\
1.93 & 0 & 0 & 0 & 0 & 0 & 0 & 0 & 1.93 & 0 & 0 & 0 & 0 & 0 & 0 & 0 \\
-0.09 & 0 & -0.09 & 0 & -0.09 & 0 & -0.09 & 0 & -0.09 & 0 & -0.09 & 0 & -0.09 & 0 & -0.09 & 0 \\
0.93 & 0 & 0 & 0 & 0.93 & 0 & 0 & 0 & 0.93 & 0 & 0 & 0 & 0.93 & 0 & 0 & 0 \\
0.43 & 0 & 0.43 & 0 & 0.43 & 0 & 0.43 & 0 & 0.43 & 0 & 0.43 & 0 & 0.43 & 0 & 0.43 & 0
\end{bmatrix}$$
[DWT 9/7 CDF],

what will be the interleaved transformed matrix $\mathbf{y}^{(4)}$ of

$$\mathbf{x} = \mathbf{y}^{(0)} = \begin{array}{cccccccccccccccc}
16 & 15 & 14 & 13 & 12 & 11 & 10 & 9 & 8 & 7 & 6 & 5 & 4 & 3 & 2 & 1 \\
16 & 15 & 14 & 13 & 12 & 11 & 10 & 9 & 8 & 7 & 6 & 5 & 4 & 3 & 2 & 1 \\
16 & 15 & 14 & 13 & 12 & 11 & 10 & 9 & 8 & 7 & 6 & 5 & 4 & 3 & 2 & 1 \\
16 & 15 & 14 & 13 & 12 & 11 & 10 & 9 & 8 & 7 & 6 & 5 & 4 & 3 & 2 & 1 \\
16 & 15 & 14 & 13 & 12 & 11 & 10 & 9 & 8 & 7 & 6 & 5 & 4 & 3 & 2 & 1 \\
16 & 15 & 14 & 13 & 12 & 11 & 10 & 9 & 8 & 7 & 6 & 5 & 4 & 3 & 2 & 1 \\
16 & 15 & 14 & 13 & 12 & 11 & 10 & 9 & 8 & 7 & 6 & 5 & 4 & 3 & 2 & 1 \\
16 & 15 & 14 & 13 & 12 & 11 & 10 & 9 & 8 & 7 & 6 & 5 & 4 & 3 & 2 & 1 \\
16 & 15 & 14 & 13 & 12 & 11 & 10 & 9 & 8 & 7 & 6 & 5 & 4 & 3 & 2 & 1 \\
16 & 15 & 14 & 13 & 12 & 11 & 10 & 9 & 8 & 7 & 6 & 5 & 4 & 3 & 2 & 1 \\
16 & 15 & 14 & 13 & 12 & 11 & 10 & 9 & 8 & 7 & 6 & 5 & 4 & 3 & 2 & 1 \\
16 & 15 & 14 & 13 & 12 & 11 & 10 & 9 & 8 & 7 & 6 & 5 & 4 & 3 & 2 & 1 \\
16 & 15 & 14 & 13 & 12 & 11 & 10 & 9 & 8 & 7 & 6 & 5 & 4 & 3 & 2 & 1 \\
16 & 15 & 14 & 13 & 12 & 11 & 10 & 9 & 8 & 7 & 6 & 5 & 4 & 3 & 2 & 1 \\
16 & 15 & 14 & 13 & 12 & 11 & 10 & 9 & 8 & 7 & 6 & 5 & 4 & 3 & 2 & 1 \\
16 & 15 & 14 & 13 & 12 & 11 & 10 & 9 & 8 & 7 & 6 & 5 & 4 & 3 & 2 & 1
\end{array}$$

in each of the three cases?

(4) Show that the value $\alpha = \frac{1}{2}$ is an *eigenvalue* for the transformation

$$\mathbf{x} = \mathbf{y}^{(0)} \longmapsto \mathbf{y} = \mathbf{y}^{(1)}$$

for each of the three 2D analysis filter bank versions that we consider. (You will find a common eigenvector (i.e. a 16×16 matrix) according to the indication for Exercise (3) of the preceding section.) Will the value $\alpha = \frac{1}{2}$ also be an eigenvalue of the transform $\mathbf{x} = \mathbf{y}^{(0)} \longmapsto \mathbf{y} = \mathbf{y}^{(4)}$?

(5) Consider the following four 16×16 matrix schemes:

$$\mathbf{x}_1 = \begin{array}{l}
1010101010101010 \\
0101010101010101 \\
1010101010101010 \\
0101010101010101 \\
1010101010101010 \\
0101010101010101 \\
1010101010101010 \\
0101010101010101 \\
1010101010101010 \\
0101010101010101 \\
1010101010101010 \\
0101010101010101 \\
1010101010101010 \\
0101010101010101 \\
1010101010101010 \\
0101010101010101
\end{array}
\qquad
\mathbf{x}_2 = \begin{array}{l}
1111111100000000 \\
1111111100000000 \\
1111111100000000 \\
1111111100000000 \\
1111111100000000 \\
1111111100000000 \\
1111111100000000 \\
1111111100000000 \\
0000000011111111 \\
0000000011111111 \\
0000000011111111 \\
0000000011111111 \\
0000000011111111 \\
0000000011111111 \\
0000000011111111 \\
0000000011111111
\end{array}$$

$$x_3 = \begin{matrix}
1\,1\,1\,1\,1\,1\,1\,1\,1\,1\,1\,1\,1\,1\,1\,1\\
1\,1\,1\,1\,1\,1\,1\,1\,1\,1\,1\,1\,1\,1\,1\,1\\
1\,1\,0\,0\,0\,0\,0\,0\,0\,0\,0\,0\,0\,0\,1\,1\\
1\,1\,0\,0\,0\,0\,0\,0\,0\,0\,0\,0\,0\,0\,1\,1\\
1\,1\,0\,0\,0\,0\,0\,0\,0\,0\,0\,0\,0\,0\,1\,1\\
1\,1\,0\,0\,0\,0\,0\,0\,0\,0\,0\,0\,0\,0\,1\,1\\
1\,1\,0\,0\,0\,0\,0\,0\,0\,0\,0\,0\,0\,0\,1\,1\\
1\,1\,0\,0\,0\,0\,0\,0\,0\,0\,0\,0\,0\,0\,1\,1\\
1\,1\,0\,0\,0\,0\,0\,0\,0\,0\,0\,0\,0\,0\,1\,1\\
1\,1\,0\,0\,0\,0\,0\,0\,0\,0\,0\,0\,0\,0\,1\,1\\
1\,1\,0\,0\,0\,0\,0\,0\,0\,0\,0\,0\,0\,0\,1\,1\\
1\,1\,0\,0\,0\,0\,0\,0\,0\,0\,0\,0\,0\,0\,1\,1\\
1\,1\,0\,0\,0\,0\,0\,0\,0\,0\,0\,0\,0\,0\,1\,1\\
1\,1\,0\,0\,0\,0\,0\,0\,0\,0\,0\,0\,0\,0\,1\,1\\
1\,1\,1\,1\,1\,1\,1\,1\,1\,1\,1\,1\,1\,1\,1\,1\\
1\,1\,1\,1\,1\,1\,1\,1\,1\,1\,1\,1\,1\,1\,1\,1
\end{matrix}
\qquad
x_4 = \begin{matrix}
1\,1\,0\,0\,0\,0\,0\,0\,0\,0\,0\,0\,0\,0\,0\,0\\
1\,1\,1\,0\,0\,0\,0\,0\,0\,0\,0\,0\,0\,0\,0\,0\\
0\,1\,1\,1\,0\,0\,0\,0\,0\,0\,0\,0\,0\,0\,0\,0\\
0\,0\,1\,1\,1\,0\,0\,0\,0\,0\,0\,0\,0\,0\,0\,0\\
0\,0\,0\,1\,1\,1\,0\,0\,0\,0\,0\,0\,0\,0\,0\,0\\
0\,0\,0\,0\,1\,1\,1\,0\,0\,0\,0\,0\,0\,0\,0\,0\\
0\,0\,0\,0\,0\,1\,1\,1\,0\,0\,0\,0\,0\,0\,0\,0\\
0\,0\,0\,0\,0\,0\,1\,1\,1\,0\,0\,0\,0\,0\,0\,0\\
0\,0\,0\,0\,0\,0\,0\,1\,1\,1\,0\,0\,0\,0\,0\,0\\
0\,0\,0\,0\,0\,0\,0\,0\,1\,1\,1\,0\,0\,0\,0\,0\\
0\,0\,0\,0\,0\,0\,0\,0\,0\,1\,1\,1\,0\,0\,0\,0\\
0\,0\,0\,0\,0\,0\,0\,0\,0\,0\,1\,1\,1\,0\,0\,0\\
0\,0\,0\,0\,0\,0\,0\,0\,0\,0\,0\,1\,1\,1\,0\,0\\
0\,0\,0\,0\,0\,0\,0\,0\,0\,0\,0\,0\,1\,1\,1\,0\\
0\,0\,0\,0\,0\,0\,0\,0\,0\,0\,0\,0\,0\,1\,1\,1\\
0\,0\,0\,0\,0\,0\,0\,0\,0\,0\,0\,0\,0\,0\,1\,1
\end{matrix}$$

Then, we are given the eight matrix schemes $y_1, y_2, \ldots, y_8$.
Each of these matrices is an interleaved version $y^{(4)}$ of $y^{(0)} = x_1, x_2, x_3, x_4$ by either the DWT 5/3 spline or the DWT 7/5 Burt or the DWT 9/7 CDF (hence, there are 12 possibilities).

Decide, with a minimum of pencil-and-paper effort, which is the decompressed version of each of the y_i, $1 \le i \le 8$ – and by which of the three DWTs at our disposal.

Example y_3 is the interleaved version $y^{(4)}$ of x_1 by *all the three* transforms.

$y_1 =$

0.24	0.1	0.17	−0.16	0.05	0.04	0	0	−0.07	0	−0.01	0	0.05	0	0.01	0
0.1	−0.13	0.07	−0.2	−0.05	0.04	0.02	0	0	0	0	0	0	0	0	0
0.17	0.07	0.33	0.05	0	−0.06	−0.03	0.02	0.01	0	−0.02	0	0	0	0	0
−0.16	−0.2	0.05	−0.14	0.04	−0.22	−0.06	0.04	0.02	0	0	0	0	0	0	0
0.05	−0.05	0	0.04	0.29	0.04	−0.01	−0.06	0.01	0.02	0.02	0	0.04	0	−0.02	0
0.04	0.04	−0.06	−0.22	0.04	−0.14	0.04	−0.22	−0.06	0.04	0.02	0	0	0	0	0
0	0.02	−0.03	−0.06	−0.01	0.04	0.41	0.04	−0.01	−0.06	−0.02	0.02	0	0	−0.04	0
0	0	0.02	0.04	−0.06	−0.22	0.04	−0.14	0.04	−0.22	−0.06	0.04	0.02	0	0	0
−0.07	0	0.01	0.02	0.01	−0.06	−0.01	0.04	0.2	0.04	−0.01	−0.06	−0.14	0.02	0.04	0
0	0	0	0	0.02	0.04	−0.06	−0.22	0.04	−0.14	0.04	−0.22	−0.06	0.04	0.02	0
−0.01	0	−0.02	0	0.02	0.02	−0.02	−0.06	−0.01	0.04	0.39	0.04	0.03	−0.06	0	0.03
0	0	0	0	0	0	0.02	0.04	−0.06	−0.22	0.04	−0.14	0.04	−0.22	−0.05	0.05
0.05	0	0	0	0.04	0	0	0.02	−0.14	−0.06	0.03	0.04	0.38	0.06	−0.22	−0.1
0	0	0	0	0	0	0	0	0.02	0.04	−0.06	−0.22	0.06	−0.12	0.07	−0.3
0.01	0	0	0	−0.02	0	−0.04	0	0.04	0.02	0	−0.05	−0.22	0.07	0.55	−0.2
0	0	0	0	0	0	0	0	0	0	0.03	0.05	−0.1	−0.3	−0.2	0.07

$y_2 =$

0.63	0	0	0	0.32	0	0.29	0.3	−0.19	−0.05	−0.11	0	−0.36	0	0	0
0	0	0	0	0	0	0	0	0	0	0	0	0	0	0	0
0	0	0	0	0	0	0	0.3	0	−0.05	0	0	0	0	0	0
0	0	0	0	0	0	0	0	0	0	0	0	0	0	0	0
0.32	0	0	0	0.13	0	0.33	0.3	−0.06	−0.05	−0.12	0	−0.15	0	0	0
0	0	0	0	0	0	0	0	0	0	0	0	0	0	0	0
0.29	0	0	0	0.33	0	0.17	0.34	−0.1	−0.06	−0.06	0	−0.3	0	0	0
0.3	0	0.3	0	0.3	0	0.34	0.18	−0.18	−0.03	−0.31	0	−0.3	0	−0.3	0
−0.19	0	0	0	−0.06	0	−0.1	−0.18	0.28	0.03	0.04	0	0.07	0	0	0
−0.05	0	−0.05	0	−0.05	0	−0.06	−0.03	0.03	0.01	0.05	0	0.05	0	0.05	0
−0.11	0	0	0	−0.12	0	−0.06	−0.31	0.04	0.05	0.02	0	0.11	0	0	0
0	0	0	0	0	0	0	0	0	0	0	0	0	0	0	0
−0.36	0	0	0	−0.15	0	−0.3	−0.3	0.07	0.05	0.11	0	0.16	0	0	0
0	0	0	0	0	0	0	0	0	0	0	0	0	0	0	0
0	0	0	0	0	0	0	−0.3	0	0.05	0	0	0	0	0	0
0	0	0	0	0	0	0	0	0	0	0	0	0	0	0	0

$$\mathbf{y}_3 = \begin{bmatrix}
0.5 & 0 & 0 & 0 & 0 & 0 & 0 & 0 & 0 & 0 & 0 & 0 & 0 & 0 & 0 & 0 \\
0 & 0.5 & 0 & 0.5 & 0 & 0.5 & 0 & 0.5 & 0 & 0.5 & 0 & 0.5 & 0 & 0.5 & 0 & 0.5 \\
0 & 0 & 0 & 0 & 0 & 0 & 0 & 0 & 0 & 0 & 0 & 0 & 0 & 0 & 0 & 0 \\
0 & 0.5 & 0 & 0.5 & 0 & 0.5 & 0 & 0.5 & 0 & 0.5 & 0 & 0.5 & 0 & 0.5 & 0 & 0.5 \\
0 & 0 & 0 & 0 & 0 & 0 & 0 & 0 & 0 & 0 & 0 & 0 & 0 & 0 & 0 & 0 \\
0 & 0.5 & 0 & 0.5 & 0 & 0.5 & 0 & 0.5 & 0 & 0.5 & 0 & 0.5 & 0 & 0.5 & 0 & 0.5 \\
0 & 0 & 0 & 0 & 0 & 0 & 0 & 0 & 0 & 0 & 0 & 0 & 0 & 0 & 0 & 0 \\
0 & 0.5 & 0 & 0.5 & 0 & 0.5 & 0 & 0.5 & 0 & 0.5 & 0 & 0.5 & 0 & 0.5 & 0 & 0.5 \\
0 & 0 & 0 & 0 & 0 & 0 & 0 & 0 & 0 & 0 & 0 & 0 & 0 & 0 & 0 & 0 \\
0 & 0.5 & 0 & 0.5 & 0 & 0.5 & 0 & 0.5 & 0 & 0.5 & 0 & 0.5 & 0 & 0.5 & 0 & 0.5 \\
0 & 0 & 0 & 0 & 0 & 0 & 0 & 0 & 0 & 0 & 0 & 0 & 0 & 0 & 0 & 0 \\
0 & 0.5 & 0 & 0.5 & 0 & 0.5 & 0 & 0.5 & 0 & 0.5 & 0 & 0.5 & 0 & 0.5 & 0 & 0.5 \\
0 & 0 & 0 & 0 & 0 & 0 & 0 & 0 & 0 & 0 & 0 & 0 & 0 & 0 & 0 & 0 \\
0 & 0.5 & 0 & 0.5 & 0 & 0.5 & 0 & 0.5 & 0 & 0.5 & 0 & 0.5 & 0 & 0.5 & 0 & 0.5 \\
0 & 0 & 0 & 0 & 0 & 0 & 0 & 0 & 0 & 0 & 0 & 0 & 0 & 0 & 0 & 0 \\
0 & 0.5 & 0 & 0.5 & 0 & 0.5 & 0 & 0.5 & 0 & 0.5 & 0 & 0.5 & 0 & 0.5 & 0 & 0.5
\end{bmatrix}$$

$$\mathbf{y}_4 = \begin{bmatrix}
0.25 & 0.13 & 0.2 & -0.19 & 0.06 & 0 & 0.06 & 0 & -0.13 & 0 & 0 & 0 & 0.18 & 0 & 0 & 0 \\
0.13 & -0.13 & 0.09 & -0.19 & -0.09 & 0 & 0 & 0 & 0 & 0 & 0 & 0 & 0 & 0 & 0 & 0 \\
0.2 & 0.09 & 0.41 & 0.09 & -0.03 & -0.09 & 0.03 & 0 & 0.03 & 0 & 0 & 0 & 0 & 0 & 0 & 0 \\
-0.19 & -0.19 & 0.09 & -0.13 & 0.09 & -0.19 & -0.09 & 0 & 0 & 0 & 0 & 0 & 0 & 0 & 0 & 0 \\
0.06 & -0.09 & -0.03 & 0.09 & 0.52 & 0.09 & -0.03 & -0.09 & 0 & 0 & 0.03 & 0 & 0.2 & 0 & 0 & 0 \\
0 & 0 & -0.09 & -0.19 & 0.09 & -0.13 & 0.09 & -0.19 & -0.09 & 0 & 0 & 0 & 0 & 0 & 0 & 0 \\
0.06 & 0 & 0.03 & -0.09 & -0.03 & 0.09 & 0.44 & 0.09 & -0.03 & -0.09 & 0.04 & 0 & 0.03 & 0 & -0.01 & 0 \\
-0.13 & 0 & 0.03 & 0 & 0 & -0.09 & -0.03 & 0.09 & 0.45 & 0.09 & -0.03 & -0.09 & -0.24 & 0 & 0.07 & 0 \\
0 & 0 & 0 & 0 & 0 & 0 & -0.09 & -0.19 & 0.09 & -0.13 & 0.09 & -0.19 & -0.09 & 0 & 0 & 0 \\
0 & 0 & 0 & 0 & 0.03 & 0 & 0.04 & -0.09 & -0.03 & 0.09 & 0.44 & 0.09 & 0.01 & -0.09 & 0.11 & 0 \\
0 & 0 & 0 & 0 & 0 & 0 & 0 & 0 & -0.09 & -0.19 & 0.09 & -0.13 & 0.09 & -0.19 & -0.09 & 0 \\
0.18 & 0 & 0 & 0.2 & 0 & 0.03 & 0 & -0.24 & -0.09 & 0.01 & 0.09 & 0.53 & 0.09 & -0.29 & -0.13 \\
0 & 0 & 0 & 0 & 0 & 0 & 0 & 0 & 0 & -0.09 & -0.19 & 0.09 & -0.13 & 0.06 & -0.25 \\
0 & 0 & 0 & 0 & 0 & 0 & -0.01 & 0 & 0.07 & 0 & 0.11 & -0.09 & -0.29 & 0.06 & 0.55 & -0.13 \\
0 & 0 & 0 & 0 & 0 & 0 & 0 & 0 & 0 & 0 & 0 & 0 & -0.13 & -0.25 & -0.13 & 0
\end{bmatrix}$$

$$\mathbf{y}_5 = \begin{bmatrix}
0.4 & -0.03 & -0.06 & 0 & -0.15 & 0 & 0.01 & 0 & -0.16 & 0 & 0 & -0.01 & 0.09 & 0.04 & 0.11 & 0 \\
-0.03 & -0.05 & 0.19 & 0 & 0.23 & 0 & 0.23 & 0 & 0.23 & 0 & 0.23 & -0.01 & 0.25 & 0.06 & 0.05 & 0.01 \\
-0.06 & 0.19 & -0.06 & 0.01 & -0.26 & 0 & 0.01 & 0 & -0.24 & 0 & 0 & 0.04 & -0.2 & -0.23 & 0.12 & -0.03 \\
0 & 0 & 0.01 & 0 & 0.02 & 0 & 0.02 & 0 & 0.02 & 0 & 0.02 & 0 & 0.02 & 0 & 0 & 0 \\
-0.15 & 0.23 & -0.26 & 0.02 & -0.07 & 0 & 0.04 & 0 & -0.26 & 0 & 0.01 & 0.05 & 0.04 & -0.28 & 0.5 & -0.03 \\
0 & 0 & 0 & 0 & 0 & 0 & 0 & 0 & 0 & 0 & 0 & 0 & 0 & 0 & 0 & 0 \\
0.01 & 0.23 & 0.01 & 0.02 & 0.04 & 0 & 0 & 0 & 0.04 & 0 & 0 & 0.05 & 0.03 & -0.28 & -0.02 & -0.03 \\
0 & 0 & 0 & 0 & 0 & 0 & 0 & 0 & 0 & 0 & 0 & 0 & 0 & 0 & 0 & 0 \\
-0.16 & 0.23 & -0.24 & 0.02 & -0.26 & 0 & 0.04 & 0 & -0.05 & 0 & 0.01 & 0.05 & 0.16 & -0.28 & 0.47 & -0.03 \\
0 & 0 & 0 & 0 & 0 & 0 & 0 & 0 & 0 & 0 & 0 & 0 & 0 & 0 & 0 & 0 \\
0 & 0.23 & 0 & 0.02 & 0.01 & 0 & 0 & 0 & 0.01 & 0 & 0 & 0.04 & 0.01 & -0.27 & 0 & -0.03 \\
-0.01 & -0.01 & 0.04 & 0 & 0.05 & 0 & 0.05 & 0 & 0.05 & 0 & 0.04 & 0 & 0.05 & 0.01 & 0.01 & 0 \\
0.09 & 0.25 & -0.2 & 0.02 & 0.04 & 0 & 0.03 & 0 & 0.16 & 0 & 0.01 & 0.05 & 0.03 & -0.3 & 0.39 & -0.04 \\
0.04 & 0.06 & -0.23 & 0 & -0.28 & 0 & -0.28 & 0 & -0.28 & 0 & -0.27 & 0.01 & -0.3 & -0.08 & -0.06 & -0.01 \\
0.11 & 0.05 & 0.12 & 0 & 0.5 & 0 & -0.02 & 0 & 0.47 & 0 & 0 & 0.01 & 0.39 & -0.06 & -0.23 & -0.01 \\
0 & 0.01 & -0.03 & 0 & -0.03 & 0 & -0.03 & 0 & -0.03 & 0 & -0.03 & 0 & -0.04 & -0.01 & -0.01 & 0
\end{bmatrix}$$

$$
y_6 = \begin{bmatrix}
0.41 & -0.04 & -0.04 & 0.01 & -0.11 & 0 & 0 & 0 & -0.18 & 0 & -0.01 & 0 & 0.06 & 0.04 & 0.11 & -0.01 \\
-0.04 & -0.09 & 0.24 & 0.02 & 0.3 & 0 & 0.3 & 0 & 0.3 & 0 & 0.3 & 0 & 0.32 & 0.09 & 0.06 & -0.03 \\
-0.04 & 0.24 & -0.03 & -0.04 & -0.17 & 0 & 0 & 0 & -0.17 & 0 & 0 & 0 & -0.14 & -0.24 & 0.09 & 0.08 \\
0.01 & 0.02 & -0.04 & 0 & -0.05 & 0 & -0.05 & 0 & -0.05 & 0 & -0.05 & 0 & -0.05 & -0.02 & -0.01 & 0.01 \\
-0.11 & 0.3 & -0.17 & -0.05 & -0.05 & 0 & -0.01 & 0 & -0.22 & 0 & -0.03 & 0 & 0.02 & -0.3 & 0.53 & 0.1 \\
0 & 0 & 0 & 0 & 0 & 0 & 0 & 0 & 0 & 0 & 0 & 0 & 0 & 0 & 0 & 0 \\
0 & 0.3 & 0 & -0.05 & -0.01 & 0 & 0 & 0 & -0.01 & 0 & 0 & 0 & -0.01 & -0.3 & 0 & 0.1 \\
0 & 0 & 0 & 0 & 0 & 0 & 0 & 0 & 0 & 0 & 0 & 0 & 0 & 0 & 0 & 0 \\
-0.18 & 0.3 & -0.17 & -0.05 & -0.22 & 0 & -0.01 & 0 & -0.06 & 0 & -0.02 & 0 & 0.11 & -0.3 & 0.52 & 0.1 \\
0 & 0 & 0 & 0 & 0 & 0 & 0 & 0 & 0 & 0 & 0 & 0 & 0 & 0 & 0 & 0 \\
-0.01 & 0.3 & 0 & -0.05 & -0.03 & 0 & 0 & 0 & -0.02 & 0 & 0 & 0 & -0.02 & -0.3 & 0.01 & 0.1 \\
0 & 0 & 0 & 0 & 0 & 0 & 0 & 0 & 0 & 0 & 0 & 0 & 0 & 0 & 0 & 0 \\
0.06 & 0.32 & -0.14 & -0.05 & 0.02 & 0 & -0.01 & 0 & 0.11 & 0 & -0.02 & 0 & -0.01 & -0.32 & 0.43 & 0.11 \\
0.04 & 0.09 & -0.24 & -0.02 & -0.3 & 0 & -0.3 & 0 & -0.3 & 0 & -0.3 & 0 & -0.32 & -0.09 & -0.06 & 0.03 \\
0.11 & 0.06 & 0.09 & -0.01 & 0.53 & 0 & 0 & 0 & 0.52 & 0 & 0.01 & 0 & 0.43 & -0.06 & -0.27 & 0.02 \\
-0.01 & -0.03 & 0.08 & 0.01 & 0.1 & 0 & 0.1 & 0 & 0.1 & 0 & 0.1 & 0 & 0.11 & 0.03 & 0.02 & -0.01
\end{bmatrix}
$$

$$
y_7 = \begin{bmatrix}
0.62 & 0 & -0.03 & 0 & 0.25 & -0.05 & 0.24 & 0.28 & -0.18 & 0.02 & -0.04 & 0 & -0.21 & 0 & 0.02 & 0 \\
0 & 0 & 0 & 0 & 0 & 0 & 0 & 0 & 0 & 0 & 0 & 0 & 0 & 0 & 0 & 0 \\
-0.03 & 0 & 0 & 0 & -0.04 & -0.05 & -0.02 & 0.28 & 0.01 & 0.02 & 0 & 0 & 0.03 & 0 & 0 & 0 \\
0 & 0 & 0 & 0 & 0 & 0 & 0 & 0 & 0 & 0 & 0 & 0 & 0 & 0 & 0 & 0 \\
0.25 & 0 & -0.04 & 0 & 0.08 & -0.04 & 0.29 & 0.26 & -0.05 & 0.02 & -0.04 & 0 & -0.07 & 0 & 0.02 & 0 \\
-0.05 & 0 & -0.05 & 0 & -0.04 & 0 & -0.05 & -0.03 & 0.03 & 0 & 0.04 & 0 & 0.05 & 0 & 0.05 & 0 \\
0.24 & 0 & -0.02 & 0 & 0.29 & -0.05 & 0.14 & 0.32 & -0.08 & 0.02 & -0.02 & 0 & -0.27 & 0 & 0.01 & 0 \\
0.28 & 0 & 0.28 & 0 & 0.26 & -0.03 & 0.32 & 0.16 & -0.17 & 0.01 & -0.27 & 0 & -0.28 & 0 & 0.01 & 0 \\
-0.18 & 0 & 0.01 & 0 & -0.05 & 0.03 & -0.08 & -0.17 & 0.26 & -0.01 & 0.01 & 0 & 0.04 & 0 & -0.01 & 0 \\
0.02 & 0 & 0.02 & 0 & 0.02 & 0 & 0.02 & 0.01 & -0.01 & 0 & -0.02 & 0 & -0.02 & 0 & -0.02 & 0 \\
-0.04 & 0 & 0 & 0 & -0.04 & 0.04 & -0.02 & -0.27 & 0.01 & -0.02 & 0 & 0 & 0.04 & 0 & 0 & 0 \\
0 & 0 & 0 & 0 & 0 & 0 & 0 & 0 & 0 & 0 & 0 & 0 & 0 & 0 & 0 & 0 \\
-0.21 & 0 & 0.03 & 0 & -0.07 & 0.05 & -0.27 & -0.28 & 0.04 & -0.02 & 0.04 & 0 & 0.06 & 0 & -0.02 & 0 \\
0 & 0 & 0 & 0 & 0 & 0 & 0 & 0 & 0 & 0 & 0 & 0 & 0 & 0 & 0 & 0 \\
0.02 & 0 & 0 & 0 & 0.02 & 0.05 & 0.01 & -0.28 & -0.01 & -0.02 & 0 & 0 & -0.02 & 0 & 0 & 0 \\
0 & 0 & 0 & 0 & 0 & 0 & 0 & 0 & 0 & 0 & 0 & 0 & 0 & 0 & 0 & 0
\end{bmatrix}
$$

$$
y_8 = \begin{bmatrix}
0.24 & 0.19 & 0.21 & -0.22 & 0.09 & 0.01 & -0.07 & 0 & -0.08 & 0 & 0 & 0 & -0.02 & 0 & 0 & 0 \\
0.19 & -0.1 & 0.07 & -0.23 & -0.12 & 0.03 & 0.01 & 0 & 0 & 0 & 0 & 0 & 0 & 0 & 0 & 0 \\
0.21 & 0.07 & 0.29 & 0.12 & 0 & -0.13 & -0.07 & 0.01 & -0.03 & 0 & 0.01 & 0 & 0 & 0 & 0 & 0 \\
-0.22 & -0.23 & 0.12 & -0.06 & 0.12 & -0.25 & -0.13 & 0.03 & 0.01 & 0 & 0 & 0 & 0 & 0 & 0 & 0 \\
0.09 & -0.12 & 0 & 0.12 & 0.27 & 0.12 & 0.04 & -0.13 & 0 & 0.01 & -0.04 & 0 & 0.01 & 0 & 0 & 0 \\
0.01 & 0.03 & -0.13 & -0.25 & 0.12 & -0.06 & 0.12 & -0.25 & -0.13 & 0.03 & 0.01 & 0 & 0 & 0 & 0 & 0 \\
-0.07 & 0.01 & -0.07 & -0.13 & 0.04 & 0.12 & 0.4 & 0.12 & 0.03 & -0.13 & -0.08 & 0.01 & -0.03 & 0 & 0.02 & 0 \\
0 & 0 & 0.01 & 0.03 & -0.13 & -0.25 & 0.12 & -0.06 & 0.12 & -0.25 & -0.13 & 0.03 & 0.01 & 0 & 0 & 0 \\
-0.08 & 0 & -0.03 & 0.01 & 0 & -0.13 & 0.03 & 0.12 & 0.22 & 0.12 & 0.03 & -0.13 & -0.12 & 0.01 & -0.05 & 0 \\
0 & 0 & 0 & 0 & 0.01 & 0.03 & -0.13 & -0.25 & 0.12 & -0.06 & 0.12 & -0.25 & -0.13 & 0.03 & 0.01 & 0 \\
0 & 0 & 0.01 & 0 & -0.04 & 0.01 & -0.08 & -0.13 & 0.03 & 0.12 & 0.4 & 0.12 & 0.04 & -0.13 & -0.07 & 0.01 \\
0 & 0 & 0 & 0 & 0 & 0 & 0.01 & 0.03 & -0.13 & -0.25 & 0.12 & -0.06 & 0.12 & -0.25 & -0.12 & 0.06 \\
-0.02 & 0 & 0 & 0 & 0.01 & 0 & -0.03 & 0.01 & -0.12 & -0.13 & 0.04 & 0.12 & 0.38 & 0.12 & -0.18 & -0.22 \\
0 & 0 & 0 & 0 & 0 & 0 & 0 & 0 & 0.01 & 0.03 & -0.13 & -0.25 & 0.12 & -0.06 & 0.1 & -0.34 \\
0 & 0 & 0 & 0 & 0 & 0 & 0.02 & 0 & -0.05 & 0.01 & -0.07 & -0.12 & -0.18 & 0.1 & 0.53 & -0.12 \\
0 & 0 & 0 & 0 & 0 & 0 & 0 & 0 & 0 & 0 & 0.01 & 0.06 & -0.22 & -0.34 & -0.12 & 0.12
\end{bmatrix}
$$

(6) Around optimal quantization.

Preliminary remark In lossy compression, there is a natural competition between two opposite criteria of quality:

(a) The compression ratio.
(b) The pictorial faithfulness of the decompressed image.

The battle-field of these two criteria is the procedure of *quantization*.

The compression ratio $\frac{nbofsourcebits}{nbofcodebits}$ increases with the brutality of quantization.

The pictorial faithfulness of the decompressed image will be guaranteed by delicate quantization.

We will have to find a satisfactory compromise.

But let us first speak of the *quantization* operation. We have to associate with a real number y (between 0 and 1, say) an 8-bit byte (a modest proposal). The natural idea is to take the eight leading positions in the binary notation of y. The obvious dequantization will yield a rather satisfactory approximation of the initial value y. This procedure is legitimate, but too rigid for our needs. We aim for a kind of "arithmetical telescope" which carries out a weighting of the considered value y.

Example Gradual quantization.

(1) $-0.31 \longmapsto (1, 01001111)$
The bit '1' in the first position indicates the presence of a negative sign (the bit '0' would indicate its absence). 01001111 is the binary notation of the integer $79 = \lfloor 0.31 \cdot 2^8 \rfloor$[17]
In other words: $0.31 = 0.01001111*$ in binary notation (the unmasked part is clearly $\frac{79}{256}$).
The dequantization: $(1, 01001111) \longmapsto -\frac{79.5}{2^8} = -0.3105$. (Dequantization "naturally" estimates the masked part of the quantized number: A real number 79.* is generically 79.5).

(2) $-0.31 \longmapsto (1, 00010011)$ 00010011 is the binary notation of the integer $19 = \lfloor 0.31 \cdot 2^6 \rfloor$
The dequantization: $(1, 00010011) \longmapsto -\frac{19.5}{2^6} = -0.3047$.

(3) $-0.31 \longmapsto (1, 00000100)$ 00000100 is the binary notation of the integer $4 = \lfloor 0.31 \cdot 2^4 \rfloor$
The dequantization: $(1, 00000100) \longmapsto -\frac{4.5}{2^4} = -0.28125$.

(4) $-0.31 \longmapsto (1, 00000001)$
00000001 is the binary notation of the integer $1 = \lfloor 0.31 \cdot 2^2 \rfloor$
The dequantization: $(1, 00000001) \longmapsto -\frac{1.5}{2^2} = -0.375$.

Let us sum up. The progression (1), (2), (3), (4) shows a going down of interest: We are looking at the 8, 6, 4, 2 most significant positions in the binary notation of $|y| = 0.31$. The leading zeros serve as padding.

For the intuitive. The more our quantization is brutal (and the dequantization imprecise), the more the ensuing compaction (of the whole of the quantized values) will be efficient – due to an important domination of the bit '0' over the bit '1'.

---

[17] $\lfloor \; \rfloor \equiv$ the integer part of a (non-negative) real number.

Consequence Look now at the situation of an interleaved scheme $\mathbf{y}^{(4)}$ obtained from a 16×16 scheme $\mathbf{x} = \mathbf{y}^{(0)}$.

Suppose that the value -0.31 there takes place at four different positions. Then the quantization will perhaps be carried out in four different ways, according to the importance that we have attributed to the considered position for the reconstruction of an acceptable image in synthesis. The more (the position of) this value will be important, the more significant bits of its binary notation will be taken into account.

The objective of this exercise is thus to get a certain feeling for the weighting of the positions (or rather of the regions) of the interleaved scheme $\mathbf{y}^{(4)}$.

LL_4 HL_4 / LH_4 HH_4	HL_3	HL_2	HL_1
LH_3	HH_3		
LH_2		HH_2	
LH_1			HH_1

We shall use the 2D DWT 5/3 spline. We choose as test matrix

$$\mathbf{x} = \mathbf{y}^{(0)} = \begin{matrix}
0 & 1.5 & 1.4 & 1.3 & 1.2 & 1.1 & 1.0 & 0.9 & 0.8 & 0.7 & 0.6 & 0.5 & 0.4 & 0.3 & 0.2 & 0.1 \\
0.1 & 0 & 1.5 & 1.4 & 1.3 & 1.2 & 1.1 & 1.0 & 0.9 & 0.8 & 0.7 & 0.6 & 0.5 & 0.4 & 0.3 & 0.2 \\
0.2 & 0.1 & 0 & 1.5 & 1.4 & 1.3 & 1.2 & 1.1 & 1.0 & 0.9 & 0.8 & 0.7 & 0.6 & 0.5 & 0.4 & 0.3 \\
0.3 & 0.2 & 0.1 & 0 & 1.5 & 1.4 & 1.3 & 1.2 & 1.1 & 1.0 & 0.9 & 0.8 & 0.7 & 0.6 & 0.5 & 0.4 \\
0.4 & 0.3 & 0.2 & 0.1 & 0 & 1.5 & 1.4 & 1.3 & 1.2 & 1.1 & 1.0 & 0.9 & 0.8 & 0.7 & 0.6 & 0.5 \\
0.5 & 0.4 & 0.3 & 0.2 & 0.1 & 0 & 1.5 & 1.4 & 1.3 & 1.2 & 1.1 & 1.0 & 0.9 & 0.8 & 0.7 & 0.6 \\
0.6 & 0.5 & 0.4 & 0.3 & 0.2 & 0.1 & 0 & 1.5 & 1.4 & 1.3 & 1.2 & 1.1 & 1.0 & 0.9 & 0.8 & 0.7 \\
0.7 & 0.6 & 0.5 & 0.4 & 0.3 & 0.2 & 0.1 & 0 & 1.5 & 1.4 & 1.3 & 1.2 & 1.1 & 1.0 & 0.9 & 0.8 \\
0.8 & 0.7 & 0.6 & 0.5 & 0.4 & 0.3 & 0.2 & 0.1 & 0 & 1.5 & 1.4 & 1.3 & 1.2 & 1.1 & 1.0 & 0.9 \\
0.9 & 0.8 & 0.7 & 0.6 & 0.5 & 0.4 & 0.3 & 0.2 & 0.1 & 0 & 1.5 & 1.4 & 1.3 & 1.2 & 1.1 & 1.0 \\
1.0 & 0.9 & 0.8 & 0.7 & 0.6 & 0.5 & 0.4 & 0.3 & 0.2 & 0.1 & 0 & 1.5 & 1.4 & 1.3 & 1.2 & 1.1 \\
1.1 & 1.0 & 0.9 & 0.8 & 0.7 & 0.6 & 0.5 & 0.4 & 0.3 & 0.2 & 0.1 & 0 & 1.5 & 1.4 & 1.3 & 1.2 \\
1.2 & 1.1 & 1.0 & 0.9 & 0.8 & 0.7 & 0.6 & 0.5 & 0.4 & 0.3 & 0.2 & 0.1 & 0 & 1.5 & 1.4 & 1.3 \\
1.3 & 1.2 & 1.1 & 1.0 & 0.9 & 0.8 & 0.7 & 0.6 & 0.5 & 0.4 & 0.3 & 0.2 & 0.1 & 0 & 1.5 & 1.4 \\
1.4 & 1.3 & 1.2 & 1.1 & 1.0 & 0.9 & 0.8 & 0.7 & 0.6 & 0.5 & 0.4 & 0.3 & 0.2 & 0.1 & 0 & 1.5 \\
1.5 & 1.4 & 1.3 & 1.2 & 1.1 & 1.0 & 0.9 & 0.8 & 0.7 & 0.6 & 0.5 & 0.4 & 0.3 & 0.2 & 0.1 & 0
\end{matrix}$$

This gives for the interleaved matrix[18]

$$
y^{(4)} =
$$

```
 0.727   0.1    0.305  -0.1    0.405   0     -0.153  0      0.186   0     -0.002  0     -0.471  0     -0.113 -0.5
-0.3    -0.3    0.2    -0.1   -0.05    0      0      0      0       0      0      0      0      0      0      0
-0.383  -0.15  -0.283   0.2    0.27   -0.05  -0.166  0     -0.077   0     -0.002  0     -0.001  0      0     -0.05
 0       0     -0.15   -0.3    0.2    -0.1   -0.05   0      0       0      0      0      0      0      0      0
-0.411   0     -0.219  -0.15  -0.197   0.2    0.271 -0.05   0.227   0     -0.077  0     -0.278  0     -0.114 -0.05
 0       0      0       0     -0.15   -0.3    0.2   -0.1   -0.05    0      0      0      0      0      0      0
 0.033   0      0.033   0     -0.211  -0.15  -0.267  0.2    0.272  -0.05  -0.164  0     -0.077  0     -0.003 -0.05
 0       0      0       0      0       0     -0.15  -0.3    0.2    -0.1   -0.05   0      0      0      0      0
-0.1     0      0.016   0     -0.231   0     -0.211 -0.15  -0.147   0.2    0.272 -0.05   0.26   0     -0.242 -0.05
 0       0      0       0      0       0      0     -0.15  -0.3     0.2   -0.1   -0.05   0      0      0      0
 0       0      0       0      0.016   0      0.033  0     -0.211  -0.15  -0.267  0.2    0.213 -0.05  -0.281 -0.05
 0       0      0       0      0       0      0      0      0      -0.15  -0.3    0.2   -0.1   -0.05   0
 0.334   0      0       0      0.114   0      0.016  0     -0.103   0     -0.199 -0.15  -0.186  0.2    0.413 -0.15
 0       0      0       0      0       0      0      0      0       0      0     -0.15  -0.3    0.2    0.15  -0.2
 0.113   0      0       0      0.113   0      0      0      0.141   0      0.056  0     -0.156 -0.15  -0.294  0.45
 0.05    0      0.05    0      0.05    0      0.05   0      0.05    0      0.05   0      0.050  0     -0.15  -0.4
```

In order to start naturally, let us first carry out a uniform quantization which saves the eight most significant positions in the binary notation of each of the 256 coefficients of $y^{(4)}$:

$$
q^{(4)} =
$$

```
 186   25   78  -25  103    0  -39    0   47    0    0    0 -120    0  -28  -12
 -76  -76   51  -25  -12    0    0    0    0    0    0    0    0    0    0    0
 -98  -38  -72   51   69  -12  -42    0  -19    0    0    0    0    0    0  -12
   0    0  -38  -76   51  -25  -12    0    0    0    0    0    0    0    0    0
-105    0  -56  -38  -50   51   69  -12   58    0  -19    0  -71    0  -29  -12
   0    0    0    0  -38  -76   51  -25  -12    0    0    0    0    0    0    0
   8    0    8    0  -54  -38  -68   51   69  -12  -42    0  -19    0    0  -12
   0    0    0    0    0    0  -38  -76   51  -25  -12    0    0    0    0    0
 -25    0    4    0  -59    0  -54  -38  -37   51   69  -12   66    0  -62  -12
   0    0    0    0    0    0    0  -38  -76   51  -25  -12    0    0    0
   0    0    0    0    4    0    8    0  -54  -38  -68   51   54  -12  -72  -12
   0    0    0    0    0    0    0    0  -38  -76   51  -25  -12    0
  85    0    0    0   29    0    4    0  -26    0  -51  -38  -47   51  105  -38
   0    0    0    0    0    0    0    0    0    0    0  -38  -76   38  -51
  28    0    0    0   28    0    0    0   36    0   14    0  -40  -38  -75  115
  12    0   12    0   12    0   12    0   12    0   12    0   12    0  -38 -102
```

The first row of $q^{(4)}$ begins as follows: 10111010, 00011001, 01001110, etc.

Now the dequantized matrix:

$$
z^{(4)} =
$$

```
 0.729   0.1    0.307  -0.1    0.404   0     -0.154  0      0.186   0      0      0     -0.471  0     -0.111 -0.4
-0.299  -0.299  0.201  -0.1   -0.049   0      0      0      0       0      0      0      0      0      0      0
-0.385  -0.15  -0.283   0.201  0.271  -0.049 -0.166  0     -0.076   0      0      0      0      0      0     -0.04
 0       0     -0.15   -0.299  0.201  -0.1   -0.049  0      0       0      0      0      0      0      0      0
-0.412   0     -0.221  -0.15  -0.197   0.201  0.271 -0.049  0.229   0     -0.076  0     -0.279  0     -0.115 -0.04
 0       0      0       0     -0.15   -0.299  0.201 -0.1   -0.049   0      0      0      0      0      0      0
 0.033   0      0.033   0     -0.213  -0.15  -0.268  0.201  0.271  -0.049 -0.166  0     -0.076  0      0     -0.04
 0       0      0       0      0       0     -0.15  -0.299  0.201  -0.1   -0.049  0      0      0      0      0
-0.1     0      0.018   0     -0.232   0     -0.213 -0.15  -0.146   0.201  0.271 -0.049  0.26   0     -0.244 -0.04
 0       0      0       0      0       0      0     -0.15  -0.299   0.201 -0.1   -0.049  0      0      0      0
 0       0      0       0      0.018   0      0.033  0     -0.213  -0.15  -0.268  0.201  0.213 -0.049 -0.283 -0.04
 0       0      0       0      0       0      0      0      0      -0.15  -0.299  0.201 -0.1   -0.049  0
 0.334   0      0       0      0.115   0      0.018  0     -0.104   0     -0.201 -0.15  -0.186  0.201  0.412 -0.1
 0       0      0       0      0       0      0      0      0       0      0     -0.15  -0.299  0.15  -0.20
 0.111   0      0       0      0.111   0      0      0      0.143   0      0.057  0     -0.158 -0.15  -0.295  0.451
 0.049   0      0.049   0      0.049   0      0.049  0      0.049   0      0.049  0      0.049  0     -0.15  -0.4
```

---

[18] With duly rounded coefficients.

We note that the maximum deviation between (the rounded version of) $\mathbf{y}^{(4)}$ and (the rounded version of) $\mathbf{z}^{(4)}$ is 0.003 (a coefficient with value -0.003 becomes 0). Finally, apply the synthesis transform of the (2D) DWT 5/3 spline to $\mathbf{z}^{(4)}$ [19]:

$$\mathbf{z}^{(0)} = \begin{matrix}
0.002 & 1.5 & 1.404 & 1.3 & 1.196 & 1.096 & 0.996 & 0.896 & 0.797 & 0.698 & 0.598 & 0.399 & 0.3 & 0.201 & 0.104 & 0.499 \\
0.101 & 0.004 & 1.503 & 1.404 & 1.301 & 1.2 & 1.098 & 0.999 & 0.899 & 0.801 & 0.703 & 0.5 & 0.4 & 0.299 & 0.202 & 0.601 \\
0.201 & 0.099 & 0 & 1.502 & 1.401 & 1.303 & 1.2 & 1.101 & 1.002 & 0.905 & 0.808 & 0.6 & 0.499 & 0.398 & 0.3 & 0.704 \\
0.303 & 0.199 & 0.098 & 0.004 & 1.501 & 1.405 & 1.303 & 1.204 & 1.102 & 1.004 & 0.906 & 0.7 & 0.598 & 0.496 & 0.398 & 0.803 \\
0.405 & 0.302 & 0.199 & 0.097 & 0 & 1.501 & 1.402 & 1.304 & 1.203 & 1.103 & 1.003 & 0.8 & 0.697 & 0.594 & 0.497 & 0.902 \\
0.505 & 0.402 & 0.3 & 0.197 & 0.096 & 0.002 & 1.501 & 1.406 & 1.306 & 1.204 & 1.1 & 0.902 & 0.802 & 0.702 & 0.604 & 1.001 \\
0.604 & 0.502 & 0.401 & 0.299 & 0.197 & 0.095 & -0.002 & 1.502 & 1.404 & 1.302 & 1.197 & 1.003 & 0.906 & 0.81 & 0.712 & 1.1 \\
0.703 & 0.603 & 0.502 & 0.4 & 0.298 & 0.196 & 0.097 & 0.005 & 1.505 & 1.407 & 1.302 & 1.103 & 1.003 & 0.904 & 0.806 & 1.203 \\
0.803 & 0.703 & 0.603 & 0.502 & 0.4 & 0.299 & 0.199 & 0.1 & 0.004 & 1.505 & 1.404 & 1.202 & 1.1 & 0.998 & 0.9 & 1.305 \\
0.903 & 0.803 & 0.703 & 0.602 & 0.501 & 0.401 & 0.3 & 0.199 & 0.1 & 0.005 & 1.502 & 1.305 & 1.202 & 1.096 & 0.999 & 1.407 \\
1.002 & 0.903 & 0.803 & 0.703 & 0.603 & 0.502 & 0.401 & 0.3 & 0.199 & 0.098 & 0 & 1.404 & 1.302 & 1.195 & 1.097 & 1.503 \\
1.102 & 1.002 & 0.903 & 0.802 & 0.702 & 0.602 & 0.502 & 0.401 & 0.3 & 0.198 & 0.098 & 1.505 & 1.407 & 1.303 & 1.204 & 0.006 \\
1.202 & 1.102 & 1.002 & 0.901 & 0.801 & 0.702 & 0.603 & 0.502 & 0.401 & 0.3 & 0.2 & 0.005 & 1.507 & 1.407 & 1.307 & 0.101 \\
1.301 & 1.201 & 1.101 & 1.001 & 0.9 & 0.802 & 0.704 & 0.603 & 0.503 & 0.402 & 0.301 & 0.103 & 0.007 & 1.504 & 1.401 & 0.201 \\
1.4 & 1.3 & 1.2 & 1.1 & 1 & 0.902 & 0.804 & 0.705 & 0.605 & 0.503 & 0.402 & 0.204 & 0.1 & 0 & 1.503 & 0.303 \\
1.498 & 1.398 & 1.298 & 1.197 & 1.097 & 1 & 0.902 & 0.802 & 0.703 & 0.601 & 0.499 & 0.301 & 0.198 & 0.099 & 0.001 & 0.4
\end{matrix}$$

We note: The maximum deviation $\max\{|\ y^{(0)}[i,j] - z^{(0)}[i,j]\ |\ 0 \le i,j \le 15\} = 0.012$.[20] What is the consequence for the faithfulness of the compression/decompression? Clearly, everything depends on the precision with which $\mathbf{x} = \mathbf{y}^{(0)}$ transcribes the values of the digital image. Suppose that we started with the 256 luminance sample values (in 8-bit bytes) of a 16×16 pixel region. The matrix $\mathbf{x} = \mathbf{y}^{(0)}$ will then be a "renormalization" of these values. Let us adopt the primitive viewpoint of a correspondence

$$1.5 \quad \longleftrightarrow 150 \longleftrightarrow \quad 10010110$$

$$0.3 \quad \longleftrightarrow 30 \longleftrightarrow \quad 00011110$$

Then the luminance matrix derived from $\mathbf{z}^{(0)}$ differs from the initial luminance matrix in 31 octets (count the critical positions!). All that for a first orientation. But let us begin with the statement of the exercise. We shall uniformly quantize on each *resolution level*.

LL_0

LL_1 $\mathcal{R}_4$: HL_1 / LH_1 HH_1

LL_2 $\mathcal{R}_3$: HL_2 / LH_2 HH_2

LL_3 $\mathcal{R}_2$: HL_3 / LH_3 HH_3

LL_4 $\mathcal{R}_1$: HL_4 / LH_4 HH_4

We shall allocate to every resolution level $\mathcal{R}_i$ a quantization factor $N_i = 2^{b_i}$ ($i = 0, 1, 2, 3, 4$), which fixes the number b_i of the significant binary positions which we shall take into account – according to the quantization/dequantization procedure already discussed.

Hence let $y = \text{sgn}(y) \cdot |y|$ be a coefficient of level $\mathcal{R}_i$.

Quantization: $q = \lfloor |y| \cdot N_i \rfloor$

---

[19] It is the $\mathbf{q}^{(4)}$ which undergoes entropy coding; hence the decompression recovers precisely $\mathbf{q}^{(4)}$ – its version to synthesize will be $\mathbf{z}^{(4)}$.

[20] $\| \mathbf{y}^{(0)} - \mathbf{z}^{(0)} \| = 0.049$. The matrix $\mathbf{y}^{(0)} - \mathbf{z}^{(0)}$ "is worth ± 0.003 per coefficient" – in quadratic mean.

Dequantization: $y' = \text{sgn}(y) \cdot \frac{q+0.5}{N_i}$

The value 0 will be quantized into 0 and dequantized into 0. The four coefficients which determine $\mathcal{R}_0 = LL_4$ as well as $\mathcal{R}_1 = (LH_4, HL_4, HH_4)$ will be "optimally" quantized: $N_0 = N_1 = 2^8$.

Philosophy: Do not touch the most primitive fragments of the image!

Concerning the other resolution levels, we shall vary.

(a) Choose the most brutal quantization: $N_2 = N_3 = N_4 = 1 = 2^0$. Determine the dequantized scheme $\mathbf{z}^{(4)}$, then the decompressed scheme $\mathbf{z}^{(0)}$. Compare $\mathbf{z}^{(0)}$ with $\mathbf{y}^{(0)}$: Maximum deviation $\max\{|\ y^{(0)}[i,j] - z^{(0)}[i,j]\ |\ 0 \le i,j \le 15\}$ and Euclidean distance $\|\ \mathbf{y}^{(0)} - \mathbf{z}^{(0)}\ \|$?

(b) Now, let us widen our horizon of interest for the details:

   First step: $N_2 = 2^8$, $N_3 = N_4 = 1 = 2^0$.

   Second step: $N_2 = N_3 = 2^8$, $N_4 = 1 = 2^0$.[21]

   Determine, in every case, the dequantized scheme $\mathbf{z}^{(4)}$, then its decompressed scheme $\mathbf{z}^{(0)}$. Compare $\mathbf{z}^{(0)}$ with $\mathbf{y}^{(0)}$: Maximum deviation $\max\{|\ y^{(0)}[i,j] - z^{(0)}[i,j]\ |\ 0 \le i,j \le 15\}$ and Euclidean distance $\|\ \mathbf{y}^{(0)} - \mathbf{z}^{(0)}\ \|$?

(c) Now pass to a graduated quantization: $N_2 = 2^6$, $N_3 = 2^5$, $N_4 = 2^4$. This seems to be natural in two aspects: On the one hand, it is logical for efficient compression – it creates the greatest number of zeros in regions with many coefficients. On the other hand, it is logical for faithful image reconstruction – it is delicate where the numerical details appear in a "small team" and it is more brutal where the numerical details appear in a "big team".

   Determine, as before, the dequantized scheme $\mathbf{z}^{(4)}$, then its decompressed scheme $\mathbf{z}^{(0)}$. Compare, once more, $\mathbf{z}^{(0)}$ with $\mathbf{y}^{(0)}$: Maximum deviation $\max\{|\ y^{(0)}[i,j] - z^{(0)}[i,j]\ |\ 0 \le i,j \le 15\}$ and Euclidean distance $\|\ \mathbf{y}^{(0)} - \mathbf{z}^{(0)}\ \|$?

(d) A last attempt, in order to verify if our viewpoint is right. Now invert the order of our quantization factors: $N_2 = 2^2$, $N_3 = 2^4$, $N_4 = 2^6$. Once more: Compute the dequantized scheme $\mathbf{z}^{(4)}$, then the decompressed scheme $\mathbf{z}^{(0)}$. Compare, for a last time, $\mathbf{z}^{(0)}$ with $\mathbf{y}^{(0)}$: Maximum deviation $\max\{|\ y^{(0)}[i,j] - z^{(0)}[i,j]\ |\ 0 \le i,j \le 15\}$ and Euclidean distance $\|\ \mathbf{y}^{(0)} - \mathbf{z}^{(0)}\ \|$?

Remark In practice, after long, honest and complicated considerations, we shall have a tendency to more or less resign in applying the following primitive formula:

   We shall fix N_0 (which gives the maximum precision) and then put $N_i = N_0 \cdot 2^{-i}$.

   [*The results* (we work with the 2D DWT 5/3 spline):

---

[21] The third step has just been discussed..

(a)

$$\mathbf{z}^{(0)} = \begin{matrix}
0.496 & 0.579 & 0.662 & 0.745 & 0.828 & 0.911 & 0.994 & 1.077 & 1.16 & 1.16 & 1.16 & 1.16 & 1.16 & 1.16 & 1.16 & 1.16 \\
0.508 & 0.582 & 0.656 & 0.729 & 0.803 & 0.877 & 0.951 & 1.025 & 1.099 & 1.099 & 1.099 & 1.099 & 1.099 & 1.099 & 1.099 & 1.099 \\
0.520 & 0.584 & 0.649 & 0.714 & 0.778 & 0.843 & 0.908 & 0.972 & 1.037 & 1.037 & 1.037 & 1.037 & 1.037 & 1.037 & 1.037 & 1.037 \\
0.531 & 0.587 & 0.642 & 0.698 & 0.753 & 0.809 & 0.865 & 0.920 & 0.976 & 0.976 & 0.976 & 0.976 & 0.976 & 0.976 & 0.976 & 0.976 \\
0.543 & 0.589 & 0.636 & 0.682 & 0.729 & 0.775 & 0.821 & 0.868 & 0.914 & 0.914 & 0.914 & 0.914 & 0.914 & 0.914 & 0.914 & 0.914 \\
0.555 & 0.592 & 0.629 & 0.666 & 0.704 & 0.741 & 0.778 & 0.815 & 0.853 & 0.853 & 0.853 & 0.853 & 0.853 & 0.853 & 0.853 & 0.853 \\
0.566 & 0.594 & 0.623 & 0.651 & 0.679 & 0.707 & 0.735 & 0.763 & 0.791 & 0.791 & 0.791 & 0.791 & 0.791 & 0.791 & 0.791 & 0.791 \\
0.578 & 0.597 & 0.616 & 0.635 & 0.654 & 0.673 & 0.692 & 0.711 & 0.729 & 0.729 & 0.729 & 0.729 & 0.729 & 0.729 & 0.729 & 0.729 \\
0.59 & 0.6 & 0.609 & 0.619 & 0.629 & 0.639 & 0.648 & 0.658 & 0.668 & 0.668 & 0.668 & 0.668 & 0.668 & 0.668 & 0.668 & 0.668 \\
0.59 & 0.6 & 0.609 & 0.619 & 0.629 & 0.639 & 0.648 & 0.658 & 0.668 & 0.668 & 0.668 & 0.668 & 0.668 & 0.668 & 0.668 & 0.668 \\
0.59 & 0.6 & 0.609 & 0.619 & 0.629 & 0.639 & 0.648 & 0.658 & 0.668 & 0.668 & 0.668 & 0.668 & 0.668 & 0.668 & 0.668 & 0.668 \\
0.59 & 0.6 & 0.609 & 0.619 & 0.629 & 0.639 & 0.648 & 0.658 & 0.668 & 0.668 & 0.668 & 0.668 & 0.668 & 0.668 & 0.668 & 0.668 \\
0.59 & 0.6 & 0.609 & 0.619 & 0.629 & 0.639 & 0.648 & 0.658 & 0.668 & 0.668 & 0.668 & 0.668 & 0.668 & 0.668 & 0.668 & 0.668 \\
0.59 & 0.6 & 0.609 & 0.619 & 0.629 & 0.639 & 0.648 & 0.658 & 0.668 & 0.668 & 0.668 & 0.668 & 0.668 & 0.668 & 0.668 & 0.668 \\
0.59 & 0.6 & 0.609 & 0.619 & 0.629 & 0.639 & 0.648 & 0.658 & 0.668 & 0.668 & 0.668 & 0.668 & 0.668 & 0.668 & 0.668 & 0.668 \\
0.59 & 0.6 & 0.609 & 0.619 & 0.629 & 0.639 & 0.648 & 0.658 & 0.668 & 0.668 & 0.668 & 0.668 & 0.668 & 0.668 & 0.668 & 0.668
\end{matrix}$$

Hence $\max\{|\ y^{(0)}[i,j] - z^{(0)}[i,j]\ |\ 0 \le i,j \le 15\} = 1.06$ and $\|\ \mathbf{y}^{(0)} - \mathbf{z}^{(0)}\ \| = 7.682.$[22]

It is clear that there is no hope to reconstruct the cyclic structure of $\mathbf{x} = \mathbf{y}^{(0)}$ by means of the four coefficients of LL_3.

(b) Actually, it is the "vicious" (highly non-pictorial) structure of $\mathbf{x} = \mathbf{y}^{(0)}$ which demands a thorough participation of the (regions of) details in an – even approximate – reconstruction of the initial scheme.

This is shown in the frustrating results of this part of the exercise. We see that the total suppression of the highest resolution level is – at least in our case – completely intolerable – and still less the simultaneous annihilation of the two highest resolution levels.

But let us go on to the results.

Begin with the first step quantization

$$\mathbf{z}^{(0)} = \begin{matrix}
0.307 & 0.66 & 1.013 & 1.366 & 1.72 & 1.471 & 1.223 & 0.975 & 0.727 & 0.631 & 0.535 & 0.439 & 0.344 & 0.344 & 0.344 & 0.344 \\
0.263 & 0.548 & 0.833 & 1.118 & 1.402 & 1.283 & 1.164 & 1.045 & 0.926 & 0.803 & 0.68 & 0.558 & 0.435 & 0.435 & 0.435 & 0.435 \\
0.22 & 0.436 & 0.653 & 0.869 & 1.085 & 1.095 & 1.105 & 1.115 & 1.125 & 0.976 & 0.826 & 0.676 & 0.526 & 0.526 & 0.526 & 0.526 \\
0.177 & 0.325 & 0.472 & 0.620 & 0.768 & 0.907 & 1.046 & 1.186 & 1.325 & 1.148 & 0.971 & 0.794 & 0.617 & 0.617 & 0.617 & 0.617 \\
0.134 & 0.213 & 0.292 & 0.371 & 0.45 & 0.719 & 0.987 & 1.256 & 1.524 & 1.320 & 1.116 & 0.912 & 0.708 & 0.708 & 0.708 & 0.708 \\
0.305 & 0.329 & 0.353 & 0.377 & 0.401 & 0.615 & 0.829 & 1.043 & 1.257 & 1.166 & 1.074 & 0.982 & 0.891 & 0.891 & 0.891 & 0.891 \\
0.477 & 0.446 & 0.415 & 0.384 & 0.353 & 0.512 & 0.671 & 0.83 & 0.99 & 1.011 & 1.032 & 1.053 & 1.074 & 1.074 & 1.074 & 1.074 \\
0.649 & 0.562 & 0.476 & 0.39 & 0.304 & 0.408 & 0.513 & 0.618 & 0.722 & 0.856 & 0.99 & 1.123 & 1.257 & 1.257 & 1.257 & 1.257 \\
0.82 & 0.679 & 0.538 & 0.396 & 0.255 & 0.305 & 0.355 & 0.405 & 0.455 & 0.701 & 0.947 & 1.193 & 1.439 & 1.439 & 1.439 & 1.439 \\
0.93 & 0.799 & 0.669 & 0.538 & 0.408 & 0.411 & 0.414 & 0.418 & 0.421 & 0.621 & 0.82 & 1.02 & 1.22 & 1.22 & 1.22 & 1.22 \\
1.039 & 0.919 & 0.8 & 0.68 & 0.561 & 0.517 & 0.474 & 0.43 & 0.387 & 0.54 & 0.693 & 0.847 & 1 & 1 & 1 & 1 \\
1.148 & 1.04 & 0.931 & 0.822 & 0.713 & 0.623 & 0.533 & 0.443 & 0.353 & 0.459 & 0.566 & 0.673 & 0.78 & 0.78 & 0.78 & 0.78 \\
1.258 & 1.16 & 1.062 & 0.964 & 0.866 & 0.729 & 0.592 & 0.455 & 0.318 & 0.379 & 0.439 & 0.5 & 0.561 & 0.561 & 0.561 & 0.561 \\
1.258 & 1.16 & 1.062 & 0.964 & 0.866 & 0.729 & 0.592 & 0.455 & 0.318 & 0.379 & 0.439 & 0.5 & 0.561 & 0.561 & 0.561 & 0.561 \\
1.258 & 1.16 & 1.062 & 0.964 & 0.866 & 0.729 & 0.592 & 0.455 & 0.318 & 0.379 & 0.439 & 0.5 & 0.561 & 0.561 & 0.561 & 0.561 \\
1.258 & 1.16 & 1.062 & 0.964 & 0.866 & 0.729 & 0.592 & 0.455 & 0.318 & 0.379 & 0.439 & 0.5 & 0.561 & 0.561 & 0.561 & 0.561
\end{matrix}$$

This yields $\max\{|\ y^{(0)}[i,j] - z^{(0)}[i,j]\ |\ 0 \le i,j \le 15\} = 0.939$ and $\|\ \mathbf{y}^{(0)} - \mathbf{z}^{(0)}\ \| = 5.157.$

---

[22] So, we get, in quadratic mean, a deviation of ± 0.5 per coefficient (do not forget that the Euclidean norm of the constant matrix of value 1 is equal to 16).

Now the second step quantization

$$\mathbf{z}^{(0)} = \begin{array}{llllllllllllllll}
0.102 & 0.953 & 1.804 & 1.476 & 1.147 & 1.072 & 0.996 & 0.896 & 0.797 & 0.698 & 0.598 & 0.499 & 0.399 & 0.288 & 0.177 & 0.177 \\
0.76 & 0.545 & 1.015 & 1.214 & 1.413 & 1.249 & 1.086 & 0.993 & 0.899 & 0.801 & 0.703 & 0.601 & 0.5 & 0.387 & 0.275 & 0.275 \\
0.05 & 0.138 & 0.226 & 0.952 & 1.678 & 1.427 & 1.176 & 1.089 & 1.002 & 0.905 & 0.808 & 0.704 & 0.6 & 0.487 & 0.373 & 0.373 \\
0.228 & 0.201 & 0.175 & 0.563 & 0.951 & 1.189 & 1.427 & 1.259 & 1.09 & 0.998 & 0.906 & 0.803 & 0.7 & 0.586 & 0.472 & 0.472 \\
0.405 & 0.264 & 0.124 & 0.174 & 0.225 & 0.952 & 1.678 & 1.429 & 1.179 & 1.091 & 1.003 & 0.902 & 0.8 & 0.685 & 0.57 & 0.57 \\
0.505 & 0.383 & 0.262 & 0.217 & 0.173 & 0.562 & 0.951 & 1.190 & 1.430 & 1.259 & 1.088 & 0.995 & 0.902 & 0.790 & 0.677 & 0.677 \\
0.604 & 0.502 & 0.401 & 0.261 & 0.121 & 0.172 & 0.223 & 0.952 & 1.681 & 1.427 & 1.173 & 1.088 & 1.003 & 0.894 & 0.785 & 0.785 \\
0.703 & 0.603 & 0.502 & 0.381 & 0.26 & 0.217 & 0.173 & 0.564 & 0.955 & 1.191 & 1.427 & 1.259 & 1.091 & 0.985 & 0.879 & 0.879 \\
0.803 & 0.703 & 0.603 & 0.502 & 0.4 & 0.262 & 0.123 & 0.176 & 0.229 & 0.955 & 1.68 & 1.429 & 1.178 & 1.076 & 0.973 & 0.973 \\
0.903 & 0.803 & 0.703 & 0.602 & 0.501 & 0.382 & 0.262 & 0.219 & 0.177 & 0.565 & 0.953 & 1.191 & 1.430 & 1.245 & 1.06 & 1.06 \\
1.002 & 0.903 & 0.803 & 0.703 & 0.603 & 0.502 & 0.401 & 0.263 & 0.124 & 0.174 & 0.225 & 0.953 & 1.681 & 1.414 & 1.147 & 1.147 \\
1.102 & 1.002 & 0.903 & 0.802 & 0.702 & 0.602 & 0.502 & 0.382 & 0.262 & 0.218 & 0.174 & 0.565 & 0.956 & 1.173 & 1.39 & 1.39 \\
1.202 & 1.102 & 1.002 & 0.901 & 0.801 & 0.702 & 0.603 & 0.502 & 0.401 & 0.263 & 0.124 & 0.177 & 0.23 & 0.932 & 1.633 & 1.633 \\
1.313 & 1.213 & 1.113 & 1.013 & 0.913 & 0.814 & 0.716 & 0.615 & 0.515 & 0.395 & 0.275 & 0.233 & 0.191 & 0.598 & 1.004 & 1.004 \\
1.425 & 1.325 & 1.224 & 1.124 & 1.024 & 0.926 & 0.829 & 0.729 & 0.629 & 0.528 & 0.426 & 0.289 & 0.152 & 0.264 & 0.375 & 0.375 \\
1.425 & 1.325 & 1.224 & 1.124 & 1.024 & 0.926 & 0.829 & 0.729 & 0.629 & 0.528 & 0.426 & 0.289 & 0.152 & 0.264 & 0.375 & 0.375
\end{array}$$

with $\max\{|\ y^{(0)}[i,j] - z^{(0)}[i,j]\ |\ 0 \le i,j \le 15\} = 1.125$ and $\|\ \mathbf{y}^{(0)} - \mathbf{z}^{(0)}\ \| = 3.165$.

We note here a disaster of the maximum deviation, which is rather impressive – and greater than at the first step. But it comes up at the boundary, and in a corner of great numerical variation. We note that the annihilation of the highest resolution level gives rise to a hook '|' of two identical rows/columns at the boundary south – boundary east of the reconstructed scheme (we remark that the annihilation of the two highest resolution levels, at the first quantization step, has created a hook of four identical rows/columns at the boundary south – boundary east of $\mathbf{z}^{(0)}$ (and the annihilation of the three highest resolution levels has created a hook of eight identical rows/columns – cf. (a)).

$$\text{(c)}\ \mathbf{q}^{(4)} = \begin{array}{llllllllllllllll}
186 & 1 & 9 & -1 & 25 & 0 & -4 & 0 & 47 & 0 & 0 & 0 & -30 & 0 & -3 & 0 \\
-4 & -4 & 3 & -1 & 0 & 0 & 0 & 0 & 0 & 0 & 0 & 0 & 0 & 0 & 0 & 0 \\
-12 & -2 & -9 & 3 & 8 & 0 & -5 & 0 & -2 & 0 & 0 & 0 & 0 & 0 & 0 & 0 \\
0 & 0 & -2 & -4 & 3 & -1 & 0 & 0 & 0 & 0 & 0 & 0 & 0 & 0 & 0 & 0 \\
-26 & 0 & -7 & -2 & -12 & 3 & 8 & 0 & 14 & 0 & -2 & 0 & -17 & 0 & -3 & 0 \\
0 & 0 & 0 & 0 & -2 & -4 & 3 & -1 & 0 & 0 & 0 & 0 & 0 & 0 & 0 & 0 \\
1 & 0 & 1 & 0 & -6 & -2 & -8 & 3 & 8 & 0 & -5 & 0 & -2 & 0 & 0 & 0 \\
0 & 0 & 0 & 0 & 0 & 0 & -2 & -4 & 3 & -1 & 0 & 0 & 0 & 0 & 0 & 0 \\
-25 & 0 & 0 & 0 & -14 & 0 & -6 & -2 & -37 & 3 & 8 & 0 & 16 & 0 & -7 & 0 \\
0 & 0 & 0 & 0 & 0 & 0 & 0 & 0 & -2 & -4 & 3 & -1 & 0 & 0 & 0 & 0 \\
0 & 0 & 0 & 0 & 0 & 0 & 1 & 0 & -6 & -2 & -8 & 3 & 6 & 0 & -9 & 0 \\
0 & 0 & 0 & 0 & 0 & 0 & 0 & 0 & 0 & 0 & -2 & -4 & 3 & -1 & 0 & 0 \\
21 & 0 & 0 & 0 & 7 & 0 & 0 & 0 & -6 & 0 & -6 & -2 & -11 & 3 & 13 & -2 \\
0 & 0 & 0 & 0 & 0 & 0 & 0 & 0 & 0 & 0 & 0 & 0 & -2 & -4 & 2 & -3 \\
3 & 0 & 0 & 0 & 3 & 0 & 0 & 0 & 4 & 0 & 1 & 0 & -5 & -2 & -9 & 7 \\
0 & 0 & 0 & 0 & 0 & 0 & 0 & 0 & 0 & 0 & 0 & 0 & 0 & 0 & -2 & -6
\end{array}$$

$$
\mathbf{z}^{(0)} =
\begin{array}{cccccccccccccccc}
0.02 & 1.463 & 1.406 & 1.271 & 1.137 & 1.084 & 1.031 & 0.916 & 0.801 & 0.697 & 0.594 & 0.49 & 0.387 & 0.277 & 0.168 & 0.168 \\
0.125 & 0.042 & 1.490 & 1.444 & 1.336 & 1.26 & 1.09 & 0.997 & 0.903 & 0.802 & 0.7 & 0.595 & 0.489 & 0.38 & 0.271 & 0.271 \\
0.23 & 0.059 & -0.051 & 1.461 & 1.348 & 1.342 & 1.149 & 1.078 & 1.006 & 0.906 & 0.807 & 0.699 & 0.592 & 0.482 & 0.373 & 0.373 \\
0.32 & 0.163 & 0.037 & 0.028 & 1.488 & 1.491 & 1.337 & 1.256 & 1.081 & 0.993 & 0.905 & 0.8 & 0.694 & 0.585 & 0.476 & 0.476 \\
0.41 & 0.299 & 0.188 & 0.064 & 0.004 & 1.483 & 1.338 & 1.341 & 1.156 & 1.08 & 1.004 & 0.9 & 0.797 & 0.688 & 0.578 & 0.578 \\
0.512 & 0.41 & 0.309 & 0.18 & 0.083 & 0.05 & 1.485 & 1.491 & 1.341 & 1.247 & 1.06 & 0.978 & 0.896 & 0.793 & 0.689 & 0.689 \\
0.613 & 0.522 & 0.431 & 0.328 & 0.225 & 0.085 & 0.008 & 1.485 & 1.338 & 1.321 & 1.116 & 1.056 & 0.996 & 0.898 & 0.801 & 0.801 \\
0.715 & 0.608 & 0.501 & 0.412 & 0.323 & 0.188 & 0.85 & 0.05 & 1.483 & 1.478 & 1.317 & 1.240 & 1.068 & 0.982 & 0.896 & 0.896 \\
0.816 & 0.694 & 0.572 & 0.497 & 0.422 & 0.323 & 0.225 & 0.083 & 0.004 & 1.479 & 1.330 & 1.329 & 1.141 & 1.066 & 0.992 & 0.992 \\
0.912 & 0.792 & 0.672 & 0.587 & 0.502 & 0.414 & 0.326 & 0.187 & 0.079 & 0.044 & 1.479 & 1.483 & 1.331 & 1.233 & 1.042 & 1.042 \\
1.008 & 0.890 & 0.771 & 0.677 & 0.582 & 0.504 & 0.427 & 0.322 & 0.217 & 0.078 & 0.002 & 1.480 & 1.334 & 1.307 & 1.092 & 1.092 \\
1.104 & 0.999 & 0.895 & 0.802 & 0.709 & 0.605 & 0.5 & 0.406 & 0.312 & 0.18 & 0.08 & 0.048 & 1.485 & 1.495 & 1.349 & 1.302 \\
1.199 & 1.108 & 1.018 & 0.927 & 0.836 & 0.705 & 0.574 & 0.49 & 0.406 & 0.313 & 0.221 & 0.085 & 0.012 & 1.527 & 1.418 & 1.324 \\
1.309 & 1.218 & 1.127 & 1.036 & 0.945 & 0.816 & 0.688 & 0.605 & 0.523 & 0.415 & 0.307 & 0.183 & 0.09 & 0.074 & 1.527 & 1.387 \\
1.418 & 1.327 & 1.236 & 1.146 & 1.055 & 0.928 & 0.801 & 0.721 & 0.641 & 0.517 & 0.393 & 0.312 & 0.23 & 0.09 & 0.012 & 1.574 \\
1.418 & 1.327 & 1.236 & 1.146 & 1.055 & 0.928 & 0.801 & 0.721 & 0.641 & 0.517 & 0.393 & 0.312 & 0.23 & 0.137 & 0.105 & 0.043 \\
\end{array}
$$

Here $\max\{|\ y^{(0)}[i,j] - z^{(0)}[i,j],|\ 0 \le i,j \le 15\} = 0.108$ and $\|\ \mathbf{y}^{(0)} - \mathbf{z}^{(0)}\ \| = 0.613$. In quadratic mean, the deviation per position is ± 0.04. But let us look more closely at the matrix $\mathbf{q}^{(4)}$. If we write the absolute values of the entries in 8-bit bytes, we obtain 1,872 zeros per 2,048 bits. In the language of the first chapter (on compaction) we thus have: $p_0 = 0.914$. The entropy of the considered binary source: $H(\mathbf{p}) = 0.423$.

Arithmetic coding will reduce to approximately 43%. This is satisfactory, since the scheme $\mathbf{x} = \mathbf{y}^{(0)}$ is pictorially chaotic.

There remains a little uneasiness concerning the low precision of the reconstruction. We shall try to find a remedy by a treatment which takes into account the irregularities at the boundaries coming from the symmetric extensions of the considered row and column vectors.

More precisely, we shall try to maintain a "frame" of 2/3/4 boundary rows/columns where the quantization will be as clement as possible; this will lower the compression ratio, but will – hopefully – increase the pictorial faithfulness of the reconstructed scheme. For a brief information, let us look quickly at the variant where the *two* boundary rows/columns will be quantized by $N_0 = 2^8$. We obtain

$$
\mathbf{z}^{(0)} =
\begin{array}{cccccccccccccccc}
-0.01 & 1.504 & 1.423 & 1.308 & 1.192 & 1.1 & 1.008 & 0.905 & 0.802 & 0.7 & 0.598 & 0.496 & 0.394 & 0.294 & 0.195 & 0.098 \\
0.108 & 0.003 & 1.494 & 1.401 & 1.286 & 1.226 & 1.073 & 0.988 & 0.903 & 0.803 & 0.703 & 0.6 & 0.496 & 0.396 & 0.296 & 0.198 \\
0.227 & 0.093 & -0.037 & 1.488 & 1.376 & 1.351 & 1.138 & 1.071 & 1.005 & 0.907 & 0.809 & 0.704 & 0.598 & 0.497 & 0.396 & 0.298 \\
0.32 & 0.182 & 0.046 & 0.044 & 1.505 & 1.497 & 1.332 & 1.252 & 1.079 & 0.993 & 0.907 & 0.804 & 0.7 & 0.598 & 0.496 & 0.399 \\
0.414 & 0.303 & 0.192 & 0.069 & 0.008 & 1.486 & 1.339 & 1.340 & 1.153 & 1.079 & 1.005 & 0.904 & 0.803 & 0.7 & 0.597 & 0.499 \\
0.503 & 0.404 & 0.306 & 0.181 & 0.086 & 0.052 & 1.486 & 1.49 & 1.339 & 1.247 & 1.062 & 0.983 & 0.903 & 0.803 & 0.703 & 0.606 \\
0.591 & 0.506 & 0.421 & 0.324 & 0.227 & 0.086 & 0.008 & 1.485 & 1.336 & 1.321 & 1.119 & 1.061 & 1.003 & 0.906 & 0.810 & 0.712 \\
0.707 & 0.603 & 0.498 & 0.411 & 0.324 & 0.189 & 0.085 & 0.05 & 1.483 & 1.48 & 1.32 & 1.245 & 1.076 & 0.988 & 0.901 & 0.803 \\
0.823 & 0.699 & 0.576 & 0.499 & 0.422 & 0.323 & 0.225 & 0.083 & 0.004 & 1.482 & 1.334 & 1.335 & 1.149 & 1.070 & 0.992 & 0.894 \\
0.916 & 0.795 & 0.673 & 0.587 & 0.501 & 0.413 & 0.326 & 0.187 & 0.079 & 0.044 & 1.477 & 1.482 & 1.332 & 1.249 & 1.073 & 0.975 \\
1.009 & 0.890 & 0.771 & 0.676 & 0.581 & 0.503 & 0.426 & 0.322 & 0.218 & 0.074 & -0.006 & 1.473 & 1.327 & 1.334 & 1.154 & 1.056 \\
1.102 & 0.997 & 0.892 & 0.8 & 0.707 & 0.603 & 0.5 & 0.406 & 0.313 & 0.177 & 0.073 & 0.044 & 1.484 & 1.461 & 1.281 & 1.182 \\
1.194 & 1.104 & 1.014 & 0.923 & 0.833 & 0.703 & 0.573 & 0.491 & 0.408 & 0.311 & 0.214 & 0.084 & 0.016 & 1.529 & 1.416 & 1.316 \\
1.293 & 1.203 & 1.113 & 1.022 & 0.932 & 0.803 & 0.674 & 0.592 & 0.51 & 0.41 & 0.31 & 0.183 & 0.088 & 0.065 & 1.499 & 1.397 \\
1.393 & 1.302 & 1.212 & 1.122 & 1.031 & 0.903 & 0.775 & 0.693 & 0.612 & 0.509 & 0.405 & 0.314 & 0.223 & 0.1 & 0.016 & 1.52 \\
1.49 & 1.4 & 1.31 & 1.219 & 1.129 & 1.001 & 0.872 & 0.791 & 0.709 & 0.606 & 0.503 & 0.412 & 0.321 & 0.199 & 0.116 & 0.018 \\
\end{array}
$$

with $\max\{|\ y^{(0)}[i,j] - z^{(0)}[i,j]\ |\ 0 \le i,j \le 15\} = 0.097$ and $\|\ \mathbf{y}^{(0)} - \mathbf{z}^{(0)}\ \| = 0.4$. Are we satisfied with this result? We shall have a modest compression rate for a sensible loss of information: 17 positions (per 256) of the initial scheme will not be correctly restored if one rounds roughly to the first decimal position. But our test scheme is pictorially very "excited", and the 16×16 format is too small in order to permit an efficient taming of the digital reflux from the boundaries, due to symmetric extension, whilst aiming at the same time at a high compression ratio.

(d)

$$
\mathbf{z}^{(0)} =
\begin{array}{llllllllllllllll}
0.238 & 1.61 & 1.357 & 1.167 & 0.977 & 0.985 & 0.994 & 1.03 & 1.066 & 1.055 & 1.043 & 1.031 & 1.02 & 0.939 & 0.859 & 0.75 \\
0.172 & 0.011 & 1.467 & 1.334 & 1.216 & 1.154 & 1.099 & 1.032 & 0.964 & 0.924 & 0.884 & 0.852 & 0.819 & 0.739 & 0.659 & 0.55 \\
0.105 & 0.052 & -0.017 & 1.524 & 1.472 & 1.330 & 1.205 & 1.033 & 0.861 & 0.793 & 0.725 & 0.672 & 0.619 & 0.539 & 0.459 & 0.35 \\
0.09 & 0.106 & 0.115 & 0.09 & 1.699 & 1.477 & 1.28 & 1.038 & 0.805 & 0.693 & 0.581 & 0.5 & 0.419 & 0.339 & 0.259 & 0.149 \\
0.074 & 0.153 & 0.231 & 0.29 & 0.332 & 1.648 & 1.37 & 1.051 & 0.748 & 0.593 & 0.438 & 0.328 & 0.219 & 0.139 & 0.059 & -0.051 \\
0.203 & 0.236 & 0.27 & 0.354 & 0.43 & 0.142 & 1.486 & 1.205 & 0.948 & 0.818 & 0.695 & 0.577 & 0.458 & 0.382 & 0.306 & 0.196 \\
0.332 & 0.32 & 0.308 & 0.41 & 0.513 & 0.268 & 0.008 & 1.383 & 1.164 & 1.051 & 0.953 & 0.825 & 0.697 & 0.625 & 0.553 & 0.443 \\
0.461 & 0.461 & 0.461 & 0.518 & 0.575 & 0.374 & 0.165 & -0.064 & 1.339 & 1.248 & 1.181 & 1.078 & 0.982 & 0.907 & 0.831 & 0.722 \\
0.59 & 0.602 & 0.613 & 0.625 & 0.637 & 0.472 & 0.307 & 0.121 & -0.08 & 1.469 & 1.424 & 1.338 & 1.268 & 1.188 & 1.109 & 1 \\
0.766 & 0.745 & 0.724 & 0.703 & 0.682 & 0.507 & 0.332 & 0.208 & 0.076 & 0.038 & 1.633 & 1.603 & 1.596 & 1.508 & 1.429 & 1.319 \\
0.941 & 0.888 & 0.834 & 0.78 & 0.727 & 0.542 & 0.357 & 0.287 & 0.217 & 0.24 & 0.248 & 1.891 & 1.939 & 1.836 & 1.748 & 1.639 \\
1.117 & 1.031 & 0.944 & 0.858 & 0.771 & 0.634 & 0.497 & 0.417 & 0.337 & 0.421 & 0.498 & 0.554 & 2.242 & 2.125 & 2.031 & 1.926 \\
1.293 & 1.174 & 1.055 & 0.936 & 0.816 & 0.727 & 0.637 & 0.547 & 0.457 & 0.595 & 0.732 & 0.85 & 0.951 & 2.438 & 2.33 & 2.229 \\
1.373 & 1.254 & 1.135 & 1.016 & 0.896 & 0.822 & 0.748 & 0.674 & 0.6 & 0.676 & 0.753 & 0.919 & 1.078 & 0.959 & 2.473 & 2.383 \\
1.453 & 1.334 & 1.215 & 1.096 & 0.977 & 0.918 & 0.859 & 0.801 & 0.742 & 0.758 & 0.773 & 0.981 & 1.189 & 1.113 & 1.021 & 2.506 \\
1.563 & 1.443 & 1.324 & 1.205 & 1.086 & 1.027 & 0.969 & 0.91 & 0.852 & 0.867 & 0.883 & 1.091 & 1.299 & 1.219 & 1.123 & 1.014
\end{array}
$$

with $\max\{|\, y^{(0)}[i,j] - z^{(0)}[i,j]\,| \mid 0 \le i,j \le 15\} = 1.023$ and $\|\mathbf{y}^{(0)} - \mathbf{z}^{(0)}\| = 5.74$.]

Lifting Structure and Reversibility

The notion of *reversible transform* distinguishes appreciably JPEG 2000 from JPEG (which does not know it). Reversibility means *exact invertibility* in *integer* arithmetic. Thus a reversible transform maps vectors (matrices) with integer coefficients onto vectors (matrices) with integer coefficients, and so does its inverse transform. Clearly, this is a fundamental device for lossless compression. Unfortunately, this notion is a priori hostile to usual matrix computations (i.e. to "geometric" linear transformations). Now, the theory which discusses the *quality* of the transform candidates for (image) compression is wavelet transform theory. So, we will search for interesting reversible transforms in the setting of *non-linear approximations* for the wavelet transforms which are optimal – according to their theory.

The crucial question: How can we find a reversible approximation for a DWT?

The answer comes from a factorization trick into elementary transformations; it is the famous lifting structure.[23]

Lifting Structure

The theme is to decompose a two channel filter bank into appropriate elementary *round transforms*, thus creating a formal analogy to the DES scheme in cryptography.

$$x[n] \quad \equiv \text{ the input sequence}$$

Situation:

$$
\begin{aligned}
& y_0[n] \\
& \quad\quad \equiv \text{ the two } \begin{array}{l} \text{low-pass} \\ \text{high-pass} \end{array} \text{ sequences} \\
& y_1[n]
\end{aligned}
$$

We will get

$$
\begin{aligned}
& y_0[n] \\
& \quad\quad \text{after } L \text{ rounds of a lifting structure in the following way:} \\
& y_1[n]
\end{aligned}
$$

---

[23] Famous for insiders – the word *lifting* comes from a method of *design* for good filter banks.

(0):
$$y_0^{\{0\}}[n] = x[2n]$$
$$y_1^{\{0\}}[n] = x[2n+1]$$

(1): For l odd $1 \leq l \leq L$

$$y_0^{\{l\}}[n] = y_0^{\{l-1\}}[n]$$
$$y_1^{\{l\}}[n] = y_1^{\{l-1\}}[n] + \sum_i \lambda_l[i] y_0^{\{l-1\}}[n-i]$$

(2): For l even $1 \leq l \leq L$

$$y_0^{\{l\}}[n] = y_0^{\{l-1\}}[n] + \sum_i \lambda_l[i] y_1^{\{l-1\}}[n-i]$$
$$y_1^{\{l\}}[n] = y_1^{\{l-1\}}[n]$$

After L lifting rounds (or "lifting steps"), we obtain

$$y_0[n] = K_0 \cdot y_0^{\{L\}}[n]$$
$$y_1[n] = K_1 \cdot y_1^{\{L\}}[n]$$

where K_0 and K_1 are two appropriate multiplicative factors (the *gain factors* of the two subbands).

Attention: The lifting structure is specified by the L sequences $\lambda_1[n], \lambda_2[n], \ldots, \lambda_L[n]$, which will be each, in all interesting cases, reduced to merely *two* non-zero numbers.

Important Observations (1) A lifting structure is trivially invertible:
$$y_{1-p(l)}^{\{l-1\}}[n] = y_{1-p(l)}^{\{l\}}[n],$$
$$y_{p(l)}^{\{l-1\}}[n], = y_{p(l)}^{\{l\}}[n] - \sum_i \lambda_l[i] y_{1-p(l)}^{\{l\}}[n-i],$$
where $p(l) = l \bmod 2$ is the parity of l.
(2) The invertibility of the structure remains unchanged, if the linear operations defined by the convolution products with the various $\lambda_l[n]$, $1 \leq l \leq L$, are replaced by *arbitrary operators*.[24]

Example Consider the two-round lifting structure given by
$$\lambda_1[-1] = -\tfrac{1}{2}, \quad \lambda_1[0] = -\tfrac{1}{2},$$
$$\lambda_2[0] = \tfrac{1}{4}, \qquad \lambda_2[1] = \tfrac{1}{4}$$

---

[24] The situation is here *the same* as with the cipher DES: two channels, every individual round only affects one of the two channels – hence there is invertibility without specification of the added "mixing function" $f(R_{i-1}, K_i)$.

(Note that the situation is already typical for most lifting structures of broader interest: At the odd step, two non-zero values for $n = -1$ and $n = 0$, in the even case, two non-zero values for $n = 0$ and $n = 1$.)

Look now at the final result of the subband transform defined this way:

$$y_1^{\{2\}}[n] = y_1^{\{1\}}[n] \; = y_1^{\{0\}}[n] - \frac{1}{2}\left(y_0^{\{0\}}[n] + y_0^{\{0\}}[n+1]\right)$$

$$= x[2n+1] - \frac{1}{2}\left(x[2n] + x[2n+2]\right)$$

$$= 2 \cdot \left(-\frac{1}{4}x[2n] + \frac{1}{2}x[2n+1] - \frac{1}{4}x[2n+2]\right).$$

$$y_0^{\{2\}}[n] = y_0^{\{1\}}[n] + \frac{1}{4}\left(y_1^{\{1\}}[n] + y_1^{\{1\}}[n-1]\right)$$

$$= x[2n] + \frac{1}{4}\Big(x[2n+1]$$

$$-\frac{1}{2}(x[2n] + x[2n+2]) + x[2n-1] - \frac{1}{2}(x[2n-2] + x[2n])\Big)$$

$$= \frac{3}{4}x[2n] + \frac{1}{4}\left(x[2n+1] + x[2n-1]\right) - \frac{1}{8}\left(x[2n+2] + x[2n-2]\right).$$

We have obtained a factorization of the DWT 5/3 spline into a two-round lifting structure
(with $K_0 = 1$, $K_1 = \frac{1}{2}$).

Remark The DWT 9/7 CDF admits a 4-round lifting structure representation, where
$\lambda_1[-1] = \lambda_1[0] = -1.586134342,$
$\lambda_2[0] = \lambda_2[1] = -0.052980118,$
$\lambda_3[-1] = \lambda_3[0] = 0.882911075,$
$\lambda_4[0] = \lambda_4[1] = 0.443506852.$
With $K = 1.230174105$ we have: $K_0 = \frac{1}{K}$ $K_1 = \frac{K}{2}$.

Exercises

(1) Consider the perfect reconstruction filter bank given by

$\mathbf{h}_0^t = (h_0^t[-3], h_0^t[-2], h_0^t[-1], h_0^t[0], h_0^t[1], h_0^t[2], h_0^t[3])$
$= (-\frac{3}{280}, -\frac{3}{56}, \frac{73}{280}, \frac{17}{28}, \frac{73}{280}, -\frac{3}{56}, -\frac{3}{280})$

$\mathbf{h}_1^t = (h_1^t[-2], h_1^t[-1], h_1^t[0], h_1^t[1], h_1^t[2]) = (-\frac{1}{20}, -\frac{1}{4}, \frac{3}{5}, -\frac{1}{4}, -\frac{1}{20}).$

Inspired by the "balanced" type of factorization in the two examples above, we aim at a lifting structure representation of the following form:
$y_0^{\{0\}}[n] = x[2n] \; y_1^{\{0\}}[n] = x[2n+1],$
$y_0^{\{1\}}[n] = y_0^{\{0\}}[n],$
$y_1^{\{1\}}[n] = y_1^{\{0\}}[n] + \lambda_1 y_0^{\{0\}}[n] + \lambda_1 y_0^{\{0\}}[n+1],$
$y_0^{\{2\}}[n] = y_0^{\{1\}}[n] + \lambda_2 y_1^{\{1\}}[n-1] + \lambda_2 y_1^{\{1\}}[n],$
$y_1^{\{2\}}[n] = y_1^{\{1\}}[n],$
$y_0^{\{3\}}[n] = y_0^{\{2\}}[n], \; y_1^{\{3\}}[n] = y_1^{\{2\}}[n] + \lambda_3 y_0^{\{2\}}[n] + \lambda_3 y_0^{\{2\}}[n+1],$
$y_0^{\{4\}}[n] = y_0^{\{3\}}[n] + \lambda_4 y_1^{\{3\}}[n-1] + \lambda_4 y_1^{\{3\}}[n], \; y_1^{\{4\}}[n] = y_1^{\{3\}}[n],$
$y[2n] = K_0 \cdot y_0^{\{4\}}[n],$
$y[2n+1] = K_1 \cdot y_1^{\{4\}}[n].$

Show that our program is realizable, while computing $\lambda_1, \lambda_2, \lambda_3, \lambda_4$ and K_0, K_1.

[*Help*: Develop first $y_1^{\{4\}}[n] = y_1^{\{3\}}[n]$ in function of a "neighbourhood of $x[2n+1]$", this will give $\lambda_1, \lambda_2, \lambda_3$ and K_1, by comparison of coefficients. Then, determine λ_4 and K_0 while developing $y_0^{\{4\}}[n]$ in function of a "neighbourhood of $x[2n]$".

You should find $\lambda_1 = 0$, $\lambda_2 = \frac{1}{5}$, $\lambda_3 = -\frac{5}{14}$, $\lambda_4 = \frac{21}{100}$, $K_0 = \frac{5}{7}$, $K_1 = \frac{7}{10}$.

This gives a "false" four-round structure – the first round being trivial.]

(2) Now let us carry out a truncation of the lifting structure above: Consider only the composition of the three first rounds (we shall skip the last round).

(a) Find the analysis filter bank thus defined (we still keep $K_0 = \frac{5}{7}$ and $K_1 = \frac{7}{10}$).

$\Big[$Solution: $\mathbf{h}_0^t = (h_0^t[-1], h_0^t[0], h_0^t[1]) = \left(\frac{1}{7}, \frac{5}{7}, \frac{1}{7}\right)$

$\mathbf{h}_1^t = (h_1^t[-2], h_1^t[-1], h_1^t[0], h_1^t[1], h_1^t[2]) = \left(-\frac{1}{20}, -\frac{1}{4}, \frac{3}{5}, -\frac{1}{4}, -\frac{1}{20}\right)\Big]$

Verify that $\alpha = \frac{1}{2}$.[25]

(b) We have found a filter bank which admits the DWT 7/5 Burt as an "extension". Compare the effect (of the 2D version) of this filter bank with that of the DWT 5/3 spline and of the DWT 7/5 Burt on the two test schemes

$$\mathbf{x} = \mathbf{y}^{(0)} = \begin{matrix} 1\,1\,1\,1\,1\,1\,1\,1 \\ 2\,2\,2\,2\,2\,2\,2\,2 \\ 3\,3\,3\,3\,3\,3\,3\,3 \\ 4\,4\,4\,4\,4\,4\,4\,4 \\ 5\,5\,5\,5\,5\,5\,5\,5 \\ 6\,6\,6\,6\,6\,6\,6\,6 \\ 7\,7\,7\,7\,7\,7\,7\,7 \\ 8\,8\,8\,8\,8\,8\,8\,8 \end{matrix} \quad \text{and} \quad \mathbf{x} = \mathbf{y}^{(0)} = \begin{matrix} 0\,0\,0\,1\,1\,0\,0\,0 \\ 0\,0\,0\,2\,2\,0\,0\,0 \\ 0\,0\,0\,3\,3\,0\,0\,0 \\ 4\,4\,4\,4\,4\,4\,4\,4 \\ 5\,5\,5\,5\,5\,5\,5\,5 \\ 0\,0\,0\,6\,6\,0\,0\,0 \\ 0\,0\,0\,7\,7\,0\,0\,0 \\ 0\,0\,0\,8\,8\,0\,0\,0 \end{matrix}$$

Remark Our "amputated" filter bank is not really in the standards of the practitioner: the low-pass analysis filter is shorter than its partner. If we try to repair this deficiency by an exchange of the analysis filter bank with the synthesis filter bank, then it is the high-pass filter which will not vanish conveniently on constant vectors ...

At any rate, the correct viewpoint is the following: Consider the elementary round transforms as a *construction kit* for possible filter banks. Then the addition of an appropriate fourth round will produce the DWT 7/5 Burt from the rudimentary version above.

[The three variants of the first transformed scheme:

$$\mathbf{y}_{5/3} = \begin{matrix} 1\ 0\ 1\ 0\ 1\ 0\ 1\ 0 \\ 0\ 0\ 0\ 0\ 0\ 0\ 0\ 0 \\ 3\ 0\ 3\ 0\ 3\ 0\ 3\ 0 \\ 0\ 0\ 0\ 0\ 0\ 0\ 0\ 0 \\ 5\ 0\ 5\ 0\ 5\ 0\ 5\ 0 \\ 0\ 0\ 0\ 0\ 0\ 0\ 0\ 0 \\ 7.25\ 0\ 7.25\ 0\ 7.25\ 0\ 7.25\ 0 \\ 0.5\ 0\ 0.5\ 0\ 0.5\ 0\ 0.5\ 0 \end{matrix} \quad \mathbf{y}_{3/5} = \begin{matrix} 1.29\ 0\ 1.29\ 0\ 1.29\ 0\ 1.29\ 0 \\ -0.1\ 0\ -0.1\ 0\ -0.1\ 0\ -0.1\ 0 \\ 3\ 0\ 3\ 0\ 3\ 0\ 3\ 0 \\ 0\ 0\ 0\ 0\ 0\ 0\ 0\ 0 \\ 5\ 0\ 5\ 0\ 5\ 0\ 5\ 0 \\ 0\ 0\ 0\ 0\ 0\ 0\ 0\ 0 \\ 7\ 0\ 7\ 0\ 7\ 0\ 7\ 0 \\ 0.7\ 0\ 0.7\ 0\ 0.7\ 0\ 0.7\ 0 \end{matrix}$$

---

[25] *Recall*: α is the determinant of the analysis matrix A.

$$\mathbf{y}_{7/5} = \begin{array}{cccccccc}
1.24 & 0 & 1.24 & 0 & 1.24 & 0 & 1.24 & 0 \\
-0.1 & 0 & -0.1 & 0 & -0.1 & 0 & -0.1 & 0 \\
2.98 & 0 & 2.98 & 0 & 2.98 & 0 & 2.98 & 0 \\
0 & 0 & 0 & 0 & 0 & 0 & 0 & 0 \\
5 & 0 & 5 & 0 & 5 & 0 & 5 & 0 \\
0 & 0 & 0 & 0 & 0 & 0 & 0 & 0 \\
7.15 & 0 & 7.15 & 0 & 7.15 & 0 & 7.15 & 0 \\
0.7 & 0 & 0.7 & 0 & 0.7 & 0 & 0.7 & 0
\end{array}$$

In this example, our "reduced" filter bank produces a rather honourable result.

Now pass to the second test scheme.

$$\mathbf{y}_{5/3} = \begin{array}{cccccccc}
0 & 0 & 0.125 & 0.25 & 1 & -0.25 & -0.125 & 0 \\
0 & 0 & 0 & 0 & 0 & 0 & 0 & 0 \\
0.375 & 0 & 0.703 & 0.656 & 3 & -0.656 & 0.047 & 0 \\
0.75 & 0 & 0.656 & -0.188 & 0 & 0.188 & 0.844 & 0 \\
4.75 & 0 & 4.781 & 0.063 & 5 & -0.063 & 4.719 & 0 \\
-1.25 & 0 & -1.094 & 0.313 & 0 & -0.313 & -1.406 & 0 \\
-0.625 & 0 & 0.359 & 1.969 & 7.25 & -1.969 & -1.609 & 0 \\
0 & 0 & 0.063 & 0.125 & 0.5 & -0.125 & -0.63 & 0
\end{array}$$

$$\mathbf{y}_{3/5} = \begin{array}{cccccccc}
0 & -0.06 & 0.18 & 0.45 & 1.1 & -0.39 & 0 & 0 \\
-0.2 & -0.01 & -0.19 & 0.04 & -0.11 & -0.03 & -0.2 & 0 \\
0.57 & -0.12 & 0.92 & 0.85 & 2.65 & -0.73 & 0.57 & 0 \\
1.15 & 0.06 & 0.99 & -0.4 & 0.16 & 0.35 & 1.15 & 0 \\
4.14 & -0.04 & 4.27 & 0.3 & 4.88 & -0.26 & 4.14 & 0 \\
-1.45 & -0.07 & -1.24 & 0.51 & -0.21 & -0.44 & -1.45 & 0 \\
0 & -0.35 & 1 & 2.45 & 6 & -2.1 & 0 & 0 \\
0 & -0.04 & 0.1 & 0.25 & 0.6 & -0.21 & 0 & 0
\end{array}$$

$$\mathbf{y}_{7/5} = \begin{array}{cccccccc}
-0.1 & -0.07 & 0.19 & 0.47 & 1.07 & -0.4 & -0.17 & 0 \\
-0.2 & -0.01 & -0.18 & 0.04 & -0.11 & -0.03 & -0.21 & 0 \\
0.75 & -0.11 & 1.23 & 0.77 & 2.69 & -0.66 & 0.63 & 0 \\
1.16 & 0.06 & 0.91 & -0.4 & 0.15 & 0.35 & 1.22 & 0 \\
4.07 & -0.05 & 4.27 & 0.32 & 4.88 & -0.28 & 4.02 & 0 \\
-1.47 & -0.07 & -1.15 & 0.51 & -0.19 & -0.44 & -1.54 & 0 \\
-0.39 & -0.37 & 1.23 & 2.61 & 6.16 & -2.24 & -0.79 & 0 \\
-0.01 & -0.04 & 0.15 & 0.25 & 0.61 & -0.21 & -0.05 & 0
\end{array}$$

We note finally: Our two test examples do not give a clear argument for the superiority of the DWT 7/5 Burt over its truncated version.

But there remains altogether the question: Why do we search to obtain a proof in favour of the DWT 7/5 Burt relative to the filter bank which is its "prefix"? The first answer is simple: If the impulse responses become longer, less localized, then the treatment of the digital material should be more refined (which explains, by the way, the non-annihilation of certain positions with respect to the action of the DWT 5/3 spline). The second answer is of ideological nature: The DWT 7/5 Burt is a true Discrete Wavelet Transform; this is not the case for its truncated version (for a filter bank which realizes a (biorthogonal) DWT you have necessarily: $\sum(-1)^k h_0[k] = 0$ – which does not hold here). Now, the principal paradigm of our theory is the supremacy of the filter banks which realize a Discrete Wavelet Transform over the filter banks which do not. The next section shall be devoted to this subject. We shall see that the arguments of excellence primarily turn around the properties of the impulse response $\mathbf{g}_0^t$ – and it is not $\mathbf{g}_0^t$ which makes the difference between the DWT 7/5 Burt and its truncated version.⌉

(3) Now we exchange, in (1), the second and the fourth round (i.e. we have $\lambda_1 = 0, \lambda_2 = \frac{21}{100}, \lambda_3 = -\frac{5}{14}, \lambda_4 = \frac{1}{5}$)[26]

(a) Determine the corresponding analysis filter bank $\mathbf{h}^t = (\mathbf{h}_0^t, \mathbf{h}_1^t)$. Compute $\sum h_0^t[k], \sum(-1)^k h_0^t[k], \sum h_1^t[k], \sum(-1)^k h_1^t[k]$.

(b) Finally, exchange the analysis filter bank with the synthesis filter bank. Determine the "dual" impulse responses $\mathbf{h}^* = (\mathbf{h}_0^*, \mathbf{h}_1^*)$: You have to write $x[2n]$ in function of a "neighbourhood" of $y[2n]$ and $x[2n + 1]$ in function of a "neighbourhood" of $y[2n + 1]$.

Compute $\sum h_0^*[k], \sum(-1)^k h_0^*[k], \sum h_1^*[k], \sum(-1)^k h_1^*[k]$.

[(a) You should obtain (putting $K_0 = K_1 = 1$):

$$\mathbf{h}_0^t = \left(h_0^t[-3], h_0^t[-2], h_0^t[-1], h_0^t[0], h_0^t[1], h_0^t[2], h_0^t[3]\right)$$
$$= \left(-\frac{3}{200}, -\frac{1}{14}, \frac{73}{200}, \frac{6}{7}, \frac{73}{200}, -\frac{1}{14}, -\frac{3}{200}\right),$$

$$\mathbf{h}_1^t = (h_1^t[-2], h_1^t[-1], h_1^t[0], h_1^t[1], h_1^t[2]) = \left(-\frac{3}{40}, -\frac{5}{14}, \frac{17}{20}, -\frac{5}{14}, -\frac{3}{40}\right).$$

Note the *synthesis* character of the coefficients: Alternating type of the denominators according to alternating membership of $\mathbf{g}_0^t$ or of $\mathbf{g}_1^t$.

(b) The "dual" case looks much more natural:

$$\mathbf{h}_0^* = (h_0^*[-3], h_0^*[-2], h_0^*[-1], h_0^*[0], h_0^*[1], h_0^*[2], h_0^*[3])$$
$$= \left(\frac{3}{200}, -\frac{3}{40}, -\frac{73}{200}, \frac{17}{20}, -\frac{73}{200}, -\frac{3}{40}, \frac{3}{200}\right)$$
$$\mathbf{h}_1^* = (h_1^*[-2], h_1^*[-1], h_1^*[0], h_1^*[1], h_1^*[2])$$
$$= \left(-\frac{1}{14}, \frac{5}{14}, \frac{6}{7}, \frac{5}{14}, -\frac{1}{14}\right).]$$

Reversibility via Lifting Structures

Recall: The invertibility of the lifting scheme does *not* depend on the linearity of the convolution term in the expression

$$y_{p(l)}^{\{l\}}[n] = y_{p(l)}^{\{l-1\}}[n] + \sum_i \lambda_l[i] y_{1-p(l)}^{\{l-1\}}[n - i].$$

Hence, the scheme will remain invertible when replacing all the convolutional terms by non-linear approximations:

$$y_{p(l)}^{\{l\}}[n] = y_{p(l)}^{\{l-1\}}[n] + \lfloor \frac{1}{2} + \sum_i \lambda_l[i] y_{1-p(l)}^{\{l-1\}}[n - i] \rfloor.$$

(where $\lfloor \cdot \rfloor$ means "integer floor"[27]).

---

[26] The synthesis filter bank of the DWT 7/5 Burt is given, in lifting structure, by $\lambda_1 = -\frac{21}{100}, \lambda_2 = \frac{5}{14}, \lambda_3 = -\frac{1}{5}, \lambda_4 = 0$. There should be some similarities.

[27] $\lfloor x \rfloor = \max\{n \in \mathbb{Z} : n \leq x\}$

By this clever trick – representation as a lifting structure and reversible approximation of the round transforms – every Discrete Wavelet Transform admits a (non-linear) reversible variant.

Remark We have decided to use the rounding function $r(x) = \lfloor \frac{1}{2} + x \rfloor$.

It differs from the usual rounding function (which is an odd function) on the arguments of the form $-(n+\frac{1}{2})$, $n \in \mathbb{N}$: The usual round of $x = -3.5$ is -4, whereas $r(-3.5) = -3$.

Exercises

(1) Let us begin with the reversible version of the DWT 5/3 spline. Show the following formulas:

(a) In analysis:

$$y[2n + 1] = x[2n + 1] - \lfloor \frac{1}{2}(x[2n] + x[2n + 2]) \rfloor$$

Disposing of these values, one computes

$$y[2n] = x[2n] + \lfloor \frac{1}{2} + \frac{1}{4}(y[2n - 1] + y[2n + 1]) \rfloor$$

(b) In synthesis:

$$x[2n] = y[2n] - \lfloor \frac{1}{2} + \frac{1}{4}(y[2n - 1] + y[2n + 1]) \rfloor$$

Then, as in analysis
$$x[2n + 1] = y[2n + 1] + \lfloor \frac{1}{2}(x[2n] + x[2n + 2]) \rfloor$$

(you have to justify the simplified expression for the floor term in high-pass).

(2) Now consider the gradation

$$
\begin{array}{cccccccc}
30 & 30 & 30 & 30 & 30 & 30 & 30 & 30 \\
60 & 60 & 60 & 60 & 60 & 60 & 60 & 60 \\
90 & 90 & 90 & 90 & 90 & 90 & 90 & 90 \\
120 & 120 & 120 & 120 & 120 & 120 & 120 & 120 \\
150 & 150 & 150 & 150 & 150 & 150 & 150 & 150 \\
180 & 180 & 180 & 180 & 180 & 180 & 180 & 180 \\
210 & 210 & 210 & 210 & 210 & 210 & 210 & 210 \\
240 & 240 & 240 & 240 & 240 & 240 & 240 & 240
\end{array}
$$

and its transformed scheme – by the DWT 5/3 spline, the DWT 7/5 Burt and the DWT 9/7 CDF:

$$
\mathbf{y}_{5/3} =
\begin{array}{cccccccc}
30 & 0 & 30 & 0 & 30 & 0 & 30 & 0 \\
0 & 0 & 0 & 0 & 0 & 0 & 0 & 0 \\
90 & 0 & 90 & 0 & 90 & 0 & 90 & 0 \\
0 & 0 & 0 & 0 & 0 & 0 & 0 & 0 \\
150 & 0 & 150 & 0 & 150 & 0 & 150 & 0 \\
0 & 0 & 0 & 0 & 0 & 0 & 0 & 0 \\
218 & 0 & 218 & 0 & 218 & 0 & 218 & 0 \\
15 & 0 & 15 & 0 & 15 & 0 & 15 & 0
\end{array}
\qquad
\mathbf{y}_{7/5} =
\begin{array}{cccccccc}
37 & 0 & 37 & 0 & 37 & 0 & 37 & 0 \\
-3 & 0 & -3 & 0 & -3 & 0 & -3 & 0 \\
89 & 0 & 89 & 0 & 89 & 0 & 89 & 0 \\
0 & 0 & 0 & 0 & 0 & 0 & 0 & 0 \\
150 & 0 & 150 & 0 & 150 & 0 & 150 & 0 \\
0 & 0 & 0 & 0 & 0 & 0 & 0 & 0 \\
215 & 0 & 215 & 0 & 215 & 0 & 215 & 0 \\
21 & 0 & 21 & 0 & 21 & 0 & 21 & 0
\end{array}
$$

$$\mathbf{y}_{9/7} = \begin{array}{llllllll} 40 & 0 & 40 & 0 & 40 & 0 & 40 & 0 \\ 4 & 0 & 4 & 0 & 4 & 0 & 4 & 0 \\ 92 & 0 & 92 & 0 & 92 & 0 & 92 & 0 \\ 0 & 0 & 0 & 0 & 0 & 0 & 0 & 0 \\ 148 & 0 & 148 & 0 & 148 & 0 & 148 & 0 \\ -3 & 0 & -3 & 0 & -3 & 0 & -3 & 0 \\ 212 & 0 & 212 & 0 & 212 & 0 & 212 & 0 \\ 13 & 0 & 13 & 0 & 13 & 0 & 13 & 0 \end{array}$$

(we have rounded the coefficients of the three results to integers, according to the spirit of the section...)

(a) Carry out the synthesis transforms on the three schemes above, then round the coefficients of the results to integers. Compare with the initial gradation.

(b) Compute, in each of the three cases, the reversible version.

[(a) In case of the DWT 9/7 CDF, we recover precisely the initial gradation; in case of the other two transforms, we obtain the same deviation on the two last rows: 210 will change to 211, and 240 will change to 241.

(b) The three transformed schemes in reversible mode[28]:

$$\tilde{\mathbf{y}}_{5/3} = \begin{array}{llllllll} 30 & 0 & 30 & 0 & 30 & 0 & 30 & 0 \\ 0 & 0 & 0 & 0 & 0 & 0 & 0 & 0 \\ 90 & 0 & 90 & 0 & 90 & 0 & 90 & 0 \\ 0 & 0 & 0 & 0 & 0 & 0 & 0 & 0 \\ 150 & 0 & 150 & 0 & 150 & 0 & 150 & 0 \\ 0 & 0 & 0 & 0 & 0 & 0 & 0 & 0 \\ 218 & 0 & 218 & 0 & 218 & 0 & 218 & 0 \\ 30 & 0 & 30 & 0 & 30 & 0 & 30 & 0 \end{array}$$

$$\tilde{\mathbf{y}}_{7/5} = \begin{array}{llllllll} 73 & 0 & 73 & 0 & 73 & 0 & 73 & 0 \\ -6 & 0 & -6 & 0 & -6 & 0 & -6 & 0 \\ 175 & 0 & 175 & 0 & 175 & 0 & 175 & 0 \\ 0 & 0 & 0 & 0 & 0 & 0 & 0 & 0 \\ 294 & 0 & 294 & 0 & 294 & 0 & 294 & 0 \\ 0 & 0 & 0 & 0 & 0 & 0 & 0 & 0 \\ 420 & 0 & 420 & 0 & 420 & 0 & 420 & 0 \\ 42 & 0 & 42 & 0 & 42 & 0 & 42 & 0 \end{array}$$

$$\tilde{\mathbf{y}}_{9/7} = \begin{array}{llllllll} 62 & 0 & 62 & 0 & 62 & 0 & 62 & 0 \\ 10 & 1 & 10 & 1 & 10 & 1 & 10 & 1 \\ 139 & -1 & 139 & -1 & 139 & -1 & 139 & -1 \\ 0 & 0 & 0 & 0 & 0 & 0 & 0 & 0 \\ 224 & -1 & 224 & -1 & 224 & -1 & 224 & -1 \\ -5 & 0 & -5 & 0 & -5 & 0 & -5 & 0 \\ 321 & 0 & 321 & 0 & 321 & 0 & 321 & 0 \\ 26 & 0 & 26 & 0 & 26 & 0 & 26 & 0 \end{array} \Big]$$

(3) In the situation of the exercise (2), replace the considered gradation by the following "gradation cross on a black background":

$$\mathbf{x} = \begin{array}{llllllll} 0 & 0 & 0 & 30 & 30 & 0 & 0 & 0 \\ 0 & 0 & 0 & 60 & 60 & 0 & 0 & 0 \\ 0 & 0 & 0 & 90 & 90 & 0 & 0 & 0 \\ 120 & 120 & 120 & 120 & 120 & 120 & 120 & 120 \\ 150 & 150 & 150 & 150 & 150 & 150 & 150 & 150 \\ 0 & 0 & 0 & 180 & 180 & 0 & 0 & 0 \\ 0 & 0 & 0 & 210 & 210 & 0 & 0 & 0 \\ 0 & 0 & 0 & 240 & 240 & 0 & 0 & 0 \end{array}$$

---

[28] Caution: We shall put everywhere $K_0 = K_1 = 1$ – this explains the surprising magnitudes of the coefficients!

(a) Compute, for each of our three standard transforms, the integer round $\breve{y}$ of the analysis transform y of x, then the integer round $\breve{z}$ of the synthesis transform z of $\breve{y}$.

(b) Compute the three reversible versions.

[(a) First the analysis transforms (with the usual rounds):

$$\breve{y}_{5/3} = \begin{array}{rrrrrrrr} 0 & 0 & 4 & 8 & 30 & -8 & -4 & 0 \\ 0 & 0 & 0 & 0 & 0 & 0 & 0 & 0 \\ 11 & 0 & 21 & 20 & 90 & -20 & 1 & 0 \\ 23 & 0 & 20 & -6 & 0 & 6 & 25 & 0 \\ 143 & 0 & 143 & 2 & 150 & -2 & 142 & 0 \\ -38 & 0 & -33 & 9 & 0 & -9 & -42 & 0 \\ -19 & 0 & 11 & 59 & 218 & -59 & -48 & 0 \\ 0 & 0 & 2 & 4 & 15 & -4 & -2 & 0 \end{array} \qquad \breve{y}_{7/5} = \begin{array}{rrrrrrrr} -3 & -2 & 6 & 14 & 32 & -12 & -5 & 0 \\ -6 & 0 & -5 & 1 & -3 & -1 & -6 & 0 \\ 22 & -3 & 37 & 23 & 81 & -20 & 19 & 0 \\ 35 & 2 & 27 & -12 & 5 & 10 & 37 & 0 \\ 122 & -1 & 128 & 10 & 146 & -8 & 121 & 0 \\ -44 & -2 & -34 & 15 & -6 & -13 & -46 & 0 \\ -14 & -11 & 37 & 78 & 185 & -67 & -24 & 0 \\ 0 & -1 & 4 & 7 & 18 & -6 & -1 & 0 \end{array}$$

$$\breve{y}_{9/7} = \begin{array}{rrrrrrrr} 5 & 1 & 11 & 9 & 35 & -12 & 2 & 3 \\ 3 & 0 & 3 & 0 & 4 & 0 & 3 & 0 \\ 22 & 1 & 34 & 19 & 83 & -23 & 15 & 7 \\ 22 & 0 & 18 & -6 & 3 & 7 & 24 & -2 \\ 123 & 0 & 127 & 7 & 145 & -8 & 121 & 2 \\ -47 & 1 & -39 & 12 & -9 & -15 & -51 & 4 \\ -5 & 4 & 32 & 58 & 183 & -72 & -25 & 20 \\ 14 & 0 & 14 & 0 & 13 & 0 & 14 & 0 \end{array}$$

Now the rounded schemes of the synthesis transforms:

$$\breve{z}_{5/3} = \begin{array}{rrrrrrrr} 0 & 0 & 0 & 31 & 30 & -1 & 0 & 0 \\ 0 & 0 & 0 & 61 & 60 & -1 & 0 & 0 \\ -1 & -1 & -1 & 91 & 90 & -1 & 0 & 0 \\ 121 & 121 & 120 & 120 & 120 & 120 & 119 & 119 \\ 151 & 150 & 149 & 151 & 150 & 149 & 151 & 151 \\ -1 & -1 & 0 & 179 & 180 & 1 & 1 & 1 \\ 0 & 0 & 0 & 210 & 211 & 0 & 0 & 0 \\ 0 & 0 & 0 & 241 & 241 & -1 & 0 & 0 \end{array} \qquad \breve{z}_{7/5} = \begin{array}{rrrrrrrr} 1 & 0 & 0 & 30 & 29 & 0 & 0 & 0 \\ 0 & 1 & 1 & 60 & 61 & 0 & 0 & 0 \\ 0 & 0 & 0 & 90 & 90 & 0 & 0 & 0 \\ 119 & 120 & 119 & 120 & 121 & 120 & 121 & 121 \\ 150 & 150 & 149 & 150 & 149 & 150 & 150 & 150 \\ 0 & 1 & 1 & 180 & 179 & 0 & 0 & 0 \\ 0 & 0 & 0 & 210 & 211 & 0 & -1 & -1 \\ 1 & 0 & 0 & 239 & 240 & 1 & 0 & 0 \end{array}$$

$$\breve{z}_{9/7} = \begin{array}{rrrrrrrr} 0 & 1 & 1 & 29 & 30 & -1 & 1 & 1 \\ 0 & 0 & 0 & 60 & 60 & 0 & 0 & 0 \\ 1 & 0 & 1 & 91 & 90 & 0 & -1 & 0 \\ 120 & 120 & 119 & 120 & 120 & 120 & 120 & 120 \\ 151 & 149 & 150 & 150 & 150 & 151 & 150 & 150 \\ 0 & 0 & 0 & 180 & 179 & -1 & 0 & 0 \\ 0 & 0 & 0 & 209 & 210 & 0 & 0 & -1 \\ 0 & 1 & 0 & 240 & 240 & 0 & 0 & 0 \end{array}$$

In order to duly appreciate the numerical values of the reversible versions to follow, recall that they differ from the numerical values of the standard versions (very, very) approximately by a factor $\frac{1}{K_0^2}$ in subband LL, by a factor $\frac{1}{K_0 \cdot K_1}$ in subbands LH and HL, and by a factor $\frac{1}{K_1^2}$ in subband HH. This makes roughly $(1, 2, 4)$ for the DWT 5/3 spline, $(2, 2, 2)$ for the DWT 7/5 Burt, and $(\frac{3}{2}, 2, \frac{8}{3})$ for the DWT 9/7 CDF.

(b) The three reversible versions:

$$\tilde{y}_{5/3} = \begin{matrix} 0 & 0 & 4 & 15 & 30 & -15 & -4 & 0 \\ 0 & 0 & 0 & 0 & 0 & 0 & 0 & 0 \\ 11 & 0 & 21 & 39 & 90 & -40 & 1 & 0 \\ 45 & 0 & 39 & -23 & 0 & 22 & 51 & 0 \\ 143 & 0 & 144 & 3 & 150 & -4 & 142 & 0 \\ -75 & 0 & -66 & 37 & 0 & -38 & -84 & 0 \\ -19 & 0 & 11 & 118 & 218 & -119 & -49 & 0 \\ 0 & 0 & 4 & 15 & 30 & -15 & -4 & 0 \end{matrix}$$

$$\tilde{y}_{7/5} = \begin{matrix} -8 & -4 & 11 & 28 & 63 & -24 & -11 & 0 \\ -13 & 0 & -11 & 3 & -7 & -2 & -13 & 0 \\ 42 & -7 & 71 & 47 & 157 & -40 & 37 & 0 \\ 70 & 3 & 54 & -25 & 9 & 21 & 73 & 0 \\ 238 & -3 & 251 & 20 & 287 & -17 & 235 & 0 \\ -89 & -5 & -69 & 31 & -11 & -27 & -93 & 0 \\ -27 & -22 & 72 & 157 & 362 & -134 & -46 & 0 \\ -1 & -2 & 9 & 15 & 36 & -13 & -3 & 0 \end{matrix}$$

$$\tilde{y}_{9/7} = \begin{matrix} 6 & 1 & 15 & 19 & 53 & -25 & 1 & 7 \\ 6 & 0 & 6 & 0 & 7 & -2 & 5 & 0 \\ 34 & 3 & 51 & 38 & 126 & -47 & 23 & 13 \\ 44 & -1 & 36 & -17 & 5 & 19 & 48 & -6 \\ 187 & 1 & 193 & 13 & 219 & -18 & 182 & 4 \\ -93 & 2 & -77 & 31 & -17 & -39 & -102 & 10 \\ -8 & 7 & 50 & 116 & 278 & -144 & -37 & 40 \\ 27 & -1 & 27 & -1 & 26 & 1 & 28 & -1 \end{matrix}$$

(4) Consider a "natural" reversible version of the DWT 5/3 spline which is obtained as follows:

First, we note: If T is the matrix of the analysis transform (of arbitrary size) of the DWT 5/3 spline, then $S = 4T$ is a matrix with *integer* coefficients. Hence, if X is an integer 2D scheme, then $Y = TXT^{\mathrm{t}} = \frac{1}{16}SXS^{\mathrm{t}}$ becomes integer if one multiplies all coefficients of Y by 16. This gives the following reversible algorithm:

Let X be an integer matrix.

(a) Transform, by the usual DWT 5/3 spline, the scheme $X' = 16X$. $Y' = TX'T^{\mathrm{t}}$ will be an integer matrix.

(b) Backwards, transform, in usual synthesis, Y' into X', then divide by 16; we obtain $X = \frac{1}{16}X'$.

Why will this mathematically correct version not be accepted by the practitioner?

(5) Do there exist integer 8×8 schemes which are *fixed points* for the 2D reversible version:

(a) of the DWT 5/3 spline
(b) of the DWT 7/5 Burt
(c) of the DWT 9/7 CDF

[*Help*: (b) You should search (and you will find) a fixed point in the integer multiples of the matrix

$$X = \begin{matrix} 1 & 0 & 1 & 0 & 1 & 0 & 1 & 0 \\ 0 & 1 & 0 & 1 & 0 & 1 & 0 & 1 \\ 1 & 0 & 1 & 0 & 1 & 0 & 1 & 0 \\ 0 & 1 & 0 & 1 & 0 & 1 & 0 & 1 \\ 1 & 0 & 1 & 0 & 1 & 0 & 1 & 0 \\ 0 & 1 & 0 & 1 & 0 & 1 & 0 & 1 \\ 1 & 0 & 1 & 0 & 1 & 0 & 1 & 0 \\ 0 & 1 & 0 & 1 & 0 & 1 & 0 & 1 \end{matrix}$$

(a) and (c) Here, the answer is probably "no" – the author counts on the mathematical cleverness of his reader.]

5.3.2 The Discrete Wavelet Transform

Until now, our exposition has given priority to two or three particular filter banks, especially the DWT 5/3 spline. Our mathematical curiosity should have become excited by the notion of a *Discrete Wavelet Transform*.

There seems to be a continuous world – that of the wavelets – which has the authority to declare certain filter banks as derived objects.

Our next objective will thus be to explain why a filter bank which realizes a Discrete Wavelet Transform is particularly interesting for the practitioner.

First, we have to introduce the wavelets.

At level zero, we have adopted an intuitive and simplifying viewpoint: *the* sin *c function is a wavelet – look at its line – and everything similar should be a wavelet.*

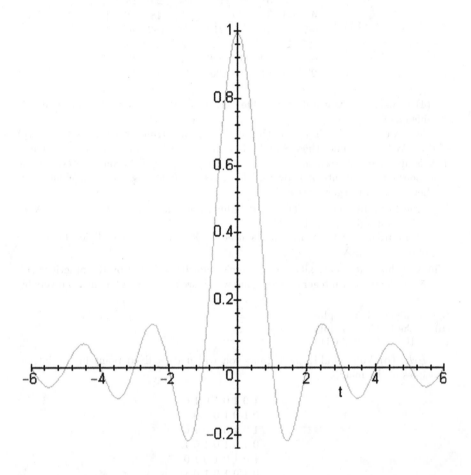

At level one, we shall insist on discretization (i.e. on coordinates). The Whittaker-Shannon theorem, with variable frequential window, introduces the translates of the dilated sin *c* function (according to the widening of the frequential window) as an orthonormal basis "in packets" of the $\mathbf{L}^2(\mathbb{R})$, allowing a more and more faithful coordinatization (by opening of the frequential horizon).

And this will also be the general situation: a *scaling function*, dominating a filtration structure, accompanied by a *mother wavelet*, which dominates the associated graded structure, will establish the setting for a more and more faithful coordinate refinement. We are in the situation of a generalized Shannon theorem: the varying frequential window will be implicitly present through the operation of (dyadic) dilation.

Now consider two successive layers of resolution (of approximation):

For $x = x(t)$ (of finite energy), its approximation $x^{(m-1)}$ at scale 2^{m-1} and its approximation $x^{(m)}$ at (wider) scale 2^m will be connected by $x^{(m-1)} = x^{(m)} + d^{(m-1)}$ (interaction of the filtration and of the grading) where the complementary term $d^{(m-1)}$ describes the "*details to the order $m-1$*". This equality – which is an algebraic banality – can be read in two directions: $x^{(m-1)} \longmapsto (x^{(m)}, d^{(m-1)})$ and $(x^{(m)}, d^{(m-1)}) \longmapsto x^{(m-1)}$.

If we now consider these operations *on the coordinates*, we actually obtain a two channel filter bank, which *does not depend on m*. This is *the Discrete Wavelet Transform* associated with a multi-resolution analysis.

The orthonormal wavelet bases (for example the basis generated by the sin c function) are mathematically too perfect in order to be really adapted to finer practical needs. So, we have to look for some compromise. Fortunately, when passing to biorthogonal wavelet bases (one cleverly splits the approximation scene into analysis filtration and into synthesis filtration) one resolves all problems which the practitioner could pose for the theoretician.

At this point, we no longer can avoid the question: Why are we interested making sure that a given filter bank is actually a Discrete Wavelet Transform?

The answer is the following:

It is the degree of *regularity* of the (synthesis) mother wavelet which controls the quality of the DWT associated with the considered biorthogonal wavelet bases. On the one hand, in analysis, we obtain the expected trivialization of very regular numerical data (they will produce no details). On the other hand, in synthesis, we obtain a neatly localized propagation of local errors.

A filter bank which is a DWT thus allows *a qualitative evaluation a priori*. Our presentation will remain essentially temporal (or rather spatial). Only for the design criteria towards "good" Discrete Wavelet Transforms we are obliged to adopt a spectral viewpoint (which is, at any rate, more technical than conceptual).

Recall   Our temporal (spatial) world will always be *real*.

Multi-Resolution Analysis and Wavelet Bases

Our formal setting will be the Hilbert space $\mathbf{L}^2(\mathbb{R})$ of (classes of) square integrable functions on $\mathbb{R}$.

Recall: The Whittaker–Shannon Theorem

Recall the principal result of our third chapter (on sampling and reconstruction of signals):

A square integrable bandlimited (hence essentially smooth) function is determined by an appropriate sequence of its values (i.e. by an equidistant sampling which is "sufficiently fine"). More precisely:

Let $T > 0$, and let $h_T(t) = \sin c(\frac{t}{T})$ $[= \frac{\sin(\pi t/T)}{\pi t/T}$, with continuous prolongation at $0]$.

Visualize: The smaller is T, the more the renormalized $\sin c$ function is "nervous"; the larger is T, the more it is "flattened".

Let $x = x(t)$ be a function which is square integrable on $\mathbb{R}$. Assume that $\hat{x}(\omega) = 0$ outside $[-\frac{\pi}{T}, \frac{\pi}{T}]$. Then $x = x(t)$ is essentially smooth (i.e. equivalent to an infinitely differentiable $x_1 = x_1(t)$). Let us suppose that $x = x(t)$ is regular from the beginning, then we have:

$$x(t) = \sum_{k \in \mathbb{Z}} x(k \cdot T) h_T(t - k \cdot T)$$

[Note that here we have a convergence in quadratic mean. But the Shannon interpolator is the (inverse) Fourier transform of a function with bounded support, hence infinitely differentiable. Our hypothesis allows an identification as *functions*.]

We insist a little bit on the ingredients of the interpolation formula: if

$$x(t) = \overline{\mathbb{F}\mathbb{F}}[x(t)],$$

then $x(k \cdot T) = \frac{1}{T}\langle x(t), h_T(t - k \cdot T)\rangle = \frac{1}{T}(x * h_T)(k \cdot T)$ [Concerning the notation: t is here an integration variable, hence logically bound ("annihilated" by $\langle,\rangle$).]

Remark For fixed $T > 0$, let $\mathbf{V}_T = \{x \in \mathbf{L}^2(\mathbb{R}) : \hat{x}(\omega) = 0 \text{ outside } [-\frac{\pi}{T}, \frac{\pi}{T}]\}$.

The *orthogonal projection* of $x \in \mathbf{L}^2(\mathbb{R})$ on $\mathbf{V}_T$ is

$$x_T = \frac{1}{T}x * h_T$$

The argument:

Let $x_T \in \mathbf{V}_T$ be the nearest element to x. Then we have, by the Plancherel formula:

$$\|x - x_T\|^2 = \frac{1}{2\pi}\int |\hat{x}(\omega) - \hat{x}_T(\omega)|^2 \, d\omega$$

$$= \frac{1}{2\pi}\int_{|\omega|>\pi/T} |\hat{x}(\omega)|^2 \, d\omega + \frac{1}{2\pi}\int_{|\omega|\leq\pi/T} |\hat{x}(\omega) - \hat{x}_T(\omega)|^2 \, d\omega.$$

This distance is minimum in case $\hat{x}_T(\omega) = \hat{x}(\omega)\mathbb{I}_{[-\frac{\pi}{T}, \frac{\pi}{T}]}(\omega) = \frac{1}{T}\hat{x}(\omega)\hat{h}_T(\omega)$. [Note that the orthogonal projection will carry out *a smoothing*: x_T is infinitely differentiable (since h_T is).]

We make $T > 0$ vary appropriately, and obtain a *filtration* of $\mathbf{L}^2(\mathbb{R})$:

Put $\mathbf{V}^{(m)} = \{x \in \mathbf{L}^2(\mathbb{R}) : \hat{x}(\omega) = 0 \text{ for } |\omega| > \frac{1}{2^m}\pi\}$. We get a chain of closed subspaces of $\mathbf{L}^2(\mathbb{R})$:

$$\cdots \subset \mathbf{V}^{(2)} \subset \mathbf{V}^{(1)} \subset \mathbf{V}^{(0)} \subset \mathbf{V}^{(-1)} \subset \mathbf{V}^{(-2)} \subset \cdots$$

For arbitrary $x \in \mathbf{L}^2(\mathbb{R})$, $m \in \mathbb{Z}$, let $x^{(m)} \in \mathbf{V}^{(m)}$ be the orthogonal projection of x onto $\mathbf{V}^{(m)}$. According to the preceding remark, we have $\|x^{(m)} - x\|^2 = \frac{1}{2\pi} \int_{|\omega| > 2^{-m}\pi} |\hat{x}(\omega)|^2 \, d\omega$ i.e. $\lim_{m \to -\infty} \|x^{(m)} - x\| = 0$. In other words

$$\overline{\bigcup_{m \in \mathbb{Z}} \mathbf{V}^{(m)}} = \mathbf{L}^2(\mathbb{R}) \quad (1)$$

On the other hand, we have $\|x^{(m)}\|^2 = \frac{1}{2\pi} \int_{|\omega| \leq 2^{-m}\pi} |\hat{x}(\omega)|^2 \, d\omega < 2^{-m} \cdot const.$ Hence $\lim_{m \to \infty} \|x^{(m)}\| = 0$. This means that

$$\bigcap_{m \in \mathbb{Z}} \mathbf{V}^{(m)} = 0 \quad (2)$$

The filtration is exhaustive (upward and downward). Now, let us look more closely at scaling operations. Since $\mathbb{F}[x(2^{-m}t)] = 2^m \hat{x}(2^m\omega)$, we have:

$$x = x(t) \in \mathbf{V}^{(0)} \Longleftrightarrow_m x = x(2^{-m}t) \in \mathbf{V}^{(m)} \quad (3)$$

Since $\mathbb{F}[x(t - k)] = e^{-ik\omega}\hat{x}(\omega)$, we have:

$$x = x(t) \in \mathbf{V}^{(0)} \Longleftrightarrow \begin{array}{l} x_k = x(t - k) \in \mathbf{V}^{(0)} \\ \text{for } k \in \mathbb{Z} \end{array} \quad (4)$$

Finally, consider the function[29] $\varphi(t) = \sin c(t)$. We have already seen:

Let $\varphi_k(t) = \varphi(t - k)$, then $(\varphi_k(t))_{k \in \mathbb{Z}}$ is an orthonormal basis of $\mathbf{V}^{(0)}$. $\quad (5)$

Consequence (of (3) and (5)) Put $\varphi_k^{(m)}(t) = \sqrt{2^{-m}}\varphi(2^{-m}t - k)$. Then $(\varphi_k^{(m)}(t))_{k \in \mathbb{Z}}$ is an orthonormal basis of $\mathbf{V}^{(m)}$ (in the usual Hilbertian sense). More explicitly (we deal with the Wittaker–Shannon theorem applied to the orthogonal projection onto $\mathbf{V}^{(m)}$): for $x = x(t) \in \mathbf{L}^2(\mathbb{R})$ we have:

$$x^{(m)}(t) = \sum_{k \in \mathbb{Z}} y_0^{(m)}[k]\varphi_k^{(m)}(t),$$

with

$$y_0^{(m)}[k] = \langle x(t), \varphi_k^{(m)}(t) \rangle = (x * \varphi_0^{(m)})(2^m k).$$

Remark How can we interpret this "scaled" version of the Whittaker–Shannon theorem?

---

[29] The change of notation $h(t) \mapsto \varphi(t)$ indicates an important change of viewpoint: an "impulse response" promotes to a "scaling function".

Accept, as a starting point, the following paradigm.

In spectral analysis of a signal (i.e. in its frequential representation) "the crude information is in the low frequencies", whereas "the fine information – the information on the details – is in the high frequencies" (for the intuitive: think of the left hand and of the right hand of a classical pianist).

We insist: The higher the frequency, the finer are the supported details. If we adopt this information hierarchy, we can perceive the approximation of $x = x(t)$ by $x^{(m)} = x^{(m)}(t)$ as a "level m resolution", which substitutes for the initial signal while neglecting the "finer details".

Let us be more precise.

On the frequential side, our resolution horizon is given, in a very suggestive manner, by a "truncation window": it is a real horizon.

On the temporal side, the resolution screen acts via "digital density": it is characterized by an equidistant sampling adapted to the demanded precision of details.

Hence, the higher (finer) is the resolution, the larger is the frequential window (the horizon), the denser is the sampling (i.e. the smaller is the sampling step).

Some *conventions of language*: In the version of the Whittaker–Shannon theorem, as cited above, the value T for the sampling step will be called *the scale*, and the reciprocal value $\frac{1}{T}$ *the resolution*.

Hence:

Small scale = high resolution
Large scale = low resolution

In our enumeration of the resolution levels, we have associated with the resolution level m the values

$2^m \equiv$ the scaling factor,

$2^{-m} \equiv$ the resolution (value).

Note finally the fundamental role of the function $\varphi(t) = \sin c(t)$ as a *coordinate generator*, by means of its translated-dilated variants, as indicated by the Whittaker–Shannon theorem.

Now we can generalize.

Multi-Resolution Analyses of $\mathbf{L}^2(\mathbb{R})$

Definition (S. Mallat) A multi-resolution analysis of $\mathbf{L}^2(\mathbb{R})$ is given by a filtration (of a chain of closed subspaces) of $\mathbf{L}^2(\mathbb{R})$

$$\cdots \subset \mathbf{V}^{(2)} \subset \mathbf{V}^{(1)} \subset \mathbf{V}^{(0)} \subset \mathbf{V}^{(-1)} \subset \mathbf{V}^{(-2)} \subset \cdots$$

with the following properties:

(MR-1) $\overline{\bigcup_{m\in\mathbb{Z}} \mathbf{V}^{(m)}} = \mathbf{L}^2(\mathbb{R})$

(MR-2) $\bigcap_{m\in\mathbb{Z}} \mathbf{V}^{(m)} = 0$

(MR-3) $x = x(t) \in \mathbf{V}^{(0)} \iff {}_m x = x(2^{-m}t) \in \mathbf{V}^{(m)}$

(MR-4) $x = x(t) \in \mathbf{V}^{(0)}$ $\iff$ $x_k = x(t - k) \in \mathbf{V}^{(0)}$ for $k \in \mathbb{Z}$

(MR-5)   There exists (one declares) $\varphi = \varphi(t) \equiv$ the *scaling function* (of the multi-resolution structure), such that, with $\varphi_k(t) = \varphi(t - k)$ we get in $(\varphi_k(t))_{k \in \mathbb{Z}}$ an orthonormal basis of $\mathbf{V}^{(0)}$.

Consequence (of (MR-3) and of (MR-5)) Put $\varphi_k^{(m)}(t) = \sqrt{2^{-m}}\varphi(2^{-m}t - k)$. Then $(\varphi_k^{(m)}(t))_{k \in \mathbb{Z}}$ is an orthonormal basis of $\mathbf{V}^{(m)}$.

Remarks Consider, for $x = x(t) \in \mathbf{L}^2(\mathbb{R})$, its orthogonal projection $x^{(m)} = x^{(m)}(t)$ onto $\mathbf{V}^{(m)}$. $x^{(m)}$ will be called the approximation

$$\begin{cases} \text{at scale} 2^m, \\ \text{of resolution} 2^{-m} \end{cases} \text{ of } x.$$

$\mathbf{V}^{(m)}$ is then the vector space of the "approximations at scale 2^m".

We note the *essentially discrete* character of the notion of multi-resolution analysis for $\mathbf{L}^2(\mathbb{R})$: The pivot of the structure is clearly the *scaling function*. It is the scaling function which supports the transition from the continuous world to the digital world: the introduction of coordinate systems allows a more and more faithful digital representation (refine the resolution) of the approximated signal. In other words, the structure of multi-resolution analysis defines the appropriate formal setting for a variable digitization via dichotomic refinement of the resolution grid. From another viewpoint, we are simply in face of a *generalized Shannon theorem*.

The dilations by the factors 2^m, $m \in \mathbb{Z}$, formalize implicitly bandwidth variation, whereas the translations of step 2^m, $m \in \mathbb{Z}$, formalize equidistant sampling grids. Note that this translation – dilation twin reflects perfectly the time – frequency coupling which promotes wavelets theory so gloriously.

Some commentaries concerning the axioms stated above:

(MR-1) means that every $x = x(t) \in \mathbf{L}^2(\mathbb{R})$ is determined by (the knowledge of) its (high) resolutions: $x = \lim_{m \to -\infty} x^{(m)}$.

(MR-2) allows a conceptual interpretation that we shall see in a moment...

(MR-3) formalizes – in quantized form – the frequential aspect of the transitions between the resolution levels.

(MR-4) formalizes – in quantized form – the "equidistant grid" aspect inside the particular resolution levels.

(MR-5) is the digitization axiom, fundamental for the notion of multi-resolution analysis. It indicates, *in which way* the $\mathbf{V}^{(m)}$ are effectively *coordinate spaces*.

Let us point out that the notion of *orthonormal basis* (for the introduction of coordinates) quickly reveals itself to be too restrictive for really interesting applications. We shall soon find the remedy.

Exercises

Piecewise Constant Approximations. Consider
$\mathbf{V}^{(m)} = \{y = y(t) \in \mathbf{L}^2(\mathbb{R}) : y(t) = y(k \cdot 2^m) \text{ For } k \cdot 2^m \leq t < (k+1) \cdot 2^m, k \in \mathbb{Z}\}.$

$\mathbf{V}^{(m)}$ is the vector space of square integrable functions which are piecewise constant on the intervals – positioned in "natural dichotomy" – of length 2^m.

It is obvious that $\cdots \subset \mathbf{V}^{(2)} \subset \mathbf{V}^{(1)} \subset \mathbf{V}^{(0)} \subset \mathbf{V}^{(-1)} \subset \mathbf{V}^{(-2)} \subset \cdots$

(1) First the formal trivialities:
   (a) Verify (MR-3) and (MR-4).
   (b) Verify (MR-5), with the scaling function $\varphi(t) = \mathbb{I}_{[0,1[}(t)$ (the unit "time-window").

(2) Consider the function $x(t) = \begin{cases} 1 & \text{for } -1 \leq t \leq 1 \\ \frac{1}{|t|} & \text{else} \end{cases}$

   (a) Compute $\|x\|$, the norm of $x = x(t) \in \mathbf{L}^2(\mathbb{R})$.
   (b) Let $x^{(m)}(t) = \sum_{k \in \mathbb{Z}} \mu_k^{(m)} \cdot \mathbb{I}_{[k \cdot 2^m, (k+1) \cdot 2^m[}(t)$.
       Find the values $\mu_k^{(m)}, \quad k, m \in \mathbb{Z}$.
   (c) Show that $\lim_{m \to -\infty} x^{(m)}(t) = x(t)$ and that $\lim_{m \to +\infty} x^{(m)}(t) = 0$ (for what type of convergence?).

(3) Let us return to the general situation of a piecewise constant approximation.
   (a) Show (MR-1).
   (b) Show (MR-2).

[Help: For real-valued $x = x(t) \in \mathbf{L}^2(\mathbb{R})$ we have:

$$x^{(m)}(t) = \sum_{k \in \mathbb{Z}} \frac{1}{2^m} \int_{k \cdot 2^m}^{(k+1) \cdot 2^m} x(t)\mathrm{d}t \cdot \mathbb{I}_{[k \cdot 2^m, (k+1) \cdot 2^m[}(t).$$

We obtain:

$$\|x - x^{(m)}\|^2 = \|x\|^2 - \|x^{(m)}\|^2 = \int_{-\infty}^{+\infty} x(t)^2 \mathrm{d}t - 2^m \cdot \sum_{k \in \mathbb{Z}} (\mu_k^{(m)})^2,$$

with $\mu_k^{(m)} = \frac{1}{2^m} \int_{k \cdot 2^m}^{(k+1) \cdot 2^m} x(t)\mathrm{d}t$.

So, we need:
$\int_{-\infty}^{+\infty} x(t)^2 \mathrm{d}t = \lim_{m \to -\infty} 2^m \cdot \sum_{k \in \mathbb{Z}} (\mu_k^{(m)})^2$.

This is a bifurcation of mathematical rigour: Either you consider this approximation of the integral as obvious (and a formal banality), or you have to prove it.]

Wavelet Bases

From the conceptual viewpoint, the notion of multi-resolution analysis (and that of scaling function) are fundamental for the organization of the arguments around the hierarchical treatment of digital information.

But from the viewpoint of mathematical formalization, it is a derived structure which is more adapted to guarantee algorithmic solutions: that's the *associated graded structure*, the concrete appearance of which is realized via *wavelet bases*.

So, let us begin with a multi-resolution analysis

$$\cdots \subset \mathbf{V}^{(2)} \subset \mathbf{V}^{(1)} \subset \mathbf{V}^{(0)} \subset \mathbf{V}^{(-1)} \subset \mathbf{V}^{(-2)} \subset \cdots$$

and consider the spaces $\mathbf{W}^{(m)} = \mathbf{V}^{(m-1)}/\mathbf{V}^{(m)}$, $m \in \mathbb{Z}$. Conceptually, here we have introduced the *spaces of the "details at scale 2^{m-1}"*. We shall realize them as the orthogonal complements of the $\mathbf{V}^{(m)}$ inside the $\mathbf{V}^{(m-1)}$:

$$\mathbf{V}^{(m-1)} = \mathbf{V}^{(m)} \oplus \mathbf{W}^{(m)}, \quad m \in \mathbb{Z}.$$

Let us iterate the decomposition:

$$\mathbf{V}^{(m-1)} = \mathbf{V}^{(m+p)} \oplus \mathbf{W}^{(m)} \oplus \mathbf{W}^{(m+1)} \oplus \cdots \oplus \mathbf{W}^{(m+p)}, \quad p \ge 0, m \in \mathbb{Z}.$$

At this point the reason for the axiom (MR-2) becomes apparent. Pass to the limit, with $p \to \infty$: $\mathbf{V}^{(m-1)} = \bigoplus_{p \ge m} \mathbf{W}^{(p)}$, i.e. every $\mathbf{V}^{(m-1)}$ is the orthogonal direct sum of the $\mathbf{W}^{(p)}, p \ge m$ (in the Hilbertian sense: as a topological closure.). Finally, with $m \to -\infty$, we obtain $\mathbf{L}^2(\mathbb{R}) = \bigoplus_{p \in \mathbb{Z}} \mathbf{W}^{(p)}$.

We note: The situation of a multi-resolution analysis now appears in a *linearized version*: the resolution refinement (let us look closer and closer) is replaced by a procedure of accumulation of details (forget only what's not needed).

Example The Shannon multi-resolution analysis.

Recall: $\mathbf{V}^{(m)} = \{x \in \mathbf{L}^2(\mathbb{R}) : \hat{x}(\omega) = 0$ for $\mid \omega \mid > 2^{-m}\pi\}$

Hence, $\mathbf{W}^{(m)} = \{x \in \mathbf{L}^2(\mathbb{R}) : \hat{x}(\omega) = 0$ outside the band $2^{-m}\pi < \mid \omega \mid \le 2^{-m+1}\pi\}$ (this is immediate, by Parseval's identity).

Exercise

Put $\psi^{(m)}(t) = \sqrt{2^{-m}} \cdot \cos\left(\frac{3}{2}\pi \cdot 2^{-m}t\right) \cdot \sin c\left(\frac{1}{2} \cdot 2^{-m}t\right)$. Then:

(a) $\hat{\psi}^{(m)}(\omega) = \begin{cases} \sqrt{2^m} & \text{for } 2^{-m}\pi < \mid \omega \mid \le 2^{-m+1}\pi \\ 0 & \text{else.} \end{cases}$

(b) $\|\psi^{(m)}\| = 1$.

$\left[\text{First verify the identity } \frac{\sin\left(\frac{2\pi t}{T}\right)}{\pi t / T} - \frac{\sin\left(\frac{\pi t}{T}\right)}{\pi t / T} = \cos \frac{3}{2} \frac{\pi t}{T} \cdot \sin c\left(\frac{1}{2} \frac{t}{T}\right)\right]$

Now consider the translates $\psi_k^{(m)}(t) = \psi^{(m)}(t - k \cdot 2^m) \; k \in \mathbb{Z}$. Then we have, for $x = x(t) \in \mathbf{L}^2(\mathbb{R})$

$$(\triangle) \qquad x^{(m-1)} - x^{(m)} = \sum_{k \in \mathbb{Z}} \langle x(t), \psi_k^{(m)}(t) \rangle \psi_k^{(m)}$$

The argument:

(i) First, $(\psi_k^{(m)})_{k \in \mathbb{Z}}$ is an orthonormal system in $\mathbf{W}^{(m)}$:

$$\hat{\psi}_k^{(m)}(\omega) = e^{-ik2^m\omega} \hat{\psi}^{(m)}(\omega) \; k \in \mathbb{Z}$$

(and Parseval's formula yields, once more, the claim).

(ii) Then, let us establish a development

$$\hat{x}^{(m-1)} - \hat{x}^{(m)} = \sum_{k \in \mathbb{Z}} \gamma_k \hat{\psi}_k^{(m)}.$$

Write

$$\hat{x}^{(m-1)} = \sum_{k \in \mathbb{Z}} \alpha_k \hat{\varphi}_k^{(m-1)}$$

$$\hat{x}^{(m)} = \sum_{k \in \mathbb{Z}} \beta_k \hat{\varphi}_k^{(m)}$$

But: $\hat{\varphi}_k^{(m-1)}(\omega) = \sqrt{2^{m-1}} \cdot e^{-ik2^{m-1}\omega} \cdot \mathbb{I}_{[-2^{-m+1}\pi, 2^{-m+1}\pi]}(\omega)$, $\hat{\varphi}_k^{(m)}(\omega) = \sqrt{2^m} \cdot e^{-ik2^m\omega} \cdot \mathbb{I}_{[-2^{-m}\pi, 2^{-m}\pi]}(\omega)$

Hence: $\hat{\psi}_k^{(m)}(\omega) = \sqrt{2} \cdot e^{-ik2^{m-1}\omega} \cdot \hat{\varphi}_k^{(m-1)}(\omega) - \hat{\varphi}_k^{(m)}(\omega)$, $k \in \mathbb{Z}$.

But $\hat{x}^{(m)} = \hat{x}^{(m-1)} \cdot \mathbb{I}_{[-2^{-m}\pi, 2^{-m}\pi]}$,

$$\hat{\varphi}_k^{(m)} = \sqrt{2} \cdot e^{-ik2^{m-1}\omega} \cdot \hat{\varphi}_k^{(m-1)} \cdot \mathbb{I}_{[-2^{-m}\pi, 2^{-m}\pi]}$$

i.e. $\hat{x}^{(m)}$ is a "restriction" of $\hat{x}^{(m-1)}$ as well as $\hat{\varphi}_k^{(m)}$ is a "modulated restriction" of $\hat{\varphi}_k^{(m-1)}$. This gives:

$$\hat{x}^{(m)} = \sum_{k \in \mathbb{Z}} \alpha_k \hat{\varphi}_k^{(m-1)} \cdot \mathbb{I}_{[-2^{-m}\pi, 2^{-m}\pi]} = \sum_{k \in \mathbb{Z}} \alpha_k \cdot \frac{1}{\sqrt{2}} \cdot e^{ik2^{m-1}\omega} \cdot \hat{\varphi}_k^{(m)} = \sum_{k \in \mathbb{Z}} \beta_k \hat{\varphi}_k^{(m)}.$$

We have thus: $\alpha_k = \sqrt{2} \cdot e^{-ik2^{m-1}\omega} \cdot \beta_k$ $k \in \mathbb{Z}$. Finally, we obtain what we want: $\hat{x}^{(m-1)} - \hat{x}^{(m)} = \sum_{k \in \mathbb{Z}} \sqrt{2} \cdot e^{-ik2^{m-1}\omega} \cdot \beta_k \cdot \hat{\varphi}_k^{(m-1)} - \sum_{k \in \mathbb{Z}} \beta_k \hat{\varphi}_k^{(m)} = \sum_{k \in \mathbb{Z}} \beta_k \hat{\psi}_k^{(m)}$.

(iii) The identification of the "coordinate" is now a formality:

$$\beta_k = \langle \hat{x}^{(m-1)} - \hat{x}^{(m)}, \hat{\psi}_k^{(m)} \rangle = \langle \hat{x} - \hat{x}^{(m)}, \hat{\psi}_k^{(m)} \rangle = \langle \hat{x}, \hat{\psi}_k^{(m)} \rangle - \langle \hat{x}^{(m)}, \hat{\psi}_k^{(m)} \rangle$$
$$= \langle \hat{x}, \hat{\psi}_k^{(m)} \rangle$$

(since $\hat{x}^{(m)}$ is in the orthogonal complement of $\hat{\mathbf{W}}^{(m)}$). So, we have established the identity

$$x^{(m-1)} - x^{(m)} = \sum_{k \in \mathbb{Z}} \langle x(t), \psi_k^{(m)}(t) \rangle \psi_k^{(m)}.$$

Let us sum up. For every $k \in \mathbb{Z}$, the system $(\psi_k^{(m)})_{k \in \mathbb{Z}}$ is an orthonormal basis of the vector space $\mathbf{W}^{(m)}$. But we know already: $\mathbf{L}^2(\mathbb{R}) = \bigoplus_{p \in \mathbb{Z}} \mathbf{W}^{(p)}$. Hence, $(\psi_k^{(m)})_{k,m \in \mathbb{Z}}$ is an orthonormal basis of $\mathbf{L}^2(\mathbb{R})$.

We insist: Every $x = x(t) \in \mathbf{L}^2(\mathbb{R})$ admits a unique approximation (a unique development)

$$x = \sum_{k,m \in \mathbb{Z}} \langle x(t), \psi_k^{(m)}(t) \rangle \psi_k^{(m)}.$$

Supplementary observation, which is rather important:
Put $\psi(t) = \cos \frac{3}{2}\pi t \cdot \sin c\left(\frac{1}{2}t\right)$

$$\left[\hat{\psi}(\omega) = \begin{cases} 1 \text{ for} \pi <| \omega |\le 2\pi \\ 0 \text{ else} \end{cases} \right]$$

Then $\psi_k^{(m)}(t) = \sqrt{2^{-m}} \cdot \psi(2^{-m}t - k)$, $k, m \in \mathbb{Z}$. In this sense, our orthonormal basis is generated by a *single function* $\psi(t)$, with the help of a "dyadic quantization" by dilation (frequential aspect) and translation (temporal aspect).

Our particular situation deserves generalization:

Definition *A wavelet basis of* $\mathbf{L}^2(\mathbb{R})$ *is given by:*

(1) A mother wavelet $\psi(t)$,

(2) A family of translates – dilates: $\psi_k^{(m)}(t) = \sqrt{2^{-m}} \cdot \psi(2^{-m}t - k)$, $k, m \in \mathbb{Z}$,

which is a basis of $\mathbf{L}^2(\mathbb{R})$*: Every* $x = x(t) \in \mathbf{L}^2(\mathbb{R})$ *admits a unique development*

$$x = \sum_{k,m \in \mathbb{Z}} y_1^{(m)}[k]\psi_k^{(m)}$$

Caution A wavelet basis is *not* necessarily orthonormal!

Wavelet Bases Generated by a Scaling Function $\varphi(t)$

Let us fix our *initial data*: We start with a multi-resolution analysis

$$\cdots \subset \mathbf{V}^{(2)} \subset \mathbf{V}^{(1)} \subset \mathbf{V}^{(0)} \subset \mathbf{V}^{(-1)} \subset \mathbf{V}^{(-2)} \subset \cdots$$

and its scaling function $\varphi(t)$.

Our objective: Find a mother wavelet $\psi(t)$ and the wavelet basis $(\psi_k^{(m)})_{k,m \in \mathbb{Z}}$ which is derived from it such that, for every $m \in \mathbb{Z}$, the sequence of the translates $(\psi_k^{(m)})_{k \in \mathbb{Z}}$ is a basis (in the Hilbertian sense) of $\mathbf{W}^{(m)}$, i.e. of the orthogonal complement of $\mathbf{V}^{(m)}$ in $\mathbf{V}^{(m-1)}$.

Strategy: Find, from the coordinates of $\varphi(t)$ in $\mathbf{V}^{(-1)}$, "appropriate orthogonal coordinates" which will define $\psi(t)$.

In detail:

(1) $\mathbf{V}^{(0)} \subset \mathbf{V}^{(-1)}$, hence $\varphi = \varphi(t)$ admits a development with respect to $(\varphi_k^{(-1)})_{k \in \mathbb{Z}}$:

$$\varphi(t) = \sqrt{2} \cdot \sum_{k \in \mathbb{Z}} g_0[k]\varphi(2t - k)$$

(2) Characteristics of $\mathbf{g}_0 = (g_0[k])_{k \in \mathbb{Z}}$
   – under the hypothesis of the *orthonormality* of $(\varphi(t - k))_{k \in \mathbb{Z}}$:
   $\langle \varphi_0, \varphi_k \rangle = 2 \int_{-\infty}^{+\infty} (\sum_i g_0[i]\varphi(2t - i) \cdot \sum_j g_0[j]\varphi(2(t - k) - j))dt =$
   $2\sum_{i,j} g_0[i]g_0[j - 2k] \int_{-\infty}^{+\infty} \varphi(2t - i)\varphi(2t - j)dt =$
   $2\sum_{i,j} g_0[i]g_0[j - 2k] \cdot \frac{1}{2}\langle \varphi_i^{(-1)}, \varphi_j^{(-1)} \rangle = \sum_i g_0[i]g_0[i - 2k] \quad k \in \mathbb{Z}$.
   We observe: $\mathbf{g}_0 = (g_0[k])_{k \in \mathbb{Z}}$ is a sequence in $\mathrm{l}^2(\mathbb{Z})$, of norm (of energy) equal to 1, which is orthogonal to all its 2-translates.
   And what about the odd (integer shift) translates? Put $\mathbf{g}_1 = (g_1[k])_{k \in \mathbb{Z}}$ with $g_1[k] = (-1)^{k+1}g_0[-k + 1]$, $k \in \mathbb{Z}$. Then $\mathbf{g}_1$ and its 2-translates fulfill their mission.

Exercise (Important)

Show that the 2-translates of $\mathbf{g}_0 = (g_0[k])_{k \in \mathbb{Z}}$ and the 2-translates of $\mathbf{g}_1 = (g_1[k])_{k \in \mathbb{Z}}$ constitute together an orthonormal basis of $\mathrm{l}^2(\mathbb{Z})$.

(The orthonormality of the considered system is easy; what we have to show is that every sequence in $\mathrm{l}^2(\mathbb{Z})$ admits a development in the 2-translates of $\mathbf{g}_0$ and of $\mathbf{g}_1$).

$\Big[$*Help*: Let then $x = (x[k])_{k \in \mathbb{Z}} \in \mathrm{l}^2(\mathbb{Z})$. We have to show that there exists $(a[n])_{n \in \mathbb{Z}}$, $(b[n])_{n \in \mathbb{Z}}$, such that $x[k] = \sum_n a[n]g_0[k - 2n] + \sum_n b[n]g_1[k - 2n]$, $k \in \mathbb{Z}$.

Recall (Two channel filter banks – exercises) A sufficient condition for an analysis filter bank (h_0^t, h_1^t) (interleaved formalism) to admit a synthesis filter bank (g_0^t, g_1^t) for a perfect reconstruction is the validity of the identity (♮) $\sum_k h_0^t[k](-1)^k h_1^t[2n-k] = \delta_{n,0}$. In this case, g_0^t and g_1^t will be the "alternating versions" of h_1^t and of h_0^t. Passing now to the conventional formalism – $h_1^t[k] = h_1[k-1]$ and $g_1^t[k] = g_1[k+1]$ – we obtain the identity (♮) $\sum_k (-1)^{k+1} g_1[k] g_0[2n+1-k] = \delta_{n,0}$.

Accept the extension of our results from $l_c^2(\mathbb{Z})$ (finite support) to $l^2(\mathbb{Z})$ (or prove it: it is precisely our strong hypothesis $\det A = 1$ which facilitates the argument …). Then our couple $(\mathbf{g}_0, \mathbf{g}_1)$ is a synthesis filter bank which admits a perfect reconstruction, and we can conclude … $\Big]$

(3) Put $\boxed{\psi(t) = \sqrt{2} \cdot \sum_{k \in \mathbb{Z}} g_1[k] \varphi(2t - k)}$

and, as usual, $\psi_k^{(m)}(t) = \sqrt{2^{-m}} \cdot \psi(2^{-m}t - k)$ $k, m \in \mathbb{Z}$. Let us show first that $(\psi_k(t))_{k \in \mathbb{Z}} = (\psi_k^{(0)}(t))_{k \in \mathbb{Z}}$ is an *orthonormal basis* of $\mathbf{W}^{(0)}$.

[*Exercise*: Show that

(a) $\varphi_i(t) = \sum_k g_0[k - 2i] \varphi_k^{(-1)}(t)$
(b) $\psi_j(t) = \sum_k g_1[k - 2j] \varphi_k^{(-1)}(t)$]

(α) Orthonormality

$\langle \psi_k, \psi_m \rangle = \sum_{i,j} g_1[i - 2k] g_1[j - 2m] \langle \varphi_i^{(-1)}, \varphi_j^{(-1)} \rangle = \sum_i g_1[i - 2k] g_1[i - 2m] = \delta_{k,m}.$

(β) Orthogonality to $\mathbf{V}^{(0)}$

$\langle \psi_k, \varphi_m \rangle = \sum_{i,j} g_1[i - 2k] g_0[j - 2m] \langle \varphi_i^{(-1)}, \varphi_j^{(-1)} \rangle = \sum_i g_1[i - 2k] g_0[i - 2m] = 0.$

Hence: $(\psi_k(t))_{k \in \mathbb{Z}}$ is an orthonormal family in $\mathbf{W}^{(0)}$, the orthogonal complement of $\mathbf{V}^{(0)}$ in $\mathbf{V}^{(-1)}$.

(γ) Density

$\mathbf{V}^{(-1)}$ is generated (in the Hilbertian sense) by $(\varphi_k(t))_{k \in \mathbb{Z}}$ and by $(\psi_k(t))_{k \in \mathbb{Z}}$. Consider $x = x(t) \in \mathbf{V}^{(-1)}$: $x = \sum_{k \in \mathbb{Z}} y_0^{(-1)}[k] \varphi_k^{(-1)}$. But, $(y_0^{(-1)}[k])_{k \in \mathbb{Z}} \in l^2(\mathbb{Z})$, and the 2-translates of $\mathbf{g}_0$ as well as the 2-translates of $\mathbf{g}_1$ constitute together an orthonormal basis of $l^2(\mathbb{Z})$. We thus can write:

$$x = \sum_k \left(\sum_i y_0^{(0)}[i] g_0[k - 2i] + \sum_j y_1^{(0)}[j] g_1[k - 2j] \right) \varphi_k^{(-1)}$$
$$= \sum_i y_0^{(0)}[i] \varphi_i^{(0)} + \sum_j y_1^{(0)}[j] \psi_j^{(0)}$$

All that is naturally inherited by the dilated versions of $\psi(t)$, which finally gives the claim:

$(\psi_k^{(m)}(t))_{k,m \in \mathbb{Z}}$ is an (orthonormal) wavelet basis of $\mathbf{L}^2(\mathbb{R})$,

generated from the mother wavelet $\quad \psi(t) = \sqrt{2} \cdot \sum_{k \in \mathbb{Z}} g_1[k]\varphi(2t - k)$
where $\mathbf{g}_1 = (g_1[k])_{k \in \mathbb{Z}}$ are coordinates orthogonal to the coordinates $\mathbf{g}_0 = (g_0[k])_{k \in \mathbb{Z}}$
of the scaling function $\varphi(t)$ in $\mathbf{V}^{(-1)}$. We insist:

Let $\varphi(t)$ be the scaling function of a multi-resolution analysis, and
let $\psi(t)$ be the mother wavelet for the associated wavelet basis.

Then the coordinates $\mathbf{g}_0 = (g_0[k])_{k \in \mathbb{Z}}$ of $\varphi(t)$ and the coordinates
$\mathbf{g}_1 = (g_1[k])_{k \in \mathbb{Z}}$ of $\psi(t)$ in $\mathbf{V}^{(-1)}$ constitute the two synthesis
filters $(\mathbf{g}_0, \mathbf{g}_1)$ of a perfect reconstruction filter bank.

Exercises

(1) Show – with the notations from above – that

(a) $\langle \varphi_m^{(0)}, \varphi_n^{(-1)} \rangle = g_0[n - 2m]$
(b) $\langle \psi_m^{(0)}, \varphi_n^{(-1)} \rangle = g_1[n - 2m]$, $m, n \in \mathbb{Z}$

(**Caution:** We suppose the system $(\varphi_k)_{k \in \mathbb{Z}}$ to be orthonormal!)

(2) Consider the Shannon approximation.
Find $\mathbf{g}_0 = (g_0[k])_{k \in \mathbb{Z}}$ and $\mathbf{g}_1 = (g_1[k])_{k \in \mathbb{Z}}$ in this case.

$\left[\begin{array}{l} Result: \\[4pt] g_0[0] = \frac{1}{\sqrt{2}} \\[10pt] g_0[2m] = 0 \qquad\qquad\qquad\qquad m \neq 0 \\[4pt] g_0[2m + 1] = \frac{\sqrt{2}}{\pi} \cdot (-1)^m \frac{1}{2m+1} \end{array} \right]$

(3) Now let us pass to the piecewise constant approximation.

(a) Find $\mathbf{g}_0 = (g_0[k])_{k \in \mathbb{Z}}$ and $\mathbf{g}_1 = (g_1[k])_{k \in \mathbb{Z}}$ in this case.
(b) Produce the wavelet basis $(\psi_k^{(m)}(t))_{k,m \in \mathbb{Z}}$.

$\Big[$ Result:

(a) $g_0[m] = \begin{cases} \frac{\sqrt{2}}{2} & m = 0, 1 \\ 0 & \text{else} \end{cases}$

$g_1[m] = \begin{cases} -\frac{\sqrt{2}}{2} & m = 0 \\ \frac{\sqrt{2}}{2} & m = 1 \\ 0 & \text{else} \end{cases}$

(b) $\psi_k^{(m)}(t) = \begin{cases} -\sqrt{2^{-m}} & k \cdot 2^m \leq t < k \cdot 2^m + 2^{m-1} \\ \sqrt{2^{-m}} & k \cdot 2^m + 2^{m-1} \leq t < (k + 1) \cdot 2^m \\ 0 & \text{else} \end{cases} \Big]$

The Discrete Wavelet Transform

The Orthonormal Situation

That's the situation that we have exclusively considered until now: The discretizations (the coordinates) are given by development with respect to orthonormal bases of the considered approximation spaces.

Let us underline the principal reason – let alone our "geometrizing instincts" – for this mathematical modeling which insists upon orthonormality: Since the norms of the functions (in $\mathbf{L}^2(\mathbb{R})$) are equal to the norms of the series of their coordinates (in $\mathbf{l}^2(\mathbb{Z})$), we get isometric coordinatizations. This frequently is called the "numerical stability of the discretizations". That's a theme which we will not abandon in the sequel.

Let us sum up:

(1) Every multi-resolution analysis gives rise to a wavelet basis.
(2) The construction of the wavelet basis generates simultaneously a two channel filter bank.
(3) The principal ingredients of the construction:

(α) Let $\varphi = \varphi(t)$ be the scaling function of the considered multi-resolution analysis. Find its coordinates in $\mathbf{V}^{(-1)}$:

$$\varphi = \sum_{k \in \mathbb{Z}} g_0[k] \varphi_k^{(-1)}$$

(β) $\mathbf{g}_0 = (g_0[k])_{k \in \mathbb{Z}}$ will be interpreted as the low-pass synthesis filter of a two channel filter bank. $\mathbf{g}_1 = (g_1[k])_{k \in \mathbb{Z}}$ is derived as the "alternate" sequence, orthogonal to $\mathbf{g}_0$, and defines

(γ) $\psi = \psi(t)$, the mother wavelet of the associated wavelet basis:

$$\psi = \sum_{k \in \mathbb{Z}} g_1[k] \varphi_k^{(-1)}$$

Remark The sequences $\mathbf{g}_0 = (g_0[k])_{k \in \mathbb{Z}}$ and $\mathbf{g}_1 = (g_1[k])_{k \in \mathbb{Z}}$ are, a priori, in $\mathbf{l}^2(\mathbb{Z})$. We shall tacitly suppose them *finite* (without always underlining it), by practical instinct and a weakness for simple reference formularies. Altogether, there are themes intrinsic to the mathematical theory (for example finite bandwidth in Shannon approximation) which are structurally hostile to finiteness of impulse responses.

Lemma *In the situation described above, we have:*

(a) $\varphi_n^{(m)}(t) = \sum_k g_0[k - 2n] \varphi_k^{(m-1)}(t)$
(b) $\psi_n^{(m)}(t) = \sum_k g_1[k - 2n] \varphi_k^{(m-1)}(t)$

This is a (trivial) generalization of a previous exercise.

Important Observation The systems of coordinates of $(\varphi_n^{(m)})_{n \in \mathbb{Z}}$ and of $(\psi_n^{(m)})_{n \in \mathbb{Z}}$ in $\mathbf{V}^{(m-1)}$ *do not* depend on m.

This is clearly a consequence of the generic character of the scaling function $\varphi = \varphi(t)$ and of the mother wavelet $\psi = \psi(t)$ for the coordinatization of the situation. We thus obtain – on the coordinate level – a two channel *filter bank* which provides the transit between consecutive resolution levels – and this *independently* of their location in the dyadic multi-scale hierarchy.

More precisely: In a coordinate-free setting, the situation is described this way: The (orthogonal) decomposition $\mathbf{V}^{(m-1)} = \mathbf{V}^{(m)} \oplus W^{(m)}$,

[More generally: $\mathbf{V}^{(m-1)} = \mathbf{V}^{(m+p)} \oplus W^{(m)} \oplus \mathbf{W}^{(m+1)} \oplus \cdots \oplus \mathbf{W}^{(m+p)}$]

reads as follows: For $x \in \mathbf{L}^2(\mathbb{R})$, the approximation $x^{(m-1)}$ at scale 2^{m-1} decomposes into the (rougher) approximation $x^{(m)}$ at scale 2^m and into $w^{(m)}$, the detail-component at scale 2^{m-1}.

[More generally: The approximation $x^{(m-1)}$ at scale 2^{m-1} decomposes into the (rougher) approximation $x^{(m+p)}$ at scale 2^{m+p} and into $(w^{(m)}, w^{(m+1)}, \ldots, w^{(m+p-1)})$, the detail-component at scales $2^{m-1}, 2^{m-2}, \ldots, 2^{m+p-1}$.]

All this is, at a first glance, a formal banality.

But let us pass to the coordinates; we shall compute the coordinates of $x^{(m-1)}$ in function of the coordinates of $x^{(m)}$ and of $w^{(m)}$:

First $x^{(m-1)}$: $x^{(m-1)} = \sum_k y_0^{(m-1)}[k]\varphi_k^{(m-1)}$. Then, $x^{(m)}$ and $w^{(m)}$:

$$x^{(m)} = \sum_n y_0^{(m)}[n]\varphi_n^{(m)},$$

$$w^{(m)} = \sum_n y_1^{(m)}[n]\psi_n^{(m)}.$$

Hence:

$$x^{(m-1)} = x^{(m)} + w^{(m)} = \sum_n y_0^{(m)}[n]\varphi_n^{(m)} + \sum_n y_1^{(m)}[n]\psi_n^{(m)}$$

$$= \sum_{k,n} (y_0^{(m)}[n]g_0[k-2n] + y_1^{(m)}[n]g_1[k-2n])\varphi_k^{(m-1)}.$$

Comparison of the coefficients:

$$y_0^{(m-1)}[k] = \sum_n y_0^{(m)}[n]g_0[k-2n] + \sum_n y_1^{(m)}[n]g_1[k-2n], \quad k \in \mathbb{Z}$$

In compact form[30]:

$$\boxed{y_0^{(m-1)} = (\uparrow 2y_0^{(m)}) * \mathbf{g}_0 + (\uparrow 2y_1^{(m)}) * \mathbf{g}_1 \qquad m \in \mathbb{Z}}$$

In other words: The reconstruction of the coordinates of an approximation at scale 2^{m-1} from the coordinates of the (rougher) approximation at scale 2^m and of the detail-coordinates at scale 2^{m-1} is carried out by means of the synthesis filters

---

[30] *Recall:* Down-sampling $\downarrow 2$ and up-sampling $\uparrow 2$ to the factor 2 are defined this way: $(\downarrow 2f)(n) = f(2n)$ and $(\uparrow 2f)(n) = \begin{cases} f(\frac{n}{2}) & \text{if } n \equiv 0 \bmod 2 \\ 0 & \text{else} \end{cases}$

$(\mathbf{g}_0, \mathbf{g}_1)$ of a two channel filter bank. The important feature is that the intervention of the filter bank is *uniform* (i.e. independent of the transit level $m \to m - 1$).

The analysis filter bank $(\mathbf{h}_0, \mathbf{h}_1)$ which produces, on the coordinate level, the decomposition $x^{(m-1)} \mapsto (x^{(m)}, w^{(m)})$ (still uniformly in m), is given by the reconstruction formulas (discussed previously in the exercises on finite impulse response filter banks):

$$h_0[n] = (-1)^n g_1[n+1] = g_0[-n]$$
$$h_1[n] = (-1)^{n+1} g_0[n+1] = g_1[-n]$$

(*Caution:* We have adopted the conventional, non-interleaved notation)
So, we have, with the notation introduced above:

$$\boxed{\begin{aligned} y_0^{(m)} &= \downarrow 2(y_0^{(m-1)} * \mathbf{h}_0) \\[2mm] y_1^{(m)} &= \downarrow 2(y_0^{(m-1)} * \mathbf{h}_1) \end{aligned}}$$

Definition *The Discrete Wavelet Transform associated with a multi-resolution analysis of* $\mathbf{L}^2(\mathbb{R})$ *is the perfect reconstruction two channel filter bank with synthesis filters* $(\mathbf{g}_0, \mathbf{g}_1)$ *defined by the coordinates (in* $\mathbf{V}^{(-1)}$*) of the scaling function* $\varphi = \varphi(t)$ *and of the mother wavelet* $\psi = \psi(t)$.

A first consequence of our definition is an invasion by the language of (digital) filters. Lowering the resolution level (increasing the scale) corresponds then to a low-pass filtering: The low-pass subband receives the lower resolution, and the high-pass subband receives the details for the reconstruction of the finer resolution.

Iterating the considered filterings makes the approximations "thinner and thinner", while "thickening" – by additive accumulation – the reconstruction details.

Look at the corresponding notation:

$$x^{(m-1)} = x^{(m+p)} + \sum_{0 \leq i \leq p} \sum_{n \in \mathbb{Z}} y_1^{(m+i)}[n] \psi_n^{(m+i)}$$

with $x^{(m+p)} = \sum_{n \in \mathbb{Z}} y_0^{(m+p)}[n] \varphi_n^{(m+p)}$

Note that, with $H_0(x) = \downarrow 2(x * h_0)$, we have $y_0^{(m+p)} = H_0^{p+1} \left(y_0^{(m-1)} \right)$.

Remark The most appreciable property of the Discrete Wavelet Transform is its uniformity: it does not depend on m, the resolution level of the considered approximation structure. This causes its simple iterative intervention.

On the other hand, we should refrain from calling it recursive; only its iterative application has the recursive aspect of every iteration (like the computation of the nth derivative of a function, for example).

Exercises

(1) Consider $\mathbf{g}_0 = (g_0[0], g_0[1], g_0[2], g_0[3])^{31}$ with $g_0[0] = -\frac{1}{10}, g_0[1] = \frac{1}{5}, g_0[2] = \frac{3}{5}$, $g_0[3] = \frac{3}{10}$.

---

[31] We only note the values of the support.

(a) Determine $\mathbf{g}_1$, $\mathbf{h}_0$ and $\mathbf{h}_1$ according to our construction rules in the orthonormal case, and verify that we thus obtain a perfect reconstruction filter bank.

(b) Find a non-trivial permutation of the four values above which gives another perfect reconstruction filter bank.

(2) Carry out, in the situation of exercise (1), an up-sampling to the order 3:

We consider

$$\mathbf{g}_0^{(3)} = (g_0^{(3)}[0], g_0^{(3)}[1], g_0^{(3)}[2], g_0^{(3)}[3], g_0^{(3)}[4], g_0^{(3)}[5], g_0^{(3)}[6], g_0^{(3)}[7], g_0^{(3)}[8], g_0^{(3)}[9])$$

with $g_0^{(3)}[0] = -\frac{1}{10}$, $g_0^{(3)}[1] = g_0^{(3)}[2] = 0$, $g_0^{(3)}[3] = \frac{1}{5}$, $g_0^{(3)}[4] = g_0^{(3)}[5] = 0$, $g_0^{(3)}[6] = \frac{3}{5}$, $g_0^{(3)}[7] = g_0^{(3)}[8] = 0$, $g_0^{(3)}[9] = \frac{3}{10}$.

(a) Verify that we obtain – with $\mathbf{g}_1^{(3)}$, $\mathbf{h}_0^{(3)}$ and $\mathbf{h}_1^{(3)}$ derived from $\mathbf{g}_0^{(3)}$ – still a perfect reconstruction filter bank.

(b) Let us admit, in the situation of exercise (1), the existence of a multi-resolution analysis which allows us to recover the considered filter bank as the associated DWT. Let $\varphi = \varphi(t)$ be the corresponding scaling function, according to the axiom (MR-5).
Put $\varphi(t) = \frac{1}{3}\varphi(\frac{t}{3})$.
Is $\varphi = \varphi(t)$ the scaling function of a multi-resolution analysis such that our "up-sampled" filter bank is the associated DWT?

(3) Let now $\mathbf{g}_0 = (g_0[0], g_0[1], g_0[2], g_0[3])$

be given by $g_0[0] = \frac{2-\sqrt{2}}{6}$, $g_0[1] = \frac{1-\sqrt{2}}{6}$, $g_0[2] = \frac{1+\sqrt{2}}{6}$, $g_0[3] = \frac{2+\sqrt{2}}{6}$.

(a) Verify that $\sum_k g_0[k] = 1$ and that $\sum_k (-1)^k g_0[k] = 0$.

(b) Determine $\mathbf{g}_1$, $\mathbf{h}_0$ and $\mathbf{h}_1$ according to the imposed identities in the orthonormal case, and verify that we thus obtain a perfect reconstruction filter bank.

(c) Consider $\hat{g}_0(\omega) = g_0[0] + g_0[1]e^{-i\omega} + g_0[2]e^{-2i\omega} + g_0[3]e^{-3i\omega}$.
Accept the following result: If $\mid \hat{g}_0(\omega) \mid > 0$ on $[-\frac{\pi}{2}, \frac{\pi}{2}]$, then $\mathbf{g}_0$ is the low-pass synthesis filter of a DWT associated with a multi-resolution analysis of $\mathbf{L}^2(\mathbb{R})$.
Verify the validity of our $\mathbf{g}_0$.

The Biorthogonal Situation

We have just seen: In the orthogonal situation $((\varphi_n(t))_{n\in\mathbb{Z}}$ is an *orthonormal basis* of $\mathbf{V}^{(0)}$), the analysis filters and the synthesis filters of the considered DWT are trivially connected:

$$(h_0[n], h_1[n]) = (g_0[-n], g_1[-n])$$

On the other hand, the two "star" wavelet transforms that we have encountered within the JPEG 2000 setting, the DWT 5/3 spline and the DWT 9/7 CDF, *do not* have this property (they are *not* orthogonal).

Actually, they present (for image processing) *practically indispensable* general characteristics, which reveal themselves to be hostile to orthogonality.
Let us explain this.

First, there is the finite support. It is unimaginable not to use *finite length* impulse responses in practical applications (except if they are theoretically indefensible; in this case, we altogether will have to live with appropriate truncations). Now, finite support alone is perfectly compatible with the orthogonality of a filter bank.

But add the demand of *symmetric* impulse responses (this means linear phase) which is a must by the very philosophy of image compression: the transition to a lower resolution should numerically be done by computation of *local means*, i.e. with the help of symmetric barycentric formulas. Unfortunately, we have the following result:

A *two* channel *orthogonal* filter bank (in the sense of the preceding section) which has impulse responses of *finite length and* of *linear phase* (i.e. they are symmetric or anti-symmetric) cannot admit more than two non-zero coefficients per impulse response

---

Hence we are obliged – if we want to work both with finite length and symmetry – to consider filter banks of a more general type.

But let us begin first with a

Remark We shall be obliged to give up the geometric paradise of orthogonality for practical reasons; but we shall try to counterbalance this loss by a maximum of formal comfort:

On the one hand, we shall try to stay as near as possible to a geometrical language (faithful to inner products "with a meaning") – this also will be dictated by practical necessities (numerical stability of the coordinatization); on the other hand, we shall strongly insist upon our principal hypothesis: the coupling of *finiteness* and *symmetry* for the considered impulse responses.

Hence: The results of the section on the (finite impulse response) filter banks will always be the basic formulary for our arguments.

Recall Let $(\mathbf{h}_0^t, \mathbf{h}_1^t)$ be the couple which defines the analysis filter bank, and let $(\mathbf{g}_0^t, \mathbf{g}_1^t)$ be the couple which defines the synthesis filter bank[32].

Then it suffices to specify $\mathbf{h}_0^t$ and $\mathbf{g}_0^t$ satisfying[33]

$$\sum_k h_0^t[k] g_0^t[2n-k] = \begin{cases} \alpha & \text{for } n = 0 \\ 0 & \text{else} \end{cases}$$

$\mathbf{h}_1^t$ and $\mathbf{g}_1^t$ are then computed by the formulas

$$h_1^t[n] = \alpha \cdot (-1)^n g_0^t[n],$$
$$g_1^t[n] = \tfrac{1}{\alpha} \cdot (-1)^n h_0^t[n].$$

If we search for a *generalization* of the orthogonal multi-resolution analysis concept, while refusing to give up the familiar formal setting – and aiming at a naturally associated DWT – we have to proceed as follows:

---

[32] $()^t \equiv$ interleaved notation.

[33] α, the determinant of the analysis matrix A, will be, in general, normalized: $\alpha = \tfrac{1}{2}, 1$ or 2.

First, we need an alleged version of the axiom (MR-5):

(MR-5)* There exists (one chooses) $\varphi = \varphi(t) \equiv$ the *scaling function* (of the multi-resolution structure), such that, with $\varphi_k(t) = \varphi(t - k)$, we have in $(\varphi_k(t))_{k\in\mathbb{Z}}$ a *Riesz basis* of $\mathbf{V}^{(0)}$.

The notion of a *Riesz basis* is a natural generalization of the notion of an orthonormal basis (in the Hilbertian sense): from the numerical viewpoint, the big advantage of the orthonormal bases comes from the *inner product coordinates*; hence we shall impose a robust condition which will maintain the continuity of the coordinate functions and which will allow the intervention of the Riesz theorem (reflexivity of Hilbert spaces). More explicitly, in our situation:

(i) Every $x = x(t) \in \mathbf{V}^{(0)}$ admits a unique development $x = \sum_{k\in\mathbb{Z}} y_0^{(0)}[k]\varphi_k$

(ii) There exists $0 < A \leq B$, only depending on $(\varphi_k)_{k\in\mathbb{Z}}$, with the estimation
$$A \cdot \|y_0^{(0)}\|^2 \leq \|x\|^2 \leq B \cdot \|y_0^{(0)}\|^2 \text{ i.e. } \tfrac{1}{B} \cdot \|x\|^2 \leq \sum_k \mid y_0^{(0)}[k] \mid^2 \leq \tfrac{1}{A} \cdot \|x\|^2$$
for all $x \in \mathbf{V}^{(0)}$.

This definition deserves a *commentary*: Visibly, we have here an equivalence of norms (on the space $\mathbf{L}^2(\mathbb{R})$ and on the space $l^2(\mathbb{Z})$). We thus obtain a homeomorphism between $\mathbf{V}^{(0)}$ and $l^2(\mathbb{Z})$ – which yields the "numerical stability" of the coordinates. Note also a finer aspect of the definition: The family $(\varphi_k)_{k\in\mathbb{Z}}$ is an *unconditional basis* of $\mathbf{V}^{(0)}$: for all sequences $(s_k)_{k\in\mathbb{Z}}$ in $l^2(\mathbb{Z})$, the sum $\sum_k s_k\varphi(t - k)$ exists in $\mathbf{V}^{(0)}$ and *does not depend on the summation mode*.

Note finally that our readjustment of the definition of a multi-resolution analysis by the relaxed condition (MR-5)* is amply sufficient for all interesting applications.

Attention The Riesz basis postulate for $\mathbf{V}^{(0)}$ puts immediately its dual basis (represented by vectors *of the vector space* $\mathbf{V}^{(0)}$) at our disposal. But it is *not* at this basis that we aim with the notion of biorthogonality ...

Now let us come back to our problem: How can we *generalize* the orthonormal situation in order to get the formal freedom for a Discrete Wavelet Transform which is described by a filter bank with *finite* and *symmetric* impulse responses?

Let us begin with splitting the multi-resolution analysis of $\mathbf{L}^2(\mathbb{R})$ into two copies:

$$(\mathbf{V}^{(m)})_{m\in\mathbb{Z}} \equiv \text{the synthesis approximation,}$$

$$(\tilde{\mathbf{V}}^{(m)})_{m\in\mathbb{Z}} \equiv \text{the analysis approximation.}$$

Let $\varphi = \varphi(t)$ and $\tilde{\varphi} = \tilde{\varphi}(t)$ be the two scaling functions for the coordinatization in both cases.

Remark It is sufficient to specify $\mathbf{V}^{(0)}$ and $\tilde{\mathbf{V}}^{(0)}$ with their Riesz bases generated by $\varphi = \varphi(t)$ and $\tilde{\varphi} = \tilde{\varphi}(t)$. All the remainder will follow:

$$x = x(t) \in \mathbf{V}^{(m)} \iff {}_0x = x(2^m t) \in \mathbf{V}^{(0)}$$

$$x = x(t) \in \tilde{\mathbf{V}}^{(m)} \iff {}_0x = x(2^m t) \in \tilde{\mathbf{V}}^{(0)}$$

Concerning the synthesis filter bank $(\mathbf{g}_0, \mathbf{g}_1)$ and the analysis filter bank $(\mathbf{h}_0, \mathbf{h}_1)$ of the associated DWT, the impulse responses will be the coordinates (in $\mathbf{V}^{(-1)}$ and $\tilde{\mathbf{V}}^{(-1)}$) of the considered scaling functions and of the mother wavelets:

$$\varphi(t) = \sqrt{2} \cdot \sum_k g_0[k]\varphi(2t - k)$$

$$\psi(t) = \sqrt{2} \cdot \sum_k g_1[k]\varphi(2t - k)$$

$$\tilde{\varphi}(t) = \sqrt{2} \cdot \sum_k h_0[-k]\tilde{\varphi}(2t - k)$$

$$\tilde{\psi}(t) = \sqrt{2} \cdot \sum_k h_1[-k]\tilde{\varphi}(2t - k)$$

[the time-reverse indexing of the analysis impulse responses with respect to the synthesis impulse responses is a characteristic feature of the filter bank formularies.]

Looking at the DWT formalism in the orthonormal case, we note: For the correct functioning of the synthesis filter bank $(\mathbf{g}_0, \mathbf{g}_1)$, we first need the identities

(a) $\varphi_n^{(m)}(t) = \sum_k g_0[k - 2n]\varphi_k^{(m-1)}(t)$
(b) $\psi_n^{(m)}(t) = \sum_k g_1[k - 2n]\varphi_k^{(m-1)}(t)$

But this is an immediate consequence of the two equations at level $m = 0$. Then, we need a direct sum decomposition

$$\mathbf{V}^{(m-1)} = \mathbf{V}^{(m)} \oplus \mathbf{W}^{(m)},$$

where $\mathbf{W}^{(m)}$ is the closure of the subspace (of $\mathbf{V}^{(m-1)}$) which is generated by the family $(\psi_k^{(m)})_{k \in \mathbb{Z}}$.

We insist: The essential point there is that $\mathbf{W}^{(m)}$ is *sufficiently large* in order to allow this decomposition of $\mathbf{V}^{(m-1)}$; on the other hand, there is no necessity for $\mathbf{W}^{(m)}$ to be orthogonal to $\mathbf{V}^{(m)}$.

Let us also point out that the identity $\mathbf{V}^{(m-1)} = \mathbf{V}^{(m)} \oplus \mathbf{W}^{(m)}$ is – under our general hypotheses – equivalent to the fact that the 2-translates of $\mathbf{g}_0$ and the 2-translates of $\mathbf{g}_1$ constitute together a *basis* of $l^2(\mathbb{Z})$ (in the Hilbertian sense).

Concerning the analysis filter bank $(\mathbf{h}_0, \mathbf{h}_1)$, our commentaries are the same: for a perfect reconstruction filter bank, the couple $(\mathbf{h}_0, \mathbf{h}_1)$ can be mirrored to carry out the reconstruction (the synthesis) for the *inverted filter bank*. So we shall have similar identities as before for the developments of the translates-dilates of $\tilde{\varphi} = \tilde{\varphi}(t)$ and of $\tilde{\psi} = \tilde{\psi}(t)$ inside the one-step higher resolution level

(c) $\tilde{\varphi}_n^{(m)}(t) = \sum_k h_0[2n - k]\tilde{\varphi}_k^{(m-1)}(t)$
(d) $\tilde{\psi}_n^{(m)}(t) = \sum_k h_1[2n - k]\tilde{\varphi}_k^{(m-1)}(t)$

as well as the need of a (not necessarily orthogonal) direct sum decomposition $\tilde{\mathbf{V}}^{(m-1)} = \tilde{\mathbf{V}}^{(m)} \oplus \tilde{\mathbf{W}}^{(m)}$, where $\tilde{\mathbf{W}}^{(m)}$ is generated by the family $(\tilde{\psi}_k^{(m)})_{k \in \mathbb{Z}}$.

Arrived at this point, we shall have a perfectly operational DWT, provided that the impulse responses (which we suppose finite and of linear phase) satisfy the required identities for a perfect reconstruction filter bank.

But where is *the biorthogonality*? And what is its interest?

Let us give a first answer: The biorthogonality is a luxury which we can allow us in the DWT world of finite and symmetric[34] impulse responses – this provides a great number of very interesting practical examples. And since biorthogonality is compatible with our objectives, we shall declare it indispensable.

Then, the arguments go this way: The essential ingredient of a multi-resolution analysis of $\mathbf{L}^2(\mathbb{R})$ is the more and more faithful *discretization* of the analog data. But: This discretization has to be *numerically stable*. In conventional mathematical terms: the operation which associates with $x = x(t) \in \mathbf{L}^2(\mathbb{R})$ its coordinates (with respect to a wavelet basis) has to "respect the norms" (in the orthonormal case, we have equality between the norm of $x = x(t) \in \mathbf{L}^2(\mathbb{R})$ and the norm of the series of its coordinates in $l^2(\mathbb{Z})$). In order to guarantee stability, we need at least – in formal analogy with the orthonormal case – both for the analysis approximation and for the synthesis approximation, developments with respect to the wavelet bases where *the coefficients are inner products*.

The following observation will serve us as a guide:

Let us consider the equations for the filtering operations of a perfect reconstruction filter bank (in conventional notation):

$$y_0[k] = \sum_n x[n]h_0[2k-n],$$
$$y_1[k] = \sum_n x[n]h_1[2k-n]$$

and

$$x[n] = \sum_l y_0[l]g_0[n-2l] + \sum_l y_1[l]g_1[n-2l].$$

Put now

$$e_{2l}[n] = g_0[n-2l] \qquad \tilde{e}_{2l}[n] = h_0[2l-n]$$
$$e_{2l+1}[n] = g_1[n-2l] \qquad \tilde{e}_{2l+1}[n] = h_1[2l-n]$$

Then:
$$y_0[k] = \langle x[n], \tilde{e}_{2k}[n] \rangle$$
$$y_1[k] = \langle x[n], \tilde{e}_{2k+1}[n] \rangle$$

and $x[n] = \sum_l \langle \mathbf{x}, \tilde{e}_{2l} \rangle g_0[n-2l] + \sum_l \langle \mathbf{x}, \tilde{e}_{2l+1} \rangle g_1[n-2l] = \sum_m \langle \mathbf{x}, \tilde{e}_m \rangle e_m[n]$.

All this seems to be a decomposition in a pair of dual bases $(e_n)_{n\in\mathbb{Z}}$ and $(\tilde{e}_n)_{n\in\mathbb{Z}}$ of $l_c^2(\mathbb{Z})$. And indeed:

Lemma *If (h_0, h_1) and (g_0, g_1) are the two couples of the analysis filters and of the synthesis filters of a perfect reconstruction filter bank, then $(e_n)_{n\in\mathbb{Z}}$ and $(\tilde{e}_n)_{n\in\mathbb{Z}}$ – as defined above – constitute a biorthogonal system (a pair of dual bases) in $l_c^2(\mathbb{Z})$.*

Proof   The relations of biorthogonality

$$\langle e_m, \tilde{e}_n \rangle = \begin{cases} 1 & \text{if } m = n \\ 0 & \text{else} \end{cases}$$

are easily verified – cf. the help for the exercise (6) on two channel filter banks).

We insist: This biorthogonality in $l_c^2(\mathbb{Z})$ does *not* depend on any particular property of the considered filter bank.

But let us try to inject it into the continuous world as a *twinning condition* between the discretization of the synthesis approximation $(\mathbf{V}^{(m)})_{m\in\mathbb{Z}}$ and the discretization of the analysis approximation $(\tilde{\mathbf{V}}^{(m)})_{m\in\mathbb{Z}}$ giving rise to a DWT, as described above.

---

[34] In interleaved version!

In this situation, we shall call the couple of synthesis and analysis approximations *biorthogonal* provided

$$\langle \varphi_k, \tilde{\varphi}_l \rangle = \delta_{k,l} \qquad k, l \in \mathbb{Z}$$

for the translates $\varphi_k = \varphi_k(t) = \varphi(t - k)$ of the synthesis scaling function $\varphi = \varphi(t)$ and for the translates $\tilde{\varphi}_l = \tilde{\varphi}_l(t) = \tilde{\varphi}(t - l)$ of the analysis scaling function $\tilde{\varphi} = \tilde{\varphi}(t)$.

Remark Let $(\varphi_k^*)_{k\in\mathbb{Z}}$ be the Riesz dual basis – in $\mathbf{V}^{(0)}$ – of $(\varphi_k)_{k\in\mathbb{Z}}$, guaranteed by $(MR - 5)^*$. Then φ_k^* is the orthogonal projection of $\tilde{\varphi}_k$ onto $\mathbf{V}^{(0)}$, for all $k \in \mathbb{Z}$. We see that $\mathbf{V}^{(0)}$ and $\hat{\mathbf{V}}^{(0)}$ are altogether rather engaged.

The target consequence of the biorthogonality on the level of the scaling functions is the biorthogonality of the (synthesis and analysis) wavelet bases:

Proposition *The biorthogonality of the scaling functions* $\langle \varphi_n, \tilde{\varphi}_{n'} \rangle = \delta_{n,n'} n, n' \in \mathbb{Z}$ *implies the biorthogonality of the associated wavelet bases*

$$\langle \psi_k^{(m)}, \tilde{\psi}_{k'}^{(m')} \rangle = \delta_{m,m'} \delta_{k,k'} \qquad m, m', k, k' \in \mathbb{Z}$$

Proof We shall reason by means of a geometric constellation. In order to establish it, we need the following

Lemma *With the hypotheses of the proposition, we have the following identities:*

(1) $\langle \varphi_n^{(m)}, \tilde{\varphi}_{n'}^{(m)} \rangle = \delta_{n,n'} \qquad m, n, n' \in \mathbb{Z}$
(2) $\langle \psi_n^{(m)}, \tilde{\psi}_{n'}^{(m)} \rangle = \delta_{n,n'} \qquad m, n, n' \in \mathbb{Z}$
(3) $\langle \varphi_n^{(m)}, \tilde{\psi}_{n'}^{(m)} \rangle = \langle \psi_n^{(m)}, \tilde{\varphi}_{n'}^{(m)} \rangle = 0 \qquad m, n, n' \in \mathbb{Z}$

Proof   (1) Immediate (change of variable in the considered integral).
     (2)

$$\psi_n^{(m)} = \sum_k g_1[k - 2n] \varphi_k^{(m-1)}$$

$$\tilde{\psi}_{n'}^{(m)} = \sum_l h_1[2n' - l] \tilde{\varphi}_l^{(m-1)}$$

$$\langle \psi_n^{(m)}, \tilde{\psi}_{n'}^{(m)} \rangle = \sum_{k,l} g_1[k - 2n] h_1[2n' - l] \langle \varphi_k^{(m-1)}, \tilde{\varphi}_l^{(m-1)} \rangle$$

$$= \sum_k g_1[k - 2n] h_1[2n' - k]$$

$$= \langle e_{2n+1}[k], \tilde{e}_{2n'+1}[k] \rangle = \delta_{n,n'}$$

(3)

$$\langle \varphi_n^{(m)}, \tilde{\psi}_{n'}^{(m)} \rangle = \sum_{k,l} g_0[k - 2n] h_1[2n' - l] \langle \varphi_k^{(m-1)}, \tilde{\varphi}_l^{(m-1)} \rangle$$

$$= \sum_k g_0[k - 2n] h_1[2n' - k] = \langle e_{2n}[k], \tilde{e}_{2n'+1}[k] \rangle = 0$$

$$\langle \psi_n^{(m)}, \tilde{\varphi}_{n'}^{(m)} \rangle = \langle e_{2n+1}[k], \tilde{e}_{2n'}[k] \rangle = 0. \quad \square$$

Observation *The biorthogonality of the scaling functions* $\langle \varphi_n, \tilde{\varphi}_{n'} \rangle = \delta_{n,n'}$ $n, n' \in \mathbb{Z}$ *implies the perfect reconstruction for the associated filter bank (which is equivalent to the biorthogonality* $\langle e_n, \tilde{e}_{n'} \rangle = \delta_{n,n'}$ *(in* $l_c^2(\mathbb{Z})$*) of the system of the synthesis translates (*e_n*) and of the system of the analysis translates (*$\tilde{e}_{n'}$*)).*

The argument: First, we have: $\left\langle \varphi_k^{(-1)}, \tilde{\varphi}_l^{(-1)} \right\rangle = \delta_{k,l}$, $k, l \in \mathbb{Z}$. *Then, write*

$$\varphi_n = \sum_k g_0[k - 2n]\varphi_k^{(-1)},$$

$$\tilde{\varphi}_{n'} = \sum_l h_0[2n' - l]\varphi_l^{(-1)}.$$

Hence, via inner products:

$$\delta_{n,n'} = \sum_{k,l} g_0[k - 2n]h_0[2n' - l]\delta_{k,l} = \sum_k g_0[k - 2n]h_0[2n' - k].$$

But this is the condition of perfect reconstruction.

But let us return to the proof of the proposition. According to the identities (3) of the lemma we have, for all $m \in \mathbb{Z}$: $\tilde{W}^{(m)}$ is orthogonal to $V^{(m)}$, and $W^{(m)}$ is orthogonal to $\tilde{V}^{(m)}$. Now, the $V^{(m)}$ as well as the $\tilde{V}^{(m)}$ constitute a chain of closed subspaces, and every $V^{(m)}$ contains all $W^{(m)}$ for $m' > m$; equally, every $\tilde{V}^{(m)}$ contains all $\tilde{W}^{(m)}$ for $m' > m$.

Finally: $W^{(m)}$ and $\tilde{W}^{(m')}$ are orthogonal for $m \neq m'$. But this is precisely – with the point (2) of the lemma – the claim of the proposition. $\square$

Remark We note how the biorthogonality (in $l_c^2(\mathbb{Z})$) of the impulse response 2-translates – which always takes place whenever we dispose of a perfect reconstruction filter bank – serves to propagate the biorthogonality from the level of the scaling functions to the level of the associated wavelet bases.

Consequence Under the hypotheses above, every $x = x(t) \in L^2(\mathbb{R})$ admits two "dual" developments

$$x = \sum_{k,m \in \mathbb{Z}} \langle x, \tilde{\psi}_k^{(m)} \rangle \psi_k^{(m)} = \sum_{k,m \in \mathbb{Z}} \langle x, \psi_k^{(m)} \rangle \tilde{\psi}_k^{(m)}.$$

We have finally got the desired coordinatization, but not yet completely: the situation effectively aimed at is that of a pair of Riesz bases *for* $L^2(\mathbb{R})$.

In other words: We need two positive constants A and $\tilde{A}$ with

$$\frac{1}{A}\|x\|^2 \le \sum_{k,m} |\langle x, \tilde{\psi}_k^{(m)} \rangle|^2 \le \tilde{A}\|x\|^2,$$

$$\frac{1}{\tilde{A}}\|x\|^2 \le \sum_{k,m} |\langle x, \psi_k^{(m)} \rangle|^2 \le A\|x\|^2.$$

Let us try to well understand the problem: A priori, the two Riesz bases $(\varphi(t - k))_{k \in \mathbb{Z}}$ and $(\tilde{\varphi}(t - k))_{k \in \mathbb{Z}}$ are "externally dual" on resolution level 0 (cf. a previous remark). This gives the same situation for $(\psi_k^{(m)})_{k \in \mathbb{Z}}$ and for $(\tilde{\psi}_k^{(m)})_{k \in \mathbb{Z}}$ *on a given resolution level* m. The problem is to make sure that the "local" Riesz constants do not explode at infinity (in m).

Let us repeat: It is the *uniformity* of the Riesz bounds – across all the resolution levels m – at which we aim. This also implies that $(\psi_k^{(m)})_{k,m \in \mathbb{Z}}$ and $(\tilde{\psi}_k^{(m)})_{k,m \in \mathbb{Z}}$ will then be unconditional bases (i.e. that the summation mode will not affect the results of the considered developments).

Design Procedures

In this last section, we shall try to answer two questions which naturally arise in application-oriented spirits:

(1) What is the *practical interest* of the fact that a (perfect reconstruction) filter bank is the DWT associated with a multi-resolution analysis (eventually split into analysis copy and into synthesis copy, with biorthogonal coupling)?
(2) Which are the *criteria* that produce realizable and practically interesting Discrete Wavelets Transforms?

On a very concrete level of reasoning: How can we justify the choice of the DWT 5/3 spline and of the DWT 9/7 CDF for JPEG 2000?

Our answers will follow – at least in certain technical details – a logic of "spectral arguments". Hence we have to pave the way.

The Frequential Formulary

Recall briefly the characteristics of a (biorthogonal) DWT:

(1) The formulary of the (perfect reconstruction) filter bank:
   The perfect reconstruction condition:

$$(\mathcal{R}) \quad \sum_k h_0[k] g_0[2n - k] = \begin{cases} 1 & n = 0 \\ 0 & \text{else} \end{cases}$$

   [Did we normalize? – cf. exercise (1) at the end of this paragraph.]
   The high-pass impulse responses (derived from the low-pass impulse responses[35]):

$$h_1^{t}[n] = h_1[n - 1] = (-1)^n g_0[n],$$

$$g_1^{t}[n] = g_1[n + 1] = (-1)^n h_0[n].$$

Caution: On this formal level, there is no (discrete) time-inversion between analysis and synthesis. The inversion shall take place in the "operational" formulas.

We suppose tacitly that all impulse responses are of finite (odd) length and symmetric. Our notation will altogether respect the signs (in discrete time) where it is demanded by general theory.

---

[35] We also shall have the other viewpoint: The synthesis impulse responses derived from the analysis impulse responses.

(2) The *scaling equations*:

(S)
$$\varphi(t) = \sqrt{2} \cdot \sum_k g_0[k]\varphi(2t - k)$$

$$\tilde{\varphi}(t) = \sqrt{2} \cdot \sum_k h_0[-k]\tilde{\varphi}(2t - k)$$

The "high-pass" equations concerning the mother wavelets $\psi(t)$ and $\tilde{\psi}(t)$ (and making appear $\mathbf{g}_1 = (g_1[k])$ and $\mathbf{h}_1 = (h_1[k])$) are analogous and logically secondary.

(3) The condition of *biorthogonality*: The condition of biorthogonality $\langle \varphi_m, \tilde{\varphi}_n \rangle = \delta_{m,n}$ on the integer translates of the two (synthesis and analysis) scaling functions has *no* discrete equivalent.

We saw previously that it implies the condition of perfect reconstruction (R) which is equivalent to the biorthogonality $\langle e_m, \tilde{e}_n \rangle = \delta_{m,n}$ (in $l^2_c(\mathbb{Z})$) of the system of synthesis translates (e_m) and of the system of analysis translates $(\tilde{e}_n)$:

$$e_{2l}[n] = g_0[n - 2l] \qquad \tilde{e}_{2l}[n] = h_0[2l - n]$$

$$e_{2l+1}[n] = g_1[n - 2l] \qquad \tilde{e}_{2l+1}[n] = h_1[2l - n]$$

Now let us pass to the reformulation of our characteristic identities in "frequential notation".

Recall: The *Fourier transform* $\hat{\mathbf{f}} = \hat{f}(\omega)$ of $\mathbf{f} = (f[k])_{k\in\mathbb{Z}}$ – which we always suppose *of finite support* – is the trigonometric polynomial

$$\hat{f}(\omega) = \sum_{k\in\mathbb{Z}} f[k]e^{-ik\omega}$$

Two important values:

$$\hat{f}(0) = \sum_k f[k]$$

$$\hat{f}(\pi) = \sum_k (-1)^k f[k]$$

Let us begin with some usual *normalized notation*:

$m_0(\omega) = \frac{1}{\sqrt{2}}\hat{g}_0(\omega)$

$m_1(\omega) = \frac{1}{\sqrt{2}}\hat{g}_1^t(\omega) = \frac{1}{\sqrt{2}}e^{i\omega}\hat{g}_1(\omega)$ \qquad (since $g_1^t[n] = g_1[n+1]$)

$\tilde{m}_0(\omega) = \frac{1}{\sqrt{2}}\hat{h}_0(-\omega)$ \qquad (we have transformed $(h_0[-n])_{n\in\mathbb{Z}}$)

$\tilde{m}_1(\omega) = \frac{1}{\sqrt{2}}\hat{h}_1^t(-\omega) = \frac{1}{\sqrt{2}}e^{i\omega}\hat{h}_1(-\omega)$ \qquad (since $h_1^t[-n] = h_1[-n-1]$)

Our "braid" relations

$$h_1^t[k] = (-1)^k g_0^t[k]$$

$$g_1^t[k] = (-1)^k h_0^t[k]$$

become

$$\hat{h}_1^t(\omega) = \hat{g}_0(\omega + \pi) \qquad \text{and} \qquad \hat{g}_1^t(\omega) = \hat{h}_0(\omega + \pi)$$

hence

$$\tilde{m}_0(\omega) = m_1(\pi - \omega) \qquad \text{and} \qquad \tilde{m}_1(\omega) = m_0(\pi - \omega).$$

Finally, there are "boundary values"

$$m_0(0) = \tilde{m}_0(0) = 1 \qquad m_0(\pi) = \tilde{m}_0(\pi) = 0$$

$$m_1(\pi) = \tilde{m}_1(\pi) = 1 \qquad m_1(0) = \tilde{m}_1(0) = 0$$

the reason for which will become apparent in a moment.
Now let us move on to our program of translating the formulary.

(1) The perfect reconstruction:

Write $m_0(\omega) = \frac{1}{\sqrt{2}} \cdot \sum_k g_0[k] e^{-ik\omega}$ [36]

$$\tilde{m}_0^*(\omega) = \frac{1}{\sqrt{2}} \cdot \sum_k h_0[-k] e^{ik\omega}$$

$$m_0(\omega + \pi) = \frac{1}{\sqrt{2}} \cdot \sum_k (-1)^k g_0[k] e^{-ik\omega}$$

$$\tilde{m}_0^*(\omega + \pi) = \frac{1}{\sqrt{2}} \cdot \sum_k (-1)^k h_0[-k] e^{ik\omega}$$

$$m_0(\omega)\tilde{m}_0^*(\omega) = \frac{1}{2} \sum_n \left(\sum_{k+l=n} g_0[k] h_0[l] \right) e^{-in\omega}$$

$$m_0(\omega + \pi)\tilde{m}_0^*(\omega + \pi) = \frac{1}{2} \sum_n \left(\sum_{k+l=n} (-1)^{k+l} g_0[k] h_0[l] \right) e^{-in\omega}$$

Hence:

$$m_0(\omega)\tilde{m}_0^*(\omega) + m_0(\omega + \pi)\tilde{m}_0^*(\omega + \pi) = \sum_n \left(\sum_k g_0[k] h_0[2n - k] \right) e^{-i2n\omega}$$

---

[36] We shall use $()^*$ for conjugation, in order to avoid a bar-invasion above the formulas.

Finally[37]:

$$m_0(\omega)\tilde{m}_0^*(\omega) + m_0(\omega + \pi)\tilde{m}_0^*(\omega + \pi) = 1$$

$$\Longleftrightarrow$$

(R) $\sum_k g_0[k]h_0[2n - k] = \delta_{n,0}$

Remark So we have found a frequential formulation of the perfect reconstruction property. This means that we have characterized the biorthogonality $\langle e_m, \tilde{e}_n \rangle = \delta_{m,n}$ of the systems of synthesis vectors and of analysis vectors (in $l_c^2(\mathbb{Z})$) associated with the considered filter bank.

We shall remain in the setting of perfect reconstruction filter banks.
Put $\omega = 0$. Then $m_0(0)\tilde{m}_0^*(0) + m_0(\pi)\tilde{m}_0^*(\pi) = 1$. Now, we face a surprising intervention of the analog world:

Test of compliance. If the considered filter bank is a DWT associated with the couple of scaling functions $(\varphi, \tilde{\varphi})$ – which we suppose to only admit a finite number of discontinuities – then we have *necessarily*

$$m_0(\pi) = \tilde{m}_0(\pi) = 0$$

(i.e. $\sum_k (-1)^k g_0[k] = \sum_k (-1)^k h_0[-k] = 0$) [without proof].

Hence, we get the following (normalized) "boundary conditions":

$m_0(0) = 1$	$\tilde{m}_0(0) = 1$	$m_1(0) = 0$	$\tilde{m}_1(0) = 0$
$m_0(\pi) = 0$	$\tilde{m}_0(\pi) = 0$	$m_1(\pi) = 1$	$\tilde{m}_1(\pi) = 1$
weak low-pass		*weak high-pass*	

(2) The scaling equations: The scaling equations $(\mathcal{S})$ become

$$\hat{\varphi}(\omega) = m_0\left(\frac{\omega}{2}\right) \cdot \hat{\varphi}\left(\frac{\omega}{2}\right)$$

$$\hat{\tilde{\varphi}}(\omega) = \tilde{m}_0(\frac{\omega}{2}) \cdot \hat{\tilde{\varphi}}\left(\frac{\omega}{2}\right)$$

---

[37] You are right: we work with $\alpha = \det A = 1$ ($A \equiv$ the analysis matrix). For certain applications, this is not the best choice.

By iteration we obtain[38]

$$\boxed{\begin{aligned} \hat{\varphi}(\omega) &= \prod_{i=1}^{\infty} m_0\left(\frac{\omega}{2^i}\right) \\ \hat{\tilde{\varphi}}(\omega) &= \prod_{i=1}^{\infty} \tilde{m}_0\left(\frac{\omega}{2^i}\right) \end{aligned}} \qquad (\mathcal{P})$$

These identities express $\hat{\varphi}(\omega)$ and $\hat{\tilde{\varphi}}(\omega)$ as infinite products of trigonometric polynomials.

We note: The constituents of the infinite products *only depend on the filter bank*, i.e. only depend on $\mathbf{g}_0 = (g_0[k])$ and on $\mathbf{h}_0 = (h_0[k])$.

Conclusion The formulas (P) allow the *(re)construction* of $\varphi(t)$ and of $\tilde{\varphi}(t)$ from $\mathbf{g}_0 = (g_0[k])$ and of $\mathbf{h}_0 = (h_0[k])$.

We have hence a $\boxed{\text{test}}$ whether a given filter bank is indeed a Discrete Wavelet Transform, i.e. if it comes from a multi-resolution analysis – copy synthesis, copy analysis:

We must guarantee that the infinite products of the trigonometric polynomials *converge properly* ($\hat{\varphi}(\omega)$ and $\hat{\tilde{\varphi}}(\omega)$ must be square integrable).

We shall take up this theme in a moment.

(3) The biorthogonality: The condition of biorthogonality (on the integer translates of the two scaling functions $\varphi = \varphi(t)$ and $\tilde{\varphi} = \tilde{\varphi}(t)$) does not translate into a "frequential identity" deduced from the impulse responses of the associated DWT.

Altogether, we dispose of a link by means of the Poisson formula.

Proposition *The* biorthogonality *condition* $\langle \varphi_n, \tilde{\varphi}_{n'} \rangle = \delta_{n,n'}$ *is equivalent to the* summation formula

$$\sum_{k \in \mathbb{Z}} \hat{\varphi}(\omega + 2k\pi)\hat{\tilde{\varphi}}(\omega + 2k\pi)^* = 1.$$

Proof We always shall suppose $\varphi = \varphi(t)$ and $\tilde{\varphi} = \tilde{\varphi}(t)$ to be square integrable. Consider the trigonometric series $S(\omega) = \sum_k \langle \varphi(t), \tilde{\varphi}(t-k) \rangle \mathrm{e}^{-ik\omega} = \sum_k \langle \hat{\varphi}(\xi), \hat{\tilde{\varphi}}(\xi) \mathrm{e}^{-ik\xi} \rangle \mathrm{e}^{-ik\omega}$ and the 2π-periodic function $T(\omega) = \sum_{k \in \mathbb{Z}} \hat{\varphi}(\omega + 2k\pi)\hat{\tilde{\varphi}}(\omega + 2k\pi)^*$.

The biorthogonality is equivalent to $S(\omega) = 1$, and the validity of our summation formula is equivalent to $T(\omega) = 1$. Hence we have to show that $S(\omega) = T(\omega)$.

The fundamental auxiliary result is the following:

Lemma *Let $F(\omega)$ be a summable function such that $G(\omega) = \sum_{k \in \mathbb{Z}} F(\omega + 2k\pi)$ is square integrable on $[0, 2\pi]$ ($G(\omega)$ is 2π-periodic).*

Then: $G(\omega) = \frac{1}{2\pi} \sum_{k \in \mathbb{Z}} \hat{F}(-k)\mathrm{e}^{-ik\omega}$.

---

[38] This is true up to a multiplicative constant, which will be usually equal to 1.

Proof By our hypotheses, $G(\omega)$ admits a development in a Fourier series. Let c_k be its kth Fourier coefficient; then

$c_k = \frac{1}{2\pi} \int_{-\pi}^{\pi} G(\omega)e^{ik\omega}d\omega$

But: $\int_{-\pi}^{\pi} G(\omega)e^{-i\xi\omega}d\omega = \int_{-\infty}^{+\infty} F(\omega)e^{-i\xi\omega}d\omega = \hat{F}(\xi)$

Hence: $c_k = \frac{1}{2\pi}\hat{F}(-k)$. $\square$

Let us return to the identity $S(\omega) = T(\omega)$.

We shall apply the foregoing lemma, with $F(\omega) = \hat{\varphi}(\omega)\hat{\tilde{\varphi}}^*(\omega)$. $F(\omega)$ is summable, since $\varphi = \varphi(t)$ and $\tilde{\varphi} = \tilde{\varphi}(t)$ are supposed to be square integrable.

We have: $T(\omega) = G(\omega)$, that's already promising. We only lack the identification of $\langle\hat{\varphi}(\xi), \hat{\tilde{\varphi}}(\xi)e^{-ik\xi}\rangle$. But $\langle\hat{\varphi}(\xi), \hat{\tilde{\varphi}}(\xi)e^{-ik\xi}\rangle = \frac{1}{2\pi}\int \hat{\varphi}(\xi)\hat{\tilde{\varphi}}^*(\xi)e^{ik\xi}d\xi = \frac{1}{2\pi}\int F(\xi)e^{ik\xi}d\xi = \frac{1}{2\pi}\hat{F}(-k)$. $\square$

Exercises

(1) $\alpha = 1$ and $\alpha = \frac{1}{2}$.[39]
   Recall: Accept first $\alpha \in \mathbb{R}^*$ arbitrary. Then the condition of perfect reconstruction will be (R) $\sum_k h_0^t[k](-1)^k h_1^t[2n-k] = \alpha \cdot \delta_{n,0}$ supplemented by the identities

$g_0^t[k] = \frac{1}{\alpha}(-1)^k h_1^t[k]$,

$g_1^t[k] = \frac{1}{\alpha}(-1)^k h_0^t[k]$.

Note that the variant for the condition of perfect reconstruction

$$(R) \qquad \sum_k h_0[k]g_0[2n-k] = \delta_{n,0}$$

is *independent of the value of α.*

(a) Show that we have: $\alpha = \frac{1}{2}(\hat{h}_0^t(0)\hat{h}_1^t(\pi) + \hat{h}_1^t(0)\hat{h}_0^t(\pi))$

$\frac{1}{\alpha} = \frac{1}{2}(\hat{g}_0^t(0)\hat{g}_1^t(\pi) + \hat{g}_1^t(0)\hat{g}_0^t(\pi))$

(b) Show that we have: $\hat{h}_0^t(0)\hat{g}_0^t(0) + \hat{h}_0^t(\pi)\hat{g}_0^t(\pi) = 2$

$\hat{h}_1^t(0)\hat{g}_1^t(0) + \hat{h}_1^t(\pi)\hat{g}_1^t(\pi) = 2$

(*independently* of the value $\alpha = \det A$!)
[In the case of a filter bank which is a DWT, we thus have

$\alpha = \frac{1}{2}\hat{h}_0^t(0)\hat{h}_1^t(\pi)$ $\qquad \frac{1}{\alpha} = \frac{1}{2}\hat{g}_0^t(0)\hat{g}_1^t(\pi)$ $\qquad$ and $\qquad \hat{h}_0^t(0)\hat{g}_0^t(0) = 2$

In other words, one will put – in theory – usually $\sum_k g_0[k] = \sum_k h_0[k] = \sqrt{2}$.]

(c) Now let us pass to a frequent renormalization:
   We shall replace $\mathbf{h}_0$ by $\mathbf{h}_0^{(1)} = \frac{1}{\sqrt{2}}\mathbf{h}_0$ (hence we shall take (the coefficients of) $\tilde{m}_0(\omega)$ as low-pass analysis filter).
   Show that we have necessarily:

---

[39] $\alpha = \det A \equiv$ the determinant of the analysis matrix for the considered filter bank.

(i) $\mathbf{g}_0^{(1)} = \sqrt{2} \cdot \mathbf{g}_0$ (we will get $2 \cdot m_0(\omega)$ as low-pass synthesis filter).

(ii) $\sum_k h_0^{(1)}[k] = 1$ and $\sum_k g_0^{(1)}[k] = 2$.

(iii) $\alpha = \frac{1}{2}$.

(2) Consider $m_0(\omega) = -\frac{1}{20}e^{2i\omega} + \frac{1}{4}e^{i\omega} + \frac{3}{5} + \frac{1}{4}e^{-i\omega} - \frac{1}{20}e^{-2i\omega}$.

  (a) Verify that $m_0(0) = 1$, $m_0(\pi) = 0$.

  (b) Find $\tilde{m}_0(\omega)$, symmetric (as $m_0(\omega)$) and of length 7, such that

$$m_0(\omega)\tilde{m}_0^*(\omega) + m_0(\omega + \pi)\tilde{m}_0^*(\omega + \pi) = 1.$$

[*Help*: You have to resolve a linear system which stems from the conditions of perfect reconstruction, with the supplementary constraint $\tilde{m}_0(\pi) = 0$.

So, you will deal with the 4×4 system coming up from the three equations $\sum_k g_0[2n - k]h_0[k] = \delta_{n,0}$, for $n = 0, 1, 2$, together with the additional condition $\sum_k (-1)^k h_0[k] = 0$.

Caution: The summation index k varies from -3 to $+3$, although you will have – by the postulated symmetry – only four unknowns $h_0[0]$, $h_0[1]$, $h_0[2]$, $h_0[3]$.]

Solution: First, we obtain – with $x_k = \frac{1}{\sqrt{2}}h_0[k]$ – the following linear system:

$$\begin{aligned}
\tfrac{6}{5}x_0 \;+\; x_1 \;-\; \tfrac{1}{5}x_2 \quad\quad &= 1, \\
-\; \tfrac{1}{5}x_2 \;+\; x_3 &= 0, \\
-\tfrac{1}{5}x_0 \;+\; x_1 \;+\; \tfrac{12}{5}x_2 \;+\; x_3 &= 0, \\
x_0 \;-\; 2x_1 \;+\; 2x_2 \;-\; 2x_3 &= 0.
\end{aligned}$$

This gives finally:

$$\tilde{m}_0(\omega) = -\frac{3}{280}e^{3i\omega} - \frac{3}{56}e^{2i\omega} + \frac{73}{280}e^{i\omega} + \frac{17}{28} + \frac{73}{280}e^{-i\omega} - \frac{3}{56}e^{-2i\omega} - \frac{3}{280}e^{-3i\omega}.$$

(3) (QMF filter banks)[40] Consider a (finite support) filter bank which is "self-orthogonal" in the following sense:

$$h_0[k] = g_0[-k],$$

$$h_1[k] = g_1[-k] = (-1)^{k+1}h_0[-(k+1)].$$

  (a) Put $m_0(\omega) = \frac{1}{\sqrt{2}}\hat{g}_0(\omega)$.

     Show that the condition of perfect reconstruction is equivalent to

$$|\, m_0(\omega)\,|^2 + |\, m_0(\omega + \pi)\,|^2 = 1.$$

  (b) Now suppose the reconstruction condition (a) and the normalization $m_0(0) = 1$ valid (one speaks then of a QMF filter bank). Suppose moreover that $\hat{\varphi}(\omega) = \prod_{i=1}^{\infty} m_0\left(\frac{\omega}{2^i}\right)$ is a square integrable function.

     Show that then our filter bank is the DWT associated to a multi-resolution analysis with scaling function $\varphi = \varphi(t) \iff \sum_{k \in \mathbb{Z}} |\, \hat{\varphi}(\omega + 2k\pi)\,|^2 = 1$.

  (c) Now pass from $m_0(\omega)$ to $m_0^{(3)}(\omega) = m_0(3\omega)$.

---

[40] QMF $\equiv$ Quadratic Mirror Filter.

Show that the filter bank defined by $m_0^{(3)}(\omega)$ (i.e. by up-sampling to the order 3) is still a QMF filter bank (if the filter bank defined by $m_0(\omega)$ is one), but, on the other hand, that it never will be a DWT (if the filter bank defined by $m_0(\omega)$ is one).

(This answers the question (b) of the exercise (2) in the section "The Orthonormal Situation").

The Filter Banks Which are Discrete Wavelet Transforms

The initial situation. A perfect reconstruction filter bank:

$\mathbf{g}_0 = (g_0[k])$ the low-pass synthesis impulse response.
$\mathbf{h}_0 = (h_0[k])$ the low-pass analysis impulse response.

(The high-pass impulse responses $\mathbf{g}_1$ and $\mathbf{h}_1$ will then be fixed). We suppose $\mathbf{g}_0$ and $\mathbf{h}_0$ to be *of finite (odd) length* and symmetric. Consider

$$m_0(\omega) = \frac{1}{\sqrt{2}} \sum_k g_0[k] e^{-ik\omega},$$

$$\tilde{m}_0(\omega) = \frac{1}{\sqrt{2}} \sum_k h_0[-k] e^{-ik\omega}$$

and

$$\hat{\varphi}(\omega) = \prod_{k=1}^{\infty} m_0\left(\frac{\omega}{2^k}\right),$$

$$\hat{\tilde{\varphi}}(\omega) = \prod_{k=1}^{\infty} \tilde{m}_0\left(\frac{\omega}{2^k}\right).$$

In order to obtain that the considered filter bank is the DWT associated with a coupled multi-resolution analysis, with scaling functions $(\varphi, \tilde{\varphi})$ (preferably biorthogonal), we need (at least) that the infinite products of the trigonometric polynomials above are convergent and that $\varphi = \varphi(t)$ and $\tilde{\varphi} = \tilde{\varphi}(t)$ are square integrable.

How can we find the appropriate conditions on $m_0(\omega)$ and $\tilde{m}_0(\omega)$ such that $\varphi = \varphi(t)$ and $\tilde{\varphi} = \tilde{\varphi}(t)$ will be *validated*?

Recall our first criterion of compliance:

If the considered filter bank is indeed a DWT, then we necessarily shall have: $m_0(\pi) = \tilde{m}_0(\pi) = 0$. Hence, only low-pass impulse responses $\mathbf{g}_0$ and $\mathbf{h}_0$ with $\sum_k (-1)^k g_0[k] = \sum_k (-1)^k h_0[k] = 0$ will be accepted.

Write then:

$$m_0(\omega) = \left(\frac{1 + e^{-i\omega}}{2}\right)^N p_0(\omega),$$

$N \equiv$ the multiplicity of the root π of $m_0(\omega)$

$$\tilde{m}_0(\omega) = \left(\frac{1 + e^{-i\omega}}{2}\right)^{\tilde{N}} \tilde{p}_0(\omega),$$

$\tilde{N} \equiv$ the multiplicity of the root π of $\tilde{m}_0(\omega)$

with $p_0(0) = \tilde{p}_0(0) = 1$.

We search for a sufficiently general result which guarantees the realization of a filter bank as a DWT.

Let us begin with an *observation* which is indispensable for the understanding of the technical hypotheses which control the central theorem of this section.

Consider

$$m_0(\omega) = \left(\frac{1 + e^{-i\omega}}{2} \right)^N p_0(\omega).$$

Now,

$$\left| \frac{1 + e^{-i\omega}}{2} \right| = \left| \cos \left(\frac{\omega}{2} \right) \right|$$

and

$$\prod_{k=1}^{\infty} \cos \left(\frac{\omega}{2^k} \right) = \frac{\sin \omega}{\omega}.$$

We hence obtain for $\hat{\varphi}(\omega) = \prod_{k=1}^{\infty} m_0 \left(\frac{\omega}{2^k} \right)$:

$$|\hat{\varphi}(\omega)| \le C(1 + |\omega|)^{-N} \cdot \prod_{k=1}^{\infty} |p_0(\frac{\omega}{2^k})|$$

In order to guarantee by simple means that $\hat{\varphi} \in \mathbf{L}^2(\mathbb{R})$, we need only that $\prod_{k=1}^{\infty} |p_0 \left(\frac{\omega}{2^k} \right)|$ is bounded by $(1 + |\omega|)^b$, where $N - b > \frac{1}{2}$.

The realization of this program demands some technical vigilance. Hence consider $B = \sup_{\omega \in [0, 2\pi]} |p_0(\omega)|$ and put $b = \mathrm{Log}_2 B$. Then we obtain the desired estimation.

Lemma $|\hat{\varphi}(\omega)| \le C \cdot (1 + |\omega|)^{-(N-b)}$

Proof It remains to show that $\prod_{k=1}^{\infty} |p_0 \left(\frac{\omega}{2^k} \right)| \le C' \cdot (1 + |\omega|)^b$.

First, $p_0(\omega)$ is a trigonometric polynomial, with $p_0(0) = 1$. Consequently

$$|p_0(\omega)| \le 1 + c|\omega| \le e^{c|\omega|}.$$

Put $q(\omega) = \prod_{k=1}^{\infty} p_0 \left(\frac{\omega}{2^k} \right)$.

$$|q(\omega)| \le \exp(c \cdot \sum_{k=1}^{\infty} \frac{|\omega|}{2^k}) = e^{c|\omega|}.$$

$|q(\omega)|$ is bounded for $|\omega| \le 1$. Let us then find an estimation for $|\omega| > 1$. Let $n_\omega \in \mathbb{N}$ with $2^{n_\omega - 1} \le |\omega| < 2^{n_\omega}$.

$$|q(\omega)| = \prod_{k=1}^{n_\omega} |p_0 \left(\frac{\omega}{2^k} \right)| \cdot \prod_{k=1}^{\infty} |p_0 \left(\frac{2^{-n_\omega}\omega}{2^k} \right)| < \prod_{k=1}^{n_\omega} |p_0 \left(\frac{\omega}{2^k} \right)| \cdot e^c \le e^c B^{n_\omega}$$

$$= e^c 2^{n_\omega \mathrm{Log}_2 B} \le e^c (2|\omega|)^{\mathrm{Log}_2 B} = e^c B |\omega|^{\mathrm{Log}_2 B} \le C' \cdot (1 + |\omega|)^b \quad \square$$

Remark *We can give a more refined version of the lemma above, when replacing the exponent b by the critical exponent β of the trigonometric polynomial $m_0(\omega)$. β is defined as follows:*

*First: $B_0 = 1$, $B_j = \sup_{\omega \in \mathbb{R}} |\prod_{k=1}^{j} p_0(2^k \omega)|$, $j > 0$. Then: $b_j = \mathrm{Log}_{2^j} B_j$, $j > 0$
Finally: $\beta = \inf_{j>0} b_j$.*

We note that b_1 is the exponent b of the foregoing lemma.

Now let us pass to the principal result of this paragraph.

Theorem *Let us start with a perfect reconstruction filter bank.*

Associate with $m_0(\omega) = \left(\frac{1+e^{-i\omega}}{2}\right)^N p_0(\omega)$ *and with* $\tilde{m}_0(\omega) = \left(\frac{1+e^{-i\omega}}{2}\right)^{\tilde{N}} \tilde{p}_0(\omega)$
the exponents b and $\tilde{b}$ (the critical exponents β and $\tilde{\beta}$).

(1) If $N - b > \frac{1}{2}$ and $\tilde{N} - \tilde{b} > \frac{1}{2}$ ($N - \beta > \frac{1}{2}$ and $\tilde{N} - \tilde{\beta} > \frac{1}{2}$), then $\varphi = \varphi(t)$ and $\tilde{\varphi} = \tilde{\varphi}(t)$ are square integrable. More precisely:

$$(D) \qquad |\hat{\varphi}(\omega)| + |\hat{\tilde{\varphi}}(\omega)| \le C \cdot (1 + |\omega|)^{-\frac{1}{2} - \epsilon}$$

with appropriate $\epsilon > 0$.

(2) If (D) holds, then the families $(\varphi_k = \varphi(t - k))_{k \in \mathbb{Z}}$ and $(\tilde{\varphi}_k = \tilde{\varphi}(t - k))_{k \in \mathbb{Z}}$ are necessarily biorthogonal and generate two biorthogonal wavelet bases $(\psi_n^{(m)})_{m,n \in \mathbb{Z}}$ and $(\tilde{\psi}_n^{(m)})_{m,n \in \mathbb{Z}}$.

Proof (1) We already treated the question.

(2) It is the inevitable biorthogonality which fascinates us. Let us show then how the (D)-controlled decreasing of $\hat{\varphi}(\omega)$ and of $\hat{\tilde{\varphi}}(\omega)$ imply the biorthogonality $\langle \varphi_k, \tilde{\varphi}_{k'} \rangle = \delta_{k,k'}$.

Thus consider the "natural" approximations of $\hat{\varphi}(\omega) = \prod_{k=1}^{\infty} m_o\left(\frac{\omega}{2^k}\right)$ and of $\hat{\tilde{\varphi}}(\omega) = \prod_{k=1}^{\infty} \tilde{m}_o\left(\frac{\omega}{2^k}\right)$, given by

$$\hat{h}_0(\omega) = \hat{\tilde{h}}_0(\omega) = \mathbb{I}_{[-\pi,\pi]},$$

$\hat{h}_n(\omega) = \prod_{k=1}^{n} m_o\left(\frac{\omega}{2^k}\right) \cdot \mathbb{I}_{[-2^n \pi, 2^n \pi]}$ and by $\hat{\tilde{h}}_n(\omega) = \prod_{k=1}^{n} \tilde{m}_o\left(\frac{\omega}{2^k}\right) \cdot \mathbb{I}_{[-2^n \pi, 2^n \pi]}$, $n \ge 1$.

First, let us show (by recursion) that $\langle h_n(t - k), \tilde{h}_n(t - k') \rangle = \delta_{k,k'}$

$$\int_{\mathbb{R}} \hat{h}_n(\omega) \hat{\tilde{h}}_n(\omega)^* e^{il\omega} d\omega = \int_{-2^n \pi}^{2^n \pi} \left(\prod_{k=1}^{n} m_0(2^{-k}\omega) \tilde{m}_0^*(2^{-k}\omega) \right) e^{il\omega} d\omega$$

$$= 2^n \int_{-\pi}^{\pi} \left(\prod_{k=0}^{n-1} m_0(2^k \omega) \tilde{m}_0^*(2^k \omega) \right) e^{i2^n l\omega} d\omega$$

$$= 2^n \int_{-\frac{\pi}{2}}^{\frac{\pi}{2}} \left(\prod_{k=1}^{n-1} m_0(2^k \omega) \tilde{m}_0^*(2^k \omega) \right) \cdot [m_0(\omega) \tilde{m}_0^*(\omega)$$

$$+ m_0(\omega + \pi) \tilde{m}_0^*(\omega + \pi)] \cdot e^{i2^n l\omega} d\omega$$

$$= 2^{n-1} \int_{-\pi}^{\pi} \left(\prod_{k=0}^{n-2} m_0(2^k \omega) \tilde{m}_0^*(2^k \omega) \right) e^{i2^{n-1} l\omega} d\omega$$

$$= \int_{-2^{n-1} \pi}^{2^{n-1} \pi} \left(\prod_{k=1}^{n-1} m_0(2^{-k}\omega) \tilde{m}_0^*(2^{-k}\omega) \right) e^{il\omega} d\omega$$

$$= \int_{\mathbb{R}} \hat{h}_{n-1}(\omega) \hat{\tilde{h}}_{n-1}(\omega)^* e^{il\omega} d\omega.$$

But: $\int_{\mathbb{R}} \hat{h}_0(\omega)\hat{\bar{h}}_0(\omega)^* e^{il\omega} d\omega = \int_{-\pi}^{\pi} e^{il\omega} d\omega = 2\pi\delta_{0,l}$. (We observe the crucial intervention of the perfect reconstruction identity.)

The remainder is easy: It is clear that $\hat{h}_n(\omega)$ and $\hat{\bar{h}}_n(\omega)$ pointwise converge towards $\hat{\varphi}(\omega)$ and $\hat{\bar{\varphi}}(\omega)$, respectively. Moreover, by the hypotheses of the theorem, the functions h_n and $\tilde{h}_n$ also verify the inequality (D) for their Fourier transforms. We only have to apply the theorem on dominated convergence in order to conclude that the sequences h_n and $\tilde{h}_n$ tend to φ and $\tilde{\varphi}$ in $L^2(\mathbb{R})$. This finally gives the biorthogonality of the sequences $(\varphi_k)_{k\in\mathbb{Z}}$ and $(\tilde{\varphi}_k)_{k\in\mathbb{Z}}$. $\square$

Remark   Let us insist upon a surprising feature of the theorem: If you want to design a DWT, starting with a candidate filter bank, and passing by the scaling equations in spectral form, then – by any natural mathematical approach – you hardly can avoid the condition (D) as a guaranty for valid scaling functions – and you will *necessarily* arrive at biorthogonal wavelet bases. All these things are interesting, but maybe too abstract for an intuitive and practical spirit. Hence we eagerly should present an (almost) playful aspect of the affair:

How to visualize $\varphi = \varphi(t)$ from $\mathbf{g}_0 = (g_0[k])$?

We shall be confronted with the

Subdivision Algorithm

Recall $\varphi(t) = \sqrt{2} \cdot \sum_k g_0[k]\varphi(2t - k)$[41]

Hence: $\varphi = \varphi(t)$ is a fixed point for the operator $(Tf)(t) = \sqrt{2}\cdot\sum_k g_0[k]f(2t-k)$. Let us then try an approximate construction of $\varphi = \varphi(t)$ by iteration. We shall begin with $u_0 = \mathbb{I}_{[-\frac{1}{2},\frac{1}{2}]}$ and shall have $u_n = T^n(u_0)$, $n \geq 1$.

Observation   The operator T transforms a step function (a piecewise constant function) into a step function – by dyadic subdivision of the step intervals.

In our case, u_n will hence be constant on intervals of length $\frac{1}{2^n}$, more precisely on the intervals of type $]2^{-n}\left(k - \frac{1}{2}\right), 2^{-n}\left(k + \frac{1}{2}\right)[$.

Exercises

(1) Let $\hat{u}_n(\omega)$ be the Fourier transform of $u_n = u_n(t)$. Show that
$$\hat{u}_n(\omega) = \prod_{k=1}^{n} m_0(2^{-k}\omega)\hat{u}_0(2^{-n}\omega)$$
[*Help*: Verify first that $(\widehat{Tf})(\omega) = m_0\left(\frac{\omega}{2}\right)\hat{f}\left(\frac{\omega}{2}\right)$]

Associate with $u_n = u_n(t)$ the sequence $\mathbf{s}_n$ of its values on the various intervals $]2^{-n}(k - \frac{1}{2}), 2^{-n}(k + \frac{1}{2})$ $[s_n[k] = u_n(2^{-n}k)]$.
The step function u_n is completely described by the sequence $\mathbf{s}_n$.

---

[41] We tacitly shall suppose $\mathbf{g}_0$ to be of finite (odd) length and symmetric.

(2) Let $S_n(\omega)$ be the Fourier transform of the sequence s_n, considered as an element of the vector space $l^2\left(\frac{1}{2^n}\mathbb{Z}\right)$.[42]
Show that
$S_0(\omega) = 1$,
$S_1(\omega) = 2m_0\left(\frac{\omega}{2}\right)$
More generally:
$S_n(\omega) = 2^n \cdot \prod_{k=1}^{n} m_0(2^{-k}\omega)$
i.e. that $S_{n+1}(\omega) = 2S_n(\omega)m_0\left(\frac{\omega}{2^{n+1}}\right)$.

(3) (The subdivision algorithm) Show that
$s_0[k] = \delta_{k,0}$,
$s_1[k] = \sqrt{2} \cdot g_0[k]$,
$s_{n+1}[k] = \sqrt{2} \cdot \sum_l g_0[k - 2l]s_n[l]$.

Accept the following result: If $\varphi = \varphi(t)$ is the valid scaling function of a (synthesis) multi-resolution analysis, then *the subdivision algorithm converges uniformly to* $\varphi = \varphi(t)$.
Note the important aspect of *graphic visualization* of $\varphi = \varphi(t)$ from $g_0 = (g_0[k])$. In other words: In a first approach, the validity of $\varphi = \varphi(t)$ (and its degree of regularity) are *visually* verified.

(4) We now insist on the symmetry of g_0: $g_0[-k] = g_0[k]$. Show that then $s_n[-k] = s_n[k]$ for $k, n \geq 0$ (and that $\varphi = \varphi(t)$ is symmetric: $\varphi(-t) = \varphi(t)$).

(5) Show, in the situation of the preceding exercise: If the support of g_0 is between $-N$ and N, then the same is true for $u_n = u_n(t)$, $n \geq 1$, hence also for $\varphi = \varphi(t)$.

(6) Consider $m_0(\omega) = \frac{1}{4}e^{i\omega} + \frac{1}{2} + \frac{1}{4}e^{-i\omega}$.
So we have: $g_0[0] = \frac{\sqrt{2}}{2}$, $g_0[1] = g_0[-1] = \frac{\sqrt{2}}{4}$.
(a) Compute s_0, s_1, s_2 and s_3.
(b) Show, by recursion: $s_n[k] = 1 - \frac{k}{2^n}$ $0 \leq k \leq 2^n$.
(c) Deduce:
$$\varphi(t) = \begin{cases} 1+t & -1 \leq t \leq 0, \\ 1-t & 0 \leq t \leq 1, \\ 0 & \text{else.} \end{cases}$$

(7) Now consider $\tilde{m}_0(\omega) = -\frac{1}{8}e^{2i\omega} + \frac{1}{4}e^{i\omega} + \frac{3}{4} + \frac{1}{4}e^{-i\omega} - \frac{1}{8}e^{-2i\omega}$.
Let us sum up: $h_0[0] = \frac{3\sqrt{2}}{4}$, $h_0[1] = h_0[-1] = \frac{\sqrt{2}}{4}$ $h_0[2] = h_0[-2] = -\frac{\sqrt{2}}{8}$.
Compute $\tilde{s}_0$, $\tilde{s}_1$, $\tilde{s}_2$ and $\tilde{s}_3$ by the subdivision algorithm for h_0.[43]
Trace the step function described by $\tilde{s}_3$.

Commentary The exercises (6) and (7) are meant to give an idea of the shape of the scaling functions $\varphi = \varphi(t)$ and $\tilde{\varphi} = \tilde{\varphi}(t)$ "behind" the DWT 5/3 spline.

---

[42] Pay attention to fractionary counting: For a recursive argumentation, we have to keep track of the dyadic sampling refinement.
[43] The symmetry of h_0 will hide the time-reversed counting "on the analysis side".

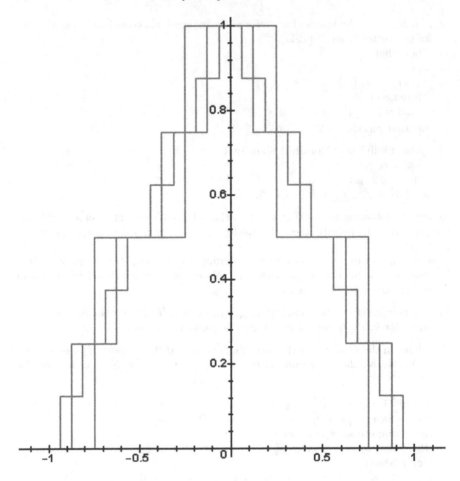

s_1, s_2, s_3, for g_0 of the DWT 5/3 spline.

(8) Let $\varphi = \varphi(t)$ be a valid scaling function – which we have approximated by the subdivision algorithm, beginning with $g_0 = (g_0[k])$.

Change the viewpoint: Carry out a "frequential" approximation of the form
$\hat{\varphi}_n(\omega) = \prod_{k=1}^{n} m_0(2^{-k}\omega)\mathbb{I}_{[-2^n\pi, 2^n\pi]}$
$\hat{\varphi}_n(\omega) \longrightarrow \hat{\varphi}(\omega)$ (in $\mathbf{L}^2(\mathbb{R})$ – take this for granted).
Put $\varphi_n(t) = \frac{1}{2\pi} \int_{-2^n\pi}^{2^n\pi} \prod_{k=1}^{n} m_0(2^{-k}\omega)e^{in\omega t}d\omega$
$\varphi_n(t) \longrightarrow \varphi(t)$ (this is immediate).
Show that the subdivision algorithm interpolates *at the same time* the sequences of functions $u_n = u_n(t)$ and $\varphi_n = \varphi_n(t)$:
$s_n[k] = u_n\left(\frac{k}{2^n}\right) = \varphi_n\left(\frac{k}{2^n}\right)$
[*Help*: On the one hand, $\prod_{k=1}^{n} m_0(2^{-k}\omega) = \frac{1}{2^n} \cdot \sum_k s_n[k]e^{-i\frac{k}{2^n}\omega}$, on the other hand

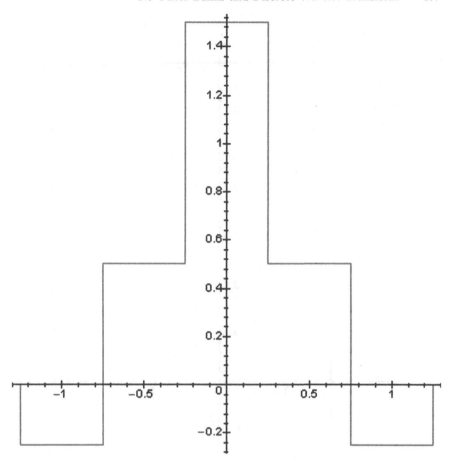

$\tilde{s}_1$ for h_0 of the DWT 5/3 spline.

$$\prod_{k=1}^{n} m_0(2^{-k}\omega)\mathbb{I}_{[-2^n\pi, 2^n\pi]} = \hat{\varphi}_n(\omega) = \frac{1}{2^n} \cdot \sum_{k} \varphi_n\left(\frac{k}{2^n}\right) e^{-\mathrm{i}\frac{k}{2^n}\omega}\mathbb{I}_{[-2^n\pi, 2^n\pi]}$$

(Fourier transform of the Shannon interpolator)]

Complement: The Subdivision Algorithm for $\psi = \psi(t)$

The identity $\psi(t) = \sqrt{2} \cdot \sum_{k} g_1[k]\varphi(2t - k)$ becomes, in frequential notation

$$\hat{\psi}(\omega) = e^{-\mathrm{i}\frac{\omega}{2}} m_1\left(\frac{\omega}{2}\right) \hat{\varphi}\left(\frac{\omega}{2}\right) = e^{-\mathrm{i}\frac{\omega}{2}} m_1\left(\frac{\omega}{2}\right) \cdot \prod_{k=2}^{\infty} m_0\left(\frac{\omega}{2^k}\right).$$

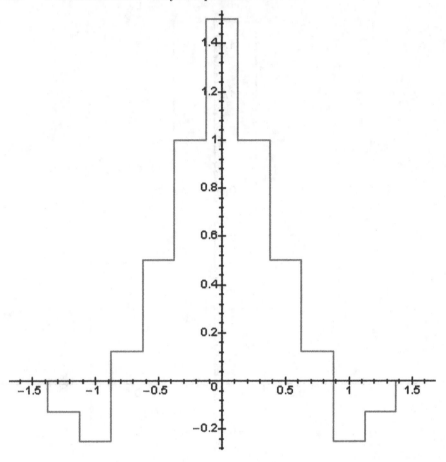

$\tilde{s}_2$ for $\mathbf{h}_0$ of the DWT 5/3 spline.

This leads us to consider the following recursive algorithm:

$$s'_0[k] = \delta_{k,0},$$
$$s'_1[k] = \sqrt{2} \cdot g_1[k],$$
$$s'_{n+1}[k] = \sqrt{2} \cdot \sum_l g_0[k - 2l]s'_n[l].$$

(9) Let $S'_n(\omega)$ be the Fourier transform of $\mathbf{s}'_n \in \mathrm{l}^2\left(\frac{1}{2^n}\mathbb{Z}\right), n \geq 0$:

$$S'_n(\omega) = \sum_k s'_n[k]e^{-\mathrm{i}\frac{k}{2^n}\omega}.$$

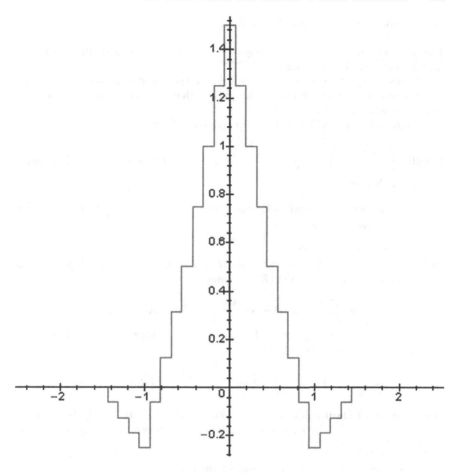

$\tilde{s}_3$ for h_0 of the DWT 5/3 spline.

Show that

$$S_0'(\omega) = 1,$$

$$S_1'(\omega) = 2e^{-i\frac{\omega}{2}} m_1\left(\frac{\omega}{2}\right),$$

$$S_n'(\omega) = 2^n e^{-i\frac{\omega}{2}} m_1\left(\frac{\omega}{2}\right) \cdot \prod_{k=2}^{n} m_0\left(\frac{\omega}{2^k}\right), \quad n \geq 2.$$

(10) Let us define an approximation $\psi_n(t) \longrightarrow \psi(t)$ by frequential truncation:

$\hat{\psi}_n(\omega) = e^{-i\frac{\omega}{2}} m_1\left(\frac{\omega}{2}\right) \cdot \prod_{k=2}^{n} m_0\left(\frac{\omega}{2^k}\right) \cdot \mathbb{I}_{[-2^n\pi, 2^n\pi]}$ (the convergence arguments are the same as in the case of $\varphi_n(t) \longrightarrow \varphi(t)$).

Show that $s_n'[k] = \psi_n\left(\frac{k}{2^n}\right)$ $k \in \mathbb{Z}, \quad n \geq 0$. (cf. the aid of exercise (8))

Construction of Biorthogonal Wavelet Bases

The star construction for biorthogonal wavelet bases comes from A.Cohen, I.Daubechies and J.-C.Feauveau.

We shall restrict ourselves to the case where the impulse responses are of finite (odd) length and of linear phase (this means that the interleaved notations are symmetric with respect to $n = 0$).

$m_0(\omega)$ and $\tilde{m}_0(\omega)$ will then be real polynomials[44] in $\cos\omega$.

Recall $\left(\frac{e^{\frac{i\omega}{2}} + e^{\frac{-i\omega}{2}}}{2}\right)^N = \left(\cos\frac{\omega}{2}\right)^N = \left(\frac{1}{2}(1 + \cos\omega)\right)^{\frac{N}{2}}$ is a polynomial in $\cos\omega$ if and only if N is even.

Hence, in our case, $m_0(\omega)$ and $\tilde{m}_0(\omega)$ must have a root of *even multiplicity* at $\omega = \pi$:

$$m_0(\omega) = \left(\cos\frac{\omega}{2}\right)^{2L} \cdot r(\cos\omega)$$

$$\tilde{m}_0(\omega) = \left(\cos\frac{\omega}{2}\right)^{2\tilde{L}} \cdot \tilde{r}(\cos\omega)$$

with $L, \tilde{L} \geq 1$, i.e. $N, \tilde{N} \geq 2$. Put now $M = L + \tilde{L}$, $R(x) = r(x)\tilde{r}(x)$. Then the perfect reconstruction condition $(\mathcal{R})$ becomes

$$(\cos^2\frac{\omega}{2})^M R(\cos\omega) + (\cos^2\frac{\omega + \pi}{2})^M R(\cos(\omega + \pi)) = 1.$$

Observing that $\sin^2\frac{\omega}{2} = \frac{1}{2}(1 - \cos\omega)$, and putting (logically) $R(x) = Q\left(\frac{1}{2}(1 - x)\right)$, we obtain the identity

$$(\cos^2\frac{\omega}{2})^M \cdot Q\left(\sin^2\frac{\omega}{2}\right) + \left(\cos^2\frac{\omega + \pi}{2}\right)^M \cdot Q\left(\sin^2\frac{\omega + \pi}{2}\right) = 1.$$

Observation (I.Daubechies) *The polynomial identity* $u^M Q(1 - u) + (1 - u)^M Q(u) = 1$ *admits a unique solution of minimal degree* $M - 1$, *which is* $Q(u) = \sum_{n=0}^{M-1} \binom{M+n-1}{n} u^n$.

So, we can (already) pass to a short summary:

At a first approach, the design of biorthogonal wavelets with associated DWT of finite support and linear phase, faces the following situation:

We shall fix the multiplicities $N = 2L$ and $\tilde{N} = 2\tilde{L}$ for the root $\omega = \pi$ of the trigonometric polynomials $m_0(\omega)$ and $\tilde{m}_0(\omega)$. Then, we shall consider $R(x) = \sum_{n=0}^{M-1} \binom{M+n-1}{n} \frac{1}{2^n}(1 - x)^n$. $M = L + \tilde{L}$.

Choosing a factorization $R(x) = r(x)\tilde{r}(x)$, we obtain

$$m_0(\omega) = \left(\cos\frac{\omega}{2}\right)^{2L} \cdot r(\cos\omega) = \frac{1}{\sqrt{2}}\sum_k g_0[k]e^{-ik\omega},$$

$$\tilde{m}_0(\omega) = \left(\cos\frac{\omega}{2}\right)^{2\tilde{L}} \cdot \tilde{r}(\cos\omega) = \frac{1}{\sqrt{2}}\sum_k h_0[-k]e^{-ik\omega}.$$

---

[44] Recall that our filter banks are real.

All this seems to be a good beginning. But there remain two fundamental questions:

---

(1) Do the equations

$$\hat{\varphi}(\omega) = \prod_{k=1}^{\infty} m_0\left(\frac{\omega}{2^k}\right) \qquad \hat{\tilde{\varphi}}(\omega) = \prod_{k=1}^{\infty} \tilde{m}_0\left(\frac{\omega}{2^k}\right)$$

define *valid* scaling functions $\varphi = \varphi(t)$ and $\tilde{\varphi} = \tilde{\varphi}(t)$, i.e. which are square integrable and give rise to *biorthogonal* wavelet bases ?

(2) Which are the criteria for the choice of the multiplicities N and $\tilde{N}$ and for an appropriate factorization of $R(x)$ into $r(x)$ and $\tilde{r}(x)$?

Let us consider the first point. The basic result will be stated in a surprising (and even enigmatic) way. Hence, we have to prepare the field.

Exercises

We consider a coupled multi-resolution analysis[45] of $\mathbf{L}^2(\mathbb{R})$, with the synthesis scaling function $\varphi = \varphi(t)$ and the analysis scaling function $\tilde{\varphi} = \tilde{\varphi}(t)$.

(1) Show the following properties:
  (a) If the support of $\mathbf{g}_0 = (g_0[k])_{k\in\mathbb{Z}}$ is between N_1 and N_2 and the support of $\mathbf{h}_0 = (h_0[k])_{k\in\mathbb{Z}}$ is between $\tilde{N}_1$ and $\tilde{N}_2$, then the supports of $\varphi = \varphi(t)$ and of $\tilde{\varphi} = \tilde{\varphi}(t)$ are in the intervals $[N_1, N_2]$ and $[\tilde{N}_1, \tilde{N}_2]$, respectively.
  (b) If $\mathbf{g}_0$ and $\mathbf{h}_0$ (have an odd number of non-zero coefficients and) are symmetric with respect to $k = 0$, then $\varphi = \varphi(t)$ and $\tilde{\varphi} = \tilde{\varphi}(t)$ are symmetric with respect to $t = 0$.

(2) Now associate with $\varphi = \varphi(t)$ and $\tilde{\varphi} = \tilde{\varphi}(t)$ the two functions

$$\alpha(\omega) = \sum_{k\in\mathbb{Z}} |\hat{\varphi}(\omega + 2\pi k)|^2 \quad \text{and} \quad \tilde{\alpha}(\omega) = \sum_{k\in\mathbb{Z}} |\hat{\tilde{\varphi}}(\omega + 2\pi k)|^2,$$

(which are real and strictly positive). Show that

$$\alpha(\omega) = \sum_k \langle \varphi(t), \varphi(t-k) \rangle e^{-ik\omega},$$

$$\tilde{\alpha}(\omega) = \sum_k \langle \tilde{\varphi}(t), \tilde{\varphi}(t-k) \rangle e^{-ik\omega}.$$

[Hence we deal with strictly positive real trigonometric polynomials – provided $\mathbf{g}_0$ and $\mathbf{h}_0$ are real, of odd length and symmetric with respect to $n = 0$]
Help: You should use the lemma which served to establish the summation formula characterizing biorthogonality – now with $F(\omega) = |\hat{\varphi}(\omega)|^2$.

(3) (The point)
Consider the operators P_0 and $\tilde{P}_0$ on trigonometric polynomials, defined by

$$(P_0 f)(\omega) = |m_0\left(\tfrac{\omega}{2}\right)|^2 \cdot f\left(\tfrac{\omega}{2}\right) + |m_0\left(\tfrac{\omega}{2} + \pi\right)|^2 \cdot f\left(\tfrac{\omega}{2} + \pi\right),$$

$$(\tilde{P}_0 f)(\omega) = |\tilde{m}_0\left(\tfrac{\omega}{2}\right)|^2 \cdot f\left(\tfrac{\omega}{2}\right) + |\tilde{m}_0\left(\tfrac{\omega}{2} + \pi\right)|^2 \cdot f\left(\tfrac{\omega}{2} + \pi\right).$$

Show that the polynomial $\alpha(\omega)$ is a *fixed point* for the operator P_0, and that the polynomial $\tilde{\alpha}(\omega)$ is un fixed point for the operator $\tilde{P}_0$.

---

[45] We know what is meant.

Help: Try to really understand the equalities

$$(P_0\alpha)(\omega) = \mid m_0\left(\frac{\omega}{2}\right)\mid^2 \cdot \sum_{k\in\mathbb{Z}}\mid\hat{\varphi}\left(\frac{\omega}{2}+2k\pi\right)\mid^2$$

$$+\mid m_0\left(\frac{\omega}{2}+\pi\right)\mid^2\cdot\sum_{k\in\mathbb{Z}}\mid\hat{\varphi}\left(\frac{\omega}{2}+\pi+2k\pi\right)\mid^2$$

$$=\sum_{k\in\mathbb{Z}}\mid\hat{\varphi}(\omega+4k\pi)\mid^2+\sum_{k\in\mathbb{Z}}\mid\hat{\varphi}(\omega+2\pi+4k\pi)\mid^2=\alpha(\omega).$$

Remark In case of a biorthogonal coupled multi-resolution analysis of $\mathbf{L}^2(\mathbb{R})$, we can show that $\alpha(\omega)$ and $\tilde{\alpha}(\omega)$ are, up to a non-zero scalar multiple, the *only fixed points* of the operator P_0 and $\tilde{P}_0$, respectively. All this is (mathematically) very satisfactory. But for a practical spirit, it seems to be rather academic stuff. So it is really surprising that the following fundamental theorem – the statement of which is highly arid – actually yields a simple and efficient *test* which validates a perfect reconstruction filter bank as a DWT associated with a couple of biorthogonal wavelet bases.

So, let us state one of the principal results of the theory of biorthogonal wavelet bases, which provides – despite its theoretical look – the indispensable practical tool for an efficient resolution to our problem: How to sort the filter banks which are the candidates for "biorthogonal wavelet transforms"?

Theorem (Cohen–Daubechies–Feauveau)   *Fixed Point Criterion.*

Let $\mathbf{g}_0 = (g_0[k])$ and $\mathbf{h}_0 = (h_0[k])$ be two real finite support sequences (eventually symmetric with respect to $k = 0$).

Suppose that the perfect reconstruction condition (R) holds.

Let us define the trigonometric polynomials $m_0(\omega)$, $\tilde{m}_0(\omega)$ by their usual formulas from $\mathbf{g}_0$ and $\mathbf{h}_0$, and $\hat{\varphi} = \hat{\varphi}(\omega)$ and $\hat{\tilde{\varphi}} = \hat{\tilde{\varphi}}(\omega)$ by their product formulas.

Then the following two statements are equivalent:

(1) $\varphi = \varphi(t)$ and $\tilde{\varphi} = \tilde{\varphi}(t)$ are square integrable, and $\langle\varphi(t-k),\tilde{\varphi}(t-l)\rangle = \delta_{k,l}$[46]
(2) There exist two strictly positive (real) trigonometric polynomials f_0 and $\tilde{f}_0$ such that $P_0 f_0 = f_0$ and $\tilde{P}_0\tilde{f}_0 = \tilde{f}_0$, and f_0 and $\tilde{f}_0$ are unique (up to a non-zero scalar multiple).

[Without proof]

Note: The property of the *unique fixed point* (in the projective sense) for the two operators P_0 and $\tilde{P}_0$ is *characteristic* for the existence of *valid* scaling functions φ and $\tilde{\varphi}$ as well as for the biorthogonality of their $\mathbb{Z}$ – translates.

What is important for the designer – as we already underlined – is that this austere theorem actually gives a test whether two candidates $\mathbf{g}_0$ and $\mathbf{h}_0$ (satisfying the condition of perfect reconstruction $(\mathcal{R})$) define the two low-pass filters of a DWT associated with a couple of biorthogonal wavelet bases. In detail:

Observations   concerning the compliance test for $\mathbf{g}_0 = (g_0[k])$[47]:

---

[46] Recall that the variable t is *bound* – by integration – and that the two integers k and l are free.
[47] For $\mathbf{h}_0 = (h_0[k])$ the situation is similar.

(1) The terms $\mid m_0(\omega)\mid^2$ and $\mid m_0(\omega+\pi)\mid^2$ are modulation invariant; thus we can suppose that

$$m_0(\omega) = \frac{1}{\sqrt{2}} \cdot \sum_{k=0}^{N} g_0[k] e^{-ik\omega}$$

and consequently, $\mid m_0(\omega)\mid^2$ and $\mid m_0(\omega+\pi)\mid^2$ will be trigonometric polynomials of degree $\leq N$.

(2) If $f_0 = f_0(\omega)$ is a trigonometric polynomial such that $P_0 f_0 = f_0$, then we have necessarily $f_0(\omega) = \sum_{k=-N}^{N} f_0[k] e^{-ik\omega}$, i.e. $f_0 = f_0(\omega)$ is a trigonometric polynomial of degree $\leq N$.

(3) Let then $\mathbb{E}_N$ be the vector space of the trigonometric polynomials of degree $\leq N$.

$\mathbb{E}_N$ is stable under the operator P_0.

Let $\mathbb{M}_0$ be the matrix of P_0 relative to the basis $\{e^{-ik\omega}, -N \leq k \leq N\}$. Then

$$\mathbb{M}_0 = (g_{mn})_{-N \leq m,n \leq N} \quad \text{with} \quad g_{mn} = \sum_{k} g_0[k] g_0[k+n-2m]$$

(cf. the exercises below).

(4) The condition of the (strictly positive) "unique" fixed point of the operator P_0 now is verified as follows:

   (a) Triangulation of the $2N+1 \times 2N+1$ matrix $\mathbb{M}_0 - \mathbb{I}_{2N+1}$ in order to make sure that $\mathbb{M}_0$ admits the value 1 as a *simple* eigenvalue.
   (b) In the affirmative case, we now dispose of a triangular system of $2N$ equations in $2N+1$ unknowns in order to determine "the" eigenvector which is fixed for the action of $\mathbb{M}_0$.
   (c) Let $f_0(\omega)$ be the (real) trigonometric polynomial the coefficients of which (in complex notation) are the components of this eigenvector. Remains to verify that $f_0(\omega)$ is strictly positive.

Exercises

(1) (Matrix representation of the operator P_0). Let us fix the notation:

$$m_0(\omega) = \frac{1}{\sqrt{2}} \cdot \sum_{k} g_0[k] e^{-ik\omega},$$

$$f(\omega) = \sum_{n} f[n] e^{-in\omega},$$

$$(P_0 f)(\omega) = \mid m_0\left(\frac{\omega}{2}\right)\mid^2 \cdot f\left(\frac{\omega}{2}\right) + \mid m_0\left(\frac{\omega}{2}+\pi\right)\mid^2 \cdot f\left(\frac{\omega}{2}+\pi\right)$$

$$= \sum_{m} (P_0 f)[m] e^{-im\omega}.$$

We will search the coordinates $((P_0 f)[m])$ of the complex notation of the trigonometric polynomial $P_0 f$ in function of the coordinates $(f[n])$ of the complex notation of the trigonometric polynomial f[48]

---

[48] The increasing order will be that of the "tap values", i.e. of the powers of $X = e^{-i\omega}$.

(a) Show that

$$m_0\left(\frac{\omega}{2}\right)m_0^*\left(\frac{\omega}{2}\right)f\left(\frac{\omega}{2}\right)=\frac{1}{2}\cdot\sum_{m\in\frac{1}{2}\mathbb{Z}}(\sum_n(\sum_k g_0[k]g_0[k+n-2m])f[n])e^{-im\omega}.$$

(b) Show that

$$m_0\left(\frac{\omega}{2}+\pi\right)m_0^*\left(\frac{\omega}{2}+\pi\right)f\left(\frac{\omega}{2}+\pi\right)$$

$$=\frac{1}{2}\cdot\sum_{m\in\frac{1}{2}\mathbb{Z}}(-1)^{2m}\left(\sum_n\left(\sum_k g_0[k]g_0[k+n-2m]\right)f[n]\right)e^{-im\omega}.$$

(c) Show that

$$\mid m_0\left(\frac{\omega}{2}\right)\mid^2 f\left(\frac{\omega}{2}\right)+\mid m_0\left(\frac{\omega}{2}+\pi\right)\mid^2 f\left(\frac{\omega}{2}+\pi\right)$$

$$=\sum_{m\in\mathbb{Z}}\left(\sum_n\left(\sum_k g_0[k]g_0[k+n-2m]\right)f[n]\right)e^{-im\omega}.$$

(d) Deduce the matrix representation of the operator P_0: Suppose that the support of g_0 is between 0 and N. Then the transformation of the co-ordinates of $f(\omega)=\sum_{n=-N}^N f[n]e^{-in\omega}$ into coordinates of $(P_0 f)(\omega)=\sum_{m=-N}^N (P_0 f)[m]e^{-im\omega}$ is done by the following matrix multiplication:

$$\begin{pmatrix} P_0 f[-N] \\ P_0 f[-N+1] \\ \vdots \\ P_0 f[N] \end{pmatrix}=\begin{pmatrix} g_{-N,-N} & g_{-N,-N+1} & \cdots & g_{-N,N} \\ g_{-N+1,-N} & g_{-N+1,-N+1} & \cdots & g_{-N+1,N} \\ \vdots & \vdots & & \vdots \\ g_{N,-N} & g_{N,-N+1} & \cdots & g_{N,N} \end{pmatrix}\begin{pmatrix} f[-N] \\ f[-N+1] \\ \vdots \\ f[N] \end{pmatrix},$$

where $g_{mn}=\sum_k g_0[k]g_0[k+n-2m]$ for $-N\leq m,n\leq N$.

(2) (Matrix representation of the operator $\tilde{P}_0$).

$$\tilde{m}_0(\omega)=\frac{1}{\sqrt{2}}\cdot\sum_k h_0[-k]e^{-ik\omega},$$

$$f(\omega)=\sum_n f[n]e^{-in\omega},$$

$$(\tilde{P}_0 f)(\omega)=\mid\tilde{m}_0\left(\frac{\omega}{2}\right)\mid^2\cdot f\left(\frac{\omega}{2}\right)+\mid\tilde{m}_0\left(\frac{\omega}{2}+\pi\right)\mid^2\cdot f\left(\frac{\omega}{2}+\pi\right)=\sum_m(\tilde{P}_0 f)[m]e^{-im\omega}.$$

Show that

$$(\tilde{P}_0 f)[m]=\sum_n\left(\sum_k h_0[-k]h_0[2m-n-k]\right)f[n].$$

(3) Let $\mathbf{g}_0 = (g_0[0], g_0[1], g_0[2])$ have support $\{0, 1, 2\}$.
Put $a = g_0[0]g_0[2]$, $b = g_0[0]g_0[1] + g_0[1]g_0[2]$, $c = g_0[0]^2 + g_0[1]^2 + g_0[2]^2$.
Show that the matrix representation of the operator P_0 on the trigonometric
polynomials of degree ≤ 2 is given by

$$
\begin{pmatrix} (P_0 f)[-2] \\ (P_0 f)[-1] \\ (P_0 f)[0] \\ (P_0 f)[1] \\ (P_0 f)[2] \end{pmatrix} = \begin{pmatrix} a & 0 & 0 & 0 & 0 \\ c & b & a & 0 & 0 \\ a & b & c & b & a \\ 0 & 0 & a & b & c \\ 0 & 0 & 0 & 0 & a \end{pmatrix} \begin{pmatrix} f[-2] \\ f[-1] \\ f[0] \\ f[1] \\ f[2] \end{pmatrix}.
$$

(4) Let $\mathbf{h}_0 = (h_0[0], h_0[1], h_0[2])$ have support $\{0, 1, 2\}$. Put $a = h_0[0]h_0[2]$,
$b = h_0[0]h_0[1] + h_0[1]h_0[2]$, $c = h_0[0]^2 + h_0[1]^2 + h_0[2]^2$.
Show that the matrix representation of the operator $\tilde{P}_0$ on the trigonometric
polynomials of degree ≤ 2 is given by

$$
\begin{pmatrix} (\tilde{P}_0 f)[-2] \\ (\tilde{P}_0 f)[-1] \\ (\tilde{P}_0 f)[0] \\ (\tilde{P}_0 f)[1] \\ (\tilde{P}_0 f)[2] \end{pmatrix} = \begin{pmatrix} a & 0 & 0 & 0 & 0 \\ c & b & a & 0 & 0 \\ a & b & c & b & a \\ 0 & 0 & a & b & c \\ 0 & 0 & 0 & 0 & a \end{pmatrix} \begin{pmatrix} f[-2] \\ f[-1] \\ f[0] \\ f[1] \\ f[2] \end{pmatrix}.
$$

(**Attention** This is a consequence of the exercise (3); but by what argument?)

(5) (A perfect reconstruction filter bank which is not a DWT).
Consider $m_0(\omega) = -\frac{1}{2} + \frac{1}{2}e^{-i\omega} + e^{-2i\omega}$ and $\tilde{m}_0(\omega) = -e^{i\omega} + \frac{1}{2} + \frac{3}{2}e^{-i\omega}$.
(a) Show that $m_0(0) = \tilde{m}_0(0) = 1$, $m_0(\pi) = \tilde{m}_0(\pi) = 0$,   and that

$$
m_0(\omega)\tilde{m}_0^*(\omega) + m_0(\omega + \pi)\tilde{m}_0^*(\omega + \pi) = 1.
$$

(b) Hence we are facing a perfect reconstruction filter bank. Verify that we
have, for the analysis filters $\mathbf{h}_0$ and $\mathbf{h}_1$ and for the synthesis filters $\mathbf{g}_0$ and
$\mathbf{g}_1$:

$(h_0[-1], h_0[0], h_0[1]) = \left(\frac{3\sqrt{2}}{2}, \frac{\sqrt{2}}{2}, -\sqrt{2} \right)$, $(g_0[0], g_0[1], g_0[2]) = \left(-\frac{\sqrt{2}}{2}, \frac{\sqrt{2}}{2}, \sqrt{2} \right)$

$(h_1[-1], h_1[0], h_1[1]) =$ $(g_1[0], g_1[1], g_1[2]) =$

$(h_1^t[0], h_1^t[1], h_1^t[2]) =$ $(g_1^t[-1], g_1^t[0], g_1^t[1]) =$

$\left(-\frac{\sqrt{2}}{2}, -\frac{\sqrt{2}}{2}, \sqrt{2} \right)$ $\left(-\frac{3\sqrt{2}}{2}, \frac{\sqrt{2}}{2}, \sqrt{2} \right)$

(c) Let $\mathbb{M}_0$ be the matrix of the operator P_0 restricted to trigonometric polynomials of degree ≤ 2. Show that

$$
\mathbb{M}_0 = \begin{pmatrix} -1 & 0 & 0 & 0 & 0 \\ 3 & \frac{1}{2} & -1 & 0 & 0 \\ -1 & \frac{1}{2} & 3 & \frac{1}{2} & -1 \\ 0 & 0 & -1 & \frac{1}{2} & 3 \\ 0 & 0 & 0 & 0 & -1 \end{pmatrix}.
$$

Show that the trigonometric polynomial $f_0(\omega) = 1 - 4\cos\omega$ is "the" solution
of the equation $P_0 f_0 = f_0$. Deduce that the identity

$$\hat{\varphi}(\omega) = \prod_{k=1}^{\infty} m_0\left(\frac{\omega}{2^k}\right)$$

cannot define a valid (i.e. square integrable) scaling function $\varphi = \varphi(t)$ (else $\alpha(\omega) = \sum_{k \in \mathbb{Z}} |\hat{\varphi}(\omega + 2\pi k)|^2$ would be proportional to $f_0(\omega)$.)

(d) Let $\tilde{\mathbb{M}}_0$ be the matrix of the operator $\tilde{P}_0$ restricted to trigonometric polynomials of degree ≤ 2. Show that

$$\tilde{\mathbb{M}}_0 = \begin{pmatrix} -3 & 0 & 0 & 0 & 0 \\ 7 & \frac{1}{2} & -3 & 0 & 0 \\ -3 & \frac{1}{2} & 7 & \frac{1}{2} & -3 \\ 0 & 0 & -3 & \frac{1}{2} & 7 \\ 0 & 0 & 0 & 0 & -3 \end{pmatrix}.$$

Show that the trigonometric polynomial $\tilde{f}_0(\omega) = 1 - 12\cos\omega$ is "the" solution of the equation $\tilde{P}_0 \tilde{f}_0 = \tilde{f}_0$.

Deduce that the identity

$$\hat{\tilde{\varphi}}(\omega) = \prod_{k=1}^{\infty} \tilde{m}_0\left(\frac{\omega}{2^k}\right)$$

cannot define a valid scaling function $\tilde{\varphi} = \tilde{\varphi}(t)$.

A final *remark*: The simplicity of our example (two impulse responses of length 3) demanded the sacrifice of symmetry: otherwise, the lengths of the two initial impulse responses should differ by an odd multiple of 2.

(6) (The DWT 5/3 spline)

Consider

$$m_0(\omega) = \frac{1}{4}e^{i\omega} + \frac{1}{2} + \frac{1}{4}e^{-i\omega} \qquad \tilde{m}_0(\omega) = -\frac{1}{8}e^{2i\omega} + \frac{1}{4}e^{i\omega} + \frac{3}{4} + \frac{1}{4}e^{-i\omega} - \frac{1}{8}e^{-2is\omega}$$

(a) Verify that $\quad m_0(0) = \tilde{m}_0(0) = 1 \quad m_0(\pi) = \tilde{m}_0(\pi) = 0 \quad$ and that

$$m_0(\omega)\tilde{m}_0^*(\omega) + m_0(\omega + \pi)\tilde{m}_0^*(\omega + \pi) = 1.$$

The perfect reconstruction filter bank thus defined is already familiar to us. We want to verify that we actually have a DWT associated with a couple of biorthogonal wavelet bases.

(b) Let $\mathbb{M}_0$ be the matrix of the operator P_0 on the trigonometric polynomials of degree ≤ 2. Show that

$$\mathbb{M}_0 = \begin{pmatrix} \frac{1}{8} & 0 & 0 & 0 & 0 \\ \frac{3}{8} & \frac{1}{4} & \frac{1}{8} & 0 & 0 \\ \frac{1}{4} & \frac{1}{2} & \frac{3}{4} & \frac{1}{2} & \frac{1}{4} \\ 0 & 0 & \frac{1}{8} & \frac{1}{2} & \frac{3}{8} \\ 0 & 0 & 0 & 0 & \frac{1}{8} \end{pmatrix}.$$

Find "the" fixed point $f_0 = f_0(\omega)$ of the operator P_0 and verify that it is strictly positive.

[*Answer*: $f_0(\omega) = 2 + \cos\omega$]

(c) Let $\tilde{\mathbb{M}}_0$ be the matrix of the operator $\tilde{P}_0$ on the trigonometric polynomials of degree ≤ 4. Show that

$$\tilde{\mathbb{M}}_0 = \begin{pmatrix} a\,0\,0\,0\,0\,0\,0\,0\,0 \\ c\,b\,a\,0\,0\,0\,0\,0\,0 \\ e\,d\,c\,b\,a\,0\,0\,0\,0 \\ c\,d\,e\,d\,c\,b\,a\,0\,0 \\ a\,b\,c\,d\,e\,d\,c\,b\,a \\ 0\,0\,a\,b\,c\,d\,e\,d\,c \\ 0\,0\,0\,0\,a\,b\,c\,d\,e \\ 0\,0\,0\,0\,0\,0\,a\,b\,c \\ 0\,0\,0\,0\,0\,0\,0\,0\,a \end{pmatrix}$$

with

$$a = \frac{1}{32}, \quad b = -\frac{1}{8}, \quad c = -\frac{1}{4}, \quad d = \frac{5}{8}, \quad e = \frac{23}{16}.$$

Find "the" fixed point $\tilde{f}_0 = \tilde{f}_0(\omega)$ of the operator $\tilde{P}_0$ and verify that it is strictly positive.

[*Answer:* $\tilde{f}_0(\omega) = 308 - 201\cos\omega + 36\cos 2\omega + \cos 3\omega$]

(d) Conclude by the theorem of Cohen–Daubechies–Feauveau: The filter bank given by $m_0(\omega)$ and $\tilde{m}_0(\omega)$ (i.e. by the low-pass synthesis filter $\mathbf{g}_0$ and by the low-pass analysis filter $\mathbf{h}_0$) is a DWT associated with a biorthogonally coupled multi-resolution analysis .

(7) Consider the perfect reconstruction filter bank given by[49]

$$m_0(\omega) = -\frac{1}{20}e^{2i\omega} + \frac{1}{4}e^{i\omega} + \frac{3}{5} + \frac{1}{4}e^{-i\omega} - \frac{1}{20}e^{-2i\omega},$$

$$\tilde{m}_0(\omega) = -\frac{3}{280}e^{3i\omega} - \frac{3}{56}e^{2i\omega} + \frac{73}{280}e^{i\omega} + \frac{17}{28} + \frac{73}{280}e^{-i\omega} - \frac{3}{56}e^{-2i\omega} - \frac{3}{280}e^{-3i\omega}.$$

(a) Show, using the Fixed Point Criterion, that we get a DWT associated with a couple of biorthogonal wavelet bases.

(b) Trace the approximations s_3 of $\varphi = \varphi(t)$ and $\tilde{s}_3$ of $\tilde{\varphi} = \tilde{\varphi}(t)$ by the subdivision algorithm.[50]

[*Help* For (a):

Concerning $\mathbb{M}_0$, the 9×9 matrix of the operator $\mathbf{P}_0$ on the trigonometric polynomials of degree ≤ 4, we are in the same situation as in part (c) of the preceding exercise, now with $a = \frac{1}{200}$, $b = -\frac{1}{20}$, $c = \frac{1}{200}$, $d = \frac{11}{20}$, $e = \frac{98}{100}$.

Observing that our linear system is invariant by inversion of the order of the variables – and that the two outer equations are trivial – we know that we can search for a solution of the form $(0, X_3, X_2, X_1, X_0, X_1, X_2, X_3, 0)$.

The final triangular system: $\begin{aligned} X_0 &- 10X_1 - 199X_2 + 110X_3 = 0 \\ &\quad 90X_1 - 396X_2 \qquad\qquad = 0 \\ &\qquad\qquad X_2 \quad - 210X_3 = 0 \end{aligned}$

---

[49] Cf. exercise (2) of the initial section on design procedures; we deal with the star in the family of Burt filters.

[50] Cf. the exercises at the end of the section "The Filter Banks Which are Discrete Wavelet Transforms".

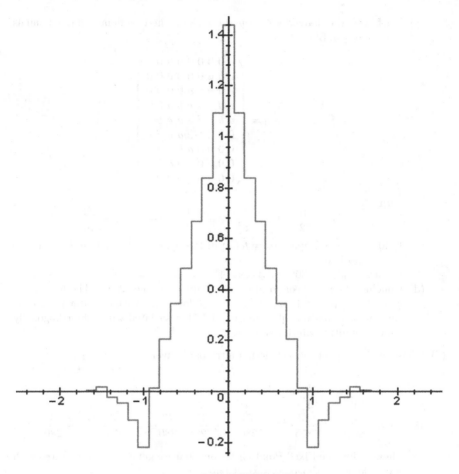

$\mathbf{s}_3$ for $\mathbf{g}_0$ of the DWT 7/5 Burt.

This yields our ("unique") strictly positive polynomial: $f_0(\omega) = 25460 + 924\cos\omega + 210\cos 2\omega + \cos 3\omega$ (corresponding to $X_3 = \frac{1}{2}$).

As to $\tilde{\mathbb{M}}_0$, the 13×13 matrix of the operator $\tilde{\mathbf{P}}_0$ on the trigonometric polynomials of degree ≤ 6, it now depends – by formal analogy with the preceding one – on the 7 parameters

$$a = \frac{18}{280^2}, \quad b = \frac{180}{280^2}, \quad c = -\frac{426}{280^2},$$

$$d = -\frac{6420}{280^2}, \quad e = -\frac{418}{280^2}, \quad f = \frac{45440}{280^2}, \quad g = \frac{80052}{280^2}.$$

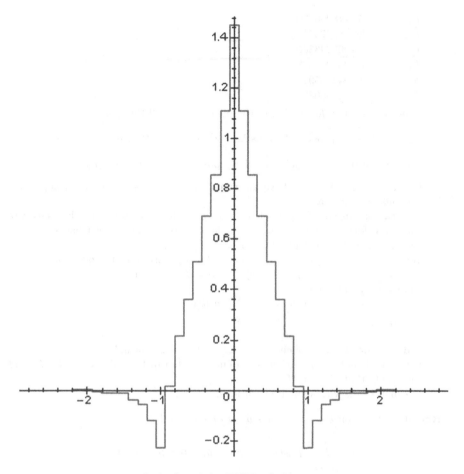

$\tilde{s}_3$ for $\mathbf{h}_0$ of the DWT 7/5 Burt.

Once more, we can search, for the homogeneous system defined by $\tilde{\mathbb{M}}_0 - \mathbb{I}_{13}$, a symmetric solution $(0, X_5, X_4, X_3, X_2, X_1, X_0, X_1, X_2, X_3, X_4, X_5, 0)$. This leads us to the following reduced system:

$$
\begin{array}{rrrrrrl}
-39110X_5 & +9X_4 & & & & & = 0 \\
-3210X_5 & -39413X_4 & +90X_3 & +9X_2 & & & = 0 \\
22720X_5 & -209X_4 & -42410X_3 & -213X_2 & +90X_1 & +9X_0 & = 0 \\
22720X_5 & +40026X_4 & +22720X_3 & -39400X_2 & -3120X_1 & -213X_0 & = 0 \\
-3210X_5 & -200X_4 & +22810X_3 & +39813X_2 & -19690X_1 & -209X_0 & = 0 \\
90X_5 & -213X_4 & -3210X_3 & -209X_2 & +22720X_1 & +413X_0 & = 0
\end{array}
$$

The solutions are the scalar multiples of the vector $(\tilde{F}_0, \tilde{F}_1, \tilde{F}_2, \tilde{F}_3, \tilde{F}_4, \tilde{F}_5)$ given by:

$$\tilde{F}_0 = 5445171178000580$$
$$\tilde{F}_1 = -99025982875717$$
$$\tilde{F}_2 = -20991909243300$$
$$\tilde{F}_3 = 1050836279598$$
$$\tilde{F}_4 = -21544955910$$
$$\tilde{F}_5 = -550881$$

"The" fixed point $\tilde{f}_0 = \tilde{f}_0(\omega)$ of the operator $\tilde{P}_0$ will then be

$$\tilde{f}_0(\omega) = \tilde{F}_0 + 2\tilde{F}_1\cos\omega + 2\tilde{F}_2\cos 2\omega + 2\tilde{F}_3\cos 3\omega + 2\tilde{F}_4\cos 4\omega + 2\tilde{F}_5\cos 5\omega.$$

It is immediate that $\tilde{f}_0 = \tilde{f}_0(\omega)$ is strictly positive (cf. the size of $\tilde{F}_0$!).]

Let us sum up: We just have answered the first question of the designer for biorthogonal wavelet bases:

How can one decide if a perfect reconstruction filter bank is actually a Discrete Wavelet Transform associated with a couple of biorthogonal wavelet bases?

There remains the second question:

Which are the criteria for the choice of the appropriate factorization of
$$R(x) = \sum_{n=0}^{M-1}\binom{M+n-1}{n}\frac{1}{2^n}(1-x)^n, \quad M = L + \tilde{L}, \quad \text{into}$$
$R(x) = r(x)\tilde{r}(x)$, where
$$m_0(\omega) = (\cos\tfrac{\omega}{2})^{2L}\cdot r(\cos\omega) = \tfrac{1}{\sqrt{2}}\sum_k g_0[k]e^{-ik\omega}$$
$$\tilde{m}_0(\omega) = (\cos\tfrac{\omega}{2})^{2\tilde{L}}\cdot \tilde{r}(\cos\omega) = \tfrac{1}{\sqrt{2}}\sum_k h_0[-k]e^{-ik\omega}$$

and what do the multiplicities $N = 2L$ and $\tilde{N} = 2\tilde{L}$ mean?

Let us first look at the information encapsulated in the value of $N = 2L$ (and symmetrically that of $\tilde{N} = 2\tilde{L}$).

We shall begin with some preparations.

Recall A function $\tilde{\psi} = \tilde{\psi}(t)$[51] has p *zero moments* if

$$\int_{-\infty}^{+\infty} t^k\tilde{\psi}(t)\mathrm{d}t = 0 \qquad \text{for} \quad 0 \le k \le p-1$$

(hence $\tilde{\psi}$ is orthogonal to every polynomial of degree $\le p-1$)

Exercises

We shall work in the setting of a coupled multi-resolution analysis, with our notations introduced above. The considered filter bank is, clearly, of finite type.

(1) Show that the following three statements are equivalent:
   (a) The analysis mother wavelet $\tilde{\psi} = \tilde{\psi}(t)$ has p zero moments.
   (b) $\hat{\tilde{\psi}} = \hat{\tilde{\psi}}(\omega)$ is zero for $\omega = 0$, as well as its $p-1$ first higher derivatives.
   (c) $\hat{g}_0(\omega)$ and its $p-1$ first higher derivatives are zero for $\omega = \pi$, i.e. $m_0(\omega)$ admits a root to the order $\ge p$ in $\omega = \pi$.

---

[51] We shall tacitly suppose it to be square integrable and of compact support.

$$\left[\text{Help: (a) } \Leftrightarrow \text{ (b): } \hat{\tilde{\psi}}^{(k)}(0) = \int_{-\infty}^{+\infty} (-it)^k \tilde{\psi}(t)\,dt\right.$$

$$\left.\text{(b) } \Leftrightarrow \text{ (c): } \hat{\tilde{\psi}}(2\omega) = \tilde{m}_1(\omega)\hat{\tilde{\varphi}}(\omega) = m_0(\pi - \omega)\hat{\tilde{\varphi}}(\omega) \quad (\text{and } \hat{\tilde{\varphi}}(0) \neq 0)\right]$$

(2) In the situation of the preceding exercise, show that
$\tilde{\psi} = \tilde{\psi}(t)$ has p zero moments $\Longleftrightarrow$ the high-pass analysis filter $\mathbf{h}_1 = (h_1[k])$
annihilates, for all polynomials $q = q(X)$ of degree $\leq p - 1$, the sequence of the
sample values $\mathbf{q} = (q[k])_{k \in \mathbb{Z}}$.
($\mathbf{h}_1$ is thus a kind of "discrete derivative to the order p").

[*Help:* $\tilde{m}_1^{(n)}(0) = \sum_{k \in \mathbb{Z}} k^n h_1[k]$ $\quad 0 \leq n \leq p - 1$]

Let us sum up: At first sight, we note that the multiplicity $N = 2L$ of the root in
$\omega = \pi$ for the trigonometric polynomial $m_0(\omega)$ reflects the *annihilation degree* on
very regular data by the analysis tools $\tilde{\psi} = \tilde{\psi}(t)$ and $\mathbf{h}_1 = (h_1[k])$.

On the one hand, for very regular (i.e. locally almost polynomial) $f = f(t)$, the
wavelet coefficients $\langle f, \tilde{\psi}_k^{(m)} \rangle$ will be very small in absolute value (for fine scales),
since then the Taylor polynomials of f will be annihilated according to the number
of zero moments of $\tilde{\psi}$.

On the other hand, on digital data, the high-pass analysis filter $\mathbf{h}_1 = (h_1[k])$
will annihilate all "polynomial regularities" of degree $\leq N - 1$. But there is also a
dual aspect, concerning $\psi = \psi(t)$, the synthesis mother wavelet and $\mathbf{g}_0 = (g_0[k])$,
the low-pass synthesis filter.

The key result of the affair is the following (A.Cohen, I.Daubechies,
J.C.Feauveau):

Let $\psi = \psi(t)$ and $\tilde{\psi} = \tilde{\psi}(t)$ be the (synthesis and analysis) mother wavelets
defined – recall! – from the scaling functions $\varphi = \varphi(t)$ and $\tilde{\varphi} = \tilde{\varphi}(t)$ by

$$\psi(t) = \sqrt{2} \cdot \sum_k g_1[k+1]\varphi(2t-k) \qquad \text{and} \qquad \tilde{\psi}(t) = \sqrt{2} \cdot \sum_k h_1[-k-1]\tilde{\varphi}(2t-k)$$

If $\langle \psi_k^{(m)}, \tilde{\psi}_{k'}^{(m')} \rangle = \delta_{m,m'}\delta_{k,k'}$, then we have:

$\psi = \psi(t)$ m times continuously differentiable $\quad \Longrightarrow \quad$ $m_0(\omega)$ is divisible by
$(1 + e^{-i\omega})^{m+1}$.

$\tilde{\psi} = \tilde{\psi}(t)$ $\tilde{m}$ times continuously differentiable $\quad \Longrightarrow \quad$ $\tilde{m}_0(\omega)$ is divisible by
$(1 + e^{-i\omega})^{\tilde{m}+1}$.

The argument is the following: The biorthogonality and the differentiability to
the order $p - 1$ of one of the wavelets imply that the other has p zero moments.
Then, one concludes by means of the exercise (1) above. We insist: The regularity
of $\psi = \psi(t)$ (to be continuously differentiable to the order m) "makes pressure"
on the multiplicity N of the root $\omega = \pi$ of the trigonometric polynomial $m_0(\omega)$:
$N \geq m + 1$.

There remains the question: Do we also have the inverse effect, i.e. does an
increasing of N make increase the regularity of $\psi = \psi(t)$?

The answer is yes, but somehow at low voice. More precisely: First, since the
degree of regularity of $\psi = \psi(t)$ is the same as that of $\varphi = \varphi(t)$, we only need discuss
the later.

In order to obtain that $\varphi = \varphi(t)$ is continuously differentiable to the order m, we need that $| \hat{\varphi}(\omega) |$ is faster decreasing than $| \omega |^{-m-1}$ (we need that $\omega^m \hat{\varphi}(\omega)$ is summable).

Recall: In the antechamber of the problem, the situation was like that: Beginning with the decompositions

$$m_0(\omega) = \left(\frac{1+e^{-i\omega}}{2} \right)^N p_0(\omega) \quad \text{and} \quad \tilde{m}_0(\omega) = \left(\frac{1+e^{-i\omega}}{2} \right)^{\tilde{N}} \tilde{p}_0(\omega)$$

and with the product formulas

$$\hat{\varphi}(\omega) = \prod_{k=1}^{\infty} m_o \left(\frac{\omega}{2^k} \right) \quad \text{and} \quad \hat{\tilde{\varphi}}(\omega) = \prod_{k=1}^{\infty} \tilde{m}_o \left(\frac{\omega}{2^k} \right),$$

the validity of $\varphi = \varphi(t)$ and of $\tilde{\varphi} = \tilde{\varphi}(t)$ with the biorthogonality as a supplement) was guaranteed by

$$N > \frac{1}{2} + \log_2 B \quad \text{with} \quad B = \sup_{\omega \in [0, 2\pi]} | p_0(\omega) |$$

$$\tilde{N} > \frac{1}{2} + \log_2 \tilde{B} \quad \text{with} \quad \tilde{B} = \sup_{\omega \in [0, 2\pi]} | \tilde{p}_0(\omega) |$$

A more refined version was obtained when replacing the term $\log_2 B$ by the critical exponent β of $m_0(\omega)$ (and the term $\log_2 \tilde{B}$ by the critical exponent $\tilde{\beta}$ of $\tilde{m}_0(\omega)$).

If we were to introduce exigencies of regularity to the order m and $\tilde{m}$, respectively, we should await conditions of the type

$$N > m + \beta,$$

$$\tilde{N} > \tilde{m} + \tilde{\beta}.$$

And actually these are the correct constraints.

From another perspective, i.e. by asymptotical arguments, we shall have $N \sim 5m$ (hence the regularity of $\varphi = \varphi(t)$ will be, in general, of considerable influence on the number of zero moments of $\tilde{\psi} = \tilde{\psi}(t)$).

We finally arrive at the question: What does the regularity of the scaling function $\varphi = \varphi(t)$ (and of the mother wavelet $\psi = \psi(t)$) mean for the quality of the synthesis filter bank of the associated DWT?

In order to well understand this point, recall the *the subdivision algorithm* which starts with $\mathbf{g}_0 = (g_0[k])$ (or with $\mathbf{g}_1 = (g_1[k])$, approximating $\varphi = \varphi(t)$ (and $\psi = \psi(t)$). More precisely: The subdivision algorithm, applied in n steps, starts with the sequence $y_0^{(n)}[k] = \delta_{0,k}$ and it synthesizes it n times – neglecting, at every step, the high-pass component: all details are uniformly set to zero.

After n steps, we obtain $y_0^{(0)}[k] = \varphi_n \left(\frac{k}{2^n} \right) \quad k \in \mathbb{Z}$.

We underline: The result of a "pure" synthesis of n steps of the unit impulse is the uniform approximation to the order n of the scaling function $\varphi = \varphi(t)$.

This manner to interpret the subdivision algorithm has a rather surprising *consequence*:

> The scaling function $\varphi = \varphi(t)$ describes the "profile of error propagation" in pure synthesis (i.e. with all high-pass details set to zero).

More concretely: Consider a finite support sequence $\mathbf{y}_0^{(n)} = (y_0^{(n)}[k])$, which we aim to synthesize in n steps. Admit an error of value ϵ for $k = 0$.

The error sequence: $\mathbf{z}_0^{(n)} = \mathbf{y}_0^{(n)} + \epsilon \cdot \delta_{0,k}$. After n synthesis steps ("without details") we obtain

$$z_0^{(0)}[k] = y_0^{(0)}[k] + \epsilon \cdot \varphi_n\left(\frac{k}{2^n}\right).$$

Thus the local error of value ϵ at $k = 0$ has been propagated to an (almost global) error $\left(\epsilon \cdot \varphi_n\left(\frac{k}{2^n}\right)\right)_{k \in \mathbb{Z}}$.

Example The DWT 5/3 spline $(m_0(\omega) = \frac{1}{4}e^{i\omega} + \frac{1}{2} + \frac{1}{4}e^{-i\omega})$.

We saw, in an earlier exercise: $\varphi_n\left(\frac{k}{2^n}\right) = \begin{cases} 1 - \frac{|k|}{2^n} & \text{for } -2^n \leq k \leq 2^n \\ 0 & \text{else} \end{cases}$

Here, the propagation of a local error at $k = 0$ is linearly decreasing, then zero.

Exercise

Show that a local error of ϵ at $k = k_0$ is propagated as follows:

$$\mathbf{z}_0^{(n)} = \mathbf{y}_0^{(n)} + \epsilon \cdot \delta_{k_0,k} \quad \Longrightarrow \quad z_0^{(0)}[k] = y_0^{(0)}[k] + \epsilon \cdot \varphi_n\left(\frac{k - k_0}{2^n}\right).$$

Now, we are ready to discuss the importance of the regularity of $\varphi = \varphi(t)$ for the stability of the synthesis filter bank of the considered DWT. Since we suppose $\mathbf{g}_0 = (g_0[k])$ of finite (odd) length and symmetric, the synthesis scaling function $\varphi = \varphi(t)$ will be even and of finite support. If $\varphi = \varphi(t)$ is differentiable to the order m, $m \geq 1$, then $\varphi = \varphi(t)$ will admit horizontal tangents to the order m at the boundaries of its support. In other words: The more $\varphi = \varphi(t)$ is regular, the more its profile (its graph) will be flattened towards the boundary of the support.

A *great regularity* thus means (more or less) a guaranty of *localized propagation of a local error*.

It is in this sense that the regularity of $\varphi = \varphi(t)$ implies the stability of low-pass synthesis. But the regularity of $\varphi = \varphi(t)$ is inherited by $\psi = \psi(t)$; hence, by the same arguments (via subdivision towards $\psi = \psi(t)$), we see that the stability of high-pass synthesis (on the details) is equally guaranteed.

Let us sum up. The regularity of $\varphi = \varphi(t)$ (hence also of $\psi = \psi(t)$) guarantees the stability of the *synthesis* filter bank, in the sense made precise above. It is correlated – via the multiplicity $N = 2L$ of the root $\omega = \pi$ of the trigonometric polynomial $m_0(\omega) = \frac{1}{\sqrt{2}}\hat{g}_0(\omega)$ – with the number of zero moments of $\tilde{\psi}(t)$, i.e. of the *analysis* mother wavelet. This number indicates to what degree polynomial sample values do not leave a trace as details (i.e. in high-pass).

But the situation is symmetric. Only, a great multiplicity $\tilde{N} = 2\tilde{L}$ of the root $\omega = \pi$ of the trigonometric polynomial $\tilde{m}_0(\omega) = \frac{1}{\sqrt{2}}\hat{h}_0(-\omega)$, which would have the discussed effect on the regularity of $\tilde{\varphi} = \tilde{\varphi}(t)$ and on the number of zero moments of $\psi = \psi(t)$, is (conceptually) difficult to justify.

This finally leads us to the criteria of *optimal factorization* of $R(x) = r(x) \cdot \tilde{r}(x)$. The *first* viewpoint is the following: Everything for the regularity of $\varphi = \varphi(t)$ (and the zero moments of $\tilde{\psi} = \tilde{\psi}(t)$). Hence, we must choose $N = 2L$ relatively large, and put $r(x) = 1$, i.e. $\tilde{r}(x) = R(x) = \sum_{n=0}^{M-1} \binom{M+n-1}{n} \frac{1}{2^n}(1 - x)^n$ with $M = L + \tilde{L}$.

Note that a priori this does not impose any constraint on the value $\tilde{N} = 2\tilde{L}$; but in practice, the lengths of the considered impulse responses should be minimally distinct[52]; hence $\tilde{N}$ should be small.

We thus obtain the *DWT splines*. The *second* viewpoint is a kind of tribute to orthonormality: Everything for a similarity between $\varphi = \varphi(t)$ and $\tilde{\varphi} = \tilde{\varphi}(t)$. We look for similar profiles (of the graphs). The design strategy is then the following:

We shall choose $N = \tilde{N}$ (i.e. $L = \tilde{L}$). Hence the degree $M - 1$ of $R(x)$ is odd; we factorize $R(x)$ as equitable as possible: the factor $r(x)$ will have degree $\frac{M}{2} - 1$, the factor $\tilde{r}(x)$ will have degree $\frac{M}{2}$.

We thus obtain the *DWT CDF*[53]. The two viewpoints are accepted. We are already familiar with the two most elementary members of the two families: On the one hand, the DWT 5/3 spline, on the other hand, the DWT 9/7 CDF.

Investiture of the DWT 5/3 spline and of the DWT 9/7 CDF

First, we shall be concerned with the DWT 5/3 spline. As we pointed out, at the end of the preceding paragraph, the family of Discrete Wavelet Transforms spline comes up by the choice $m_0(\omega) = (\cos\frac{\omega}{2})^N$, $N = 2L \geq 2$. In this case, we will have

$\hat{\varphi}(\omega) = \prod_{k=1}^{\infty}(\cos\frac{\omega}{2^{k+1}})^N = \left(\frac{\sin\frac{\omega}{2}}{\frac{\omega}{2}}\right)^N = \operatorname{sin} c\left(\frac{\omega}{2\pi}\right)^N$ But: $\operatorname{sin} c\left(\frac{\omega}{2\pi}\right)$ is the Fourier transform of the function $\mathbb{I}_{[-\frac{1}{2},\frac{1}{2}]}$. Hence, for $N = 2L \geq 2$, $\varphi = \varphi(t)$ will be the $N-1$-times iterated convolution product $\mathbb{I}_{[-\frac{1}{2},\frac{1}{2}]} * \cdots * \mathbb{I}_{[-\frac{1}{2},\frac{1}{2}]}$ $\left(N \text{ factors } \mathbb{I}_{[-\frac{1}{2},\frac{1}{2}]}\right)$.

In this case, $\varphi = \varphi(t)$ will have its support in $[-L, L]$, will be a polynomial of degree $N - 1$ on every interval $[K, K+1]$, $-L \leq K \leq L - 1$, and will be $(N - 2)$ times differentiable at the (integer) nodes of interpolation. In other words: $\varphi = \varphi(t)$ will be a B-spline of order N.

Now consider the simplest case: $N = 2$, i.e. $L = 1$.

First: $m_0(\omega) = \cos^2\frac{\omega}{2} = \frac{1}{2}(1 + \cos\omega) = \frac{1}{4}e^{i\omega} + \frac{1}{2} + \frac{1}{4}e^{-i\omega}$.

Then: $\hat{\varphi}(\omega) = \left(\operatorname{sinc}\left(\frac{\omega}{2\pi}\right)\right)^2$

Hence: $\varphi(t) = \mathbb{I}_{[-\frac{1}{2},\frac{1}{2}]} * \mathbb{I}_{[-\frac{1}{2},\frac{1}{2}]} = \begin{cases} 1 - |t| & -1 \leq t \leq 1 \\ 0 & \text{else} \end{cases}$

For $\tilde{m}_0(\omega)$, the options vary according to the choice of $\tilde{N} = 2\tilde{L}$. We shall choose the simplest case: $\tilde{L} = 1$, i.e. $\tilde{N} = 2$.

$M = L + \tilde{L} = 2$, hence $R(x) = Q\left(\frac{1}{2}(1 - x)\right)$ is of degree 1: $R(x) = 1 + 2 \cdot \frac{1}{2}(1 - x) = 2 - x$. This gives:

$$\tilde{m}_0(\omega) = \cos^2\frac{\omega}{2} \cdot \tilde{r}(\cos\omega) = \cos^2\frac{\omega}{2} \cdot R(\cos\omega) = \cos^2\frac{\omega}{2} \cdot (2 - \cos\omega)$$

$$= \left(\frac{1}{4}e^{i\omega} + \frac{1}{2} + \frac{1}{4}e^{-i\omega}\right)\left(2 - \frac{1}{2}e^{i\omega} - \frac{1}{2}e^{-i\omega}\right) = -\frac{1}{8}e^{2i\omega} + \frac{1}{4}e^{i\omega}$$

$$+ \frac{3}{4} + \frac{1}{4}e^{-i\omega} - \frac{1}{8}e^{-2i\omega}.$$

We have indeed recovered our DWT 5/3 spline.

---

[52] We note that our hypotheses (odd length and symmetry for the impulse responses) force these lengths to differ by an odd multiple of 2.

[53] CDF $\equiv$ Cohen–Daubechies–Feauveau

Note that in our spectral formalism, the impulse responses $\mathbf{g_0}$ and $\mathbf{h_0}$ would have irrational coefficients (the coefficients of $m_0(\omega)$ and of $\tilde{m}_0(\omega)$ are multiplied by $\sqrt{2}$). But for practice applications one renormalizes:

$$\mathbf{g_0} \longmapsto \sqrt{2} \cdot \mathbf{g_0},$$
$$\mathbf{h_0} \longmapsto \tfrac{1}{\sqrt{2}} \cdot \mathbf{h_0}$$
$$(\text{or} \quad \mathbf{g_0} \longmapsto \tfrac{1}{\sqrt{2}} \cdot \mathbf{g_0}, \quad \mathbf{h_0} \longmapsto \sqrt{2} \cdot \mathbf{h_0}).$$

Exercises

(1) Consider $m_0(\omega)$ and $\tilde{m}_0(\omega)$ – after having fixed $N = 2L$ and $\tilde{N} = 2\tilde{L}$.

  (a) Show that then $\quad m_0(\omega) = \sum_{k=-L}^{L} \frac{1}{2^N}\binom{N}{k+L}e^{-ik\omega}$.

  (b) In real notation, $\quad \tilde{m}_0(\omega) = (\cos^2\frac{\omega}{2})^2 \cdot \sum_{k=0}^{M-1}\binom{M-1+k}{k}(\sin^2\frac{\omega}{2})^k$.

  Find the complex notation of the trigonometric polynomial $\tilde{m}_0(\omega)$.

(2) Let us keep $m_0(\omega) = \frac{1}{4}e^{i\omega} + \frac{1}{2} + \frac{1}{4}e^{-i\omega}$ and choose now $\tilde{N} = 2\tilde{L} = 4$.

  (a) Find $\tilde{m}_0(\omega)$.

  $\Big[\text{Answer: } \tilde{m}_0(\omega) = \frac{3}{128}e^{4i\omega} - -\frac{3}{64}e^{3i\omega} - \frac{1}{8}e^{2i\omega} + \frac{19}{64}e^{i\omega} + \frac{45}{64} + \frac{19}{64}e^{-i\omega} -$

  $\frac{1}{8}e^{-2i\omega} - \frac{3}{64}e^{-3i\omega} + \frac{3}{128}e^{-4i\omega}.\Big]$

  (b) For the fans of computation: Show, by means of the Fixed Point Criterion applied to the operator $\tilde{P}_0$, that the product formula for $\hat{\tilde{\varphi}}(\omega)$ indeed gives a valid analysis scaling function $\varphi(t)$ (and that we hence obtain biorthogonal wavelet bases).

  (Attention: In principle, we have to triangulate a 17×17 linear system the first and the last row of which are trivial and which is invariant with respect to order inversion of the variables; so we finally only deal with an 8×8 system).

(3) A deviation to the case of odd multiplicities N and $\tilde{N}$ (and of impulse responses of even length).

We have:
$$m_0(\omega) = e^{-i\frac{\omega}{2}}(\cos\tfrac{\omega}{2})^N$$
$$\tilde{m}_0(\omega) = e^{-i\frac{\omega}{2}}(\cos\tfrac{\omega}{2})^{\tilde{N}} \cdot \sum_{k=0}^{M-1}\binom{M-1+k}{k}(\sin^2\tfrac{\omega}{2})^k \qquad {}^{54}$$
with $M = \frac{1}{2}(N+\tilde{N})$.

(a) Choose $N = 3$ and $\tilde{N} = 1$.
  Show that then $\quad m_0(\omega) = \frac{1}{8}e^{i\omega} + \frac{3}{8} + \frac{3}{8}e^{-i\omega} + \frac{1}{8}e^{-2i\omega}$
  $$\tilde{m}_0(\omega) = -\frac{1}{4}e^{i\omega} + \frac{3}{4} + \frac{3}{4}e^{-i\omega} - \frac{1}{4}e^{-2i\omega}$$

(b) Show, by means of the Fixed Point Criterion, that $\tilde{\varphi} = \tilde{\varphi}(t)$ is *not* valid ($\tilde{\varphi}(t)$ is not square integrable).

(c) Now pass to $N = \tilde{N} = 3$. Show that we obtain
  $$\tilde{m}_0(\omega) = \frac{3}{64}e^{3i\omega} - \frac{9}{64}e^{2i\omega} - \frac{7}{64}e^{i\omega} + \frac{45}{64} + \frac{45}{64}e^{-i\omega} - \frac{7}{64}e^{-2i\omega} - \frac{9}{64}e^{-3i\omega} + \frac{3}{64}e^{-4i\omega}$$

---

[54] The factor $e^{-i\frac{\omega}{2}}$ creates a symmetric situation with respect to $t = \frac{1}{2}$.

Once more: We could show, with the help of the Fixed Point Criterion applied to the operator $\check{P}_0$, that the product formula for $\hat{\check{\varphi}}(\omega)$ gives a valid analysis scaling function $\varphi(t)$ (and that we hence obtain biorthogonal wavelet bases). Note that $\tilde{\varphi} = \check{\varphi}(t)$ is valid *without* satisfying the condition (D) of controlled decreasing.

Now pass to the DWT 9/7 CDF.

The situation of the designer is the following: He has chosen $N = 2L = \tilde{N} = 2\tilde{L}$, and he now must factor
$$R(x) = \sum_{k=0}^{M-1} \binom{M-1+k}{k} \frac{1}{2^k}(1-x)^k \qquad \text{(with } M = L + \tilde{L} = N = \tilde{N}\text{)}.$$
into two factors $r(x)$ and $\tilde{r}(x)$ of degrees $L-1$ and L, respectively.
This gives
$$m_0(\omega) = (\cos \tfrac{\omega}{2})^N \cdot r(\cos\omega)$$
$$\tilde{m}_0(\omega) = (\cos \tfrac{\omega}{2})^N \cdot \tilde{r}(\cos\omega)$$
The first typical representative[55] of this family is obtained for $L = \tilde{L} = 2$, i.e. for $N = \tilde{N} = 4$.

Consider $\quad R(x) = \sum_{k=0}^{3} \binom{3+k}{k} \frac{1}{2^k}(1-x)^k = 8 - \frac{29}{2}x + 10x^2 - \frac{5}{2}x^3$.
$R(x)$ admits the following three roots:

$$R_1 = \frac{4}{3} + \left(\frac{S}{15} - \frac{7}{3S}\right)$$

$$R_2 = \frac{4}{3} - \frac{1}{2}\left(\frac{S}{15} - \frac{7}{3S}\right) + i\frac{\sqrt{3}}{2}\left(\frac{S}{15} + \frac{7}{3S}\right)$$

$$R_3 = \frac{4}{3} - \frac{1}{2}\left(\frac{S}{15} - \frac{7}{3S}\right) - i\frac{\sqrt{3}}{2}\left(\frac{S}{15} + \frac{7}{3S}\right)$$

with $S = \sqrt[3]{(350 + 105\sqrt{15})}$.
We get[56]
$$m_0(\omega) = (\cos \frac{\omega}{2})^4 \cdot \left(\frac{R_1 - \cos\omega}{R_1 - 1}\right)$$

$$\tilde{m}_0(\omega) = \left(\cos \frac{\omega}{2}\right)^4 \cdot \left(\frac{R_2 - \cos\omega}{R_2 - 1}\right) \cdot \left(\frac{R_3 - \cos\omega}{R_3 - 1}\right).$$

In complex polynomial notation we hence shall have:

$$m_0(\omega) = -\frac{1}{32(R_1 - 1)}e^{3i\omega} - \frac{2 - R_1}{16(R_1 - 1)}e^{2i\omega} + \frac{8R_1 - 7}{32(R_1 - 1)}e^{i\omega} + \frac{3R_1 - 2}{8(R_1 - 1)}$$
$$+\frac{8R_1 - 7}{32(R_1 - 1)}e^{-i\omega} - \frac{2 - R_1}{16(R_1 - 1)}e^{-2i\omega} - \frac{1}{32(R_1 - 1)}e^{-3i\omega}$$

Numerical evaluation of these irrational coefficients gives the following approximate notation:

$$m_0(\omega) = 0.557543526229 + 0.295635881557(e^{i\omega} + e^{-i\omega})$$
$$-0.028771763114(e^{2i\omega} + e^{-2i\omega}) - 0.045635881557(e^{3i\omega} + e^{-3i\omega}).$$

Concerning $\tilde{m}_0(\omega)$, we shall treat it in the following

---

[55] For $N = 2$, we still have the DWT spline.
[56] The normalizing denominators guarantee $m_0(0) = \tilde{m}_0(0) = 1$

Exercise

(a) Show that

$$\frac{1}{(R_2 - 1)(R_3 - 1)} = \frac{5}{2}(R_1 - 1).$$

(b) Compute $\tilde{m}_0(\omega)$ in function of R_1, R_2 and R_3.

$\Big[$Answer: $\tilde{m}_0(\omega) = \frac{5}{128}(R_1 - 1)\Big(C_0 + C_1 \cos \omega + C_2 \cos 2\omega + C_3 \cos 3\omega + C_4 \cos 4\omega \Big)$

    with

$$C_0 = 14 + 24R_2R_3 - 16(R_2 + R_3)$$

$$C_1 = 24 + 32R_2R_3 - 28(R_2 + R_3)$$

$$C_2 = 16 + 8R_2R_3 - 16(R_2 + R_3) \quad \Big]$$

$$C_3 = 8 - 4(R_2 + R_3)$$

$$C_4 = 2$$

Numerical evaluation gives here

$$\tilde{m}_0(\omega) = 0.602949018236 + 0.266864118443(e^{i\omega} + e^{-i\omega})$$

$$-0.078223266529(e^{2i\omega} + e^{-2i\omega}) - 0.016864118443(e^{3i\omega} + e^{-3i\omega})$$

$$+0.026748757411(e^{4i\omega} + e^{-4i\omega}).$$

This is the definition of the DWT 9/7 CDF[57].

We shall not deal with the validity of the scaling functions $\varphi = \varphi(t)$ and of $\tilde{\varphi} = \tilde{\varphi}(t)$.

Remark on a renormalization (that we already encountered with the DWT 5/3 spline):

According to our conventions, $m_0(\omega) = \frac{1}{\sqrt{2}} \cdot \hat{g}_0(\omega)$ and $\tilde{m}_0(\omega) = \frac{1}{\sqrt{2}} \cdot \hat{h}_0(-\omega)$.

But in practical implementation for JPEG 2000, one prefers to take the coefficients of $\tilde{m}_0(\omega)$ as the coordinates of $\mathbf{h}_0$[58]. This corresponds to a change $\mathbf{h}_0 \longmapsto \frac{1}{\sqrt{2}} \cdot \mathbf{h}_0$, which forces the rescaling $\mathbf{g}_0 \longmapsto \sqrt{2} \cdot \mathbf{g}_0$. The renormalized coefficients of $\mathbf{g}_0$ will then be the coefficients of $2 \cdot m_0(\omega)$.

The renormalized impulse responses $\mathbf{h}_1^{\mathrm{t}}$ and $\mathbf{g}_1^{\mathrm{t}}$ are given by

$$(h_1^{\mathrm{t}}[0], h_1^{\mathrm{t}}[1], h_1^{\mathrm{t}}[2], h_1^{\mathrm{t}}[3]) = (0.557543526229, -0.295635881557, -0.028771763114,$$

$$0.045635881557) \quad (g_1^{\mathrm{t}}[0], g_1^{\mathrm{t}}[1], g_1^{\mathrm{t}}[2], g_1^{\mathrm{t}}[3], g_1^{\mathrm{t}}[4]) = 2 \cdot (0.602949018236,$$

$$-0.266864118443, -0.078223266529, 0.016864118443, 0.026748757411).$$

---

[57] 9/7 for the lengths of $\mathbf{h}_0$ and of $\mathbf{g}_0$.

[58] One aims at $\sum_k h_0[k] = 1$. Thus the low-pass analysis filter computes local means.

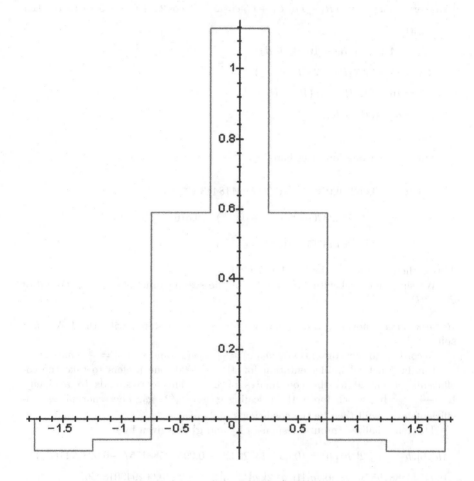

s_1 for g_0 of the DWT 9/7 CDF.

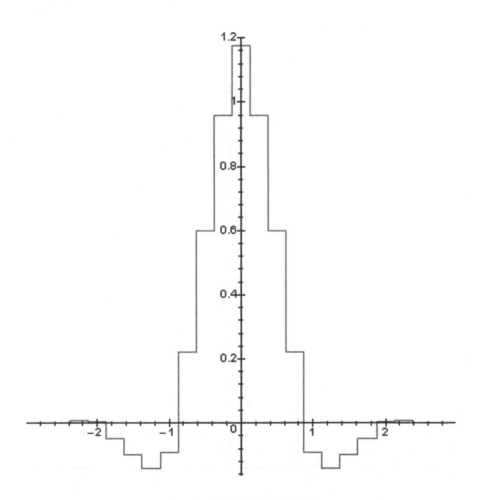

s_2 for g_0 of the DWT 9/7 CDF.

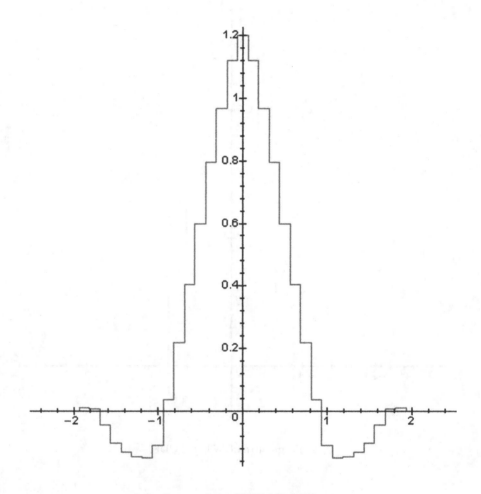

s_3 for $\mathbf{g}_0$ of the DWT 9/7 CDF.

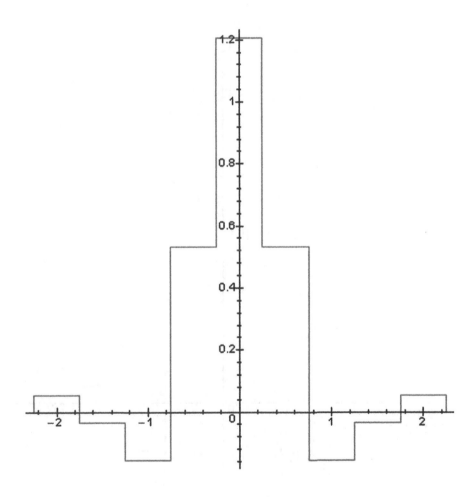

$\tilde{\mathbf{s}}_1$ for $\mathbf{h}_0$ of the DWT 9/7 CDF.

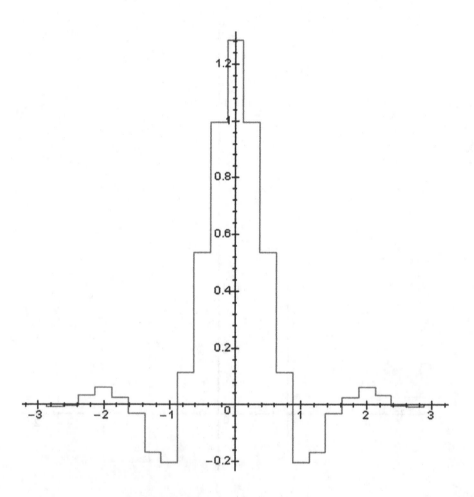

$\tilde{\mathbf{s}}_2$ for $\mathbf{h}_0$ of the DWT 9/7 CDF.

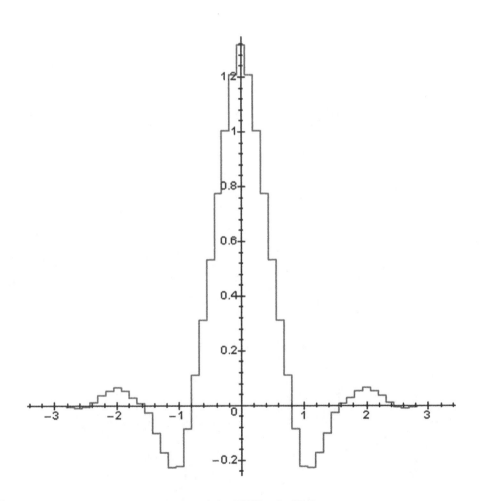

$\tilde{\mathbf{s}}_3$ for $\mathbf{h}_0$ of the DWT 9/7 CDF.

References

Digital Signatures Using Reversible Public Key Cryptography for the Financial Services Industry (rDSA). American National Standards Institute X9.31, 1998.

Digital Signatures Using Reversible Public Key Cryptography for the Financial Services Industry: The Elliptic Curve Digital Signature Algorithm (ECDSA). American National Standards Institute X9.62, 1998.

R.E. Blahut. *Theory and Practice of Error Control Codes*. Addison-Wesley, 1983.

R.E. Blahut. *Digital Transmission of Information*. Addison-Wesley, 1990.

P. Brémaud. *An Introduction to Probabilistic Modeling*. UTM, Springer, Berlin Heidelberg New York, 1988.

D.M. Bressoud. *Factorization and Primality Testing*. Springer, Berlin Heidelberg New York, 1989.

R. Calderbank, I. Daubechies, W. Sweldens, and B. Yeo. Wavelet transforms that map integers to integers. *Applied and Computational Harmonic Analysis*, 5(3):332–369, July 1998.

A. Cohen, I. Daubechies, and J.-C. Feauveau. Biorthogonal bases of compactly supported wavelets. *Communications on Pure and Applied Mathematics*, 45(5):485–560, June 1992.

D.A. Cox. *Primes of the form $x^2 + ny^2$*. Wiley, 1989.

J. Daemen and V. Rijmen. *The Design of Rijndael. AES - The Advanced Encryption Standard*. Springer, Berlin Heidelberg New York, 2002.

I. Daubechies. Orthonormal bases of compactly supported wavelets. *Communications on Pure and Applied Mathematics*, 41:909–996, November 1988.

W. Diffie and M.E. Hellman. New directions in cryptography. *IEEE Transactions on Information Theory*, IT-22:644–654, 1976.

T. ElGamal. A public key cryptosystem and a signature scheme based on discrete logarithms. *IEEE Transactions on Information Theory*, IT-31:469–472, 1985.

D.F. Elliott and K.R. Rao. *Fast Transforms. Algorithms, Analyses, Applications*. Academic, 1982.

P.M. Farrelle. *Recursive Block Coding for Image Data Compression*. Springer, Berlin Heidelberg New York, 1990.

D.A. Huffman. A method for the construction of minimum redundancy codes. *Proceedings of the IRE*, 40:1098–1101, 1952.

A.K. Jain. *Fundamentals of Digital Image Processing*. Prentice-Hall, Englewood Cliffs, NJ, 1989.

C.B. Jones. An efficient coding system for long source sequences. *IEEE Transactions on Information Theory*, IT-27:280–291, 1981.

N. Koblitz. *A Course in Number Theory and Cryptography*. Springer, Berlin Heidelberg New York, 1987.

G. Kraft. A device for quantizing, grouping, and coding amplitude modulated pulses. MS Thesis, Department of Electrical Engineering, Massachusetts Institute of Technology, Cambridge, MA, 1949.

G.-L. Lay and H.G. Zimmer. Constructing elliptic curves with given group order over large finite fields. In *Proceedings of ANTS I, LNCS 877*: 250–263, 1994.

S. Mallat. A theory for multiresolution signal decomposition; the wavelet representation. *IEEE Transactions on Pattern Analysis and Mathematical Intelligence*, 11(7):674–693, 1989.

B. McMillan. Two inequalities implied by unique decipherability. *IRE Transactions on Information Theory*, IT-2:115–116, 1956.

A.J. Menezes. *Elliptic Curve Public Key Cryptosystems*. Kluwer Academic, Boston Dordrecht London, 1993.

Data Encryption Standard (DES). National Bureau of Standards FIPS Publication 46, 1977.

DES modes of operation. National Bureau of Standards FIPS Publication 81, 1980.

Secure Hash Standard. National Bureau of Standards FIPS Publication 180-1, 1995.

Digital Signature Standard (DSS). National Bureau of Standards FIPS Publication 186-2, 2000.

H. Nyquist. Certain topics in telegraph transmission theory. *AIEE Transactions*, 47:617–644, 1928.

W.B. Pennebaker and J.L. Mitchell. *JPEG: Still Image Data Compression Standard*. Van Nostrand Reinhold, New York, 1992.

K.R. Rao and P. Yip. *Discrete Cosine Transform*. Academic, New York, 1990.

I.S. Reed and G. Solomon. Polynomial codes over certain finite fields. *Journal of the Society of Industrial Applied Mathematics*, 8:300–304, 1960.

R.L. Rivest, A. Shamir, and L. Adleman. A method for obtaining digital signatures and public key cryptosystems. *Communications of the ACM*, 21:120–126, 1978.

J. Rissanen and G.G. Langdon, Jr.. Arithmetic coding. *IBM Journal of Research and Development*, 23:149–162, 1979.

J. Rissanen and G.G. Langdon, Jr.. Universal modeling and coding. *IEEE Transactions on Information Theory*, IT-27:12–23, 1979.

C.E. Shannon. A mathematical theory of communication. *Bell Systems Technical Journal*, 27:379–423 (Part I), 623–656 (Part II), 1948.

D. Stinson. *Cryptography - Theory and Practice*. CRC, Boca Raton, 1995.

R. Strichartz. *A Guide to Distribution Theory and Fourier Transforms*. CRC, Boca Raton, 1994.

D.S. Taubman and M.W. Marcellin. *JPEG2000. Image Compression Fundamentals, Standards and Practice*. Kluwer Academic, Boston Dordrecht London, 2002.

H. Triebel. *Theory of Function Spaces*. Birkhäuser Verlag, Boston, 1992.

M. Vetterli and J. Kovačević. *Wavelets and Subband Coding*. Prentice-Hall, NJ, 1995.

A.J. Viterbi. Error bounds for convolutional codes and an asymptotically optimum decoding algorithm. *IEEE Transactions on Information Theory*, IT-13:260–269, 1967.

J. Whittaker. *Interpolatory function theory*. Cambridge Tracts in Math. and Math. Physics, 33, 1935.

J. Ziv and A. Lempel. A Universal Algorithm for Sequential Data Compression. *IEEE Transactions on Information Theory*, 23(3):337–343, 1977.

Index

Page numbers followed by n indicate foot notes.